RESTORING YOUR HISTORIC HOUSE

The Comprehensive Guide For Homeowners

SCOTT T. HANSON

with photography by David J. Clough

Down East Books

Published by Down East Books
An imprint of Globe Pequot
Trade division of The Rowman & Littlefield Publishing Group, Inc.
4501 Forbes Boulevard, Suite 200
Lanham, Maryland 20706
www.rowman.com
www.downeastbooks.com

Distributed by NATIONAL BOOK NETWORK

First Down East Books edition, 2023

ISBN: 978-1-68475-116-7 (hardcover)
ISBN: 978-0-88448-695-4 (ebook)

Library of Congress Control Number: 2019948645

Text and jacket design by Frame25 Productions

Printed in China

CONTENTS

PART FOUR: THE EXTERIOR ENVELOPE

PART FIVE: INTERIOR FINISHES

Introduction

Another book on renovating old houses? Really? Is it needed?

Yes, it is. While there are other books on renovating old houses, I have not found a single one that prioritizes the identification and preservation of a historic house's character-defining features as a starting point in the process. The purpose of this book is to describe and illustrate a best-practices approach for updating historic homes for modern life in ways that do not turn old houses into new ones. The book also suggests many ways to save money along the way without settling for cheap or inappropriate solutions.

In my professional world—the world of architectural history and building rehabilitation using state and federal historic tax credits—*rehabilitation* is defined as "the act or process of making possible a compatible use for a property through repair, alterations, and additions while preserving those portions or features which convey its historical, cultural, or architectural values."[1] This ponderous definition sounds a lot like what old-house lovers mean by *restoration*, as in the title of this book.

In the professional world, restoration has a more precise, circumscribed definition. Specifically, it means returning a building to its condition and appearance at a particular time in the past. Restoring an 1820 house under this definition, for example, would mean removing its central heat, electricity, and running water, since it would have lacked such conveniences originally. Naturally, very few homeowners—and not even the majority of historic home museums—will want to go to this extreme. In a sensitive rehabilitation, however, it is often possible and desirable to restore elements of the house that have been altered or even removed. This is how the words restore and restoration are used in this book—as a part of the rehabilitation process. When you rehabilitate an old house in a way that maintains its historic character, you are practicing *preservation*.

The approaches set forth in this book come from multiple sources, including my professional experience with commercial historic tax credit rehabilitations and museum house restorations and my personal experience with rehabilitating houses. More than 40 years of working on, in, and around historic buildings in multiple capacities has allowed me to observe the work of masters and the work of hacks, historic and present-day. It has given me the background to judge what is historically significant and what is not. And it has given me ample opportunity to learn from my mistakes.

Preservation is a frame of mind, an attempt to understand how a historic building came to be what it is and to identify the elements that make it historic. Rehabilitation makes creative use of the preservation

[1] *Secretary of the Interior's Standards for Rehabilitation*

mindset to adapt a historic building to the needs of the contemporary world while retaining the most important historic elements. Every project will require compromises, but with the right frame of mind it is possible to retain the historic character of an old house while adapting it to modern life.

This book is not intended to repeat basic information that is readily available in many standard do-it-yourself how-to home remodeling books. Good how-to books on carpentry, wiring, plumbing, and other tasks are available from bookstores, home improvement centers, and online retailers. The purpose of this book is to adapt those DIY skills to the specialized needs of a historic house.

The one basic guide to the renovation of houses that I most recommend owning is *Renovation: A Complete Guide*, by Michael W. Litchfield. First published in 1982, the book is now in its fifth edition (Taunton Press, 2019) and remains the most complete and understandable basic text on the subject. It was the required textbook for a continuing-education program in renovation technology that I took in 1985, and I still have that well-worn edition and three subsequent editions as well. If you are rehabbing a house, buy this book.

The featured homes presented between chapters of this book represent a range of periods and architectural styles spread across the United States. Some of them underwent extensive rehabilitation; others were rehabbed or updated in less dramatic ways. Some were rehabbed by do-it-yourself homeowners alone, some by do-it-yourselfers working with hired contractors, and a few by professionals working without hands-on owner participation. Two of the houses have been a labor of love for their owners for more than thirty years. Others were rehabbed in a year or two. The majority are year-round family homes. Several are operated as bed-and-breakfast inns, one is a summer cottage, and another is a winter home in the Southwest. One house accommodates visiting artists at a summer art school. What they all have in common is that their owners have approached their rehabilitation with sensitivity and a desire to preserve or restore as much historic character as possible. While none of the houses represents *all* of the recommendations made in the following pages, they illustrate many of them.

PART ONE

Project Planning

Too many people learn too late that good planning is an essential prerequisite for a manageable house restoration project. The first five chapters aim to help you avoid being one of those people.

Part One will help you identify what kind of house is right for you; understand its stylistic identity and its important character-defining features; learn how to preserve those features in your restoration design; and navigate assorted bureaucracies to obtain the required permitting. Much of my work as a historic preservation consultant is focused on these issues, helping my clients plan a successful project that can be completed on time and on budget. Time spent on these issues up front always saves time and money down the line.

While this part of the book focuses on what you want and need from a house, it also addresses what the house needs from you. Every owner of a historic house wishes at times that the walls could talk. What stories they could tell! Underlying this book is the idea that every owner of a historic house is as much steward as owner. We inherit something of value—materials, workmanship, and character—from a succession of prior owners. And with that legacy we inherit an obligation to preserve it for future owners while meeting our own needs and those of our families as long as we occupy the house. Ideally, we will pass along a house that is better prepared to serve the stewards who follow while respecting those who came before.

This duality of purpose, meeting our needs while acknowledging and honoring the temporary status of our ownership, is what makes restoring a historic house different from merely renovating an old house.

CHAPTER ONE

Finding the Right Historic House

Taos Pueblo in New Mexico is among America's oldest historic houses. Inhabited for a thousand years, it continues to provide housing to descendants of the builders. Few people in America will ever live in a thousand-year-old house, but the mere existence of a dwelling this old provides useful context for our efforts to preserve much younger houses. It also makes a mockery of the sadly common idea that a house should be demolished simply because it is old. *David Clough photo*

A successful restoration begins with the house you choose, and the discussion that follows can help you make that choice. Even if you already own the house you're going to restore, you'll find helpful planning information here.

When Is an Old House a Historic House?

The answer to this seemingly simple question depends very much on the respondent's perspective. In my historic preservation work, I occasionally hear someone say, "It's just an old house. It isn't historic." When I ask what that means, I generally hear, "No one famous lived here" or "It wasn't designed by a famous architect" or "It's just like lots of other old houses, nothing special." From perspectives like these, only a small number of American houses would be considered historic.

These might include National Historic Landmark (NHL) houses, for example. Just over 2,500 NHLs were scattered across the United States in 2017—not a large number—and many of these are historic sites with buildings other than houses (such as churches and city halls) or without buildings at all, further reducing the number of homes that are "historic" under such a restrictive definition.

Thomas Jefferson's architectural masterpiece, Monticello, is a National Historic Landmark in Charlottesville, Virginia.

A slightly broader definition includes houses listed on the National Register of Historic Places (NRHP). These can be of national, state, or local significance; of the approximately 65,000 National Register properties, perhaps half are residential buildings. All NRHP buildings are recognizably old, and many are outstanding examples of an architectural style. Adding these to the NHLs gets us to perhaps 35,000 homes that could be considered historic by strict standards.

Another perspective on historic homes is provided by the requirements for inclusion as a "contributing" building in a local or National Register historic district. Inclusion depends on a building's age and its "architectural integrity"—the degree to which it is an intact or recognizable representative of the "period of significance" for the district's "area (or areas) of significance." For example, a neighborhood of Greek Revival and Italianate homes might be significant in the areas of architecture and community development, with a period of significance from 1835 to 1875 if the neighborhood was largely developed in those years. A house within the district that remains recognizable from that time would be considered to contribute to the historic character of the district and would therefore be historic itself.

A Ranch-style house built a century later in that Greek Revival/Italianate neighborhood would not be considered as contributing or historic, but it could well be a contributing building if built in a post–World War II neighborhood development of Ranch houses that is a designated historic district in the area of architecture and community planning for the period 1945 to 1965. As a general rule, fifty years is considered the minimum age necessary for a building to acquire historic significance.

Not all historic districts were developed in such a brief period. Many include buildings that, although built in numerous styles over a long period, are

The Reuben Coburn House (Pittston, Maine) is individually listed on the National Register of Historic Places.

related by their area(s) of significance. Also, not all local historic districts meet the same requirements as National Register districts. There are approximately 13,600 National Register historic districts and 2,500 locally designated districts in the United States. Collectively, these districts add tens of thousands of houses to the nation's inventory.

There are, of course, many historic houses that are not included in historic districts but *could* be. There are currently more than 100 million houses in the United States, nearly half of which are more than 50 years old.[2]

Many of these have been altered to such a degree that little historic character remains, and it would be difficult to see them as historic houses under the definitions explored above. But if only forty percent of these houses retain enough integrity to be considered contributing buildings in a historic district, there could be nearly 20 million historic houses in the United States.

As these examples make clear, there can be many ways for a house to be considered historic. For the purposes of this book, a house will be considered historic if it is fifty or more years old, is recognizably of that age, and is considered historic by its owner or potential owner.

The Lewis-Wheeler House is a "contributing" building in a local and National Register Historic District.

This postwar Modern Movement house built in the 1960s would be "non-contributing" in a historic district of nineteenth-century houses but "contributing" in a neighborhood of postwar houses.

Is This One the Right One?

The goal, of course, is not just to find a historic house but to find one that is right for you, or at least one that can be *made* right for you with minimal damage to its historic character. If you care about historic houses, don't buy one you have to destroy to adapt. Although that may seem obvious, it happens all the time. Let the answers to the following questions guide you to the house you need and want.

Does it match your lifestyle?

Most historic houses were built to accommodate lifestyles that were very different from a typical modern American household. If you favor open floor plans

[2]There were 37 million houses in the U.S. in 1940, and 57 million as of 1960. See www.census.gov/hhes/www/housing/census/historic/units.html.

These nineteenth-century Queen Anne–style houses are not on the National Register or located in a local or National Register Historic District but are clearly historic.

This nineteenth-century vernacular house was the home of my great-great-grandparents, Civil War veteran W. H. Hanson and his wife, Lucy, and was the birthplace of my paternal grandfather in 1901. It is not on the National Register or located in a historic district, but it is historic to me.

with minimal divisions between the kitchen, living room, and dining room, you will find yourself at odds with the separated and enclosed spaces of most historic houses. Focus on those few historic house styles that tend toward more openness, such as Shingle Style houses of the late nineteenth century and Craftsman homes of the early twentieth century. Prairie-style houses and many post–World War II houses feature open plans as well.

Middle-class houses of the mid- to late-nineteenth century were often designed on an assumption that servants would do the cooking, cleaning, and laundry. In consequence, kitchens and laundry rooms were often segregated from owner-occupied spaces. Few of us have servants today, and kitchens are often the center of household activity. Fortunately, there are ways to resolve such issues sensitively, and these will be explored in later chapters.

Before buying an old house or rehabilitating one you already own, consider how you and those with whom you share your home live currently. Do you wish for more privacy or less? Does television noise drive you crazy while you are trying to read a book? Do you entertain formally, informally, or not at all? How do you celebrate holidays? How many people need to use the bathroom, eat breakfast, and get out the door by 8:00 a.m.? The answers to these and other such questions will help you select an appropriate house.

Beyond functional requirements, what are the elements of character that say "home" to you and your family? Are you comfortable in formal settings, or do you prefer cozy and casual? Do high ceilings, molded plaster ornamentation, and hardwood floors make you happy, or do you prefer low ceilings and wide board floors?

Does it have the features that are essential to you?

We all have essential requirements for a home that will suit our lives, and it is important to identify yours. What is the minimum space you and your household need to coexist happily? What is the maximum space you want to care for? Are you a cook who needs a serious kitchen, or someone who eats out regularly and mostly microwaves at home? Do you have hobbies that need their own spaces away from the dining room table? Does anyone need a dedicated space for working from home? If personal mobility is an issue, a house with bedrooms on the first floor might be essential. Do you have children who need a large yard to play in? Are you or other members of the household involved in sports or other recreational activities that require storage space for equipment? What about the needs of your pets? Do friends or family visit regularly or for extended periods? Figuring out which requirements are essential will help you find the right house and rehabilitate it in a way that meets your needs.

Does it have features that are desirable to you?

Few of us consider a working fireplace essential, but many find it desirable. If you enjoy growing houseplants, a good southern exposure might be important. Multiple bathrooms might be desirable though not essential, or perhaps you consider it desirable but not essential that your historic house have undergone

This nineteenth-century Greek Revival house with a Second Empire addition was the home of my great-great-great-grandparents in the 1850s, and it's the house in which my father grew up in the 1940s and 1950s. It is one of the millions of historic homes in the United States that are not officially designated as such.

minimal changes prior to your ownership. Taking the time to think about such factors and discuss them with your household prior to looking for a house—or at least prior to designing changes—will save time, money, and possibly relationships.

Does it offer one or more rental units to help with the mortgage?

Many historic homes have been divided into apartments or contain underutilized spaces that could be converted into long-term rental units. It is also possible to generate significant income from short-term rentals using online platforms like Airbnb, though this will be more work than renting on a long-term basis. Many towns regulate short-term rentals, and some restrict them; find out how your town handles this before committing yourself to the idea. Rental income often makes the difference between an affordable mortgage and one that is out of reach, but you have to be ready to be a landlord.

Is the location good?

According to the old saying, the three most important factors in choosing a house are location, location, and location. All the usual considerations apply to the purchase of a historic home—property taxes, commuting distances, children's schooling, long-term property values, and others—but there are additional considerations too. Is the house surrounded by other historic homes, providing a historic context for the property? Is the house in a local or National Register historic district? Have nearby historic homes been rehabilitated, indicating a neighborhood with an appreciation of the houses? As a general rule, homes in historic neighborhoods, particularly historic district neighborhoods, hold their value better during economic downturns and appreciate more during good times than other houses. A neighborhood where good historic housing stock is being sensitively rehabbed can be a good place to invest. A neighborhood where vinyl siding salesmen are having a good year may not be quite as promising if your goal is to rehab a historic house in a careful and thoughtful manner.

CHAPTER TWO

What Are the Character-Defining Features of Your House?

A classic Second Empire house.

As mentioned in the Introduction, a *restoration* project returns a historic building to its original structure and form; it does *not* adapt a building to new uses and new ways of living. Museum buildings are often restored, but you are much more likely to *rehabilitate* your historic house—incorporating modern functionality with historic character—than to restore it. A rehabilitated building offers a value to the community without which it would likely be neglected and eventually lost. *Historic preservation* is a term that applies to either restoration or rehabilitation. To ensure that your rehabilitation project is a sensitive historic preservation, *it is essential to understand the character-defining features of your house.*

Evaluating Your House

To begin, what is the age of your house, what is its architectural style, and how has it changed over time? Few historic houses stand today exactly as they were built. Most have had kitchens and bathrooms added or updated, at minimum. Often, the areas that have been changed present the best opportunities for further changes, but sometimes earlier changes have acquired significance of their own. To sort out these questions, you can either do the research yourself or hire an architectural historian to do it for you.

Hiring a professional

The greatest advantage of hiring a professional is that the work can be done quickly. An architectural historian will approach the project with a thorough knowledge of historic styles and years of experience looking for evidence of past changes. He or she will know where to find local records, historic photographs, and maps that can provide important information about your house. You can request that the conclusions be delivered in a written report with captioned photographs of character-defining features that will make future discussions with an architect or builder clearer and more concise. Miscommunication between owner and contractor is often at the root of problems on a rehab construction site, and lack of a common vocabulary frequently plays a part. A determined amateur can acquire sufficient knowledge and ferret out local historic archives eventually, but a professional can do it faster.

For example, my firm was hired by a couple who wanted to donate their property, including an early-nineteenth-century house, to a local land trust. The trust was more interested in the acreage than the house and would likely sell off the house on a small piece of land at some future time. Having spent thirty years rehabilitating the house, the owners wanted to protect its key features in a preservation easement that would remain with the house when the trust sold it. My job was to document the house's character and condition and identify the features to be preserved and "areas of opportunity" where future owners could make whatever changes they wished.

In a six-hour site visit, I examined and photographed the house, ell, and barn inside and out. I discovered interesting moldings for a house of its time and place, and I took molding profiles from exterior trim and principal rooms. I interviewed the owners about the work they had done and what they knew about changes to the house before their ownership, and I took away with me scanned copies of their deed research and several historic photos.

The deed research revealed when the original owner—a farmer and bootmaker—purchased the land, and this, along with corroborating physical evidence, enabled me to date the house. The house had been built with heating fireplaces in the principal

The historic Ezekiel and Mariam Record farm.

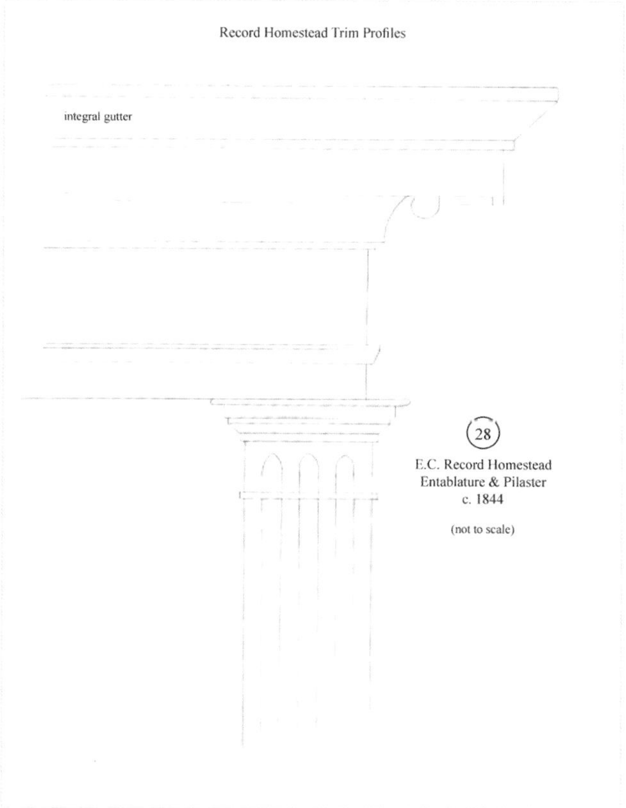

The eave, frieze, cornice, and corner board trim of the Record House documented in a sketch.

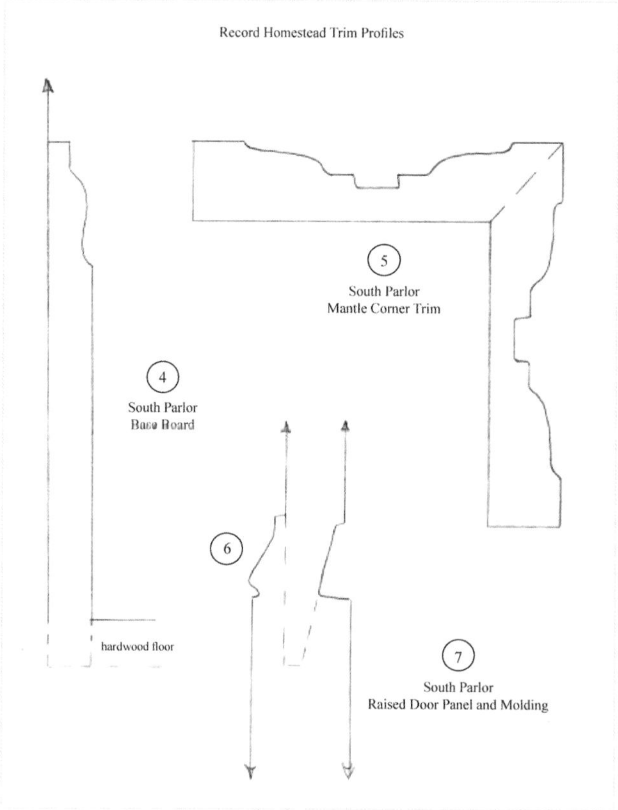

Molding profiles taken with a profile gauge and recorded on paper.

rooms but no cooking hearth and brick oven in the kitchen. There was a narrow time window when cast iron stoves had replaced hearths and brick ovens for cooking but had not yet replaced fireplaces for heating, and that narrow window corresponded with the land purchase date. This dated the house to about 1844.

The Greek Revival style was up to date when it was built, and some features of the house could be traced to an Asher Benjamin builder's guide of the period. Other features were interesting local variations on Greek Revival elements that could also be found on several nearby houses of the same period. It was clear that the original owner, who was descended from one of the oldest families in his rural community, intended the house to make a strong stylistic statement without being large or ostentatious.

Subsequent alterations ranged from barely discernible old changes, like a relocated cellar door, to obvious changes like the glassing in of the open porch. Historic photos documented the original appearance of the porch. The windows of the house had been replaced three times. The original six-over-six double-hung windows had been replaced by two-over-two windows in the early twentieth century. After purchasing the house in the 1970s, the current owners had salvaged historic sashes of the right size from the local dump and restored the windows to their original six-over-six configuration. While aesthetically

Open fireplaces for heating the principal rooms and the absence of a cooking hearth and brick oven in the kitchen are clues that the Record House was built in the early nineteenth century. A cast iron heating stove similar to this one was likely installed in this fireplace not long after the house was built, due to the greater efficiency of stoves for heating.

successful, the reused windows from other buildings never fit well, causing energy efficiency and comfort issues in the winter. Eventually, the owners replaced these repurposed windows with compatible new six-over-six wood windows with insulated glass—retaining their historic appearance while enhancing their efficiency and comfort. They also installed a modern kitchen and two bathrooms and remodeled the attic space in the ell to serve their home business. Historic photos and exposed framing on the interior of the ell showed minor changes to doors and windows over time. Although newer than the house, the barn was largely intact from the time of its construction, with one side partitioned off for the owner's woodshop.

All this and more was detailed in our report, which recommended areas to preserve and identified areas of opportunity for future owners. In short, the largely intact and most significant historic rooms—the two parlors, front hall, and principal bedrooms—were worthy of preservation and protection, while the kitchen, baths, and ell contained little of historic significance and were open to future changes. The view of the house from the roadways—its exterior form, materials, and setting—was character defining and worthy of protection, but changes would be acceptable on the elevations of the house, ell, and barn that were not visible from the roads. The documentation created for the preservation easement found additional use when the owners chose to have the house listed on the National Register of Historic Places. I wrote the nomination from the same materials.

In summary, the advantages of hiring a professional architectural historian to assess your house are several:

- An experienced professional has already looked at hundreds of houses of many periods and can place yours within a wide context of styles and typical alterations over time.
- He will be able to tell you what style the house is and what characteristics of that style are present.
- She will have a trained eye for spotting the often-subtle indications of changes that have occurred over time.
- If prior restoration work has been done, the architectural historian will often be able to identify which features are historic and which are not.

- Often, a professional will have experience and knowledge of restoration craftspeople, materials, and other sources that can be included in the report.

Doing it yourself

While hiring a professional to assess your house has advantages, so does doing the work yourself. This approach is likely to take longer but will be more educational than reading a report prepared by someone else. The broader understanding you will gain of architectural styles and materials will help your rehab project in many ways going forward. Understanding how a house was lived in and functioned in the past can suggest ways to live in it today that you might not otherwise have thought of. For example, traditional passive systems for controlling heat, light, and cooling can unite your historic house with modern concepts of green and sustainable living. Assessing

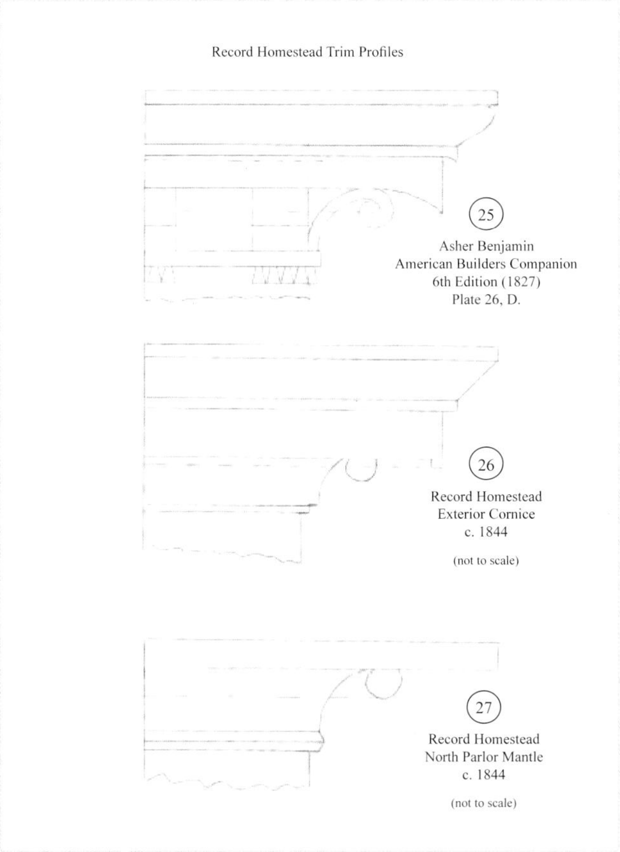

A sketch showing design elements from Asher Benjamin's *American Builders Companion*, 1827, and two simplified versions of the design used in building the Record House c. 1844.

The Record House corner board, frieze, and eave trim is quite elaborate for a modest rural home of the period. The level of detail seen in the trim inside and outside the house tells something about the people who built it.

Late-nineteenth-, mid-twentieth-, and early-twenty-first-century photos document changes to the Record House over time.

the historic character of your house will also acquaint you with its condition by forcing you to examine it closely from basement to attic, inside and out. The resultant inventory of the existing condition of parts and features will save you time and money on every stage of the rehab project that follows.

The structure of your house contains a great deal of information about its history (that's why its preservation matters!), but so do documentary sources. Deed research is a good place to start. Nearly every property leaves a paper trail of its ownership through the years, generally in a county or municipal registry of deeds office. Some jurisdictions have lost early records to fire or other causes, complicating the research, but unless you encounter this particular piece of bad luck, you can trace the property back to its first owner. Deeds contain a written description of the property, which generally notes whether buildings were present. Often you will find that the lot was originally part of a much larger lot for a house that is not your house. Read the descriptions carefully and watch for references to "improvements" and dramatic changes in the price paid for the property. A large increase in price over a short time often indicates that a house has been built or added to. A large decrease in price could indicate that the lot was divided or that a building burned or was moved off the lot.

Deeds will also identify all the owners of your property through the years. This opens the door to researching their economic status, which can be useful in trying to determine when changes were made. Changes are frequently made shortly after a house is purchased, as the new owners make updates and improvements to meet their family's needs. This was as true in 1819 as in 2019.

Another primary source of information on your house and its owners is found in probate court records. Like the registry of deeds, this is usually a county function. The probate court oversees the execution of wills and transfer of property when a person dies intestate (without a valid will). Generally, these records will include an inventory taken to document the value of an estate at the death of its owner. Such an inventory will describe land and buildings in detail, and often lists the contents of the house. With a bit of genealogical research, probate inventories of the parents or grandparents of former owners can be located and may identify sizable inheritances. Knowing when a former owner received a large sum of money can help you date a remodeling or expansion project.

Local property tax records are also a valuable source of information, since taxes are based on the assessed value of the land and buildings. Sometimes the local tax assessor has these records, sometimes they've been transferred to a historical society, and unfortunately, sometimes they are lost. Tax maps and other historic maps can provide useful information. If your house is in a developed neighborhood, it may be included on Sanborn fire insurance maps. The Sanborn company began documenting built-up areas of towns and cities with detailed maps showing building footprints marked with symbols to indicate a building's height, materials, uses, etc. Starting around the time of the Civil War, these maps were updated periodically to show changes since the last map. These are among the most important sources of historic documentation for buildings. Unfortunately, rural and suburban buildings are seldom documented on Sanborn maps.

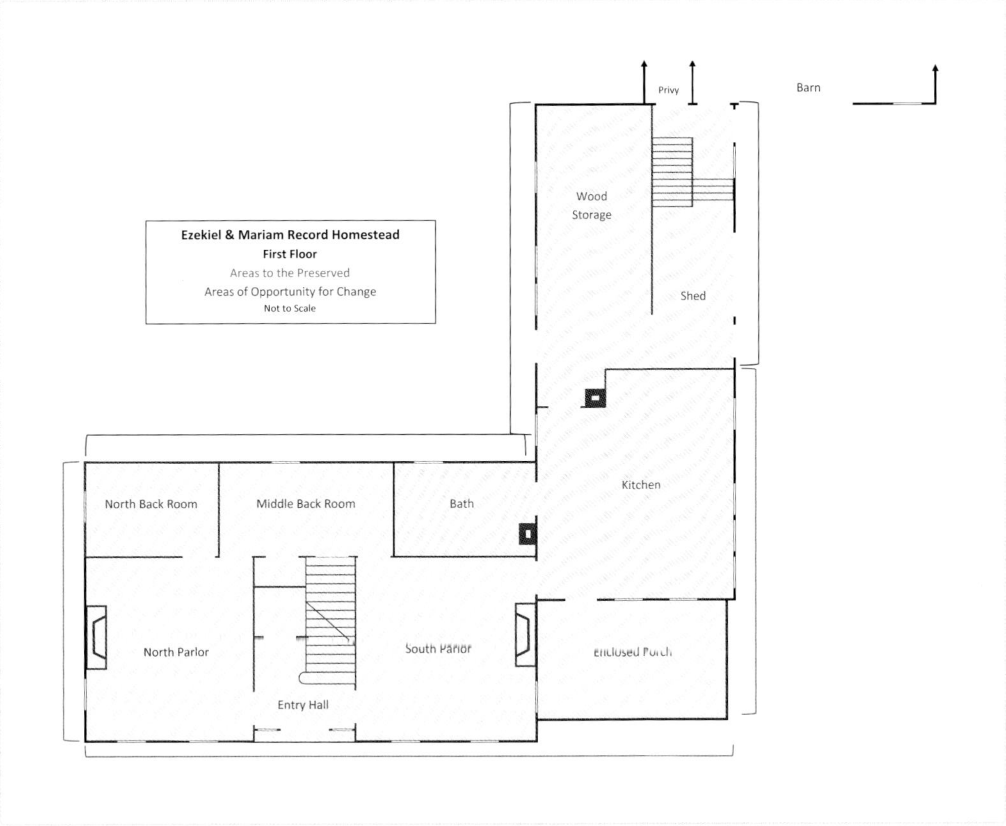

First-floor plan of the Record House, including the ell and shed, showing areas to preserve in a rehabilitation and areas where future alterations can be made without affecting the most important character-defining features of the house.

Doing the genealogical research to learn about past owners has become far easier with the availability of Ancestry.com and other online sites that provide fast and easy tools for locating historic census data and other records. Many local libraries have subscriptions to these sites and allow free access to them.

Identifying the Age and Style of Your House

Most houses in America are relatively easy to date—at least approximately—because styles and construction technologies changed over time. If later alterations have not greatly obscured original features, a historic house will likely exhibit enough characteristics of its place and time of building to allow an educated estimate of its age. You'll need a basic knowledge of architectural styles and when major changes in construction technologies occurred. A number of well-written and well-illustrated guidebooks cover historic American architectural styles in considerably more detail than I can include here. I recommend Virginia Savage McAlester's *Field Guide to American Houses (Revised): The Definitive Guide to Identifying and Understanding America's Domestic Architecture* (Knopf, 2013). See Chapter 20, Resources, for additional suggestions.

A brief overview of historic American residential architectural styles will give you the baseline knowledge to understand the style references in this book. It is important to note that stylistically "pure" houses are fairly rare. The shift from one style to another often happened incrementally, with some elements of a new style mixed with elements from a style the builder was already accustomed to working in. The homes of the wealthy were more likely to be a pure expression of a new style than the homes of the middle or working classes. But a builder who acquired new molding plane profiles to build an up-to-date home for a well-to-do customer could then use those profiles on subsequent homes for less wealthy customers. Then too, later but still historic alterations often impact the stylistic purity of a house.

The major styles in American domestic architecture are presented here in roughly chronological order. Some were regional, while others appeared across the country. The photos illustrate some regional variations but can only begin to show the great variety created within each style by thousands of architects and builders across the nation and over the decades.

There is often confusion about the difference between a house "type" versus an architectural style. A house's type is generally based on its form and sometimes on the materials it is constructed from. There is no "Cape Cod" *style* of house. The Cape Cod, or cape, is a *type* of house with a distinctive form that appears in a variety of styles—Georgian, Federal, Greek Revival, etc. Among the houses featured in this chapter, the Dow Farm is a Federal-style cape and the Record House is a Greek Revival–style cape. Other common types include the saltbox and bungalow, each with a distinctive form. Making the matter more confusing, some types *are* essentially styles. These include adobe, rustic, and vernacular houses, which are appended to the Styles feature that follows.

"Type" distinctions are not well understood. Much of the terminology familiar to the public comes from the real estate industry, not architectural history, and that industry often uses type interchangeably with style. In this book, I have attempted to keep the distinction clear except where common usage of a technically incorrect term is so deeply ingrained that it would only create more confusion to use the correct term.

Colonial

First Period Colonial (1600–1740)

Sometimes called "Post-Medieval," these houses are rare and most often survive as much-altered portions of larger houses that resulted from later additions. Any intact features from the seventeenth century should be considered of great significance.

1. Close to the ground (typical in the northern U.S.)
2. Elevated basement (typical in the South)
3. Steeply pitched roof
4. Cross-gabled roof
5. Saltbox roof
6. Exposed half-timbering (typical in the mid-Atlantic states)
7. Cruciform chimney
8. Center chimney (North)
9. End chimneys (South)
10. Overhanging second story
11. Minimal roof overhang
12. Minimal ornament
13. Board doors with exposed nailheads
14. Casement windows
15. Diamond-pane windows (often leaded)
16. Small glass panes in windows

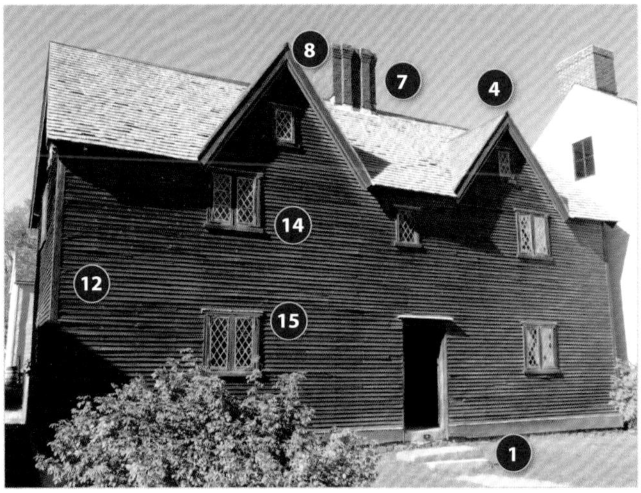

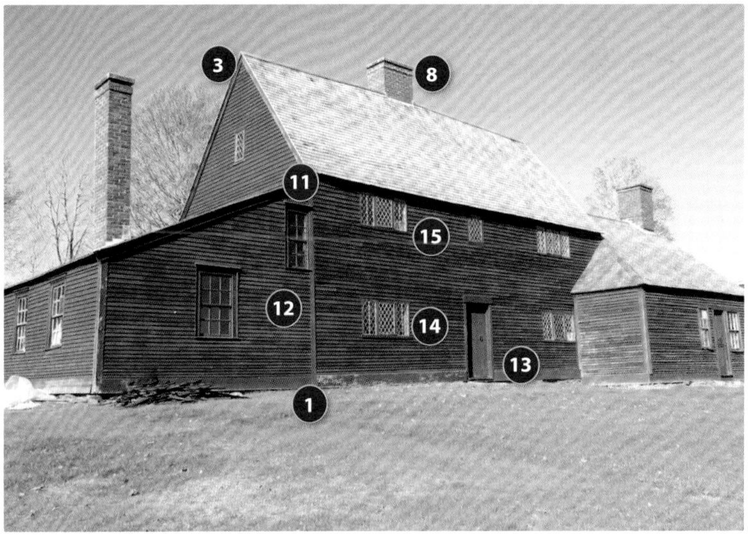

Courtesy of Wikimedia

Courtesy of Brian C. Rainville

Courtesy of Wikimedia

Courtesy of Andres Verzosa

Courtesy of Elmer Shelton

Georgian Colonial (1730–1790)

There are English, Dutch, French, and German variations on the Georgian theme, reflecting the construction technology carried by European settlers to various regions of the United States. These houses utilize Classical architectural forms and details revived across Europe during the Renaissance. Spanish Colonial often uses a different vocabulary and different materials, drawing on the Native American building traditions and technology of the Southwest while incorporating some European architectural ideas.

1. Symmetrical façade
2. Close to ground (typical North)
3. Elevated basement (typical South)
4. High hipped roof
5. High gabled or gambrel roof
6. Pent roof
7. Pedimented door surround
8. Portico
9. Pedimented dormers
10. Center chimney (typical North)
11. End chimneys (typical South)
12. Gambrel roof
13. Arched transom
14. Classical detailing
15. Small-paned windows
15. Wood siding imitating ashlar stone
16. Plank-frame windows (New England)

Courtesy of Wikimedia

Georgian Colonial (1730–1790)

Courtesy of Wikimedia

Courtesy of Brian C. Rainville

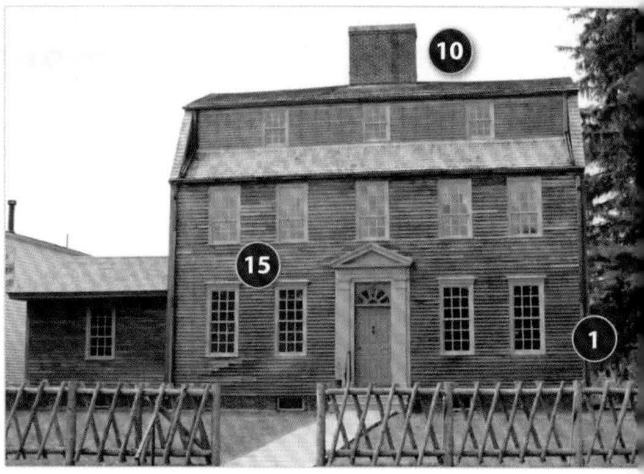

Georgian Colonial (1730–1790)

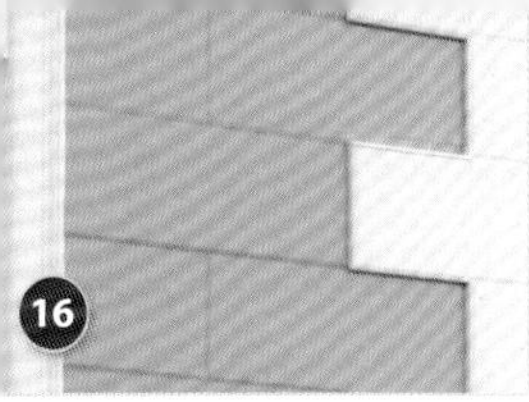

Early National

Federal (also known as Early Classical Revival) (1790–1830)

These houses draw primarily on English precedents but are considered the first "American" style because they post-date the adoption of the U.S. Constitution. There is somewhat less regional variation than in the earlier colonial period styles.

1. Symmetrical façade
2. Elevated site (large houses)
3. Low hipped roof
4. Low gabled roof
5. Palladian window
6. Attenuated molding profiles and columns
7. Portico
8. Triple- or double-hung windows (shorter on upper stories)
9. Balustrade around roof
10. Elliptical arch transom or fan
11. Transoms and sidelights often leaded
12. Chimneys generally at outside walls (North and South)
13. Minimal roof overhang
14. Balconies with iron railings (urban settings)

10
6

13
1
6

3
13
10
6

11
6

12
10
13
1

6

13
1
14

3
12
13
10
1

6
10
7

4
12
9
13
8
6
14
10
1

3
13
6
1
2

12
4
13
6
14
8
1
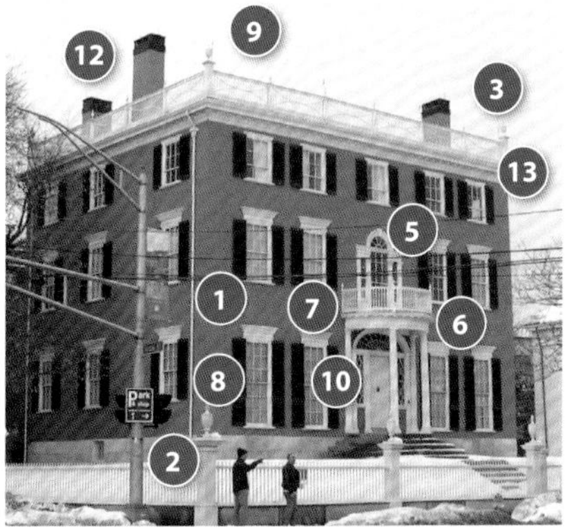
12
9
3
13
5
1
7
6
8
10
2

12
13
10
1
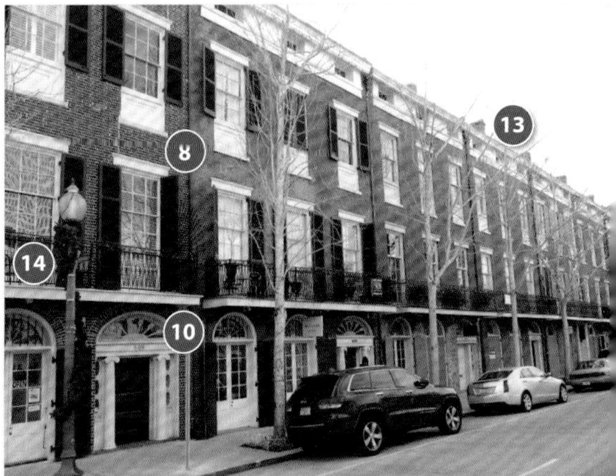
13
8
14
10

Greek Revival (1820–1860)

Greek Revival was the first truly national style in the United States, with similar houses built in nearly all the settled parts of the country. Houses built from the popular Builder's Guide books can be virtually identical from Maine to Oregon. Southern examples are frequently set on higher basements and often have more or larger porches than Northern and Midwestern examples. Greek Revival detailing can even be found on vernacular adobe buildings in the Southwest.

1. Emphasis on pedimented gable
2. Gable end to street
3. Temple-form portico
4. Monumental columns or pilasters
5. Bold molding profiles
6. Tall frieze
7. Wide corner boards
8. Doric and Ionic capitals
9. Classical motifs in carved ornament
10. Classical motifs in iron ornament
11. Porches and balconies
12. "Greek peak" lintels
13. Pilastered door surround with entablature
14. Columned door surround
15. Transoms and/or sidelights
16. Recessed entry

2
1
8
4
11
3

5
10
11
3

1
8
11
4
3
7
3

12

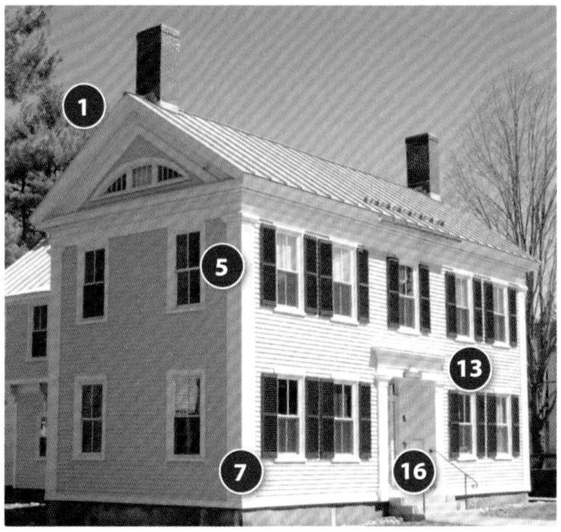
1
5
13
7
16

13
6
14
15

11
10
8
14
16

8
3

2
1
8
4
11
3

6
10
8
5
4
10
14
16
16
PINE ST
PENOBSCOT ST

1
6
7
3

6
7
14

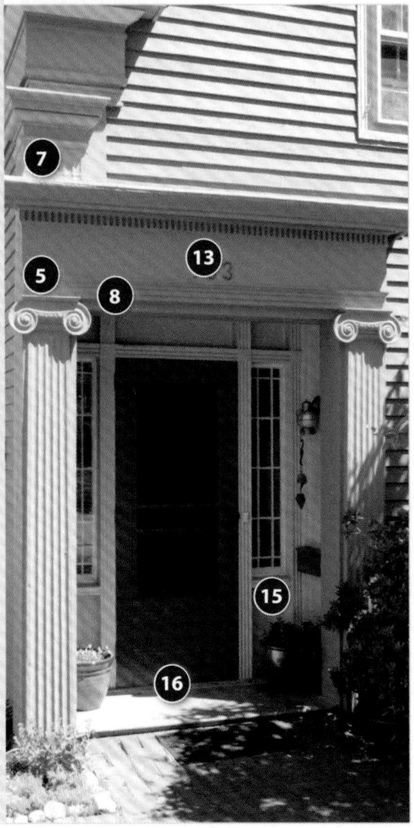
7
13
5
8
15
16

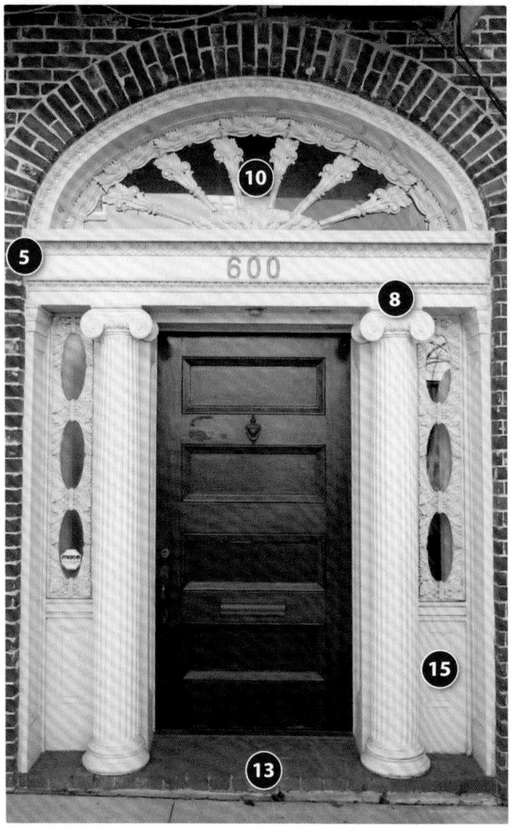
10
5
600
8
15
13

11
5
8
16
10

6
15

Greek Revival (1820–1860)

Courtesy of Brian C. Rainville

Early Victorian

Gothic Revival (1830–1870)

Romanticism swept the United States in the early nineteenth century, first in literature and art and then in architecture. The Gothic Revival was inspired by the Waverly novels of Sir Walter Scott and by a rediscovery and new appreciation of Gothic architecture in Europe. It was popularized in America by the house design books that were increasingly common from the mid-1830s onward. The leading authors of these books were Alexander Jackson Davis and Andrew Jackson Downing. Downing's *Rural Residences* (1837), *Cottage Residences* (1842), and *The Architecture of Country Houses* (1850) were particularly influential.

1. **Steeply pitched gabled roofs**
2. **Cross gables**
3. **Pointed arch doors and windows**
4. **Board-and-batten siding**
5. **Wood "ashlar stone" siding**
6. **Carved or scroll-sawn bargeboards at eaves**
7. **Crockets, trefoil, quatrefoil, and other Gothic detailing**
8. **Porches**
9. **Bay window**
10. **Occasional tower**

Courtesy of Wikimedia

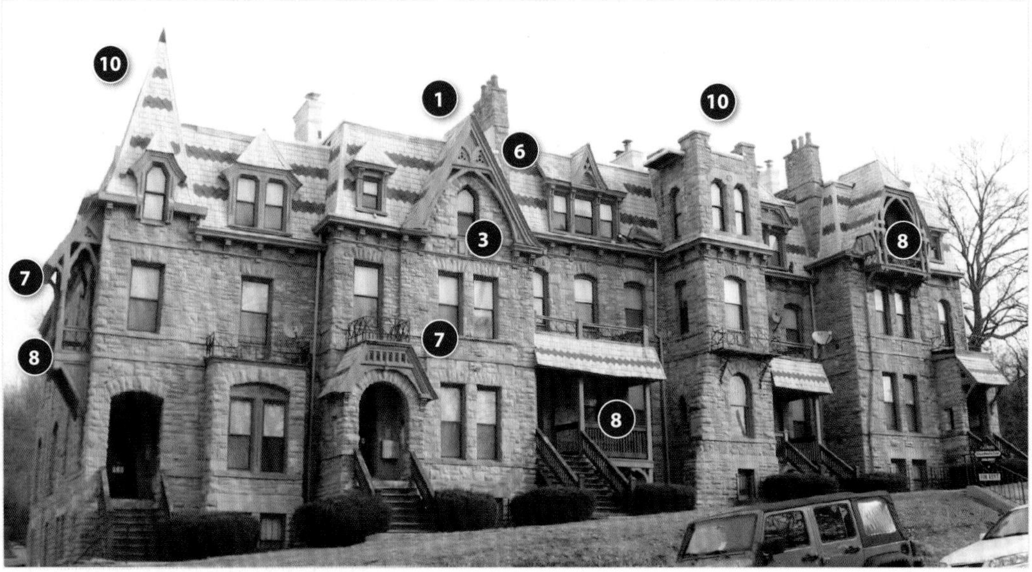

Courtesy of Wikimedia

Courtesy of Wikimedia

Courtesy of Wikimedia

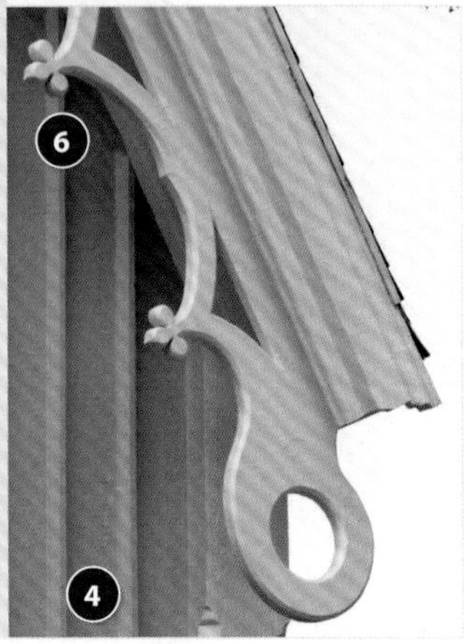

Italianate (1850–1875)

The Italianate style, like most popular American architectural styles, started in Britain and spread to the U.S. via popular magazines and architectural publications. Queen Victoria's country residence on the Isle of Wight off the south coast of England, Osborne House, completed in 1851, played a significant role in popularizing the style. This style was also spread by the numerous house-plan books being published by Andrew Jackson Downing and others.

1. Asymmetrical form
2. Low hipped roof
3. Gabled roof
4. Flat roof
5. Round/segmental arched openings
6. Bracketed eaves and porches
7. Deep overhanging eaves
8. Double doors
9. Bay windows
10. Square "campanile" tower
11. Cupola/belvedere on roof
12. Often brownstone
13. Quoins at corners
14. Wide paneled corner boards
15. Porches

Italianate (1850–1875)

Italianate (1850–1875)

Later Victorian

Second Empire (1860–1900)

These mid- to late-Victorian styles were popular at a time of great growth and expansion of the nation and its wealth, resulting in many examples being built. The Second Empire and Queen Anne styles were truly national, with examples from Maine to Hawaii. The Stick and Shingle styles were more regional, with most (though certainly not all) examples located on the East Coast.

1. **Mansard roof**
2. **Dormers**
3. **Patterned slate roofing**
4. **Round/segmental arched openings**
5. **Bracketed eaves/porches**
6. **Classically inspired detailing**
7. **Double doors**
8. **Mansard tower**
9. **Cupola/belvedere**
10. **Quoins at corners**
11. **Wide paneled corner boards**
12. **Bay windows**
13. **Porches**
14. **Iron cresting**

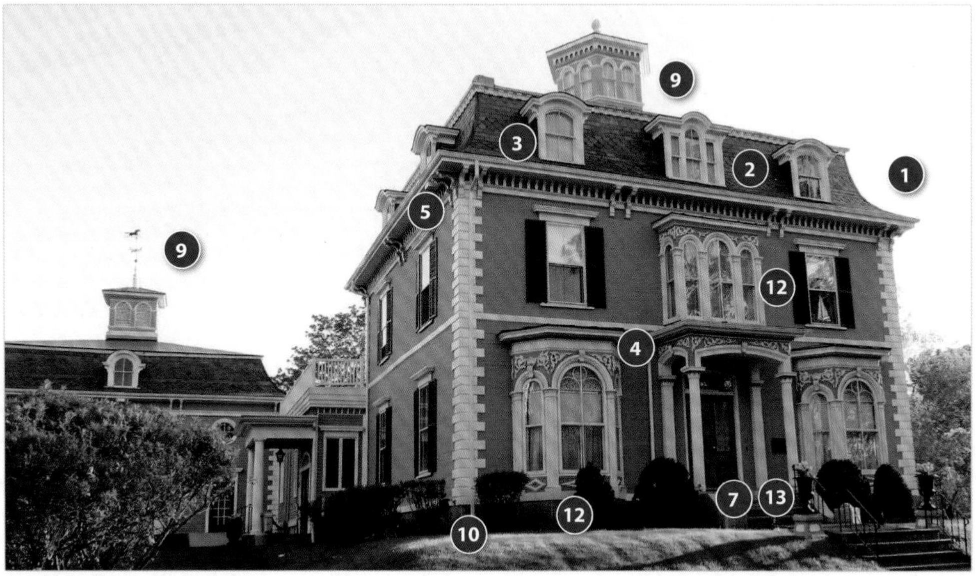

14
8
3
2
1
4
5
5
7
13
8

1
2
5
12
6
8

1
2
5
10
8
7
13

3
1
2
10
L'Aiglon
1309
4
6

1
2
3
4
5
7
7

3
2
1
6
5
13
3
8

2
3
1
5
4
5
6
12
13
7

1
2
5
8

2

8
2
1
5
6
4
13
10

4
13
7
6

1
5
7
8

2
9
1
12
6
7

1
2
4
5

1
2
4
12

1
2

2
1
5
13

14
8
3
2
1
5
4

3
2

1
3

14
3

14
1
2
6
7
8

8
1
2
5
13
8

3
5

1
2
5
12
7
7

4
13
12

Queen Anne (1860 – 1900)

1. Asymmetrical form
2. Multiple roof forms
3. Round and octagonal towers
4. Wraparound/multiple porches
5. Turned porch posts
6. Spindlework and gingerbread trim
7. Varied window sizes and configurations
8. Leaded/stained glass
9. Multiple siding/trim materials
10. Cast terra cotta ornament
11. Carved wood/stone ornament
12. Elaborate chimneys

Courtesy of Wikimedia

Courtesy of Wikimedia

Queen Anne (1860–1900)

3
8
9

12
2
6
7
4

9
7
5

11
4

12
6

6
5

12
9
7

6
4

11
6
9

Queen Anne (1860–1900)

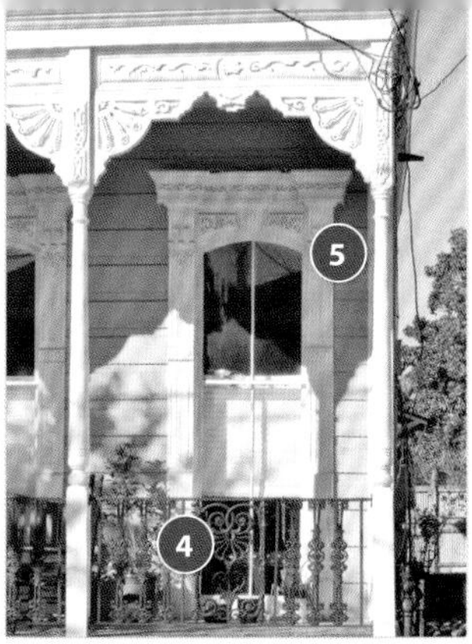

Courtesy of Wikimedia

Stick Style (1860–1900)

1. Often complex form
2. Wall surfaces often divided into panels
3. Elaborate eave and porch brackets
4. Trim may have exotic influences
5. May share characteristics with Italianate style
6. Vertical boarding often an element in trim

6
2
4
5
1

4
3
3

4

6
4

3
4

4
3

3
1

2
4
1

4
3
6

4

Shingle Style (1860–1900)

1. Complex forms
2. Brick or stone base
3. Gambrel roof
4. Saltbox roof
5. Double gabled roof
6. Multiple roof forms
7. Recessed porches
8. Wood shingle "skin" covering wall and roof surfaces
9. Flared "skirts" at floor plates
10. Towers
11. Dormers
12. Small-paned double-hung windows
13. Diamond-paned windows

11
13
4
8
9
64
7
1

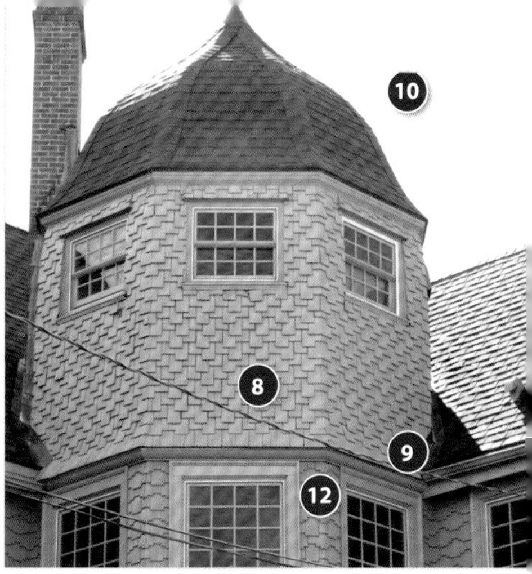
10
8
9
12

8
13
9

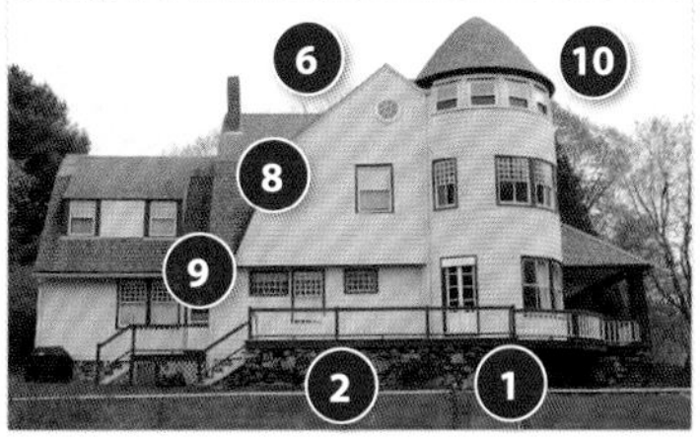
6
10
8
9
2
1

5
8
7
1
12
2

10
11
8
9
1

4
10
8
1

13
9

3
8
7
1

10
12
8
1

Richardsonian Romanesque (1880–1900)

Largely developed by Boston architect Henry Hobson Richardson, and bearing his name, this style looked toward the Romanesque period that preceded the development of Gothic during the Middle Ages. Typically executed in stone and brick, the style is massive and heavy and usually features wide, rounded arches in its construction.

1. Often complex form
2. Typically masonry
3. Heavy and often massive in character
4. Round Romanesque arches
5. Broad arches at entries
6. Short columns, often grouped
7. Rock-faced stone walls
8. Rusticated stone trim
9. Round towers
10. Recessed entrances
11. Sometimes heavy-timber porches
12. Carved stone trim

Courtesy of Wikimedia

Courtesy of Wikimedia

Courtesy of Wikimedia

Courtesy of Wikimedia

Courtesy of Wikimedia

Courtesy of Wikimedia

Eclectic and Exotic (1830–1930)

The Victorian taste for romanticism was sometimes expressed in fanciful works of architecture that defy categorization into the common styles. These often draw design influences from multiple sources and combine them in unexpected ways. There are plenty of examples created in the styles of the twentieth century as well. These homes are often unique, and every effort should be made to preserve their character-defining features. When one unique home is lost, they are all lost.

1. **Widely varied in shape and size**
2. **Middle Eastern design elements**
3. **Medieval castle design elements**
4. **Freely mixed elements from different eras/styles/places**

Courtesy of Wikimedia

Courtesy of Brian A. Bartlett

Courtesy of Wikimedia

Late-Nineteenth- and Early-Twentieth-Century Revival

The late-Victorian revivals again looked back to earlier periods for inspiration but tended toward more serious and less romanticized versions of the past, particularly as time went on through the last decades of the nineteenth century.

Colonial Revival (1880–present)

The Colonial Revival style emerged from the Shingle Style, which in turn borrowed architectural motifs from early Colonial period dwellings at first, incorporating increasingly formal Colonial elements as time went on. America's 1876 Centennial celebration awakened the nation to its own architectural history, and a definable Colonial Revival style emerged by 1890 and grew more scholarly and "correct" in the early twentieth century. It remains popular today.

1. First Period Colonial elements
2. Georgian Colonial elements
3. Dutch/German Colonial elements
4. Federal elements
5. Greek Revival elements
6. Mixed elements from Colonial and Victorian styles
7. Architectural use of elements from Colonial furniture
8. Has elements not found on original style examples (bay windows, sunporches, garages, etc.)

2
4
8
4

8
2

4
2

7

2

5

3

1
8

1
4

7
8
6

90
4

3
5
6

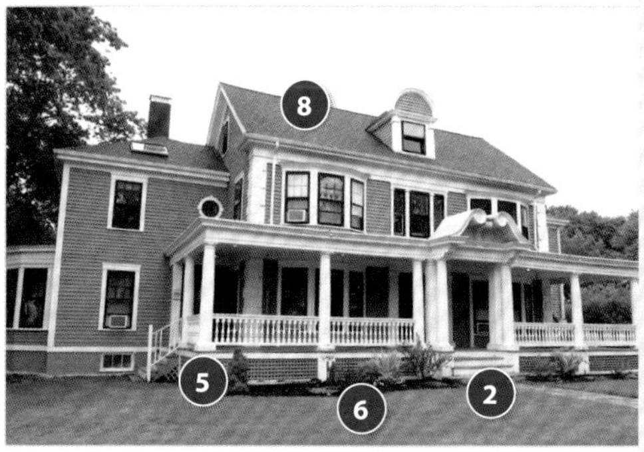
8
5
6
2

2
6

2

3
8

3
8
2

2

4

Neo-Classical Revival (1880–1940)

The Neo-Classical Revival style looked past the Colonial styles of the past to their European inspirations, using the architectural vocabulary of the Renaissance as expressed primarily in Italy, France, and England. These houses were formal and often grand. The style is more commonly found in urban settings but was also used for country houses of the wealthy.

1. **Grand and formal in style and character**
2. **English Classical elements**
3. **French Classical elements**
4. **Neo-Federal elements**
5. **Greek Revival elements**
6. **Monumental columns**
7. **Neo-classical ornamentation**

Courtesy of Wikimedia

Courtesy of Wikimedia

Spanish Colonial Revival (1880–1930)

Unlike the more popular Colonial Revival style—which looked to English, Dutch, French, and German buildings of the eighteenth and early nineteenth centuries for inspiration—this style looked toward Southern European architecture of the same period, often incorporating elements from those regions with elements of the more common Georgian Colonial Revival. A variation on the Spanish Colonial Revival is the Spanish Mission/Pueblo style (1900–1930)—often a Southwest and Californian variation on the Tudor, ranging from fantastical to grand and serious—which was popular among the rich and famous of 1920s Hollywood. Other variations were truer to Southwestern Native and Colonial building traditions and can be found blended with Craftsman forms in bungalows of the early twentieth century.

1. **Spanish Colonial elements**
2. **Spanish Colonial Mission elements**
3. **Georgian Colonial elements**
4. **Fanciful "Spanish" elements**
5. **Wrought ironwork**
6. **Tile roofs**
7. **Slate roofs**
8. **Stucco walls**
9. **Arched openings**
10. **Casement windows**

8
5
10
9
1

6
8
10

7
5
9
8
10
1

8
4

6
8
4

9
10
8

4
5
8

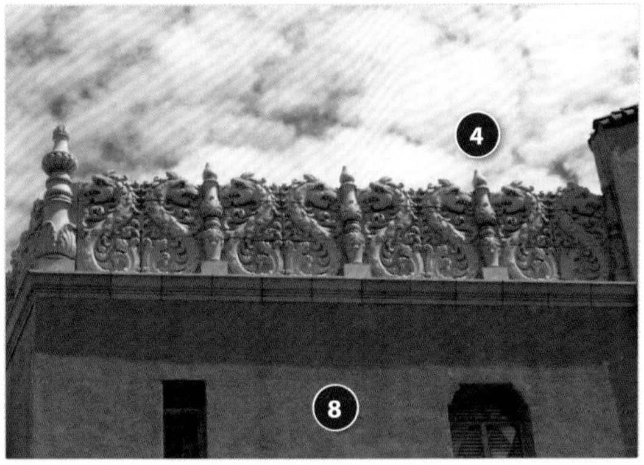
4
8

6
9
2
8

Tudor Revival (1890–1930)

The Tudor Revival style looked toward Queen Elizabeth I's England for inspiration in houses that tended toward elements of asymmetry in form. These ranged from "storybook" cottages to grand mansions, with the former a popular option in mail-order kit house catalogs in the early twentieth century.

1. **Asymmetrical forms**
2. **"Storybook" character**
3. **English Renaissance character**
4. **Brick**
5. **Stone**
6. **Stucco**
7. **Half timbering**
8. **Mixed materials**
9. **Arched entry doors**
10. **Leaded or steel windows**
11. **Casement windows**
12. **Slate roofs**
13. **Tile roofs**
14. **Expressive chimneys**

Courtesy of Wikimedia

Courtesy of Wikimedia

Courtesy of Wikimedia

Courtesy of Wikimedia

Courtesy of Wikimedia

Courtesy of Wikimedia

Tudor Revival (1890–1930)

Courtesy of Wikimedia

Early Twentieth Century

American Foursquare (1900–1930)

The American Foursquare house was perhaps the most popular catalog and plan-book house of the early twentieth century. Available in a range of sizes and with a variety of optional features, these houses could express Colonial Revival or "Victorian" characteristics, or an Arts & Crafts design aesthetic of the late nineteenth and early twentieth centuries, depending on the applied trim and details. Underneath the trim and details, the houses are all cubical two-story boxes with pyramidal roofs. Whether ordered from a catalog or site-built, this is possibly the most common form of historic house in the United States.

1. **Simple form**
2. **Pyramidal/hipped roof**
3. **Colonial Revival character trim**
4. **Queen Anne character trim**
5. **Craftsman character trim**
6. **Mixed clapboard and shingle**
7. **Front porch**
8. **Hipped dormers**
9. **Flared "skirt" between floors**
10. **Bay windows**
11. **Sunporch**

2
8
7
3
1
10

2
8
3
7
1

2
8
9
6
4
7
10
1

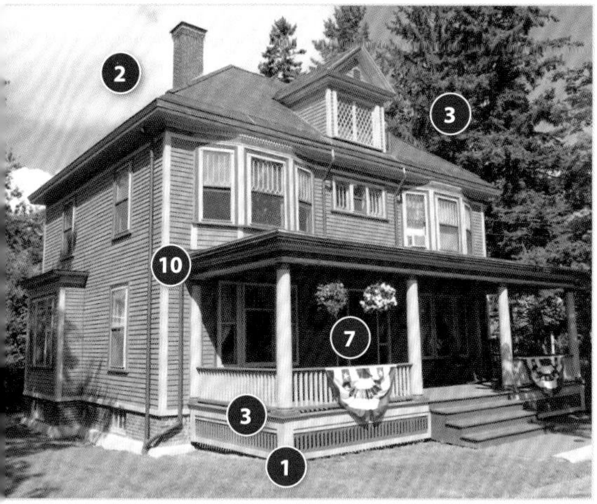
2
3
10
7
3
1

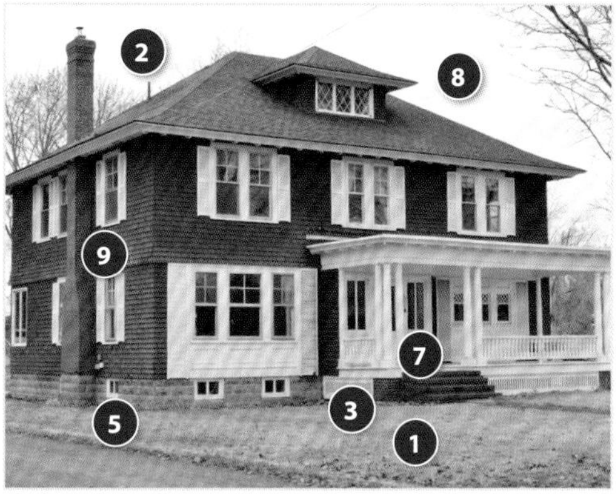
2
8
9
7
5
3
1

4
188

2
8
9
1
5

2
8
7
1
11

2
8
7
4
10
1

2
7
3
1

2
8
11
10
1
5

Craftsman (1880–1940)

At the end of the nineteenth century, there was a turn away from the highly decorative styles of the late Victorian period. Machine-made and mass-produced ornamentation drew particular criticism, and a return to a simpler, hand-crafted aesthetic developed. In home furnishings, this resulted in mass-produced, machine-made "Mission" furniture that delivered the look (sort of) but not the ethic of the Arts & Crafts design movement. In architecture, the movement found expression in several types of houses that were built in great numbers in all parts of the nation. The low-slung Bungalow form came into use after 1900 and was enormously popular. Loosely based on, and named for, a type of dwelling in India, the Bungalow was another very popular catalog house style. Bungalows were built in a wide range of materials, with discernible regional preferences. Craftsman-style detailing can be found on other house forms of the period, too—particularly American Foursquare houses.

1. Low "bungalow" form
2. Gabled roof
3. Pyramidal/hipped roof
4. Uncommonly Gambrel roof
5. Shingle siding
6. Stucco siding
7. Occasional brick
8. Occasional fieldstone
9. Bay windows
10. Dormers
11. Porches (often glazed)
12. Tapered piers on porches
13. Doubled columns on porches
14. Multiple-light-over-one-light window configurations
15. Pergolas

2
10
14
11
12
1
7

2
11
12
15

3
13
7
11
9
1

2
10
5
12
14
11
9
1

3
14
6
15

10
2
5
11
9
7
1

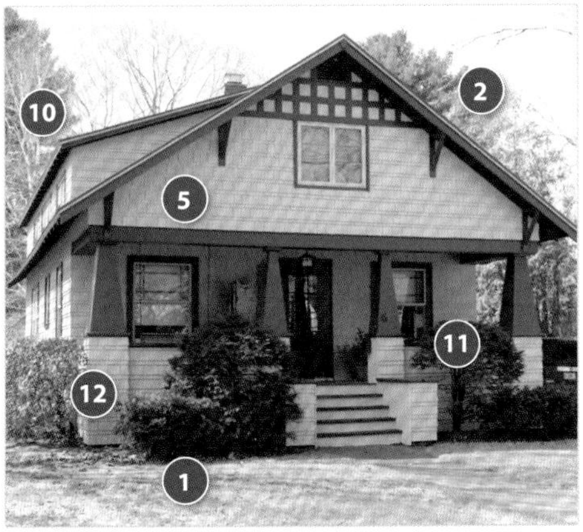
10
2
5
11
12
1

2
10
14
9
6

3
11
13
7
1

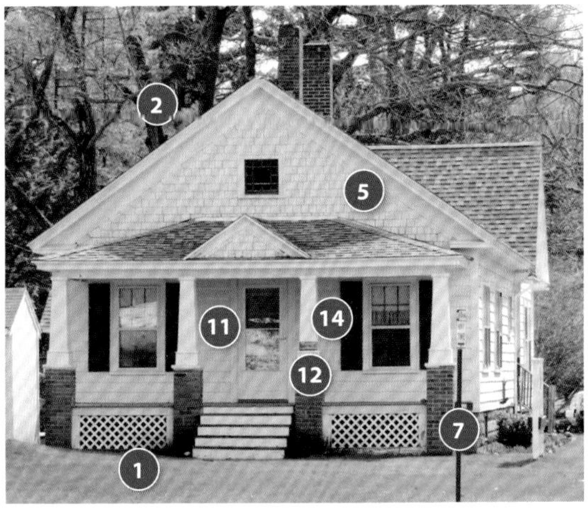
2
5
11
14
12
7
1

2
6
628
13
11
8
1

3
14
11
5
11
6
9
12

5
6
11

2
6
13
11
8
1

11
13
6

2
9
12
11
5

11
6
12

2
6
13
11
7

10
2
5
1
11

2
11
13
6
1

4
5
10
14
9
11
6

10
2
5
13
11

10
2
11
6
1

Prairie (1910–1940)

The Prairie style is most associated with the architect who developed it, Frank Lloyd Wright. In its most common form, it is a flat-roofed, one-story, open-plan house with a massive hearth and fireplace and large glazed sections of exterior wall. Exposed wood beams and wall paneling are common features. The Prairie style merges elements of the Shingle Style (open plan) with the Bungalow style (one-story living) and anticipates the International style (flat roof, large areas of glazing) that followed in the 1940s. Two-story Prairie-style houses were also built, and emphasize the horizonal dimension with overhanging eaves, bands of windows, and often a demarcation of the stories via different building materials.

1. Emphasis on horizontal lines
2. Usually masonry
3. Clear division between floors
4. Large chimneys as vertical element
5. Windows arranged in bands
6. Flat roofs
7. Low-pitched hipped roofs
8. Masonry base/plinth for house

Courtesy of Wikimedia

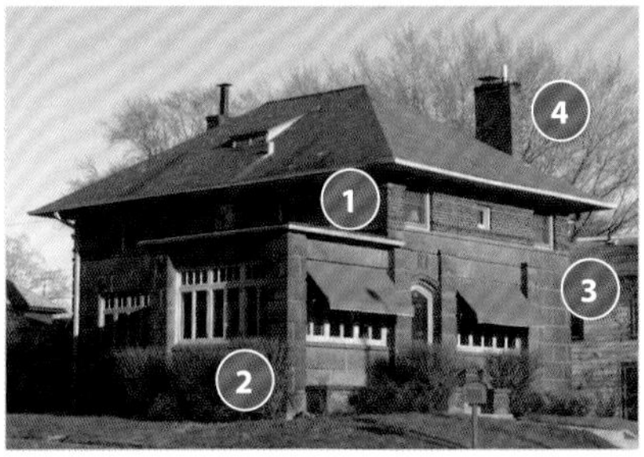

Courtesy of Wikimedia

Courtesy of Wikimedia

Courtesy of Wikimedia

Courtesy of Wikimedia

Courtesy of Wikimedia

Mid-Twentieth Century

Modernist (1940–1970)

The Modernist style arrived from Europe with architects escaping Nazi Germany. Although most associated with skyscrapers and large suburban office complexes, the style also appeared in residential architecture across the U.S. Substantial concentrations of such houses can be found in Palm Springs, Los Angeles, and other regions of the country. These houses tend to be one story, flat-roofed, and highly glazed, and are often found in suburban neighborhoods.

1. **Nontraditional forms**
2. **Flat roof**
3. **Low-pitched roof**
4. **Exposed structural steel**
5. **Concrete walls**
6. **Vertical board siding**
7. **Large areas of glazing**
8. **Trapezoidal windows**
9. **Glass block windows/walls**
10. **Corner windows**
11. **Window bands**
12. **Integral garage**

Courtesy of Wikimedia

Ranch (1945–1980)

Ranch-style houses were widely popular from the postwar period through the 1970s. They have much in common with Modernist-style houses in plan, but they are dressed in more traditional materials and generally have low-pitched gabled roofs. An attached garage is not uncommon.

1. One-story
2. Simple form
3. Low hipped roof
4. Low-pitched gable roof
5. Mixed wood and veneer masonry siding
6. Wide-exposure clapboard siding
7. Mixed shingle, vertical board, and/or clapboard siding
8. No chimney
9. Broad, low chimney
10. Recessed pedestrian entries
11. Picture or grouped windows
12. Bow or bay windows
13. Built-in planters
14. Aluminum or steel ornamental posts
15. Prominent integral garage

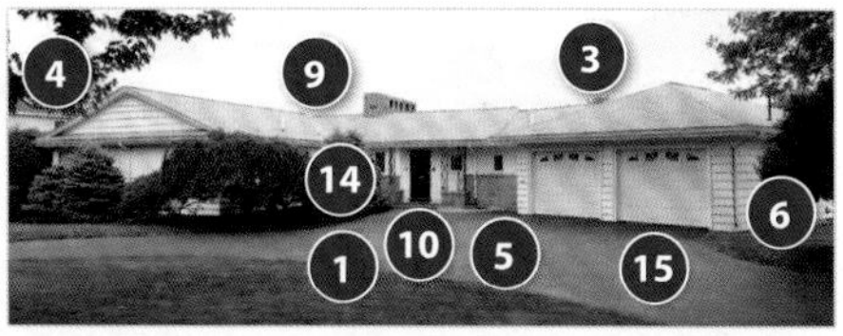

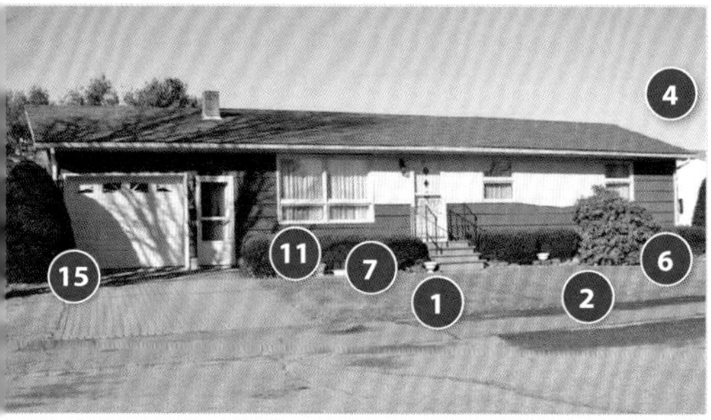

Courtesy of Sutherland Conservation & Consulting

Courtesy of Sutherland Conservation & Consulting

Raised Ranch/Split Level Variant (1950–1980)

This version of a Ranch-style house is raised on a high basement, with the front door placed at a landing mid-height between the basement and first floor. The footprint is typically smaller than a single-level ranch because several bedrooms are placed in the basement. The upper floor generally contains the living/dining room, kitchen, bathroom, and master bedroom. Often, a garage is incorporated into one end of the basement level if the house is built on a sloping site. A Split-Level house is a Ranch house built on a sloping site so that part of the main floor is placed somewhat lower than another part. Sometimes the lower level is mid-height to a basement level under the upper portion of the house. The main entrance generally enters the lower portion of the house.

1. **Two-story with mid-level entry**
2. **Simple form**
3. **Low hipped roof**
4. **Low-pitched gable roof**
5. **Mixed wood and veneer masonry**
6. **Wide-exposure clapboard/shingle siding**
7. **Mixed shingle/clapboard/ vertical siding**
8. **Recessed pedestrian entries**
9. **Picture or grouped windows**
10. **Bow or bay windows**
11. **Prominent integral garages**

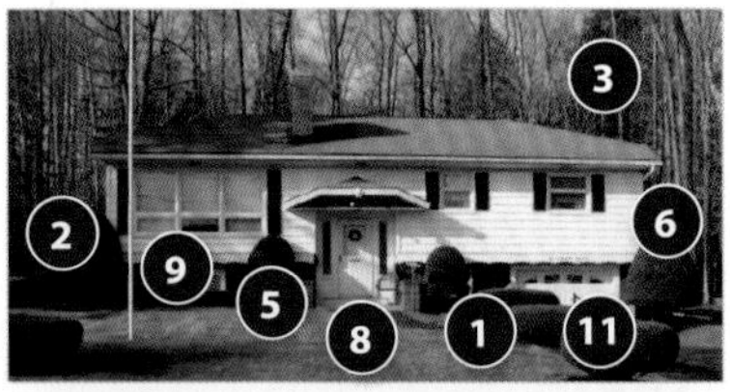

Courtesy of Sutherland Conservation & Consulting

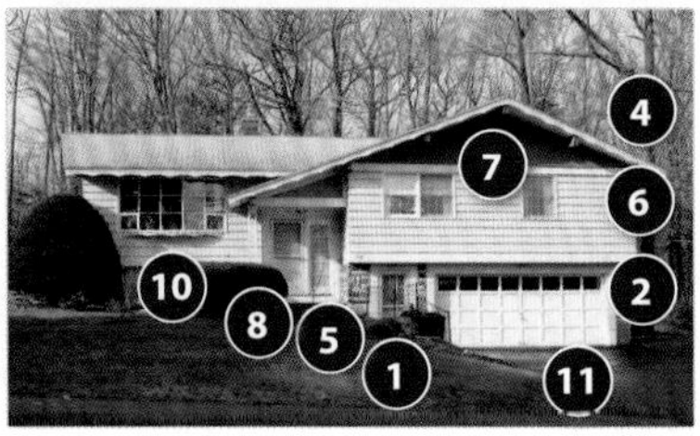

Courtesy of Sutherland Conservation & Consulting

The Postwar Cape, or "Capelette"—as much a type as a style—first appeared during the war to provide housing for people employed in war production work. It is a small house that typically features a living room, kitchen, entry/stair hall, bath, and master bedroom on the first floor, with two small bedrooms tucked under the roof. These houses were built in great numbers after the war to house returning soldiers and sailors and their young families, often lined up like Monopoly houses on small lots in planned developments. Other traditionalist houses built during the postwar period include the "Garrison" raised ranch with an overhanging second story and Ranch houses with Colonial Revival detailing.

1. **"Capelette" with Colonial Revival character**
2. **"Garrison" raised ranch with Colonial Revival character**

Courtesy of Sutherland Conservation & Consulting

Types that are "Styles"

Adobe (1600–present)

Adobe is both a style and a type (i.e., method of construction) featuring relatively soft, unfired adobe bricks made of clay and straw. A key characteristic is the soft, rounded edges of the walls and parapets. The stucco that protects the adobe bricks is applied by hand in a thick and typically uneven sculptural coat. Adobe buildings are seldom more than two stories tall, and most have a single story due to the limitations of the material. Most have flat roofs. Historically the stucco was made of clay and had to be reapplied annually, but the cement-based stucco applied to most adobe houses since the early twentieth century is much longer lasting. These houses—once limited to the dry Southwest due to the water-solubility of clay stucco and adobe bricks—have a unique character. Detailing of window and door trim, parapets, and porches can impart a touch of other styles, such as Greek Revival or Italianate. Modern construction with cement stucco over concrete block is often used to mimic the style of historic adobe buildings.

1. **Asymmetrical sculptural forms**
2. **Adobe constructionm**
3. **Usually one story**
4. **Flat roofs**
5. **"Territorial" period brick parapet**
6. **Projecting log or beam vigas**
7. **Projecting canaletas to drain flat roofs**
8. **Low-pitched tin roofs**
9. **Porch or gallery along street**
10. **Greek Revival peak on windows/doors**
11. **Italianate arched windows/doors**
12. **Simply carved brackets and balusters**
13. **Walled courtyards**
14. **Gated entries**

4
2
13
1

4
11
2

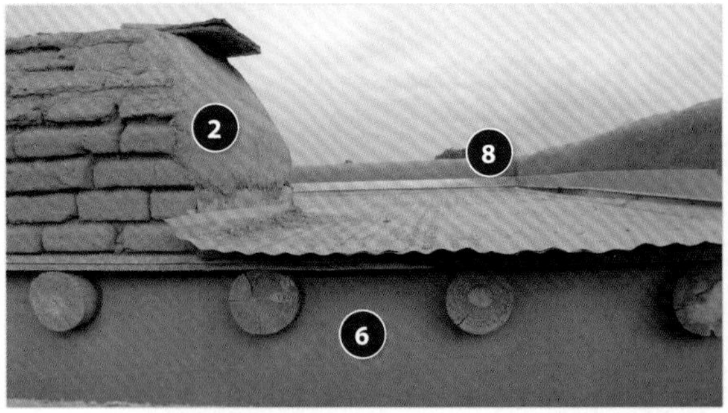
2
8
6

4
2
6
13
14
1

12

4
5
2
13
1

13
2
1

1

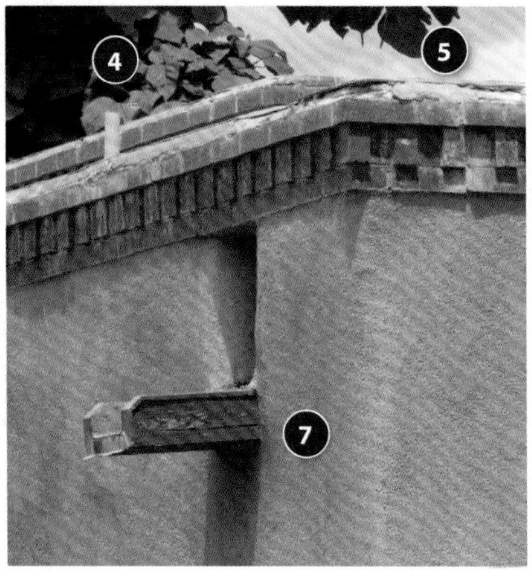
4
5
7

5
4
7
9

2

14

10

4
7
2
1

5
4

545
14

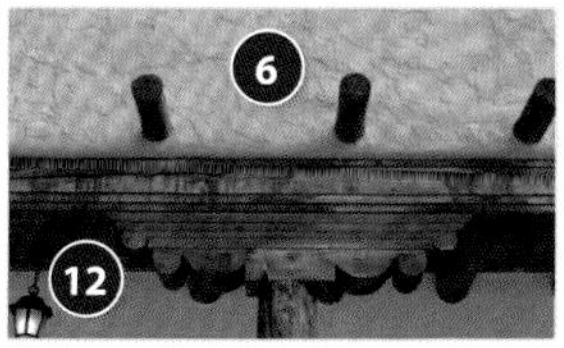
6
12

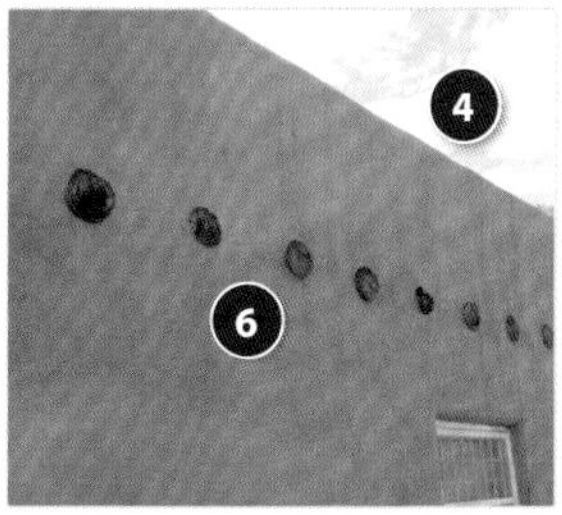
4
6

4
2
9

5
6

6
10
13
1

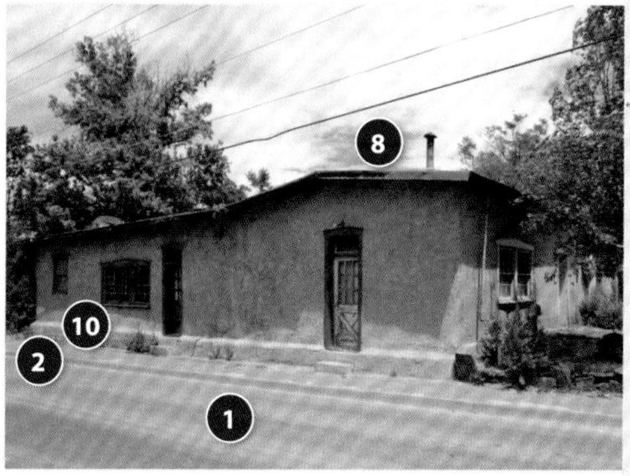
8
10
2
1

7
7
4
12

Rustic or Log Cabin (1700–1940)

Rustic buildings range from crudely built log camps for itinerant logging crews, hunters, and fishermen to grand Adirondack-style recreational lodges. Mixed into the type are log houses built for year-round habitation, often following the common forms used for masonry or timber-framed houses. Many of these houses are or were covered in clapboard or other siding to disguise their log construction. Year-round log homes date from the eighteenth to the late nineteenth centuries, depending on location. Logging camp cabins date from the 1860s to the 1940s. Adirondack-style camps and cabins, often architect designed, generally date between 1880 and 1930.

1. **Round log construction**
2. **Squared log construction**
3. **Mud or mortal chinking**
4. **Simple gabled roofs**
5. **Shingled gables**
6. **Boarded gables**
7. **Log gables**
8. **Brick chimneys**
9. **Stone chimneys**
10. **Porches with log posts and railings**
11. **Ornamental shaped branch detailing**

Courtesy of Wikimedia

Courtesy of Wikimedia

Courtesy of Wikimedia

Courtesy of Arlene Morris

Vernacular (1750–1960)

Vernacular architecture exemplifies the commonest techniques, materials, and (usually) minimal decorative elements of a particular historical period, region, or community. It is architecture that is not trying to be architecture. Often there are hints of bygone architectural styles, but these are insufficient to identify the building with any particular style. These plain buildings can be as historically significant as architect-designed homes, since they, too, represent and document the history of their community.

1. **Locally common forms**
2. **Minimal detailing**
3. **Few, simplified elements of recognizable styles**

Courtesy of Wikimedia

Courtesy of Wikimedia

Courtesy of Wikimedia

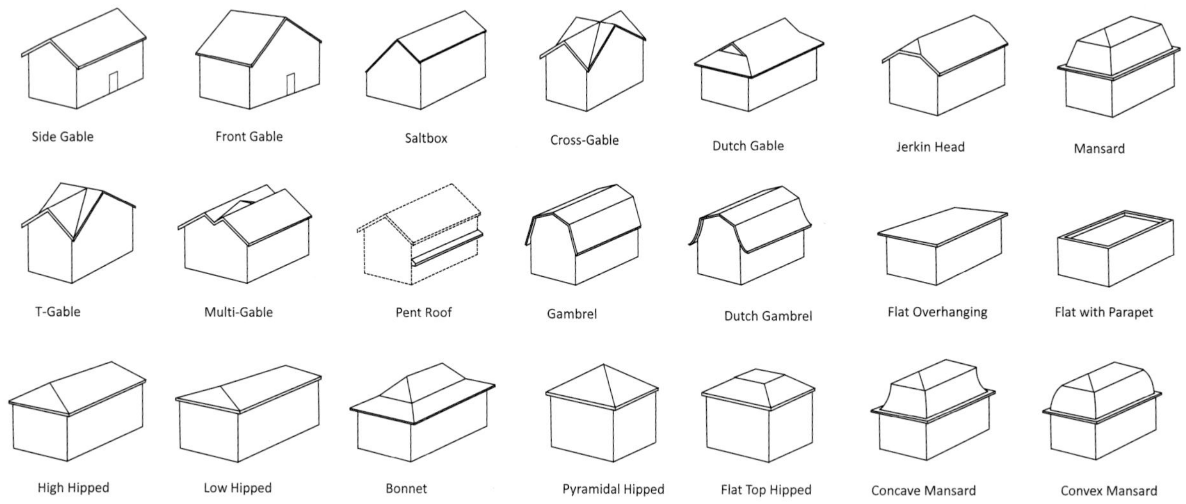

Roof Forms and Architectural Styles Chart

KEY

Commonly Found with Style ◍

Less Common but Possible ●

	Colonial Period					Early Republic		Early Victorian Period			Later Victorian Period					Late 19th Early 20th Century Revivals				Early 20th Century			Mid-20th Century			Types that are Styles		
Roof Form:	First Period	Georgian	French	Dutch/German	Spanish	Federal	Greek Revial	Gothic Revival	Italianate	Second Empire	Queen Anne	Stick Style	Shingle Style	Romanesque	Exotic/Eclectic	Colonial Revival	Spanish Col. Rev.	Tudor Revival	Neo-Classical	Craftsman	Am. 4 Square	Prairie Style	Deco/Moderne	Mod Tradition	Ranch	Adobe	Rustic-Log	Vernacular
Side-Gable	◍	◍	◍	◍		●	◍	◍	●		●	◍	◍	◍	●	◍	◍	●	◍	◍			●	◍	◍		◍	◍
Front-Gable	◍	◍	◍	◍		●	◍	◍	●		●	◍	◍	◍	●	◍		●	◍	◍			●	◍	◍		◍	●
Saltbox	◍	◍				●							◍			◍	●			●								
Cross gable	●	●	●	●			●	●	◍		◍	◍	◍		◍	●		◍		●								●
Dutch-Gable				◍							●		◍							●								
T-Gable	●	●	●	●		●	●	◍			◍	●	◍		●	◍	●	◍	●	◍				●	●			◍
Multi-Gable											●		◍			●												
Pent roof (attached to wall)				◍												●												
Gambrel	●	●		◍		●							◍			◍											●	
Dutch Gambrel				◍							●				●	◍				●				◍				
High-Hipped		◍	◍		●						◍	●		◍		◍	◍		◍									●
Low-Hipped						◍			◍							◍	◍		◍	●		◍	◍					●
Bonnet			◍	◍	◍											●				●								
Pyramidal Hipped															●			●			◍							●
Flat Top Hipped									●		●			●	●				●									
Jerkin Head											◍	●	●		●						●							
Mansard										◍					●				●									
Convex/Concave Mansard										◍					●													
Shed																											●	●
Flat Overhanging					◍																	◍	◍			◍		
Flat with Parapet											●	●					●			●								
Conical (on tower)											◍			◍	◍			●										
Candle Snuffer (on tower)											◍				●													
Hipped (on tower)								◍			●				●		●											

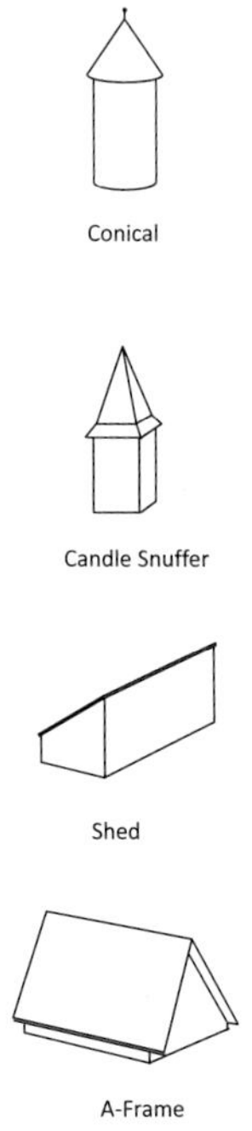

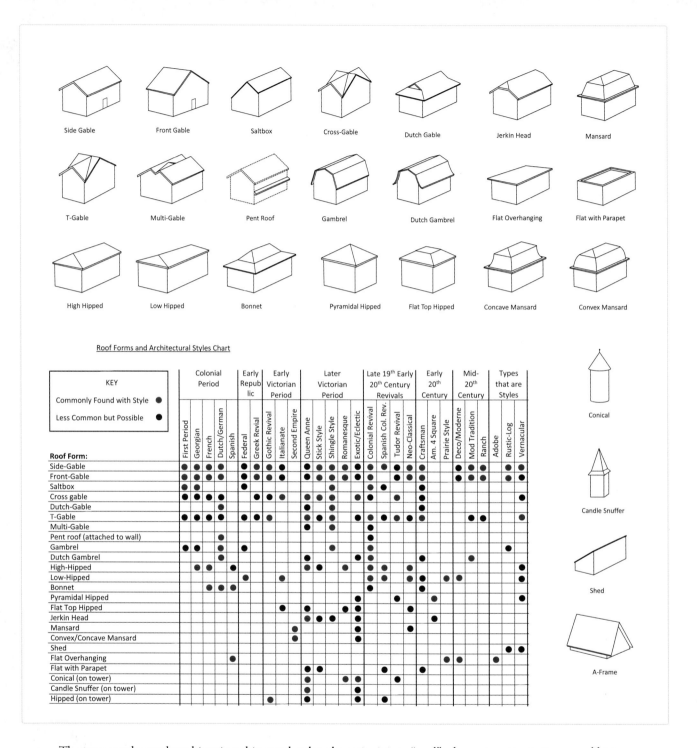

The more you know about historic architectural styles, the easier it is to "read" a house to estimate its age and history of alterations. Among the most important character-defining features that help to identify a building's period and style are roof forms. While some roof forms, such as the Side-Gable roof, are common to many styles, others are closely identified with specific styles, like the Mansard roof that is the most characteristic feature of the Second Empire style. The most common roof forms and the architectural styles they are typically found with are illustrated here.

Identifying Changes Over Time

Some changes over time are relatively easy to identify. A picture window in a Federal house is clearly an alteration, since picture windows did not exist in the early nineteenth century. A bracketed Italianate door hood of the 1860s on a Federal house of 1802 is a somewhat less obvious alteration unless you understand the characteristics and chronology of styles. An alteration done within decades of the original construction of a house can be particularly hard to detect, having been done with virtually identical materials and tools and possibly even by the same builder.

Whitten House: A Case Study

Whitten House, in midcoast Maine, was built in 1827–28 as a classic New England center-chimney Federal-style Cape with kitchen ell. Approximately twenty-five years later, a full second story and finished attic were

(continued on page 86)

An Italianate door hood installed c. 1860 on a Federal house built in 1802.

This nineteenth-century Federal house today might seem little changed from when it was built, but it was built in 1803 with a low-hipped roof, as shown in the altered photo at top left, and received an up-to-date Greek Revival gabled roof in about 1840.

Here's another Federal house with a Greek Revival roof alteration. The close-up photo of the eave area and lower corner of the pediment shows a high level of detail in the Federal-style eave cornice with much simpler detailing in the pediment. Had they been built at the same time, they would likely have received similar or identical detailing. Another clue to its original roof form can be seen in the roof of the kitchen ell, which is low and hipped as the original roof would have been. In this case, documentation for the change was found in a letter written by Charity Mustard to her brother George in January 1838. She wrote, "[your friend] Purrington has sailed for Liverpool—he bought Cyrus' part of that house, put on a steep roof, put an addition on the back part, painted and blinded it, and it looks very handsome."

This Gothic Revival confection, known as the Wedding Cake House, is actually a formal five-bay brick Federal with an amazing "wrapper" of wooden Gothic Revival trim added in 1852. The characteristic Federal-style entryway with leaded sidelights and fan light and the Palladian window in the second story remain visible, framed with Carpenter Gothic elements. A Gothic Revival shed and carriage house are attached.

The origin of this Second Empire house is not obvious without other documentation. Historic photos show that this house began life as a five-bay Federal in the early nineteenth century. Altered trim, an altered roof, and an added tower and porch have left virtually no evidence of the original exterior appearance.

This modest house appears to be a common Greek Revival of the 1840s with a recessed corner porch. Indeed, it is similar to a nearby house with the same entry porch that is documented to have been built in 1849. Deed research showed this house to have been built well before the Greek Revival style came into use, however. A Google aerial view finally provided the clue that brought history into focus, revealing a subtle line across the roof where the porch ends. The line indicates that the house was extended toward the street to add the porch and a new front room on each story. Knowing this, it becomes clear that the "side" elevation was originally the front of a five-bay Federal-style Cape. The center door remains there, as do three of the four window openings. A later nineteenth-century bay window has replaced the fourth window bay.

This Queen Anne house is an example of a once-common method of expanding an existing house by jacking it up nine or ten feet and building a new first story under it. In this case the front door was left in place to serve a new second story porch. A more steeply pitched roof allowed for some Queen Anne detailing to be added.

Here's another house that was enlarged by lifting it into the air and adding a new first story. This alteration was done by Civil War General Joshua Lawrence Chamberlain when he became president of Bowdoin College in 1871. Chamberlain had already had the 1824 transitional Federal to Greek Revival house moved and rotated ninety degrees in 1867 while serving as governor of Maine. While leaving the original house largely intact, the 1871 first story combines elements of the Italianate and Gothic Revival styles.

added to the main block of the house, and a half-story was added to the ell. Between 1827 and 1852, however, a shift in architectural styles had occurred in the region, with the Greek Revival style replacing the Federal. Thus, when the house was enlarged, it also received a stylistic update on the exterior, with wide paneled corner boards (i.e., corner boards built up with narrow boards at the sides, top, and bottom to create a recessed flat panel at the center), a pilastered door surround with broad entablature, and a wide frieze under the eaves. More than 160 years later, the house appears to have been built as a two-and-a-half story Greek Revival in the 1840s or 1850s, not as a Federal in the 1820s. No photos or other images of the building before 1850 could be located. Understanding its developmental history required careful study.

The most apparent hint of a significant post-construction change was the mixture of Federal and Greek Revival profile moldings inside the house. Federal-style moldings tend to be delicate—thin and attenuated—while Greek Revival moldings are robust and bold. This stylistic difference suggested that the moldings had been installed at different times.

A related but subtler clue was that the original thumb-latch hardware on the doors to the main rooms on the first floor had been replaced at an early date with knob-operated morticed-latch mechanisms.

The house's central chimney was common in houses built in the region prior to 1840, but quite uncommon for a Greek Revival house of a couple of decades later. It left little space for an entry/stair hall inside the front door, so the addition of a second story

Photos of Whitten House taken in 1940 and 2004.

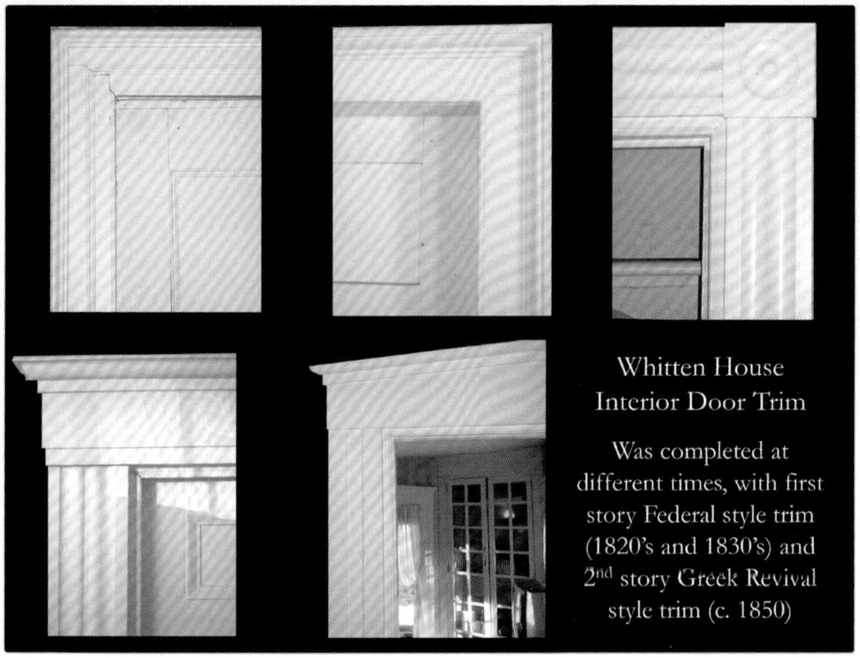
Whitten House
Interior Door Trim
Was completed at
different times, with first
story Federal style trim
(1820's and 1830's) and
2nd story Greek Revival
style trim (c. 1850)

Whitten House door trim. *Top left*: sitting room, Federal forms and moldings, 1827; *top center*: original kitchen, simple Federal forms and moldings, 1827; *top right*: parlor, Federal forms and Greek Revival moldings, c. 1850; *bottom left*: front hall, Federal forms and Greek Revival moldings, c. 1850; *bottom center*: upper hall and bedrooms, Greek Revival forms and moldings, c. 1850.

required a dramatic spiral stair that is exceptionally narrow and steep to fit the hall's available space. Under the stairs, the floor shows clear patches that document a narrower original entry hall with a closet against the chimney. Building even a narrow stair required widening the hall into the adjoining sitting room with a curved wall. The door threshold to the sitting room is installed atop the pine floorboards, while that on the parlor door, on the other side of the entry hall, is set into the pine flooring. Floor patches suggest that the original threshold between the entry and sitting room had also been set into the floor.

The exterior surround of the front door lacks the sidelights that are a ubiquitous feature of the Greek Revival style, and it has curious flat pilasters set inside half-round columns, with a transom over the six-panel door—which is typical of a Federal-style entry door. The moldings on the pilasters and door are thin and attenuated, like Federal-style moldings, while the columns and entablature are robust and bold, like Greek Revival elements. Close examination revealed that the original door and surround remained in place when the columns and entablature were added during the 1850 remodeling. As built, the transom would have been tucked up against the eaves of the cape, like several nearby examples from the 1820s and 1830s.

A small Dutchman patch in the wood of the door between the front hall and parlor is evidence of the original thumb latch being replaced by the knob seen below the patch as part of a stylistic updating and expansion of the house c. 1850.

Front stairs of Whitten House, installed c. 1850. The French Gothic Revival hand block printed wallpaper was hung in 1852.

Back inside the house, several additional clues emerged. The plaster ceiling and side walls in the two attic bedrooms had previously been removed to allow strengthening of the roof framing. Because the attic was not being used, the framing had been left exposed and was clearly from the mid-nineteenth-century transitional period from traditional pegged, timber-framed roofs—with widely spaced large rafter beams and horizontal purlins to support vertical sheathing—to nailed, closely spaced, 2-inch-thick rafters overlain by horizontal sheathing. This roof had nailed, 4-inch by 6-inch rafters with moderately wide spacing overlain by horizontal sheathing—an 1850s roof system.

When new work in a back corner of the house required the removal of a 10-foot-wide section of plaster and lath on the interior and corresponding clapboards and sheathing on the exterior, fully exposing the framing on both stories from foundation to eaves, several additional things of interest were discovered. The lath on the first floor was 1-inch-thick accordion lath, while that on the second floor was ½-inch-thick sawn lath. There was also a subtle difference in the plastering, with the second floor appearing somewhat more uniformly flat, perhaps due to the more consistent sawn lath used.

First-floor clapboards had hand-planed scarf joints, while second-floor clapboards were

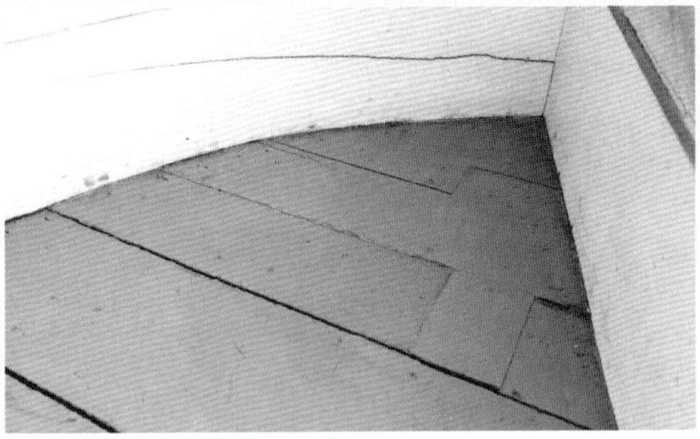

Patched area of floor under the spiral stair installed c. 1850. The patch documents the original width of the hall and the location of a closet against the chimney (on the left) before the curved wall was built with the staircase.

This Federal-style door surround is in a house built nearby and within a few years of Whitten House. The flat pilasters flanking the door extend to the low eaves of the cape and frame the transom above the door, as on the original door surround on Whitten House.

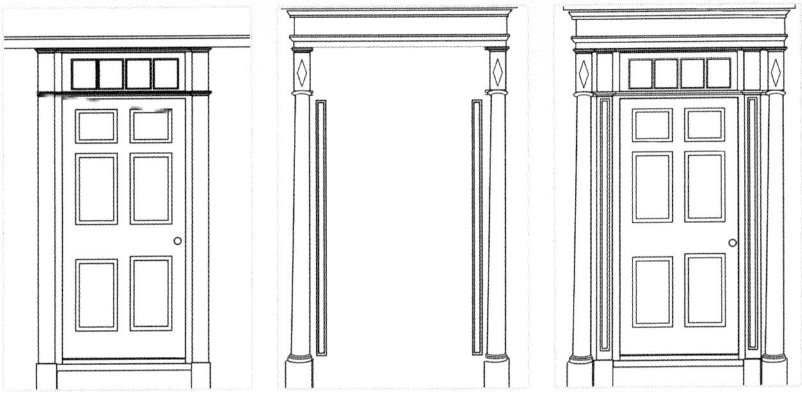

These three drawings illustrate how the 1827 Federal-style door surround of Whitten House was modified by the addition of half-columns and a bold frieze and entablature to update it to the Greek Revival style and increase its size to be more in scale with the enlarged two-story house.

The c. 1850 remodeled door surround on Whitten House.

butt-joined. The structural corner post of the house was not continuous from bottom to top; rather, a top-plate beam separated the first- and second-story por-

View in the Whitten House attic showing darker c. 1850 framing and the lighter framing added to strengthen the roof in the 1980s. The original nailed 4-inch x 6-inch rafters spaced approximately 30 inches apart are typical of mid-nineteenth-century roof framing in the region and represent a transitional phase between traditional pegged timber framing and later stick-built framing.

tions. Studs and corner braces differed subtly between the floors, and the first-story sheathing boards were noticeably wider than those on the second floor.

Whitten House was vinyl sided when the rehab project began, hiding all exterior paint. The removal of this siding from the rear of the house resulted in the "self-removal" of much of the built-up old paint that the vinyl had been covering for nearly thirty years. (This often happens, probably because the heat and moisture trapped behind vinyl siding loosen the bond between the first layer of paint and the wood siding.) The paint residue on the exposed bare clapboards showed that the first floor had originally been painted red, the common red oxide of the early nineteenth century, but the second story had always been painted white, the classic color for mid-century Greek Revival houses.

Another clue that the second story had been added some time after the house was built was a fragment of plaster with attached wallpaper discovered beneath the second-story floor. Changes had been made to this area of the house in the 1940s, but there is no evidence that the floor had been lifted during that work, so the plaster chunk could not have gotten below the floor then. Differences in the flooring between the south and north bedrooms, and an awkward area that looked patched or pieced together in a back room near where the wallpaper fragment was found, indicated much earlier changes to the floors. The wallpaper on the fragment was an identifiable c. 1830 pattern, suggesting that it dated from the house's original construction. It seems likely that the plaster fragment was part of the original finish in the attic bedrooms of the cape, which were removed when the second story was added.

One-inch-thick lath on the first floor and ½-inch-thick lath on the second floor gave evidence of two periods of construction.

Photographs from about 1940 showed the exterior of the house prior to the installation of the vinyl siding and vinyl window shutters in the 1980s. The original wooden shutters, visible in the photos, differed between floors, with the lower ones having a single panel of louvers and the upper ones being divided into two panels. This represents another shift in construction styles between early- and mid-nineteenth century.

First- and second-floor wall framing had subtle differences in stud size and design and in the size of the angled corner brace. The wider sheathing boards used on the first floor are typical of earlier construction in the region; the narrower boards on the second floor are typical of later construction.

Whitten House provides a good example of the many types of evidence that, together, can document changes and reveal a house's history. Sometimes an alteration of this type is much more obvious. Within a mile of Whitten House are at least three other capes that became full two-story houses later in the nineteenth century. Three of them were jacked into the air and lowered onto a newly constructed first story. Two of these even retained the front door, now on the second story, to access a porch or balcony. In all three cases, the second-floor interior finishes clearly read as early nineteenth century, while the first stories are clearly later in their trim and detailing.

Removing the vinyl siding from the rear of Whitten House exposed the historic wood clapboards. Several decades of heat and moisture trapped behind the vinyl had loosened much of the paint, and additional exposure while work continued caused much of the paint to fall from the house. The 1827 clapboards on the first story retained evidence of original red paint, which was commonly used for exteriors in New England before white became the most popular color for houses with the spread of the Greek Revival style. Also in evidence on the first story are scarfed joints on the clapboard ends, a labor-intensive joint that went out of use in the region around the middle of the nineteenth century. On the second story, the clapboards show no evidence of red paint, having always been painted white, and have butt joints.

Identifying Character-Defining Features

Identifying what may have changed over time requires careful examination of a building's features as just described, and this scrutiny can also identify the house's character-defining features—i.e., those features that make your historic house

A wallpaper fragment on plaster found under the second-story floorboards of Whitten House. An example of this paper printed in different colors can be found in Historic New England's on-line database of wallpaper patterns, where it is described as an "Ashlar pattern of blocks with foliate in center and framed by interlace in the corners. Printed in green, brown and dark tan on a light tan ground" and dated 1825–1835.

historic. The features that define a Queen Anne–style house of the 1880s will be very different from those of a postwar Ranch of the 1950s. Two houses of closely related styles, such as Italianate and Second Empire, will show less dramatic differences in their character-defining features and may be similar in most respects.

Character-defining features tell the "origin story" of your house: when, where, and how it was built, and when and how it changed over time. Mid-eighteenth-century life was quite different from the mid-nineteenth century, which in turn was different from life in the mid-twentieth century. Homes from these periods reflect the lives lived in them, each telling part of the story. Collectively, the character-defining features of a community's buildings tell the community's story, a narrative recorded in bricks, wood, stone, plaster, and all the other materials that have been used to build, change, and repair buildings over time.

Knowing the key historic features of your house will tell you what can or should not be changed when you adapt the house to suit your life. Character-defining features can be spatial as well as physical. If plans for the house exist, make copies you can take notes on; otherwise, sketch a simple floor plan and make copies to write on. Properly scaled and measured drawings are not necessary at this point but will save time later. If you intend to involve an architect in your project, have them start by producing the "existing condition" drawings as soon as possible.

It is generally easiest to start by recognizing the architectural features that define the house. Beginning outside, note the siding, trim, doors, windows, porches, roofing materials, chimneys, or other historic features. Inside features may include flooring, wall materials, trim, ceilings, staircases, fireplaces, light fixtures, built-in furniture and cabinetry, stained glass windows, and other historic materials. Note any observable

This detail from a 1940 photo of Whitten House shows earlier single-panel louvered shutters on the first story and later two-panel shutters on the second story, further evidence of the house being built in two stages several decades apart.

Like the two examples shown on page 86, this two-story house began as a single-story cape. Like those two, this one was lifted into the air to receive a new first story. This one was more carefully masked on the exterior, with new wide-paneled corner boards, frieze under the eaves, added overhangs on the gable ends, and flush boarding on the façade making the change virtually unnoticeable to the casual viewer, as on Whitten House. Here, however, the entire second-story interior is Federal in its detailing, and the entire first story is Italianate. All four of these early-nineteenth-century Federal-style capes—one Greek Revival, one mixed Greek Revival and Italianate, one mixed Gothic Revival and Italianate, and one Queen Anne style—are within a mile and a half of each other.

differences between the features of the main entrance and principal rooms and the secondary spaces at the back of the house. There may also be differences in the level of detail and quality of materials between the first and upper stories. If the attic has any finishes, they may be much plainer than those in the first and second stories. Inventory all of these with notes on a floor plan or in a spreadsheet. Also note where obviously modern materials are present.

Spatial features are often as important as physical features to the historic character of a house. A room is basically a box, but the size, shape, and height of the

In the room that appeared to have been the original kitchen of Whitten House, there was no visible evidence of a hearth (1), brick oven (2), fireplace (3), early mantel (4), or chimney cupboard (5)—all features to be expected in an 1827 kitchen in the region. The existing mantel was mounted over a chimney thimble for a woodstove and had a Greek Revival profile, suggesting a mid-nineteenth-century date. The room also had obvious modern alterations including a half bath where the brick oven should have been and a missing wall portion, opening the room into an adjacent room that had been added in the 1960s.

space can make it grand, cozy, or oppressive. These are character-defining features. Spend time in each room at different times of day and night, noting how each room feels. How staircases create transitions from floor to floor and how rooms flow from one to another or are separated by doors and walls can impact the historic character of a house.

Natural light is an important feature. Many architects of the past created intentional effects with light, such as inducing it to flow down a staircase from a large window on a landing.

Spatial character is also important outside a house. How a building sits on its lot, its setback from the street, how the entrances are approached, and even how landscape features frame the house can all be character-defining spatial features.

Once you've identified the features of a house, you need to assign them to architectural styles. Your goal is to prioritize features in order of their importance to the historic character of the house; the more information you have about the style of the house, the easier that task will be. You want to be able to say, "These features are what make this house an example of this style." Those are the features you will want to preserve or restore as you make necessary changes.

Identifying Missing and Added Interior Features

Important historic features of a house may have been lost along the way. Lowered ceilings may have altered the spatial character of significant rooms. A large house may have been converted into a multi-unit apartment building in which added partitions subdivide formerly large rooms or halls into smaller spaces.

The advent of central heating in the twentieth century rendered formerly essential fireplaces superfluous, and many were removed or closed up and their mantels removed. Later in the century, as the television became the ubiquitous focal point of a living room, some owners removed mantels to create wall space.

In Whitten House, the original kitchen fireplace, brick oven, hearth, and mantel were gone from the

The Whitten House 1827 kitchen after restoration of its missing features. See page 138, Restoring a Historic Kitchen, for the story of this transformation.

kitchen ell, replaced by a single-flue "stove stack" chimney and a mantel shelf mounted to the wall above a stove thimble. This was a common arrangement in the later nineteenth century, when cast iron heating and cooking stoves had replaced open fireplaces and brick ovens for heating and cooking. Many houses built after about 1850 had mantels or mantel shelves but no fireplaces. Whitten House was built earlier than that, however, and would have needed masonry cooking facilities before iron stoves came along. None of the fireplaces in the central chimney of the main block of the house showed evidence of an iron crane for cooking, nor was there any visible evidence that a brick oven had ever been present.

The only visible evidence for the original chimney mass in the 1827 kitchen was an unmistakable brick-and-stone chimney arch in the cellar. The presence of this 6-foot by 8-foot chimney base was proof that the sixteen-inch-square chimney it had been supporting since before 1940 (as confirmed by a photo of the house taken then) was not the original chimney. Two vertical trim boards flanking the original chimney location, together with a narrow door to the right, appeared to be original features and provided crucial information for reconstructing the missing chimney, oven, and mantel during the restoration of the 1827 kitchen. That restoration process is described and shown in the example project starting on page 138.

Door and window trim in a stair hall and adjoining bedroom, respectively. They look the same, and I looked at them daily for nearly a decade before wondering whether, indeed, both were installed at the same time. I knew the bedroom had been expanded in the 1940s by removing a wall to incorporate a smaller adjacent bedroom into the space. When that was done, the simple trim in the smaller bedroom was altered to match the more elaborate trim in the bigger room it was joining. This change was evident as an added peak above the flat board of the original trim, as seen in the window on the far wall in the photo below.

Added features are generally easier to identify than missing ones. To start with, there is something to look at and analyze. Common added features include doors, windows, and changes to the floor plan, which are often easy to identify because they do not match surrounding features in materials or quality of workmanship. For instance, a single door with flat-stock trim in a room that otherwise has matching molded trim on all doors and windows was probably added later. Sometimes, however, the craftsmanship of an added feature is more carefully done; an added door might have matching trim that was moved from elsewhere in the house. Clues to this might include old nail holes that don't have nails in them (poke with an awl to find out), nails that do not match the others in the room (modern nails are round, whereas historic nails are generally square or rectangular in section), or joints where the reused trim was pieced together. An added door may not match the others or may not have the same threshold style. Added windows can generally be identified by similar clues. When a wall has been removed between two rooms, the trim in one may have been altered to match the other; this can be detected through careful observation.

Less evident was the fact that the carpenter who combined the two bedrooms matched the more elaborate trim in the front room to the window and doors in the stair hall. He accomplished this by adding a cap molding to the original "Greek peak" window trim in the front bedroom. The giveaways were fewer layers of paint on the caps and modern, round nail holes.

Amy Cole Ives of Sutherland Conservation & Consulting is taking paint samples with a scalpel. The owners wondered whether the two corner cabinets were original to this eighteenth-century room. A paint analysis by Susan Buck confirmed that the cabinets were not original and identified the relative times of other alterations made to the room. *(Courtesy Susan L. Buck, Ph.D.)*

Paint samples from the wainscot trim at left and the cabinet at right show clear differences under a microscope at 400x magnification. *(Gari Melchers Home and Studio, University of Mary Washington, Fredericksburg, VA)*

Sometimes a room was "fancied up" by adding to trim to make it more detailed. Well-executed added features can be nearly impossible to identify without documentary evidence such as historic photos. Lacking such photos, you can sometimes identify newer work because it will be covered by fewer layers of paint than older work. Work of the same age relocated from another room will often show different underlying paint colors.

It is not uncommon for modern materials to be layered over historic materials. For example, plywood sheet paneling was considered a cheap and easy way to cover damaged plaster walls in the 1970s and 1980s. Often the paneling was attached with tube adhesives and a few small nails right over wallpaper (which makes its removal quite easy). Trim was often left in place. In more recent years, people looking for fast and cheap solutions to damaged walls have often opted for layering ⅜- or ½-inch gypsum board (sheetrock) over the plaster and finishing it in the standard manner. This yields decent-looking walls but reduces the projection of the trim elements from the wall surface by half or more, a giveaway that something has been altered. In cases where the original trim had minimal projection beyond the plaster, moldings are sometimes applied to build out the trim from the new wall surface. Although subtle, such changes can affect the historic character of a space and should be noted. Because modern trim is often thinner and skimpier than historic trim, layering new material over plaster can make adjacent trim look skimpy and modern even when it isn't.

Identifying Missing and Added Exterior Features

Porches are exposed to the weather and often deteriorate faster than other parts of a building, resulting in their removal. Sometimes they are replaced with a new porch possessing a much different character, and sometimes they simply disappear.

The replacement window industry has convinced a great many American homeowners that their old windows absolutely must be replaced. This issue is explored elsewhere in the book; here it is enough to note that original windows are all too commonly lost, and the loss is usually easily identified.

This applied stone veneer is a recent addition to this historic house. Although not historically accurate, it is in character with the original Tudor Revival style and more effective than most such applications.

This former window opening was partially infilled and partially extended to become a door opening, and more recently the door was removed. The result is two generations of infill, both carefully toothed in but identifiable with close observation.

Missing door and window openings are more or less obvious, depending on the care with which they have been covered. Easiest to spot are those for which the trim was left in place and the opening filled with siding that matches the rest of the wall. When the trim was removed and the new siding butted to the old, the missing window or door can be traced by its telltale vertical lines. The loss of the opening can be difficult to spot, however, when the carpenter or mason removed sections of surrounding wall and "wove in" clapboards or "toothed in" bricks, leaving no vertical lines. The best hope for spotting one of these on the exterior of a clapboard building is to look for any obvious breaks in the fenestration pattern (arrangement of openings in a wall). If a house has a perfectly symmetrical façade except for one missing opening, there is reason to suspect that an opening may once have been there.

This infilled window is obvious.

This infilled window is less obvious but still identifiable.

Another clue to well-disguised lost openings is found in paint buildup. If the house was painted many times before the opening was removed and has only been painted a few times since,

This simple mid-nineteenth-century cottage, built in 1851–52, appears to be quite unchanged from its original appearance, except for some porch details that may be from later in the century. However, it provides a good example of how the tiniest, fuzziest detail in a faded photo can disclose lost elements of an original design.

This photo from around the turn of the twentieth century supports the idea that the house is largely intact, with even the same porch detailing.

there will be a noticeable difference in the paint texture on the newer clapboards or shingles.

Even carefully filled openings in brick and stone are generally identifiable, because few masons go to the trouble of exactly matching old and new bricks. Even subtle differences in brick size or color will be noticeable with careful observation. Early masonry openings were generally built with brick arches or stone or brick headers to carry the load across the opening. When these are left in place, they are an obvious giveaway. Later masonry openings often had iron or steel lintels across the top of the opening, which many masons leave in place. A thin, dark line of metal in a wall at the right height for the top of a door or window is a giveaway for a missing opening.

One earlier photo exists, perhaps from the 1860s. In spite of its faded condition, the photo clearly shows an earlier porch and a dormer on the ell that was gone by the time the later photo was taken.

On exteriors, as on interiors, modern materials are often layered atop historic ones. Vinyl siding is undoubtedly the most common example, but there is a long history of such materials made of cement asbestos board, asphalt, aluminum, and galvanized steel, which commonly attempt to replicate the look of clapboard, brick, or other traditional materials. The original siding material is unlikely to have been removed before the new siding was applied.

Cement-based stucco and fake stone sidings are an exception in this respect. These were generally applied to a wire mesh mounted on laths and nailed or stapled to the house, and the greater

Another lost detail is just barely visible in the earliest photograph.

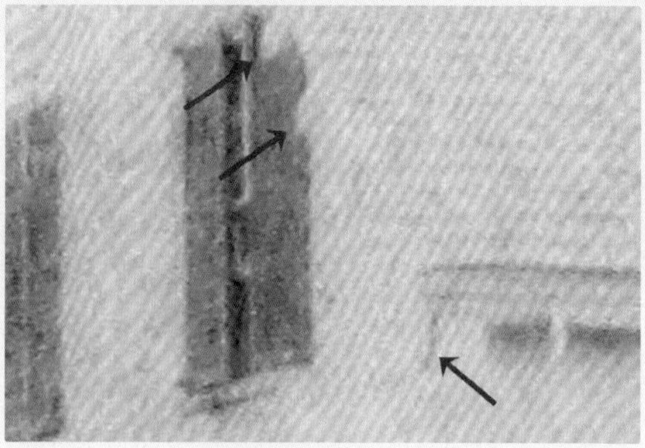

The detail is a sawn Gothic Revival bargeboard along the gable, just visible as white painted crockets standing out against the dark shutter on the second story (two upper arrows). At the bottom of the bargeboard, the shadowed edge of a terminating loop is visible (lower arrow). Along the remainder of the gable eaves, the white bargeboard is lost against the white paint of the house.

thickness of this siding made it desirable to remove the original siding, exposing the full thickness of the trim boards to run up against.

No matter what modern siding is used, it is common for the historic trim to be thickened with modern moldings where it meets the siding. In more than one case I have been involved in, there were

At the north gable of the house is an awkwardly sawn arch that does not match the quality of trim work on the rest of the exterior. It is what is left of the bargeboard, the remainder having been sawn off flush with the soffit. The awkwardness of the arch cut is the result of having to saw it in place.

This illustration of the Henry Boody House—designed by Gervase Wheeler, built in 1849, and considered an important example of the Gothic Revival style—was published in Andrew Jackson Downing's influential book *The Architecture of Country Homes* in 1850. The bargeboard with crockets along the eaves and the terminating loop at the bottom of the eaves are clearly shown. The Boody House is only two miles from the cottage in these photos and was built just a year earlier. (Author's collection)

This 1924 photo provided valuable information about the historic appearance of this nineteenth-century Greek Revival building, which had been altered during the twentieth century.

Here is the same building in about 2000, before efforts were made to rehab it and return much of its historic character.

After rehabilitation based on the 1924 photo, the building closely resembles its historic appearance.

two layers of added siding: a 1940s layer of asbestos shingles and a 1980s layer of vinyl siding.

Documentary Evidence

Documentary evidence, especially historic photos, can be enormously helpful in identifying missing elements. Historic interior photos are relatively rare, but exterior photos often show trim elements, chimneys, roofing, landscape elements, or porches that have been lost. Local historical societies usually have historic photo collections that include images of buildings. Neighbors may have historic photos of their houses that include part or all of yours. Some cities and towns had photos taken of every building at given points in history, usually for tax assessment purposes. Portland, Maine, did so in 1924 and again in the 1950s. The 1924 Portland photos have recently been digitized and made available online. Other communities across the country are undertaking similar initiatives.

Final Notes

Please note that the purpose of identifying these changes, some of which have minimal impact on historic character or even add an interesting chapter, is to understand the house's story. Whether or not any of these changes should be removed or changed further will be discussed later in the book.

This chapter has assumed that you are assessing a house that is relatively intact, limiting your ability to look for changes in the structural framing members. When you work on the house, you will probably expose framing and may find additional clues. Add new information to your notes, drawings, or spreadsheets as you get it.

S. Nevin Hench House, 1887

Queen Anne Style

This Queen Anne home was built in 1887 by industrialist S. Nevin Hench in York, a city in south-central Pennsylvania. His business partner and cousin, Walker A. Dromgold, built an identical house next door.

The house remained in the Hench family until 1923. It was renovated by a later owner in the 1940s, at which time a bathroom was added on the first floor and a substantial amount of steel reinforcement was installed in the basement. Well maintained for years, the house entered a period of deferred maintenance and neglect by the 1980s. The current owners, Jim and Jean, who lived nearby, watched the house slowly disappear behind untamed shrubbery and trees as paint peeled and roof slates fell. The then-owner eventually hired a contractor to repair the roof but abandoned the job after a large area of slate was removed, leaving tarps to cover the roof. When the tarps succumbed to the weather, water began to pour into the house, a condition that continued for twenty years. Finally, in 2006, the city condemned the house and evicted its resident.

Recognizing that this distinctive historical house was likely to be demolished, Jim and Jean decided to sell the nearby home they had spent years returning to its 1890s character and take on saving the Hench House. Recently retired, they had the time and energy for what they knew would be a big job.

Before any work could take place on the house itself, the jungle around it had to be cleared. Nine thousand dollars' worth of tree and brush removal exposed the house's true condition, which was bad. Undaunted, Jim and Jean dug in and started working. The first priority was a new roof. Recovering the entire roof with slate would have been too costly, but they were able to repair the slate on the highly visible corner tower roof. Elsewhere they chose an asphalt roofing shingle that mimicked the color and pattern of slate.

Inside the house, the hoarding of the last resident had filled every room waist-high with piles and piles of refuse, which, fortunately, had served as a giant sponge, sopping up much of the water that had come through the roof for twenty years. Removing the soggy mess revealed hardwood floors in remarkably good condition. In fact, the high-quality materials used in the original construction, including old-growth woods, had survived their long neglect remarkably well.

Jim and Jean decided to return the exterior from white (peeled and weathered) to its original colors. Jim spent two years on a rented lift, stripping all the remaining paint, repairing trim and siding, replacing pieces where necessary, and painting the exterior from the gables down. Rebuilding the porches was the last part of the exterior restoration. The new owners retained as much original material as possible; when new materials were needed, they were selected or milled to match what had to be replaced. To save costs near the end of the project,

Jim decked the porches with pressure-treated wood, a decision he later regretted, as it does not hold paint well. Since the Hench House was repainted, twenty nearby houses have also been returned to period colors, transforming the neighborhood of Victorian homes.

Inside the house, all the plaster on the first two floors and all the flooring and trim were saved. Gypsum board was used only where missing walls were rebuilt and in the attic, which had suffered the most water damage. (The plaster there was already on the floor when Jim and Jean bought the house.) Trim for the rebuilt walls was custom milled to match what it replaced, and missing pocket doors were replaced with architectural salvage doors from another building. Jim and Jean finished the principal rooms with Bradbury & Bradbury wall and ceiling papers, which they installed themselves. Bedrooms received a simpler treatment with wallpaper from York Wallcoverings, which is located nearby, and borders from Bradbury & Bradbury and other specialty reproduction companies.

Going into the project, Jim and Jean were unfamiliar with building codes and at the mercy of their local code enforcement officials. These officials—apparently unaware of the chapters in the BOCA code that treat historic buildings differently from new construction—insisted that the homeowners bring the house up to new construction standards, adding significantly to the cost. In fact, the only new element added to the house was a bathroom; all the other work was either restoration or system updates. The entire electrical and plumbing systems were replaced. Because the owners planned to operate a bed and breakfast in the house, a sprinkler system had to be installed.

The Hench House in 1889 with Hench's coach and coachman in the foreground. Hench's first son is on the tricycle. This photo provided vital clues to the original paint scheme.

The house after Jim and Jean bought it, cleared the site of overgrowth, and fixed the roof.

Jim and Jean operate the restored house as a B&B,
which helps to offset the costs of restoration.

The entry hall's restored floor was nearly ruined by decades of water infiltration through a failed roof. The staircase, bay window, and stained-glass transoms are characteristic of the Queen Anne style.

An angled bay in the corner of the parlor is at the base of the tower. It creates the kind of cozy nook beloved by Victorians.

The front parlor is set off by Bradbury & Bradbury's Anglo-Japanese roomset, which cover the walls and ceiling with a profusion of Aesthetic Movement patterns and colors. The mantel's ceramic tile surround and the striped floor (two contrasting species of wood) are original to the room.

Aesthetic Movement motifs distinguish the cast brass hardware on the restored pocket doors.

A local artist captured the interest and excitement when Jim and Jean started their restoration. Note the hole in the roof, which had caused enormous water damage.

The porch roof pediment over the main entrance includes shaped shingles and spindlework detailing.

Looking into the kitchen from the back parlor. The large cabinet unit came from Jim and Jean's previous house.

The restored dining room with the table set for the guests' breakfast. The wallpaper frieze is by Bradbury & Bradbury.

This bedroom corner detail, looking up at the ceiling, shows elements of the Aesthetic Movement including a sunflower motif and Japonesque strapwork.

This cast iron implement was made in the Hench & Dromgold factory two blocks from the cousins' identical houses.

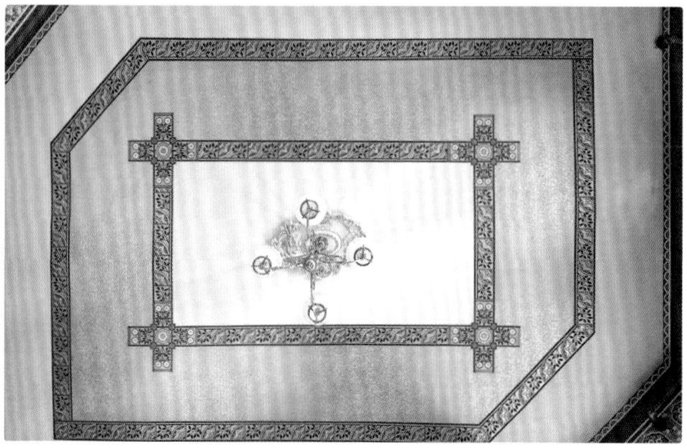

The design of the Bradbury & Bradbury reproduction ceiling papers in the restored parlor highlights the shape of the room.

A profusion of architectural elements—porches, towers, bay windows, dormers of multiple shapes, clapboards, shingles, brackets, and moldings—characterizes the Queen Anne style, the most "Victorian" of the Victorian styles. Four of the seven porches on the Hench House are visible in this view.

A detail view of one of the stained-glass transoms in the bay window off the entry hall.

A bedroom decorated with a Bradbury & Bradbury frieze and wallpaper from the York Wallcovering factory near the house.

The rounded bay window off the front entry hall features scalloped shingles, clapboards, and beadboard siding, with stained-glass transoms over the windows.

The main stair continues to the attic, winding its way to the top of the house.

Detail of one of the Bradbury & Bradbury wallpaper friezes installed by Jim.

Japonesque railings and Aesthetic Movement spindlework on the front porch. The overhead sprinkler pipe (required by code for a B&B) is painted to blend with the wall behind it.

The attic was the only area in which the plaster was damaged beyond repair (left) and had to be replaced (right).

This bedroom has a more restrained ceiling treatment with a reproduction border around a painted field. The locally produced wallpaper evokes the Victorian era without being an exact reproduction.

This bath was added in the 1920s under the main staircase, bumping out into a small addition. The original tilework and fixtures were preserved during the restoration.

This bathroom was redone in the 1940s, and Jim and Jean preserved the wonderful tile floor and wainscot, bathtub, and sink of this period.

The Work

Despite decades of neglect, enough of the home's original materials survived in good condition—even on the exposed porches—to make restoration possible.

Looking down the attic stairs after the wallpaper was stripped from the walls.

Jim spent countless hours on scaffolding restoring original plaster. Only the attic and a rebuilt missing wall needed new gypsum board surfaces.

All remaining exterior paint was stripped before repainting to ensure a lasting finish after so many decades of neglect.

Another exterior view at the start of the project.

Plaster restoration in the second-story hallway.

A new roof to halt decades of water ingress was the first priority.

This photo was taken after the vegetation that had hidden the house was cleared. Sodden trash was removed by the dumpster load, and windows were left open to allow the interior to dry out.

Starting at the top, Jim spent two years on a rented lift stripping old paint, repairing trim and siding, and repainting the house in its original colors.

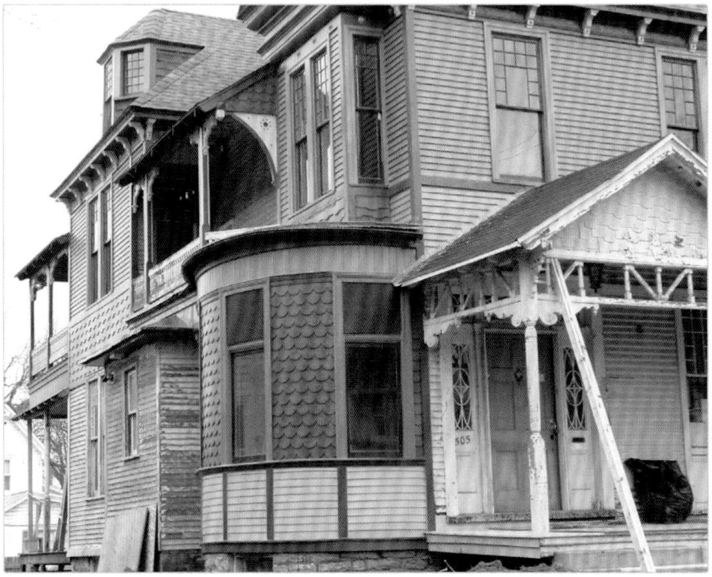

Enough of the home's original paint remained under many subsequent layers of white paint to allow the scheme to be recreated.

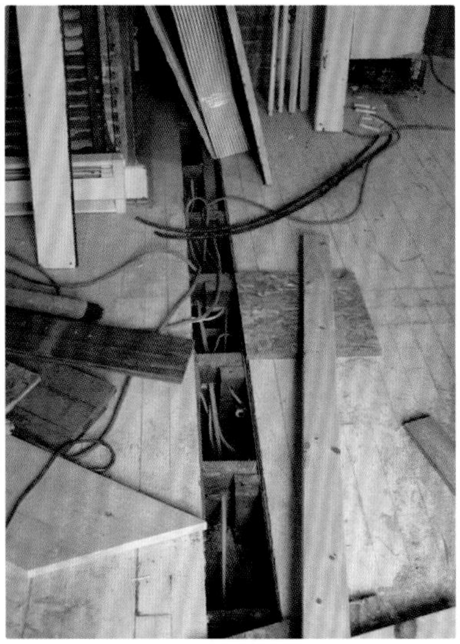

New electrical and plumbing lines were installed under temporarily lifted floorboards.

This photo suggests how much effort went into repainting hundreds of trim elements in multiple colors.

A gypsum board patch and plaster washers in a ceiling repair.

CHAPTER 3

Developing Your Project Design

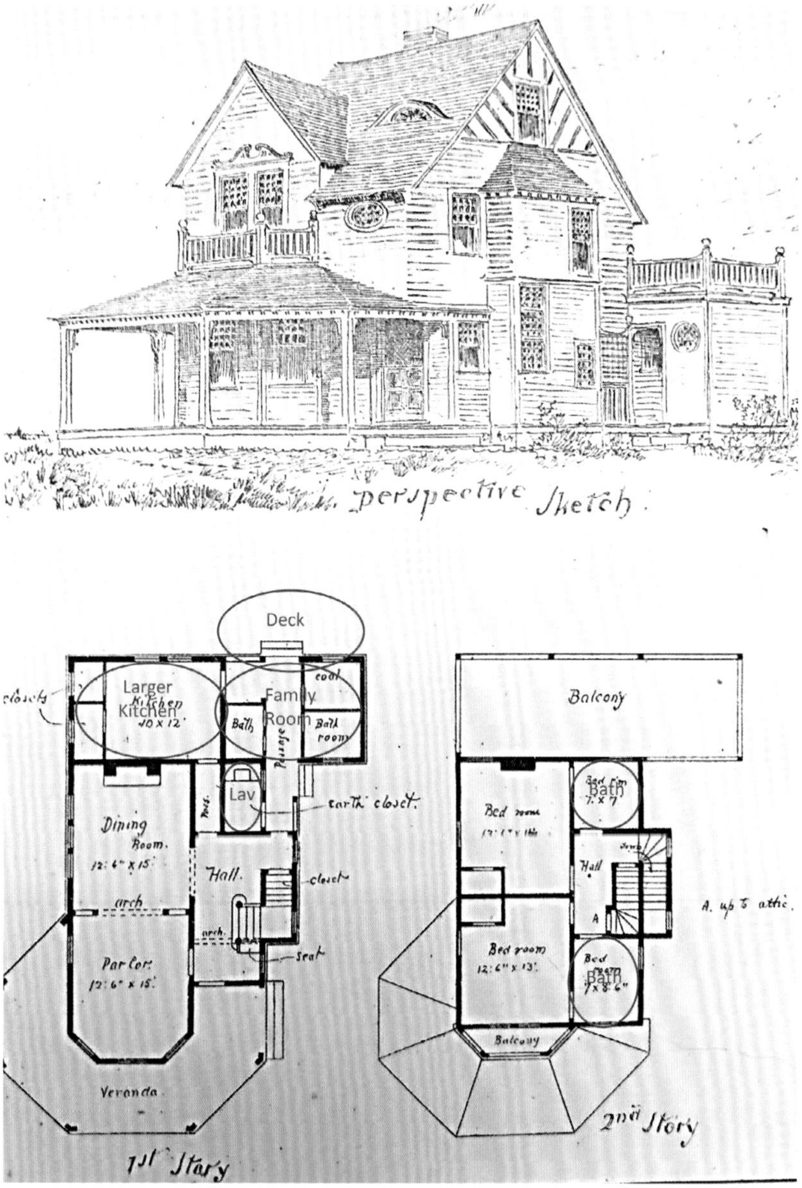

A careful study of your house will give you a prioritized inventory of its character-defining features. Important exterior features generally include:

- The overall form of the building, including porches, towers, dormers, etc.
- Siding and trim materials
- Windows and doors
- Roofing materials
- The relationship of the building to the street
- Historic outbuildings

Important interior features include:

- Primary public rooms such as the parlor, dining room, sitting room, and library
- Major circulation spaces, especially the front hall and front staircase
- Fireplaces and other visual focal points
- Doors, windows, trim, and flooring
- Wall and ceiling plaster, including plaster moldings
- Rare surviving features such as period wallpaper or decorative painting

Your prioritized list of features to be preserved will guide you (by the process of elimination) to the areas of the house where further changes can be accommodated with least impact.

Identifying Areas of Opportunity

Generally speaking, spaces and features apart from those listed above are likely to represent *areas of opportunity*—i.e., rooms or features that you can change without degrading the house's historic character. It is usually helpful to define a house's *period of significance*—the time after all the major character-defining features were in place and before subsequent (often unsympathetic or incompatible) changes were introduced. My own house provides a case study.

Whitten House Case Study

For Whitten House, the period of significance is approximately 1850–1950: after the second story was added and the house took its present form, but before modern alterations were introduced. The last Whitten, Sarah, lived there from her birth in 1845 until her death in 1941 and maintained the house as her parents had left it. She bequeathed it to the town to be used as a public library, and the library made few changes prior to 1950 beyond lining the walls with bookcases, removing a wall on the second floor to merge two bedrooms into a single meeting room, and making a few careful second-floor trim alterations that were in keeping with the house's character. Several more drastic changes occurred in the 1960s, including the installation of a central heating system and a half bath (the first indoor plumbing in the house) and the demolition of the shed to make way for a small addition to the ell. Vinyl siding was installed and the roof framing strengthened in the 1980s.

Several things had to be added to Whitten House to make it a twenty-first-century home, starting with a modern kitchen and at least one full bathroom. Sarah Whitten's wood-fired cast iron cook stove had disappeared with the shed, which had also contained the "modern" kitchen created around 1900. The only surviving kitchen features when the house was sold by the town in 2003 to help pay for a new library were a five-foot laminate counter with sink and one 1940s wall cabinet reused in the room added in the 1960s.

An examination of the house's character-defining features helped to identify "areas of opportunity" for a new kitchen and bathrooms. In plan, the house has two large front rooms with intervening stair hall on each of its two main stories. At the rear of both stories, a series of small rooms was interspersed with hallways and the back stair, which continues to the attic, where two bedrooms had historically been located. The one-and-a-half-story kitchen ell extending from the rear of house has one room on each floor, and the 1960s addition continues the ell into a single one-story room. Since it had no historic character to maintain, stood next to the dining room (the original kitchen), and occupies the space where the shed housing the c. 1900 "modern" kitchen formerly stood, the 1960s addition was an obvious location for a new kitchen.

The existing half bath occupied the location of the original brick oven and cooking fireplace in the original kitchen ell. Since I intended to restore those features, that bath would have to go. Since the small rooms at the rear of the house were too small for most uses, and since their historic trim and finishes were much simpler than in the large rooms, these provided the best candidate locations for new bathrooms. The design that eventually emerged created a bath and laundry room at the center rear of the first floor and two baths in the rear of the second floor, one at either end. No alterations in the significant historic spaces were needed to accommodate these changes. The small room just in front of the original kitchen ell appeared to have been a pantry (now stripped of all cupboards and shelving) and would serve well for displaying antique china and other dining-related objects once the pantry cabinetry was recreated. The remaining small room, at the southwest corner of the house, would become an office/library.

Although Whitten House was more intact than most historic houses, the spaces it contains are typical of old houses. Larger rooms with finer finishes appear toward the front, with diminishing detail as one moves up or back from the front door. Character-defining interior feature locations tend to follow the same pattern, and unless major spaces are already

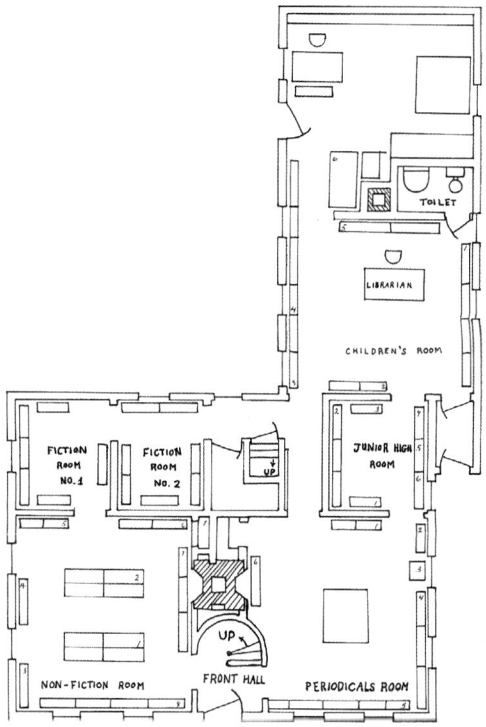

Whitten House first-floor plan, c. 1988, while it was serving as the Topsham (Maine) Public Library.

altered and stripped of historic character, it is better to make changes in the less significant spaces. In most houses predating indoor plumbing, bathrooms and a kitchen with range, refrigerator, cabinets, and sink have already been added. Occasionally, a house will have a particularly intact and interesting secondary space, such as a bathroom installed in 1890 and never updated, that should be considered as significant as the major spaces. Such a bathroom might not be ideal for daily use in the twenty-first century, but it would be fine for a guest room or other occasional use.

Sheds and outbuildings can provide other areas of opportunity. In the case of Whitten House, reconstructing the shed that had been demolished in the 1960s would allow a larger modern kitchen connecting the house to a new barn, which likewise recreated what had stood on the property historically. The barn would contain garage spaces, a workshop, and, on its second floor, additional living space. Many historic homes have similar secondary structures attached or in close proximity; these are areas of opportunity for adding modern spaces. Unfinished attics can also be converted to usable space, and sometimes basements can be made functional. Before thinking about adding on, look for underutilized spaces that can be repurposed.

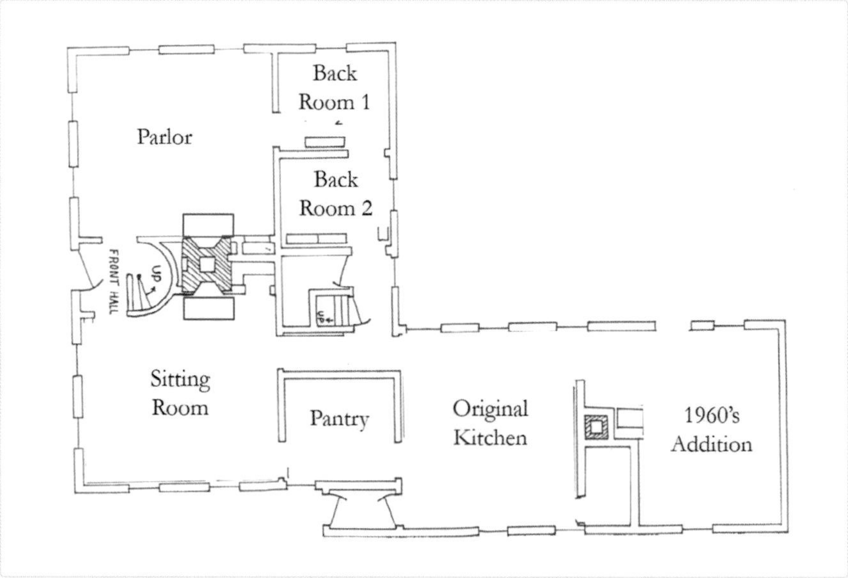

Whitten House with rooms labeled for their original uses before the house became a public library.

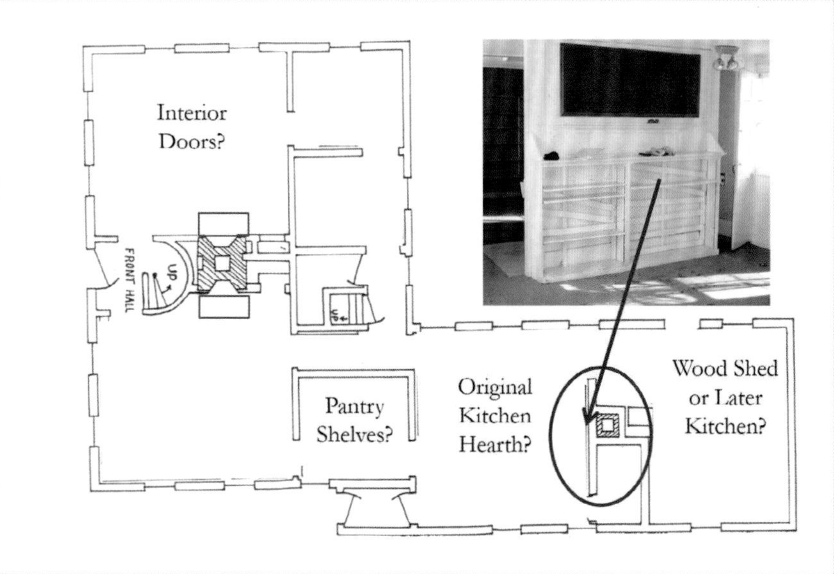

Whitten House with missing historic features labeled.

Ground Rules for Changes

It isn't always as easy as in Whitten House to identify where changes can be made without impacting character-defining features. Here are a few suggestions:

- Do nothing to fundamentally alter the exterior historic character of the house as viewed from the street(s).
- To the extent possible, preserve and repair existing historic materials.

- When materials cannot reasonably be repaired, replace them with matching material of the same quality.
- Design your alterations and additions to be compatible with the house in form, scale, placement, and materials.
- Fenestration patterns (i.e., the window layout) for any additions should not distract from the historic house.
- Concentrate your changes in areas of lesser importance to the house's character.
- Where possible, restore lost character-defining features for which sufficient documentation exists to do so accurately.
- When lost features are not well documented, make your replacements compatible with the character of the house.

Local Historic District Regulations and Design Standards

If your house is in a locally designated historic district, there will be local standards or guidelines to follow. (No restrictions or mandatory reviews of changes are imposed by a listing on the National Register.) Local historic district ordinances vary considerably but are always designed to preserve a community's historic character. In general, they are only concerned with the portions of a building that can be seen from a public right of way.

If your house is protected under a local historic preservation ordinance, you will want to read the ordinance and note the standards for review. If possible, you should meet with the administrator of the ordinance—usually someone in the municipal planning or code enforcement department—before you invest time, effort, and money in developing plans. Review boards or commissions are often made up of town residents with limited expertise in architectural history or recommended preservation practices. The standards of review they follow may be the only reference materials they have to work with. The better you understand those standards, and the more knowledge you have about best practices being used elsewhere, the more likely you are to have a positive experience with the local review board. This subject is explored more fully in Chapter 5.

Do You Need an Architect?

If you are not planning extensive alterations and are able to create reasonably clear plans to communicate your intentions to review boards, building permit departments, and contractors, you probably don't need an architect. An architect may be advisable, however, for structural or masonry repairs beyond the most basic. For serious structural issues, at least get the input of a structural engineer with experience working on historic buildings. If stamped drawings are needed by the local code-enforcement office, a structural engineer will sometimes review and stamp your drawings for a fee, provided the drawings show solutions that meet applicable codes. Some structural issues that are common to old buildings—such as sill rot—can be addressed by a good contractor or competent DIY property owner and do not require the services of an architect or engineer. DIY structural repairs are addressed in Chapter 7.

If you decide to hire an architect, it is essential to hire one with experience in historic preservation work. Most architecture schools do not teach historic preservation, and the majority of registered architects have no understanding of traditional construction

These charts will help you sort the character-defining features of your house into three categories: Primary, Secondary, and Lowest-Priority Features. Within each category, numerical scores further define the priorities based on condition, with higher-priority features earning a higher number.

Prioritizing Preservation and Restoration of Historic Interior Character-Defining Features

Alterations more than 50-75 years old may be considered historic if they are of a quality consistent with the original work and/or contributing to the current historic character of the house.

	PRESERVE					RESTORE	
	Intact	Damaged or Deteriorated	Replaced Appropriately	Replaced Inappropriately	Missing	Accurately Restored	Compatibly Replaced
Primary Features:							
Plan	4	3	2	1	0	4	3
Primary room dimensions/proportions	4	3	2	1	0	4	3
Circulation spaces	4	3	2	1	0	4	3
Front staircases	4	3	2	1	0	4	3
Primary room and circulation space flooring	4	3	3	1	0	4	3
Primary room and circulation space fireplaces/mantles	4	3	3	1	0	4	3
Primary room and circulation space doors	4	3	3	1	0	4	3
Interior stained or leaded glass	4	3	3	1	0	4	3
Primary room and circulation space trim/paneling/wainscot	4	3	3	1	0	4	3
Decorative painting on walls, floors, ceilings, or trim	4	2	3	1	0	4	3
Historic paper or textile wallcovering	4	2	3	0	0	4	3
Historic tile	4	3	3	0	0	4	3
Primary room and circulation space ceiling plaster	3	2	3	0	0	3	2
Primary room and circulation space wall plaster	3	2	3	0	0	3	2
Historic pantry built-ins	3	2	3	1	0	3	2
Historic kitchen built-ins	3	2	3	1	0	3	2
Historic bathroom finishes and fixtures	3	2	3	1	0	3	2
Secondary Features:							
Back staircases	3	2	3	1	0	3	2
Secondary space flooring	3	2	3	0	0	3	2
Secondary space fireplaces/mantles	3	2	3	1	0	3	2
Secondary space doors	3	2	3	1	0	3	2
Secondary space Trim including paneling/wainscot	3	2	3	1	0	3	2
Secondary space dimensions/proportions	2	1	2	0	0	2	2
Secondary space ceiling plaster	2	1	2	0	0	2	1
Secondary space wall plaster	2	1	2	0	0	2	1
Lowest Priority Features:							
Unfinished shed or attic space	1	1	1	0	0	1	1

Any rare, uniquely designed or crafted, or locally/regionally distinctive element or exceptional historic work/materials must be considered highest priority regardless of condition. Damaged, inappropriately replaced, or missing highest-priority features should be restored/replicated if at all possible.

Prioritizing Preservation and Restoration of Historic Exterior Character-Defining Features

Alterations more than 50-75 years old may be considered historic if they are of a quality consistent with the original work and/or contributing to the current historic character of the house.

	PRESERVE					RESTORE	
Primary Features:	Intact	Damaged or Deteriorated	Replaced Appropriately	Replaced Inappropriately	Missing	Accurately Restored	Compatibly Replaced
Primary elevation fenestration pattern	4	3	3	1	0	4	3
Primary elevation siding	4	3	3	1	0	4	3
Primary elevation trim	4	3	3	1	0	4	3
Primary elevation porches	4	3	3	0	0	4	3
Slate, clay, or historic ornamental metal roofing	4	3	3	2	0	4	3
Roof towers/cupolas	4	3	3	2	0	4	3
Primary elevation historic dormers	4	3	3	1	0	4	3
Primary view Chimneys	4	3	3	1	0	4	3
Primary elevation windows	4	3	3	1	0	4	3
Primary elevation doors	4	3	3	1	0	4	3
Stained glass or leaded windows	4	3	3	1	0	4	3
Secondary Features:							
Secondary elevation fenestration pattern	3	2	2	0	0	3	2
Secondary elevation siding	3	2	2	0	0	3	2
Secondary elevation trim	3	2	2	1	0	3	2
Secondary elevation porches	3	2	2	1	0	3	2
Secondary elevation doors	3	2	2	1	0	3	2
Secondary elevation windows	3	2	2	1	0	3	2
Secondary elevation dormers	3	2	2	0	0	3	2
Secondary chimneys	2	1	1	0	0	2	1

Any rare, uniquely designed or crafted, or locally/regionally distinctive element or exceptional historic work/materials must be considered highest priority regardless of condition. Damaged, inappropriately replaced, or missing highest-priority features should be restored/replicated if at all possible.

Any primary element inappropiately replaced/altered in the last 50 years moves to the bottom of the list unless it can be restored/replicated based on documented historic appearance or with compatible new design.

Missing features should be restored/replicated if at all possible. Moves to top if restored.

Any secondary element inappropiately replaced/altered in the last 50 years moves to the bottom of the list.

Missing secondary-priority features may be restored/replicated if desired.

methods and materials. In most communities there is at least one architect who works regularly on historic buildings, and your local preservation organization or historical society may be able to tell you who this is. If your town has a preservation ordinance, the municipal planning staff or a member of the board can tell you which architects have brought good projects before their board. Neighbors or friends who have done sensitive rehabilitation projects on their own houses may have suggestions. Take heed of their warnings as well as their recommendations.

The firm I work for does a lot of commercial historic rehab projects using state and federal historic tax credits as part of their financing. These credits require that the project meet National Park Service standards for historic rehabilitation, which ensure that the historic character of a building is maintained. Nearly all such projects use architects, and they seem to come in two basic varieties: those who work *with* a building's historic features and those who work *against* them. You want the former. If you sense that an architect's primary interest is in making a "statement" of some sort, find another architect. The house *is* the statement, and the architect needs to serve the house—not the other way around.

Drawing Your Own Plans

There are many useful tools to help you produce workmanlike drawings and specifications. Whether you are using a pencil on graph paper or a CAD program on a computer, the essential first step is to measure the house accurately. You will need a rough sketch on paper to start with. This can be as simple as a rectangle representing the footprint of the building with lines drawn within it in the approximate locations of walls, stairways, and chimneys.

Start outside and measure the exterior wall lengths, working your way around the house and writing down measurements as you go. Whether you use a traditional tape measure or a modern laser measuring device, the process will be much easier with an assistant. After completing your first circuit around the house, make a second one to place the first-story doors and windows on your sketch. Record their distances from corners and from each other, as well as the widths of the openings. Also measure the vertical distance from the top of the foundation to the bottom of each window opening, and the height of the opening itself. Write these measurements in the appropriate locations on the sketch and circle them to distinguish them from horizontal measurements.

Measure the height from the top of the foundation to the eaves and note this on the sketch. If different sections of the house have different heights, note all of them. If the house has more than one story, measure the vertical drop from the eave to the top of each upper-story window opening and the height and width of each opening, recording them on a second sketch of the building's footprint. If there are more than two stories, measure the openings in the middle stories from the foundation or eaves (noting which) and record on additional footprint sketches.

Finally, measure the height to the peak of the gable or top of the hip roof. Lacking a lift, it can be difficult to get exact measurements high off the ground. It is generally acceptable to calculate roof heights by other means, such as measuring from the attic floor to the tip of a hipped roof or counting clapboards in a gable and calculating their combined height, then adding any frieze and cornice at the eaves. Unless you are planning complicated alterations to the roof, exact measurements are not critical.

Now move inside to a corner room on the first floor. Work your way around the room, measuring overall dimensions first and then window and door sizes and locations. Include fireplaces, built-in cabinetry, closets, etc. Open a window and measure the thickness of the exterior wall (not including the thickness of interior trim). Open a door and measure the thickness of an interior wall (not including the thickness of trim). Measure the floor-to-ceiling height. Once you have measured and recorded all the dimensions needed to represent the room accurately in drawings, move to the next room and repeat the process. Continue through the house until every space is recorded. Be aware that you may find different door and window sizes and even different wall thicknesses in different parts of the house. Ceiling heights often differ between stories and between the front and rear areas of a large house.

Once you've recorded all the dimensional information on your sketches, you can use it to create scale drawings with pencil and paper or a computer-based architectural drawing program. If you are not an experienced draftsperson and are working with paper and pencil, I suggest working on tracing paper over a 1/16-inch grid paper using a metal straightedge and fine lead mechanical pencil. A good eraser is critical.

A computer-aided-drawing (CAD) program makes it simple to convert floor plans into elevations and often into three-dimensional renderings. This can be very helpful for visualizing how a proposed alteration, such as a building addition, will look in three dimensions, and for conveying such information to review boards or building inspectors. Professional CAD programs are expensive, hard to master, and probably unnecessary. Inexpensive house-design software—available from multiple sources—is easier to learn and will answer your needs. A free downloadable architectural drawing program that I use frequently is Sketchup. I strongly recommend that you read user reviews before purchasing any program.

As you draw by hand or computer, start with your largest dimensions to outline the footprint, exterior wall thickness, and then interior rooms and walls. Once the footprint is in place, add doors, windows, and other features. This provides an opportunity to double-check your measurements, as the smaller dimensions need to add up to match the larger ones. Discrepancies are normal at this stage; remeasure as necessary. There may be hidden spaces or other details of the building's construction that are not readily apparent but become obvious through the measurements. You may discover hidden chases for plumbing or dumbwaiters, obsolete chimneys that no longer go through the roof, or even staircases that have been sealed off during past alterations. Any such spaces need to be added to your drawings.

Once you have accurate plans of the existing house, you are ready to move ahead to the alterations you have in mind. Always do this work on copies of the existing-condition plans, not the originals! It seems obvious, but even experienced people can forget and start altering an existing-condition drawing without saving or copying it first—trust me.

Whether working in pencil or on a computer, I recommend using pencil and tracing paper over existing-condition drawings to sketch initial concepts. "Bubble drawings" can be useful for roughing out the placement of rooms and uses within a house. With a red marker, draw circles or bubbles to show the approximate sizes and locations of changes you have in mind, working with the existing plan as much as possible. The fewer walls you change, the lower the cost and the more historic character you will retain. Grouping bathroom and kitchen bubbles close to each other creates a possibility of cost savings on plumbing, but character-defining features should not be sacrificed to accomplish this. Add bubbles outside the existing footprint to conceptualize where additions might go.

Tracing paper is cheap; try different possibilities on fresh sheets of paper. Once you have one or more conceptual plans that seem to meet your needs and do not do violence to character-defining features, let the drawings sit while you walk through and around the house, visualizing the concepts you have sketched. Consider how daylight will reach various spaces at different times of day, and whether there are views that should be visible from particular rooms. A western view that captures sunsets might be wasted from bedrooms that will seldom be occupied at that time of day.

Do not invest time in careful drawings until you are comfortable with your concepts. Does each bubble approximate a reasonable amount of space for the anticipated use? A bubble representing a space four feet square is fine for a small lavatory but not for a bedroom. Are related uses located in proximity to each other? You do not want your dining room two rooms away from your kitchen. Are private spaces located away from public spaces? A lavatory opening off the dining room could be awkward for guests during a dinner party. Try to imagine day-to-day living in the house as you have conceptualized it.

Once you have a conceptual plan that works for your life and your house, draw the proposed alterations carefully and to scale. If you are using pencil and paper, work on tracing paper over the existing-condition drawings. If you are using a computer program, make your alterations to copies of the existing-condition drawings. In moving from concept to scale drawings, it is common to discover issues and conflicts that were not apparent in the conceptual view. Make adjustments as necessary to accommodate these. As you develop the scale drawings, determine where mechanical systems are going and leave space for necessary ducts, vents, pipes, etc.

If the changes you're planning will affect the exterior appearance of the house via additions, new windows, new doors, and so forth, you should draw elevations to see how those changes will look from the exterior. What works in plan does not always work as well in three dimensions, and drawing the elevations is the only way to see how it plays out. New window openings should be sized and placed to match existing window sizes and placements for compatibility. Roof forms for additions need to be compatible with existing roof forms and generally no taller than the existing roof heights. Elevation drawings give you an opportunity to work out these important aspects of the design. If you are working in a computer-based design program, elevations can be easily or even automatically produced from plans, depending on the program. It is a little more work on paper, but well worth the effort.

Tips and Tricks

If changes in the plan require eliminating a window, it is possible to simply block it in from the interior with a black painted surface (plywood or MDF) set just inside the glass. From the exterior, it will not be noticeable that the window is no longer visible from the interior and the historic window remains in place if a future owner wants to reverse the change.

Do not shorten historic window openings to accommodate kitchen counter heights. It is preferable to run the cabinets right past the lower openings and install a black painted panel to cover the exposed back portion of the cabinets. The fenestration pattern will remain unchanged from the exterior and the windows remain as they were if a future owner decides on a different kitchen cabinet configuration.

Documenting Your Work

Most people do not give much thought to documenting their work until it is done. However, it is much easier in the long run to set up a digital or paper filing system and document as you go. Documenting what you have learned about your house and the work you do on it is important for several reasons. First, future maintenance will be easier if you have good records of when and what you did during rehab. Second, good records document the investment you've made, demonstrating increased value if you want to refinance or sell. Documentation can also be very helpful if the house is damaged or destroyed and you have to file an insurance claim. Third, documenting your historic research will make it easier to check details later when it all gets a bit foggy in your memory; it will ensure that the information is archived and can be passed to the home's next owners.

Working with What You've Got

Existing conditions in your house provide the take-off point for your project plan. What to do starts with what you've got. A few examples will suggest how necessary changes can be accommodated without destroying historically significant features and character.

Using Areas of Opportunity

My own house, Whitten House, had been in use as a public library for forty years when I bought it (see plan on page 122) and, as the plans show, had no modern kitchen or full bath. It had intact historically significant spaces at the front of the house and a restorable significant space in the ell. Small secondary spaces in the main block of the house and a modern addition to the ell provided opportunities for bathrooms and a modern kitchen.

On the first floor, I partially opened the small center room at the rear of the main block into the walk-in closet under the back stairs, then placed a new partition through the center room to create a hallway outside this new, larger space. The closet door was moved into the new partition to provide access to the new laundry room and bath. The only historic material removal was 39 inches of wall between the closet and center room.

When the single-story ell extension had been added in the 1960s, a cellarway and chimney closet were removed to create a wide opening between the added room and the original ell. My rehab plan removed the half bath that had been added to the ell in the 1960s and returned this area to its original configuration, with a cooking hearth, brick oven, and chimney cupboard facing the original kitchen, which would become the dining room. The original cellarway space was recreated to serve as a boiler closet. The cellarway door (which had been stored in the attic since the 1960s) was moved around the corner from its original location to provide more wall space in the new kitchen. Because the rehabilitation would be a phased project, the new kitchen was designed to fit into the 1960s addition in the first phase, with a plan to double its size in the second phase.

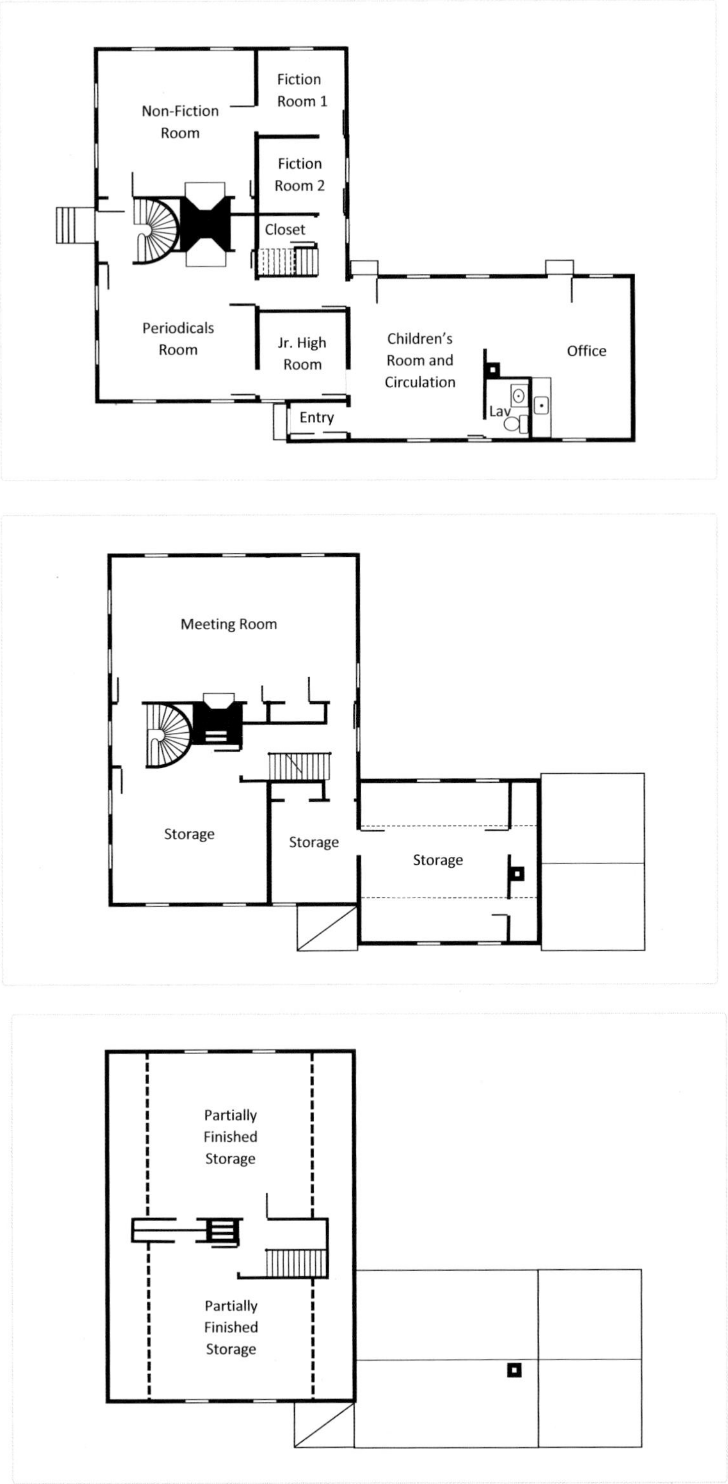

Floor plans for Whitten House at the time of purchase.

On the second floor, a wall between a front and back bedroom had been removed in the 1940s to create a large meeting room for the library. I returned this wall to its original location, creating space for a bathroom and walk-through closet next to the master bedroom where a small back bedroom had once been located. In the existing plan, it was necessary to walk through the small room at the north end of the house to get to the room in the ell attic. Walking through one room to get to another, even bedrooms, was not uncommon in houses prior to the late nineteenth century. The rehab plan inserted a new partition in the small room to create a hallway to the ell room. A linen closet fit nicely at the end of this new hallway.

The remaining space of the small room was used for a second upstairs bathroom. The only historic materials removed for these changes came from the removal of a closet in each small room. The closet doors and trim were salvaged and reused for the new door openings.

The only other change to the existing house was the creation of a two-story bay window on the southwest corner. The rooms at this corner, on both floors, were only 8 feet by 8½ feet with no windows in the west wall. The bay window made these rooms 25 percent larger and much brighter. The second phase of the Whitten House rehabilitation will recreate the shed and barn that once extended from the kitchen ell; this will accommodate a larger modern kitchen in the shed and a garage and workshop space in the barn with a large modern living space in the barn loft.

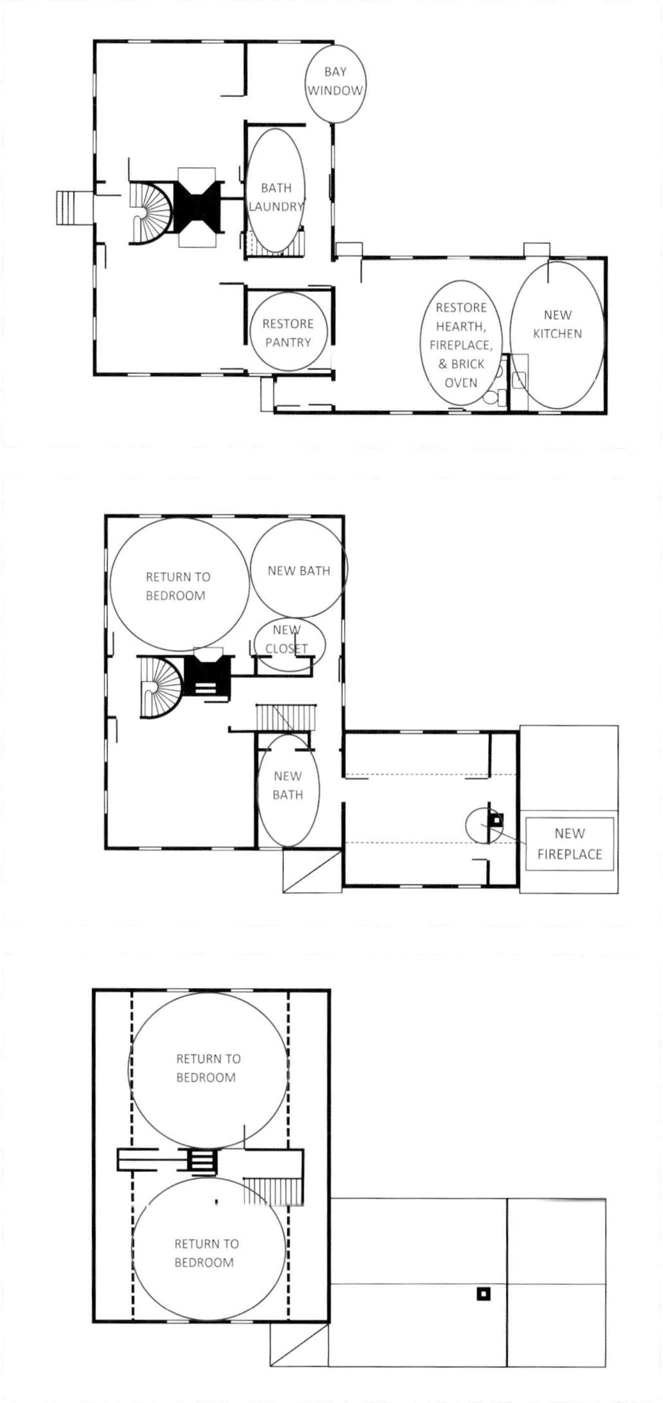

The Whitten House project was planned in two phases. These are the bubble drawings for Phase 1, showing roughly where changes were to be made during rehabilitation.

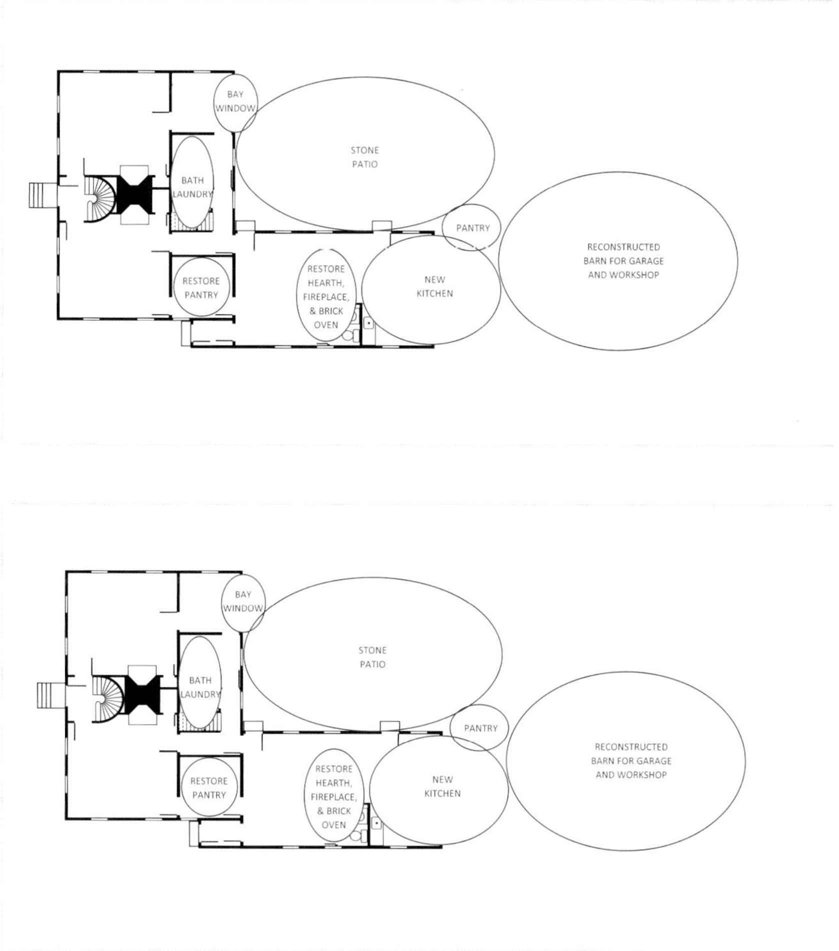

Phase 2 will remove the 1960s addition and reconstruct the missing woodshed and barn/carriage house. No additional changes are planned for the historic house and ell in Phase 2.

A Reconfigured Duplex

Another good example of sensitive changes to a house can be seen in the reconfiguration of a side-by-side double house, or duplex, into first- and second-floor units. The duplex had most recently housed professional offices in one half and a residence in the other. The owners, one of whom had his psychology practice in the building, wanted to create a one-story home in which they could "age

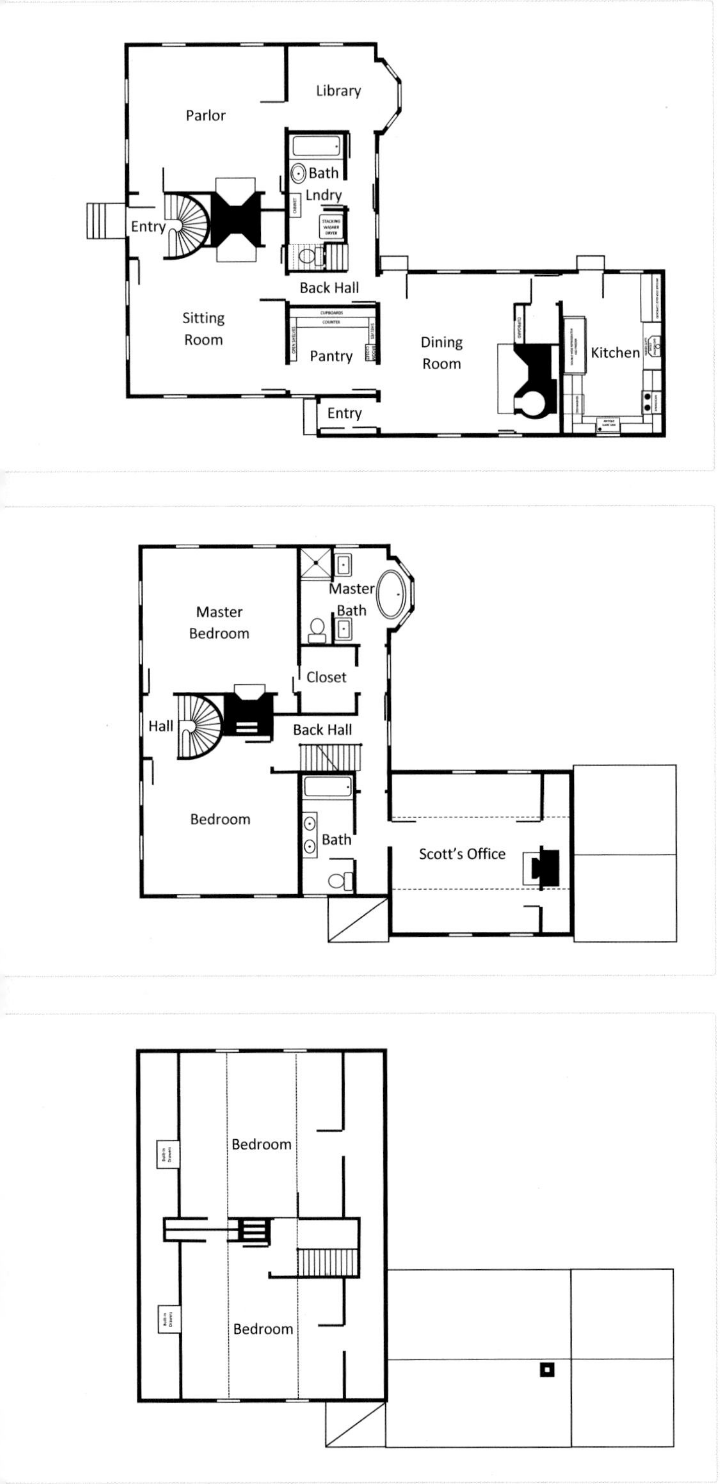

Design drawings for Phase 1 of the Whitten House rehabilitation

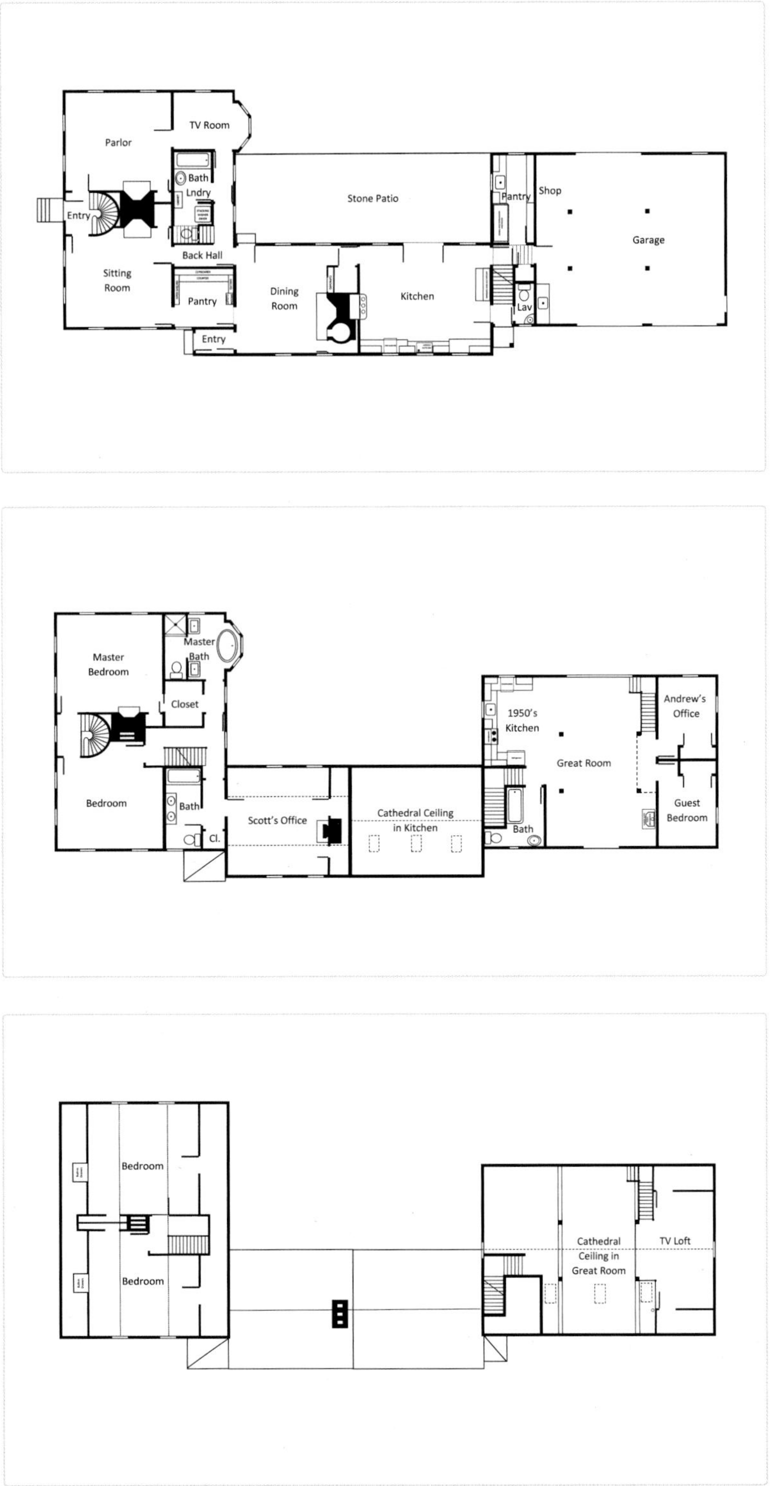

Design drawings for Phase 2 of the Whitten House rehabilitation.

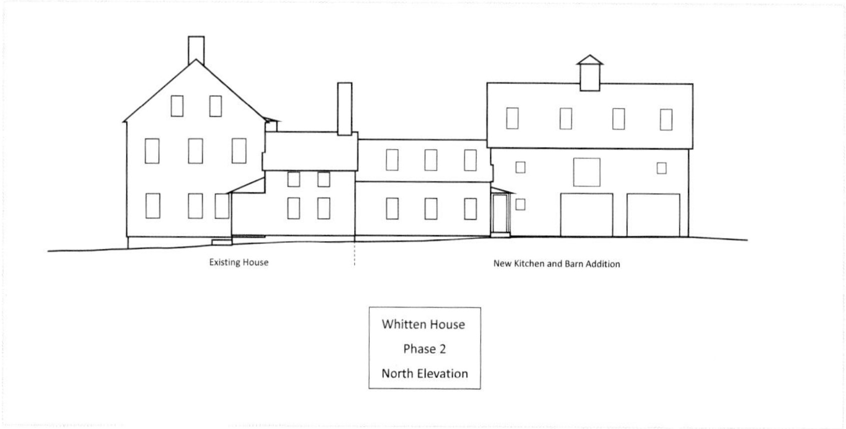

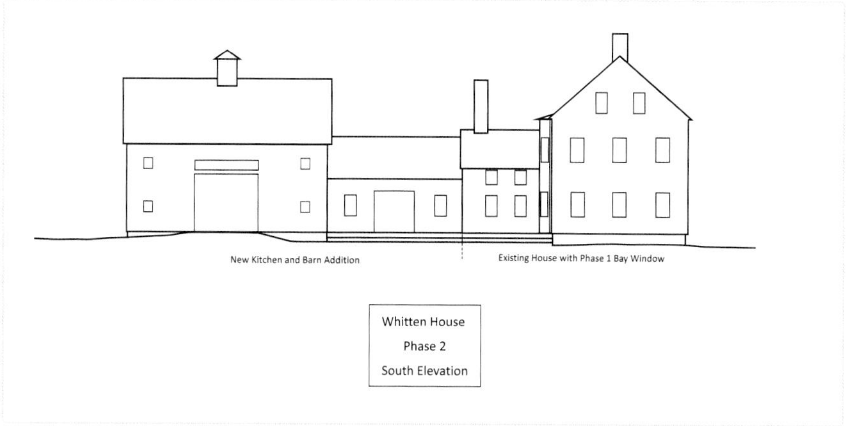

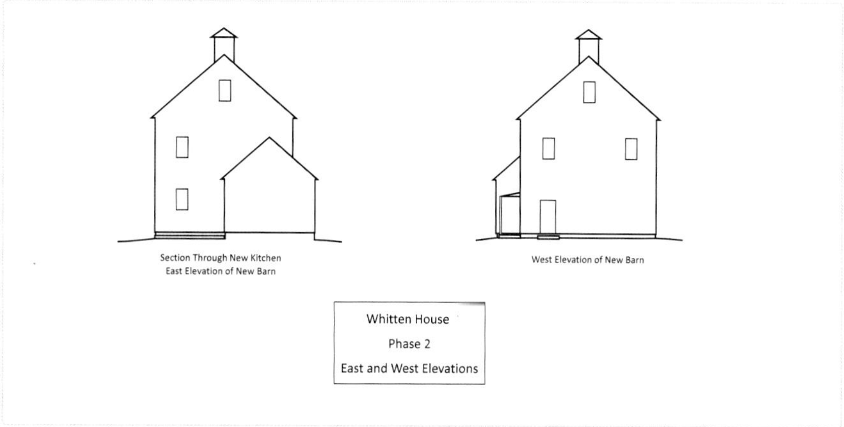

Elevation drawings for Phase 2 of the Whitten House rehabilitation. It is not always necessary to draw detailed elevations, particularly if you will be mimicking the existing materials and details of the house. These simple drawings were sufficient to see how the overall form, scale, and fenestration (doors and windows) would look. Trim for the addition is minimal, as it is a recreation of secondary buildings that would never have been as detailed as the house. There was no need to draw it on the elevations. More detailed drawings were developed for the bay window addition in Phase 1 because it involved detailed trim elements (see "Example Project: A Bay Window Addition" on Page 162).

This eighteenth-century house was moved and converted into a side-by-side double house, or duplex, in the 1830s. Recently it was reconfigured with minimal impact to intact historic spaces and materials to create one living unit on the first floor and professional office space on the second.

in place" while continuing to have income-producing office space in the building.

The building is an eighteenth-century house that was relocated in the 1830s and converted into a side-by-side "double house" with a roughly mirrored plan on each side of a center dividing wall. The original plan was preserved to the extent possible in this nineteenth-century rehab, which resulted in several minor anomalies that kept the two units from being perfect mirror images. These were small spaces that had probably been part of the front hall and the location of a central chimney prior to the relocation of the building.

The architect for the twenty-first-century rehabilitation project recognized that a small, windowless storage room on the first floor (originally occupied by the center chimney) could provide direct access to all first-floor rooms except the kitchen without passing through the two stair halls. Thus, the entire first floor could be used for a single residential unit with minimal alterations to the historic floor plan. This allowed the existing stairways to serve as the code-required two means of egress from the second story. An existing door between the two second-story stair halls, closed since the conversion to a duplex 190 years earlier, was opened to connect both sides of the second story into a single office suite with three offices and a waiting room. This also made it possible to remove a modern wood fire escape that had been built as a second egress from the second floor when the office space used only one side of the house.

New Walls

New partitions do not have to go all the way to the ceiling. If a historic space needs to be subdivided for a new plan, it is sometimes possible to stop the new partition(s) short of the ceiling so that the original space is still understandable. A good example of this is a built-in storage unit divider in a nineteenth-century Santa Fe adobe house. The small house needed storage space, and the owners wanted some separation between the kitchen and the bathroom. The new unit features two open shelving sections on the kitchen side with five narrow doors on the opposite side providing access to storage shelving, a stacked washer and dryer, and the hot water heater and whole-house water filter. A narrow corridor leads to the bathroom. The exposed vigas (log beams) and board ceiling are visible across the top of the unit, so it is clear that it was, historically, one continuous space.

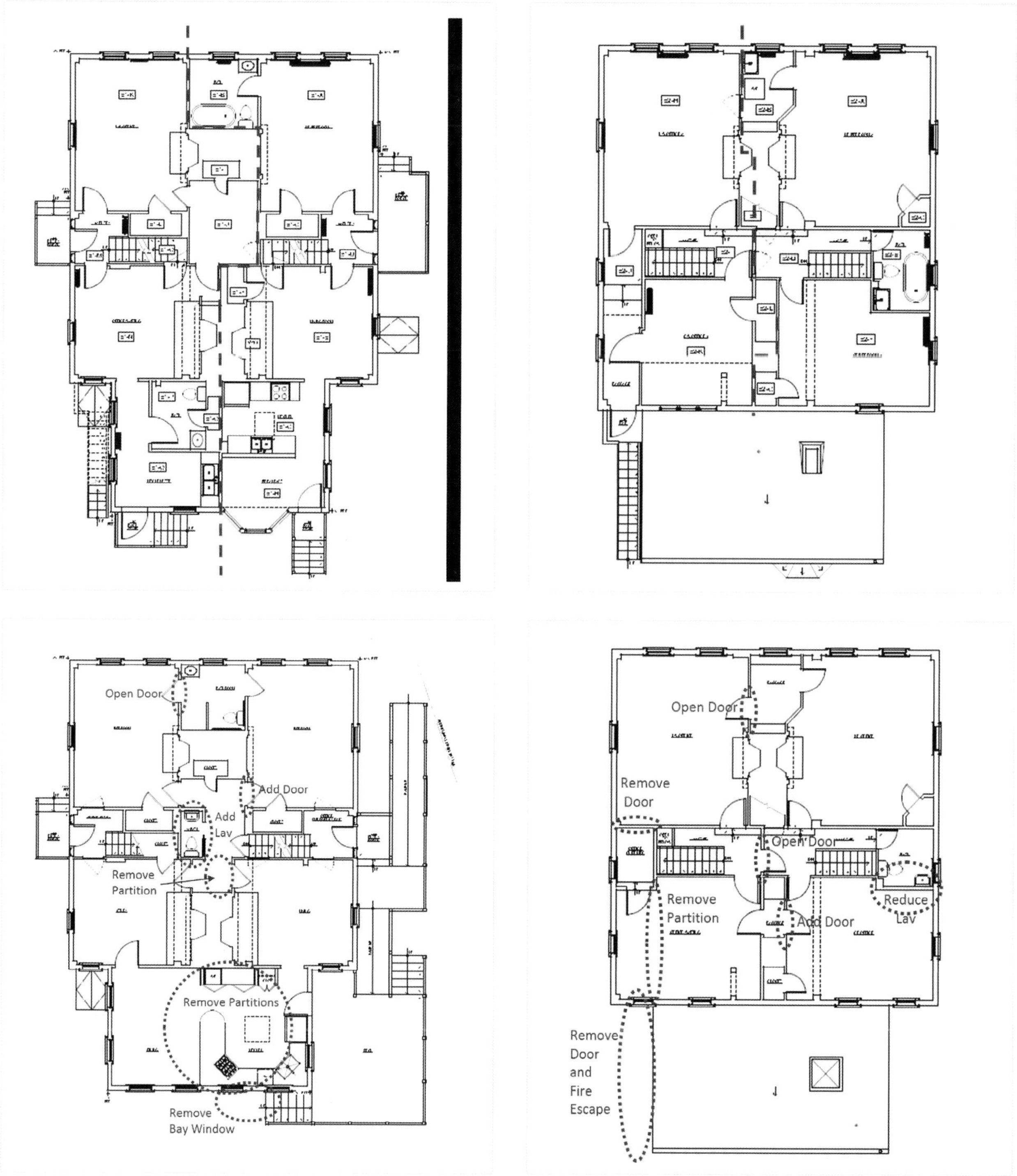

Before-and-after plans for the reconfigured duplex. With minimal change, this side-by-side two-unit building was converted to a single unit on each floor. Several historic doors were reopened that had been sealed since the original single-family house had been converted to a duplex in the 1830s, and only a few new door openings were needed. Most changes were made to areas that had already been significantly altered when half of the duplex was converted to professional office use in the 1980s. *(Courtesy of Mills–Whitaker Architects)*

How Many Bathrooms Do You Really Need?

As these examples show, it is often possible to accommodate changes with minimal alterations to a historic floor plan. Unfortunately, this degree of sensitivity is not always evident. Thousands of houses annually lose significant historic features and spaces for unnecessary gut rehabs, frequently because owners are told it is the only way to update the house. Architects and builders who are accustomed to designing and building new houses assume that the goal of rehabbing an old house is to turn it into a new house. Whatever features are currently trendy or common in new houses get shoehorned into old shells, whether or not they fit and whether or not historic character or structural integrity has to be sacrificed to accommodate them.

This frequently shows up in the assumption that every bedroom needs its own bathroom. Fifty years ago, it was typical for a middle-class family of six or eight to share a single bathroom. A really nice middle-class home might have two baths. Today, a new house with five or six bathrooms is not uncommon. This is convenient for everyone living in the house, but it also takes up an enormous amount of space. In a historic house that is largely intact, creating these small spaces often requires carving up significant larger spaces. Since it is complicated to route plumbing drains, vents, and supply lines in an existing house (especially if it is timber-framed), the owner is usually told that the house has to be gutted to install the plumbing for all those bathrooms. Before heading down this path, ask yourself how many bathrooms you *really* need. Will one bathroom with a double sink suffice for two children? Does a guest room that will be used ten days of the year need its own bath, or can another bathroom be shared for those few days? Can an intact historic bathroom that is not ideal for daily use serve a guest room? Does a home office really need its own half bath? Also ask yourself how many bathrooms you want to clean regularly or pay someone else to clean.

The new two-sided storage unit in this adobe house kitchen stops short of the ceiling to allow the original volume of the space to remain understandable.

Apply a similar skepticism to other currently "expected" modern features of new houses: professional kitchens, master suites, home gyms, great rooms, catering kitchens, etc. People have lived in these historic homes for a century or more without these currently trendy spaces, so we know they are not necessary. If some of them are desirable to you, and if those features can be accommodated in your house without unduly impacting character-defining features, by all means incorporate them in your plans. But if you need such features and cannot make room for them without substantial destruction of historic features, please find another house.

A Modern Kitchen for a Historic House

Many books and thousands of magazine articles have been written on updating kitchens. Americans love to redo their kitchens, and the "shelter publications" have fed their appetite for new (and recycled) ideas while selling acres of advertising space to cabinet, countertop, and appliance manufacturers. The kitchen is generally the area of the house most influenced by style trends. In recent years, granite countertops and stainless-steel appliances have been considered essential. Just ask a real estate agent. In the 1970s, formed laminate countertops (probably orange) with avocado-green or harvest-gold appliances were the rage. In the 1950s, gold-flecked white laminate (usually Formica) with shiny stainless-steel or aluminum edge banding and white enameled appliances were *au courant*. In the first half of the twentieth century, linoleum countertops were set off with white painted or knotty-pine wood cabinets. All these trends went through a must-have period before hitting an "oh-the-horror!" phase. Many people now appreciate the earlier trends and are excited to find intact examples. The end of the stainless and granite trend seems to be coming, though it isn't yet clear what will follow.

Few people want to cook on the hearth of an eighteenth-century house every day, however enjoyable it might be to do so occasionally. In the eighteenth and early nineteenth centuries, kitchens truly were the heart of the home in most of America. Wealthy homeowners had servants or slaves and seldom entered the kitchen except to give instructions, but the great majority of families spent a lot of time there. In New England, the kitchen was often the only regularly warm room in the house for five months of the year. In the later nineteenth century and into the twentieth, servants became common among middle-class families, who spent less time in kitchens, waiting in the next room for meals to appear. The late twentieth and early twenty-first centuries have seen families return to their kitchens, as servants have become much less common in middle-class households.

Not surprisingly, during the decades when kitchen servants were common, kitchens were not especially comfortable or attractive. They were often separated from the rest of the house by pantries and hallways to increase the privacy of family members. In the late nineteenth century, after the arrival of the Colonial Revival movement with its romanticizing of the past, many older houses built with open hearths for cooking saw those historic kitchens turned into dining or breakfast rooms and new, modern kitchens built in an adjacent woodshed or as an addition.

The roots of today's kitchen, filled with built-in cabinets and appliances, can be traced to the pantries

This 1874 chromolithograph shows a well-equipped kitchen of the late nineteenth century. It featured an iron brick-set range with a copper hot water tank (heated by the range), a built-in zinc, copper, or iron sink with hot and cold running water and a shelf and clock above, a freestanding work table with good window light, and an adjacent pantry for dishes and food storage. Not many kitchens would have been so well outfitted at the time. *("Prang's Aids for Object Teaching—The kitchen." Boston, L. Prang and Co., 1874)*

of the nineteenth century. Spare, functional pantries with base cupboards and open shelves served for food storage in modest homes, while the butler's pantries

This illustration is from *The New England Economical Housekeeper*, and *Family Receipt Book* written by Esther Allen Howland and published in 1845. It shows a kitchen where cooking is still done on a hearth and over an open fire. There is little built-in storage or workspace beyond a few wall shelves and a mantel shelf. A worktable and dry sink provide preparation and cleanup spaces (the dry sink is covered by a board to create more workspace in this view) and a "tin kitchen" roaster sits in front of the fireplace. This was published when cast iron cookstoves were starting to replace open fireplace cooking.

of the wealthy and upper middle classes frequently had built-in copper counters and hardwood cabinets with solid or glass-paned doors, often with a sink built into a copper-clad counter. Freestanding cast iron ranges, tables, and furniture cabinets provided cooking and work surfaces in the kitchen. By the early twentieth century, built-in sink cabinets had moved into the kitchen in a noticeable way. More built-in cabinets followed, and by the 1930s, electric and gas ranges that were insulated so they could stand next to wooden cabinets were becoming more common. Coal- and wood-fired stoves still had to be freestanding for fire safety. The refrigerator had arrived by this time as well. Today's kitchen is just an assortment of refinements on that theme, with the same components whether the look is rustic, contemporary, or something between.

The kitchen is one room where modern conveniences are usually considered essential. For the owner of a pre-World War II house, the challenge is to incorporate these conveniences in a way that is compatible with the house. The chances are good that your kitchen has been updated at some point in the past fifty or so years, so your starting point is not the original kitchen, but some more recent iteration. Start a new kitchen design by considering what interior trim elements exemplify the style and age of the house. Is the trim simple and painted, heavily molded and carved with a natural finish, or somewhere between those extremes? The trim in the formal rooms of the house is likely to be more elaborate than that in the less public rooms. Historically, kitchens have fallen into the less elaborate category, so a new kitchen in a historic house should not have the trim detail of the most formal rooms unless the whole house has consistently simple trim throughout. Evaluating your house's design "vocabulary" will guide your new kitchen to feel compatible with the rest of the house.

Once you've identified this vocabulary, you need to decide whether you want a modern, fully fitted, built-in kitchen or one that evokes a time before built-in cabinets and appliances became the norm. The latter can be achieved with a combination of modern cabinetry and furniture pieces, providing much of the functionality of a built-in kitchen without the continuous countertops that separate them from historic kitchens. Hybrid kitchens can be seen in several of the historic photos reproduced here; these are just a few of the more than 16,000 historic kitchen photos that can be found on the Library of Congress website, if you need inspiration.

The hybrid approach often involves building or installing cabinets along all or part of one wall, usually containing the sink, and locating smaller cabinet sections or pieces of furniture on other walls. A modern kitchen design—i.e., a work triangle of reasonable size—is as easily achieved this way as with continuous cabinetry and countertops.

If you want a fully built-in kitchen that is compatible with the house, materials and finishes become critical. If the house has painted interior trim, the cabinets should also be painted, unless you want them to appear as furniture pieces. If the trim has natural finishes, the cabinets should have a similar natural finish. Use simple cabinets in a simple house. Most cabinet companies offer a line of Shaker or "Shaker inspired" cabinets that will often work well in a simply detailed house. If your house has beadboard walls or wainscoting, choose beadboard cabinets. A house with elaborately detailed trim will give you more options while still keeping the kitchen simpler than the formal rooms.

This nineteenth-century farmhouse pantry shows one of the predecessors of today's kitchen full of built-in cabinets and counters.

This late nineteenth-century townhouse pantry shows another forebear of today's kitchen.

Historically, pantries were important parts of many kitchens. Early pantries were used primarily for food storage; later, Victorian-era pantries were for food or china storage, often with a sink for washing dishes right in the pantry. A large house of that era might have both types of pantries off the kitchen. The food-storage pantry was usually on the cooler north or east wall, and the china (or butler's) pantry was between the kitchen and dining room. Numerous nineteenth-century housekeeping books, available in reprints, include plans for kitchens and pantries. These can be a great source of ideas and solutions. *The American Woman's Home,* by Catherine Beecher and Harriet Beecher Stowe, was one of the first popular works to suggest continuous countertops when it was published in 1869. Many of the book's suggestions have been replicated in the restoration of the Harriet Beecher Stowe House in West Hartford, Connecticut. Historic house museums are paying increasing attention to

This view of President Calvin Coolidge's preserved Plymouth, Vermont, kitchen illustrates the period in which the cast iron range had replaced hearth cooking and the sink had moved from the pantry to the kitchen itself. *(Photograph by Samuel H Gottscho, 1961. Library of Congress, Prints & Photographs Division)*

Described as a "modern farm kitchen," this kitchen of the late 1920s would have been found in a prosperous middle-class home. *(Library of Congress, Prints & Photographs Division)*

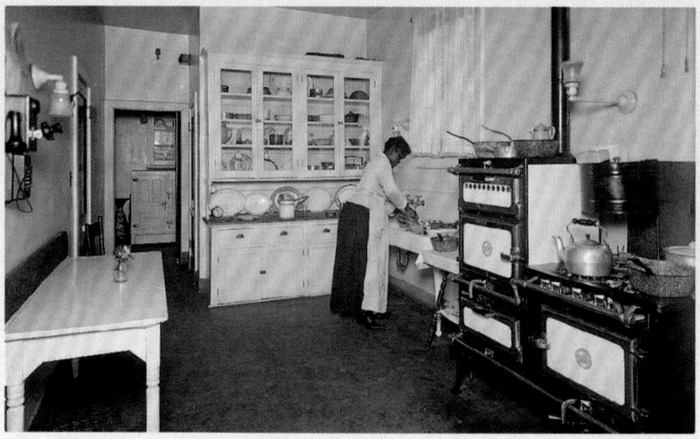

This is a good example of an early twentieth-century kitchen in an expensive home. A large gas range dominates the room, which is also equipped with an enameled cast iron sink with drainboard, a large built-in cabinet with glass doors, and a worktable with backsplash against the wall. The floor is covered with easily cleaned linoleum or cork flooring, and an icebox is visible through the pantry door. *(Photograph by Harris & Ewing, c. 1915. Library of Congress, Prints & Photographs Division)*

The home of a Lakeview, Arkansas farmer shows a modern kitchen typical of a modest home of the 1930s. Note the advanced continuous countertop with sink and backsplash. The housewife is using a hand pump at the sink. Matching cabinets flank the window over the sink, and the floor is covered with a patterned linoleum. *(Library of Congress Prints and Photographs Division, FSA/OWI Collection)*

interpreting kitchens and other service areas that used to be ignored on tours, and these can provide inspiration for modern kitchens.

Kitchens and bathrooms are the two spaces where the "period of significance" of your house might be disregarded while still evoking historic character. This approach adopts the idea that a "modern" kitchen and indoor plumbing were added to the house at some point around the turn of the twentieth century, and you are seeking to recapture the character of that period in those rooms. This is what actually happened in most houses, and there may be surviving evidence of this period kitchen. If so, try to work with it. More often, however, this kitchen has long since been replaced with a more modern one.

My approach in Whitten House was to evoke an earlier kitchen that might have been. A modern kitchen has been built around a salvaged turn-of-the-twentieth-century slate sink with double drainboard using painted beadboard cabinetry. A freestanding dropleaf table serves as an island, and there is an antique glass-doored cabinet for china storage. The bathrooms evoke the same period, with white subway tile, 1-inch hexagonal floor tile, pedestal sinks, and stained glass architectural salvage pieces. Beadboard wainscoting and cabinetry are used in the laundry room/bath. Even though the rest of the house reflects the middle of the nineteenth century, these functional spaces are attractive and feel like they belong in an old house.

Designing with Modular Cabinetry

There are two usual options for modern kitchen cabinetry: custom-built or modular. The majority of kitchens in America are created with factory-made modular cabinetry, while custom work is the norm at the upper end of the housing market.

A modern kitchen of 1938 with matching built-in steel cabinets below a soffit, electric refrigerator and range, and an enameled cast iron sink with drainboard countertop. *(Library of Congress Prints and Photographs Division, FSA/OWI Collection)*

There are wide ranges of quality and price points available in modular cabinets, the better ones having doors that are fully recessed into the cabinet-face openings, requiring close tolerances in construction and installation. Full-overlay doors cover the face frame, while partial-overlay doors are recessed but have a lip that overhangs the face frame.

Full- and partial-overlay doors allow for looser tolerances in cabinet construction and installation and are the most popular cabinet types. Frameless cabinets have doors that completely cover the front of the cabinet box but, unlike full-overlay cabinets, have no face frame. This is a contemporary look that might be appropriate for some houses built after World War II. Most lines of cabinets come with a variety of door styles and finishes, all of which use the same cabinet boxes made with frames to match the door finish.

All modular cabinets come in standard sizes: 24 inches deep for base cabinets and 12 inches deep

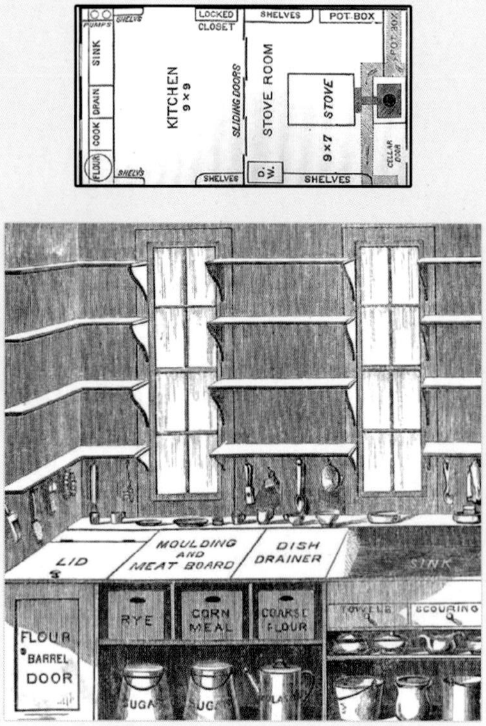

The "convenient kitchen" design published by the Beecher sisters in the mid-nineteenth century. At a time when few kitchens had any built-in cabinetry, this emphasis on organization of workspaces and the comfort of the cook (the sliding glass-pane door to separate the heat of the range from the primary work area, for instance) were ahead of their time.

for wall cabinets. There are also 24-inch-deep wall cabinets to go above refrigerators. Widths generally start at 9 inches and increase in 3-inch increments to around 42 or 48 inches, depending on the manufacturer. Base cabinets are 34½ inches tall, allowing an inch and a half of countertop thickness to bring the work surface to 36 inches above the floor. Wall cabinets range in height from 12 to 42 inches, with 30 inches being the most common. Shorter cabinets are used over ranges and refrigerators. Working around appliance sizes, usually 30 inches wide for ranges and 24 inches for dishwashers, these cabinets are fitted into the available space in an arrangement that is hopefully both functional and attractive. Filler pieces are used between cabinets to fill leftover spaces. Some cabinet companies make items like a 3-inch-wide slide-out spice rack to fill small spaces.

The more expensive the cabinet brand, the more sizes and options it will generally offer, including deeper or shallower wall and base cabinets. Nearly all brands offer some degree of size customization. Kitchen designers work with computer software that includes the catalogs of the cabinet brands they sell. As a designer works, the program inventories the cabinets and fillers used to create a design and generates a list of pieces and prices. A talented designer can create a kitchen from modular cabinets that feels custom-built. Talented designers are not the norm, however.

The two most common sources of modular cabinets are big-box home improvement retailers and locally owned lumberyards, kitchen centers, and independent kitchen designers. Most carry the same cabinets, though some brands are unique to particular companies or outlets. The brand name is usually the only difference, though sometimes the finishes also differ. Members of the preservation community sometimes assume that big-box home improvement centers carry inferior or inappropriate products for old houses, but this is not necessarily true. There are good arguments for supporting local businesses over national chains, but most big-box stores have the same special-order catalogs as the local retailers and can order more appropriate items than the bestsellers they keep in stock. These companies succeed on the volume of their sales, and an item has to sell a lot of pieces to earn a place on the shelf. You will have to wait a bit longer for a special-order item to be shipped, but you can usually get it. This is particularly true in the kitchen and bath departments, where the range of special-order products is vast.

Whether you work with a designer or design your kitchen yourself, accurate measurements of the space are essential, including all doors, windows, plumbing locations, electrical fixtures and outlets, etc. Start with a floor plan sketch and write in the room dimensions in inches, then add heights from the floor for all windows and the ceiling.

In an old house, corners are often out of square. Check for squareness with a 3-4-5 measurement in each corner where cabinetry will be placed: Measure 3 feet out from the corner and make a mark on the wall, then measure 4 feet out on the other wall and make another mark. If the corner is square, the measurement between the two marks will be 5 feet. If it is over or under that measurement, you are working with a space that is out of square, and you will have to accommodate that in the plan. This is particularly important where cabinets will be turning the corner or where a run of cabinets needs to fit tightly from wall to wall. A row of rectangular boxes exactly the same length as the wall will not fit as you want them to if the corners are not square.

Also check the level of the floor where you will be placing cabinets. A slightly out-of-level floor is easily accommodated with shims during installation. A seriously out-of-level floor can be challenging, but there are workarounds for this if it is planned for before buying the modular cabinets.

Most stores and kitchen centers will measure the space for a fee, generally under $50. If they will be designing your new kitchen, it is well worth the fee to have them accept the responsibility for the measurements. Many will work with your measurements, but any mistakes resulting from inaccurate measurements will be your responsibility. When a $400 cabinet doesn't fit, you want it to be the retailer or designer's responsibility to solve the problem.

One of the things that makes a kitchen recognizably modern is the standardization of cabinet sizes. Virtually every kitchen you walk into has 24-inch-deep base cabinets and 12-inch-deep wall cabinets with a soffit above. Counters stand 36 inches above the floor. No matter the style or finish, the base cabinets will have a 3-inch-tall toe-kick space along the bottom, and the distance between countertop and wall cabinets will be 18 inches. Varying some of these standardized elements can help to create a kitchen with a more period feel. Pre-twentieth-century kitchen and pantry cabinetry was not standardized and typically had lower counter heights than modern kitchens. Wall cabinets often went to the ceiling without a soffit above. Toe-kick spaces were uncommon. A good kitchen designer can work with standardized cabinets in a way that replicates some of these characteristics, particularly if the brand selected has a wide ranges of size options and offers some degree of size customization. A skilled installer, whether you or someone else, is important when heading into the "non-standard" zone.

Basic strategies for designing compatible kitchens in a pre-1910 historic house include:

- Break the cabinets into 3- to 6-foot sections rather than running continuous cabinets along two or three walls in an L or U shape.

- Lower the height of the base cabinets in at least one section, keeping in mind that standard dishwashers and ranges are built to go with 36-inch-high countertops. Separate cooktops, ovens, and dishwasher drawers can help get around this. Historically, counters on built-in cabinets were closer to tabletop height, in the 28- to 32-inch range.
- Reduce the depth of base cabinets to 21 or 18 inches in one area.
- Run wall cabinets to the ceiling if it is not much over 8 feet high.
- Use base cabinets with full-height doors and a separate cabinet or two containing drawers. The drawer-over-door base cabinet configuration did not become standard until after 1910.
- Extend wall-cabinet end panels down to the countertop so the wall cabinets appear to be supported by the base cabinets, as in traditional step-back cupboard furniture.
- Run baseboard over the toe-kick spaces except where you will frequently stand close to the cabinets. A shaped piece of baseboard can create the effect of a footed piece of furniture while allowing toes to extend under the cabinets.
- Stick to traditional countertop materials such as slate, wood, and copper. Honed black granite is a good substitute for slate and is available in many regions of the U.S.
- Use historic slate, copper, or enameled iron sinks, which are readily available from architectural salvage companies and classified ad sites. Reproductions are also available.
- Use different countertop materials for different cabinet sections, and avoid obviously modern formed laminate countertops.
- Avoid the standard 4-inch-tall countertop backsplash. Go higher where needed, and eliminate the backsplash altogether where it is not needed.
- Use antique furniture pieces in place of modular cabinets for some parts of the kitchen.
- Consider using an antique cooking range converted to gas or electric, or a modern reproduction cast iron range.
- If you use a standard modern kitchen range, make it either black (for pre-1910) or white (for more recent).
- There is a rationale for using stainless steel appliances in a pre-1910 historic house. Stainless steel does not precisely replicate the iron, copper, or brass once used in kitchens, but it echoes them.
- Use pantries for food storage to reduce the number of cabinets needed in the kitchen.
- Locate the refrigerator in a pantry or place it in a wooden cabinet to make it less obvious.
- Use refrigerated drawers or an under-counter refrigerator in the kitchen for regularly used items, keeping a full-sized refrigerator in the pantry for bulk storage.

- Use wood panels on dishwasher fronts or locate the dishwasher in the butler's pantry if dishes are stored there.

- Avoid cabinetry islands. Use a piece of furniture instead.

- Stick with wood flooring for houses prior to 1880. See Chapter 14 for more information.

Many of these ideas can also be used in post-1910 kitchens, and a few additional options are available as well:

- Enameled metals, usually white, came into use for sinks and appliances in the early twentieth century. Enameled cast iron cooking ranges had more color choices.

- Continuous countertops began to appear (with linoleum surfaces) by the 1920s but took several decades to become the norm.

- Linoleum flooring became common in kitchens during this period.

- Chrome appeared on faucets, countertop edging, and other kitchen hardware and accessories after World War II.

- Enameled metal cabinets also became popular after WWII.

- Refrigerators became increasingly common after the first decade of the twentieth century. While these early refrigerators were small, had tiny freezer compartments, and were inefficient by today's standards, one can be displayed in the kitchen as uncooled storage space, with a modern refrigerator and freezer in the pantry. It is possible to have an early refrigerator restored to working condition for use as a secondary unit.

- By the 1930s, cooking range manufacturers were insulating the sides of gas and electric ranges so they could stand next to cabinets with abutting countertops. Wood- and coal-fired ranges remained too hot to be placed next to flammable surfaces.

- In the postwar period, built-in kitchen nooks became popular, as did scalloped valances to hide electric lights above sinks.

- 9-inch by 9-inch vinyl floor tiles became common in the 1940s and 1950s, often laid in checkerboard or more complex patterns. Modern floor tiles are 12 inches by 12 inches but can be cut down for a more appropriate period look.

- Laminate countertops came into use around the same time, glued to plywood with either metal or laminate edging. Period patterns and colors are now being reproduced by several laminate companies.

In the early twentieth century, manufacturers were continually introducing new products and materials for kitchens while competing for the attention of American homeowners in the many print publications available to middle-class households. The expanding consumer marketplace and increasingly sophisticated advertising industry combined to create a vast number of print ads to convince people that the latest appliance would make life better. These ads are readily available today in old magazines and newspapers that can be found at flea markets, yard sales, and online marketplaces such as eBay. A number of books have also been published reproducing kitchen ads of the period. These ads are an exceptional source of design ideas for a new kitchen.

Designing with Custom Cabinetry

Custom-built cabinets are considerably more expensive than the factory-built modular cabinets described above. Nevertheless, well-designed custom cabinetry—custom-fitted to the spaces available and built using traditional methods and materials—can further enhance the character of a kitchen in a home built before the introduction of modular cabinets in the mid-twentieth century. There are a number of fine custom cabinetry companies in the U.S. specializing in kitchens for historic homes, and many contractors can also build custom cabinets. Slate, honed black granite, and white marble are typically used for countertops and sinks. The joinery is traditional, and hardware is usually brass or iron. Just as with disguised modular cabinets, the builders create a period feeling by varying dimensions from modular standards and by suggesting separate, freestanding pieces of furniture.

Custom cabinet builders do not need much design advice, but sometimes they give you too much of a good thing. Not every section of counter has to be a different height or material, and not every cabinet piece has to have a different finish. These kitchens cost a lot of money, and you may understandably want to make sure you're getting all the bells and whistles for your money, but do not let this desire blind you to what is appropriate to the house. A twenty-first-century "traditional" kitchen that outshines the house's authentic historic features is not the goal.

If you or your contractor will be custom-building kitchen cabinets, be sure they are properly designed. Architects are not usually kitchen designers, but if you are using an architect for other work, he or she should be able to recommend a custom kitchen designer. Any good cabinet designer will understand the nuances of clearances for appliance doors and other technical details, but even an excellent modular-cabinet designer may not be your best choice to design custom traditional cabinets, which are a different breed. If you are brave enough to design your own cabinetry, you might want to have an experienced designer review your drawings and identify any technical pitfalls. The last thing you want is to discover that you can't remove the drawers in your refrigerator to clean them because the door swing is too constricted. No matter who is designing your kitchen, you need to select appliances before the design is finalized to ensure compatibility.

Example Project: A Compatible Kitchen

Whitten House had never had a modern kitchen when I undertook its rehabilitation. The house had been in institutional use since the 1940s and had only a kitchen sink and one cabinet built into a 12-foot by 16-foot room that had been added to the back of the house in the 1960s. The room had a concrete slab floor, a pressed-paper tile ceiling, and wood sheet paneling on the walls. It also had a historic wood door and windows, reused from the shed that was replaced with the new room.

A reality of rehabilitating a house that lacks expected elements like a kitchen or bathroom is that banks and mortgage companies do not want to finance them. I was able to buy the house for cash thanks to an advantageous purchase and the sale of a previous home, but I needed to take out a mortgage to cover rehabilitation costs. The goal of this Phase 1 kitchen project was to create a compatible and affordable modern kitchen that would allow refinancing and function effectively until I was ready to remove the 1960s addition and replace it with a larger addition for the "ultimate" kitchen in Phase 2 of the house rehabilitation. Elements of the temporary kitchen needed to be either reusable in the ultimate kitchen or inexpensive if they would not be reused.

This low-cost temporary kitchen needed to look like an expensive high-quality kitchen for purposes of the appraisal for refinancing. Some of the approaches that accomplished this can hold down costs in a permanent kitchen rehab as well.

Walls: The sheet paneling on the walls was covered with inexpensive lining paper and painted after the paneling grooves were filled with Durabond setting plaster.

Ceiling: Half-inch drywall was installed over the existing 1960s paper tile ceiling.

Floor: Inexpensive laminate flooring was installed over the concrete slab. Laminate flooring has serious limitations in durability, particularly where it might come into regular contact with water (i.e., in kitchens and baths), but for a temporary floor to cover concrete with minimal expense and effort, it serves the purpose well.

Heating: A new gas-fired radiant heating system was being installed in the rest of the house but could not be installed in the concrete slab of this room, so a thermostat-controlled Vermont Castings gas stove was installed here. A "new old stock" thermostat (i.e., a never-used old thermostat still in its original packaging) found on eBay was used here (and elsewhere in the house).

Cabinets: The cabinets needed to be reusable in the ultimate kitchen or pantry, so these rooms had to be designed before the temporary kitchen could be designed. Unfinished poplar cabinets were ordered from a small, family-owned company through a big-box home improvement center. These custom cabinets were built to fit the c. 1900 slate sink with drainboard counters that I'd purchased from an architectural salvage shop. Because the company did not offer fully inset doors through the big-box retailer, I ordered the cabinets without doors and ordered slightly oversized doors separately, to be fitted to the cabinets once they were installed. Traditional brass hinges (mortised into the cabinet frames and doors) and reproduction catches and pulls were used.

Because the sink cabinet and slate sink would be reused in the ultimate kitchen, the unit was assembled on casters so it could be moved out of the room and returned to the new kitchen when the time came. Beadboard was selected for the cabinets because it was commonly used for kitchens around 1900, when the slate sink had been made. The company that built the cabinets offers finished sides as integral parts of end cabinets, and I used these where needed. This yields a more custom and appropriate look than the end panels applied over the basic box in most mid-price cabinets.

I purchased a large glass-doored cabinet, c. 1900, from an architectural salvage company to hold dishes. It was not deep enough for my 12-inch dinner plates, so 3 inches of

A three-dimensional view of the planned Phase 1 kitchen at Whitten House. Renderings like this—created by the kitchen design programs used by nearly all cabinet retailers—are helpful for understanding how elements will fit together and look in the space.

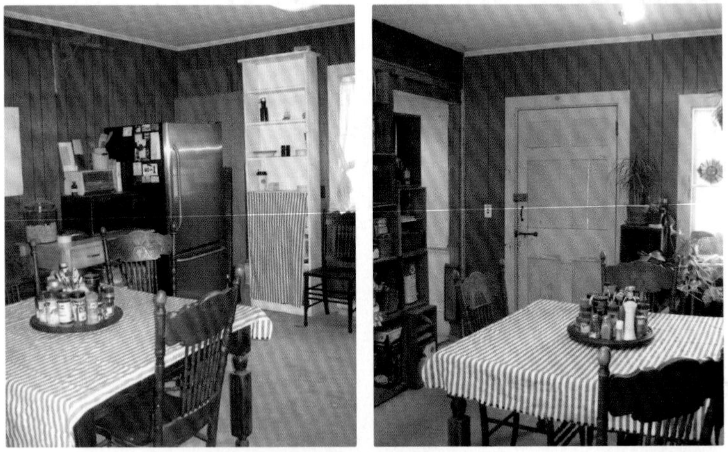

The room—shown here before installation of the new kitchen—had been added to the rear of the original kitchen ell in the 1960s after the historic shed and barn were removed. A gas range and refrigerator turned the former library break room into a rudimentary kitchen while the new one was planned.

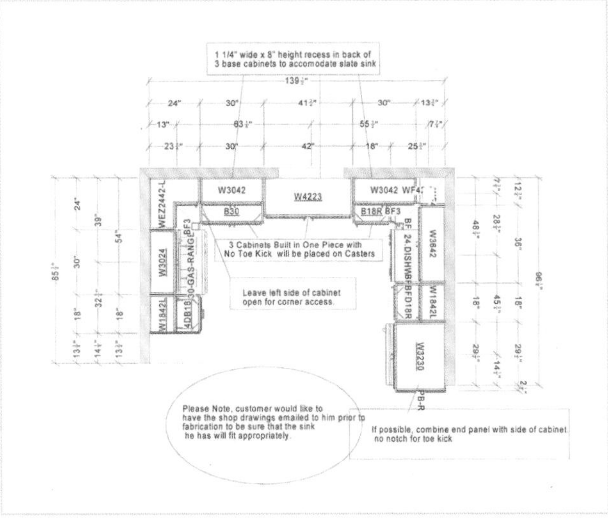

I developed the cabinet and appliance plan with a kitchen designer. Notes on the plan specify several custom elements, particularly the design of the cabinet to support the salvaged slate sink and countertop that would become a focal point of the room. Not seen on the plan are the furniture pieces that would also provide storage and workspace. The refrigerator cabinet was later changed to accommodate a much larger refrigerator/freezer unit.

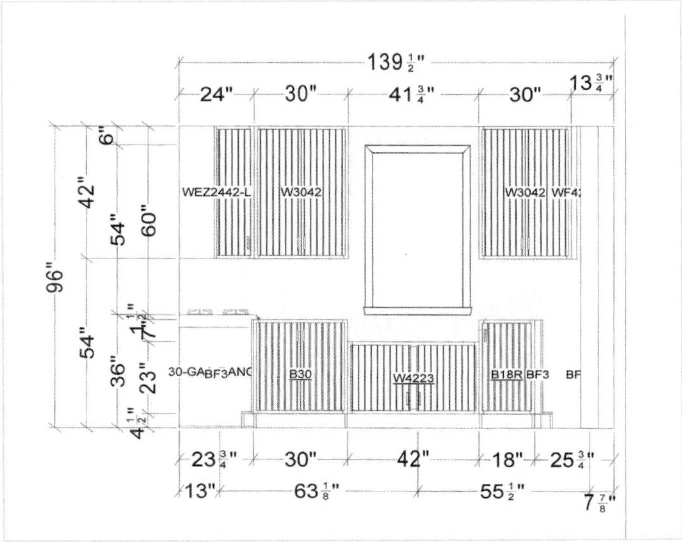

The kitchen designer's elevation drawing of the sink wall.

To hide the 1960s surfaces, gypsum board was installed over the paper tile on the ceiling and lining paper was applied over the wall paneling. Here the gypsum board has been installed and the grooves in the wall paneling have been filled with setting plaster prior to applying the lining paper. Gypsum board has not yet been installed on the exposed framing of the rebuilt wall between this room and the original kitchen.

These views show the ceiling and walls completed and painted. The finished recess behind the gas range, meant to be a small closet, was covered over when the plan was modified to accommodate the larger refrigerator/freezer unit.

The antique glass-doored china cabinet was too shallow for the largest plates it would need to accommodate, so 3 inches of depth was added by extending the cabinet frame and installing new, deeper shelves. The new wood was finished to match the old, and the original beadboard back was reinstalled. I found a length of matching crown molding at an architectural salvage shop to replace the now-too-short end piece.

The interiors of the unfinished poplar cabinet boxes were finished with two coats of clear polyurethane.

The exteriors and doors of the cabinets were painted with one coat of primer and two finish coats of Fine Paints of Europe's Hollandlac Brilliant paint for a durable finish. The cabinets were painted prior to installation so that there would be no paint bond between cabinets. This would facilitate uninstalling and reconfiguring them for Phase 2 of the kitchen rehabilitation.

The three cabinets that would support the antique slate sink were fitted to the sink while lying on their backs on the floor. The cabinets were mounted on heavy-duty casters so the whole unit could be rolled into the adjacent new kitchen once upright.

I stood up the 1,000-pound sink-and-cabinet unit with levers and a jack. Each time I jacked it up a few inches, I had to support it there with wood blocking screwed in place while I reset the jack to lift it some more. Working alone, it was a slow and somewhat nerve-wracking process.

Once the unit was upright, the slate was sanded with 220-grit sandpaper in preparation for being treated with mineral oil. Black boat caulking was applied to all seams before sanding, so slate dust would be mixed with the caulk and forced into the seam with the excess caulk sanded off. The hole for a modern faucet was drilled (slowly) into the slate with a woodworking hole saw bit.

The slate sink and cabinets are shown in place with the faucet and drain installed. The slate has been treated with mineral oil. A simple baseboard will cover the exposed casters at the base. Note that the wall cabinets were installed before the base cabinets, and the blocking on which the wall cabinets sit will be used to mount a slate backsplash 1½ inches out from the existing walls. The intervening void will allow installation of new electrical wiring and outlets above the counter without having to fish wires through walls.

Views of the cabinet installation with the first version of the refrigerator cabinet. Note the concrete floor.

Laminate flooring was chosen as an affordable temporary cover for the concrete slab, which would be removed in Phase 2 when the entire 1960s addition would be replaced with a larger timber-framed structure. The gas heating stove was installed on a thermostat because the concrete floor prevented any extension of the new radiant heating system from the rest of the house into this space.

Upon realizing that a big house allows entertaining on a big scale, it was decided that a larger refrigerator/freezer would be a good idea. Once the unit was chosen, I took the first refrigerator cabinet apart so that the end panel and doors could be reused for the new cabinet. Matching tongue-and-groove beadboard from a local lumberyard was used to increase the depth of the existing end panel, fabricate a second end panel, and make two additional doors. The necessary materials cost under $200, whereas the cabinet company had quoted $1,400 for a larger custom replacement cabinet.

Here is the nearly completed replacement cabinet. Because the large stainless refrigerator/freezer unit was contemporary in character, the cabinet was painted a different color from the other cabinets, and the doors were given invisible hardware for a cleaner look. The traditional brass hinges and latches used elsewhere in the kitchen looked out of place here. I wanted the unit to have its own character but feel related to the rest of the cabinetry and appliances.

Full-inset cabinet doors were ordered oversized and cut to fit once the cabinet boxes were installed. Because they are poplar (which is a softwood) and would be subjected to humid summers and dry winters, a fair amount of seasonal expansion and contraction had to be planned for. Tabs of blue painter's tape serve as temporary pulls on the doors that are already hung.

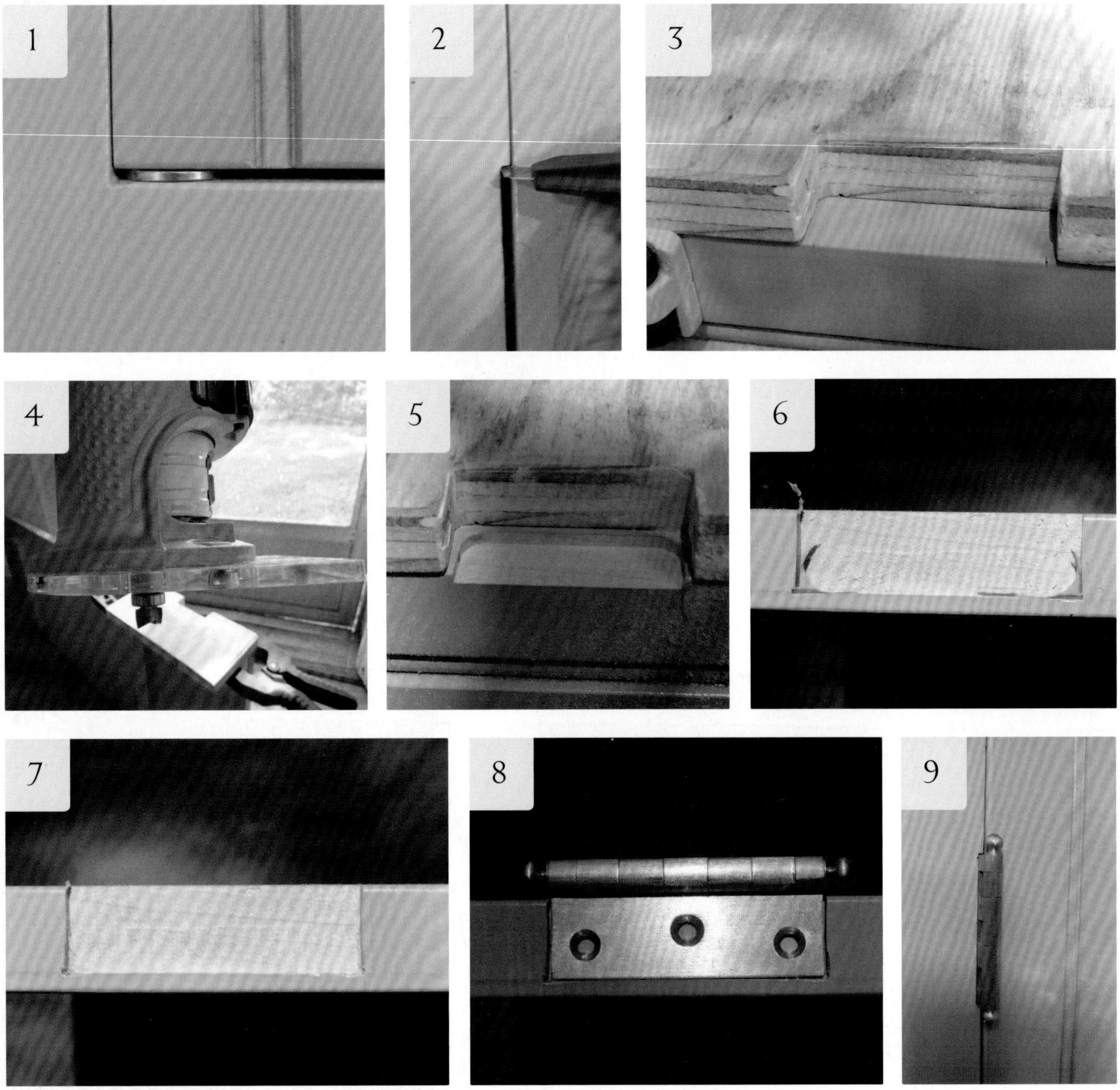

Mortising for traditional cabinet hinges. 1. A nickel is the correct thickness for spacing full-inset cabinet doors. 2. The cabinet mortises were cut before painting and installation. With the doors cut to size and held in place with spacers, the doors were marked for the corresponding mortises. 3. A jig was cut from plywood to guide the router bit. Clamps hold the jig in place. The same jig was used for the mortises on the cabinet boxes. 4. Scrap material is used to set the router bit to the correct depth for the hinges. 5. The mortise cut by the router bit. 6. Squaring up the mortise corners with a hand chisel. 7. The completed mortise. 8. Test-fitting the hinge. 9. The installed hinge.

Heavy-duty brass cabinet latches and draw pulls in a traditional turn-of-the-nineteenth-century style were selected. These are high-quality, American-made, and expensive, as are the hinges, because they need to hold up to daily use.

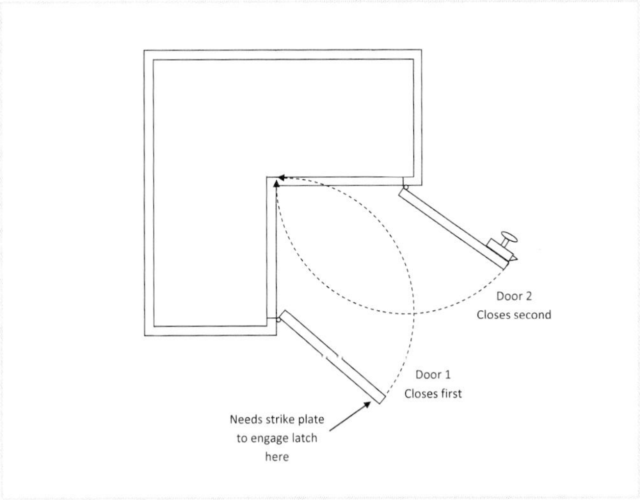

A corner wall cabinet required a custom hardware solution. Cabinets like this typically have a single L-shaped door that is hinged on one side and is somewhat awkward in appearance and function. I hinged the two parts of the door separately for a more traditional look and better cabinet access. I wanted to use the same latch as on the other doors, but the projecting keeper would not work in this configuration. This drawing was done to let the hardware manufacturer judge whether one of their flat brass strike plates would accommodate the sprung catch.

Here is the installed latch with flat strike plate to engage the catch. Once the manufacturer confirmed that the hole in the latch would fit the rectangular hole in the strike plate, I ordered the plate. Although the cost of shipping was many times the cost of the small piece of brass, I was happy with the solution (and I finished putting in all the screws eventually).

Half-inch-thick slate salvaged from a scientific laboratory was installed over a plywood base for the countertops. Painted wood edging will cover the plywood. Quarter-inch-thick slate salvaged from a school blackboard will be used to cover the backsplash area between the countertop and the cabinets above. The total investment in slate was under $500.

This open shelf unit was fabricated on-site in a day from approximately $50 in materials. A similar unit from the cabinet manufacturer would have cost several hundred dollars.

A power planer quickly removed excess material from the back of the shelf unit to scribe it into a curving wall surface.

The completed compatible kitchen. The small dropleaf table serves as a work island for food prep and serving. With leaves up, it can be combined with the matching table in the foreground to seat a larger group.

depth was added to the back. *Note: Passing inexpensive flat drawer fronts through a bead bit on a router gave them a custom appearance.*

Countertops: The ½-inch-thick countertop slate was salvaged from a scientific laboratory in New York City (via Craigslist), and the ¼-inch-thick backsplash slate came from a school classroom blackboard. The backsplash was shimmed out 1½ inches from the wall so new electrical outlets could be mounted and wired on the existing wall surface behind the slate.

Hardware: Because kitchen cabinet hardware gets heavy use, it is always worthwhile to buy the highest-quality hardware you can manage. Philips-head screws were not commonly used to install hardware before the late twentieth century, so I purchased slot-head screws separately. Small details like this cost little and help a lot for evoking historic character, even though most people will not consciously notice why. Some hardware companies will supply slot-head screws if requested. Cheap hardware will rarely last long under daily use by all members of the household.

Appliances: The ultimate kitchen will have a large cast iron "brick set" range converted to gas. This range would have been too large for the temporary kitchen, however, so a relatively inexpensive 30-inch gas range was purchased and installed along with a matching dishwasher. After building a cabinet for a 30-inch refrigerator, we realized that we needed a larger refrigerator/freezer unit. We disassembled the cabinet so the side panel could be reused, then built a cabinet for the larger unit on-site using matching beadboard from a local lumberyard.

Lighting: A period fixture was rewired for general and task lighting over the sink. Inexpensive "hockey puck" under-cabinet lights were installed prior to installing the slate backsplash, allowing the wires to be hidden. The touch-pad dimmer that came with these lights proved unreliable and was replaced with a standard wall switch with dimmer. The push-button wall switches being used throughout the house were used in the kitchen as well. Compatible and reasonably priced schoolhouse-style ceiling fixtures were purchased from a major online retailer at a fraction of the cost of high-end reproductions. Most people will never notice the difference between these $35 fixtures and $300 versions of the same style.

The kitchen remained more or less usable throughout the rehab process, which stretched over several years. Only on a handful of nights was eating out necessary.

Designing Bathrooms

Much of the kitchen advice above can be applied to bathrooms as well. As with kitchens, inspiration and design ideas for compatible bathrooms can be drawn from examples that were built into houses a century ago. Historic bathrooms can be roughly broken into three periods.

The earliest, from the late nineteenth century, typically featured naturally finished beadboard or paneled wainscoting, high-tank toilets, and sinks and tubs on legs. White marble was commonly used for sink tops and door thresholds. Floors were varnished hardwood.

By the early twentieth century, white ceramic subway tile replaced the beadboard, 1-inch hexagonal tile covered floors, and high-end tubs were skirted and built into tiled surrounds.

After World War II, 4- or 6-inch-square tile in many colors replaced the white subway tile, and matched sets of colored porcelain or porcelain enamel fixtures became popular.

Historic bathroom sink faucets typically had separate hot and cold taps, which were used to mix water to the desired temperature in the basin. For people who had always mixed hot and cold water in a ceramic basin from separate ceramic pitchers, this was the height of convenience. The ability to pull a rubber stopper and have it all drain away was a bonus after centuries of carrying the ceramic basin outside to dump the dirty water.

In the century-plus since the development of indoor plumbing, we have become accustomed to washing our hands under running water that emerges from the tap at the temperature we desire, and few of us ever fill our bathroom sink basins. There are two ways to get tap water at the desired temperature while maintaining the character of earlier plumbing fixtures. If you're using historic fixtures with only two holes for faucets, you can find a retro-styled modern faucet that has two handles connected by a tube serving a single spout at the center of the sink. These spanning faucets are available from specialty and "restoration" plumbing companies. Alternatively, you can purchase reproduction fixtures with "widespread" faucet holes. These have three holes, one on either side for the hot and cold controls and one in the center for a spout. The connections are made under the sink. Modern sink faucets with a 4-inch spread do not create the feeling of historic fixtures nearly as well. Bathroom fixtures get a lot of use, and buying the best quality you can afford will pay off in the long run. High-quality fixtures are made of cast brass, not pot metal or plastic. Chrome plating on plastic might look like the real thing for a while, but the illusion will not last long.

The finish chosen for fixtures can also help or harm your attempts to evoke historic character, whether

A bathroom of the late nineteenth century with stained and varnished V-groove walls, ceramic tile floor, ceramic and cast iron fixtures, and exposed plumbing. *(Photo c. 1906. Detroit Publishing Co. Collection, Library of Congress)*

A late-nineteenth-century bathroom fitted into a finished attic space in an ell, above the kitchen and near the second-floor bedrooms of the main house. The kitchen coal stove had an integral water-heating coil to supply hot water to a tin tub set into a wood "casket." The ceramic toilet has a stained and varnished wood tank and seat. Out of sight behind the photographer, a ceramic sink bowl is undermounted in a white marble top supported with cast iron wall brackets. *(At the Farnsworth House, Farnsworth Museum, Rockland, Maine)*

Another example of late-nineteenth-century bathroom fixtures and finishes. Here the wood wainscot is paneled, and the floor is a black-and-white marble checkerboard on the diagonal. *(At the Vaughan Estate, Hallowell, Maine)*

Early sink cabinets often resembled the dressing table and washstands they were replacing, with drawers, doors, and marble tops with backsplashes. A ceramic sink bowl would be undermounted to the marble in place of the ceramic bowl that would sit on the marble top of a furniture piece in the bedroom. Here is another tin tub mounted in a wood casket. *(At Historic New England's Eustis Estate, Milton, Massachusetts)*

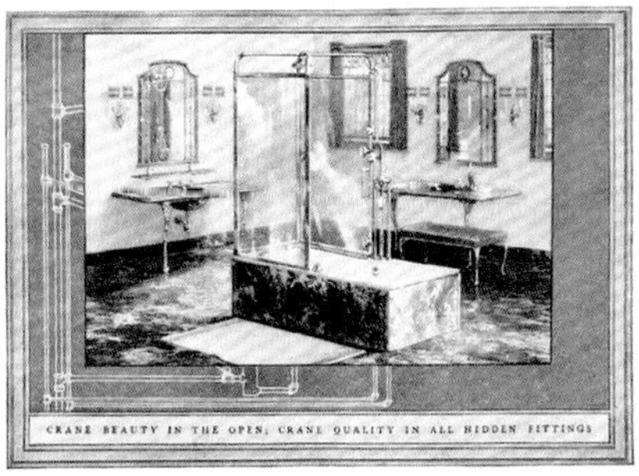

The most interesting setting a bath can be given is in the open, away from walls and corners. And now the new Crane *Crystal* shower provides the final luxury of a curtainless, splash-proof shower adapted to use in combination with the *Tarnia* bath of cream white enamel on iron. Inclosed on three sides in plate glass, framed in standards of nickeled brass, water is led to the overhead needle shower and four horizontal sprays through two of the vertical supports. The hot-and-cold mixing faucet supplies tempered water to the shower, sprays and tub. Large hand grips on main supports at both sides. All valves within easy reach. The *Tarnia* bath shown is encased in black and white marble, matching the *Neumar* lavatory and dressing table. Tiles of any color or pattern can replace the marble.

Period advertising can be an excellent source for design ideas. This magazine ad is from 1924.

This sink set into a marble-topped cabinet is located between two bedrooms. The interior window above it allows light into the adjacent tub room, which is entered from the main hall. *(At Historic New England's Eustis Estate, Milton, Massachusetts)*

Advances in the scientific understanding of germs and disease generated a movement toward sanitary surfaces in the early twentieth century, as shown in this bathroom of the period. Subway tile wainscot and hexagonal marble or ceramic tile floors became the norm. The recessed medicine cabinet and glass shelf over the sink also became common, along with nickel-plated fixtures and hardware. The toilet has been replaced with a modern unit in this bathroom.

Trim detail at edge of mirror.

This early-twentieth-century bathroom dispenses with the subway tile in favor of a simple tile base transitioning between the marble floor and plaster wall. Note how the wood molding that defines the wainscot is cut to fit the round mirror. A painted wood cabinet is built into the corner.

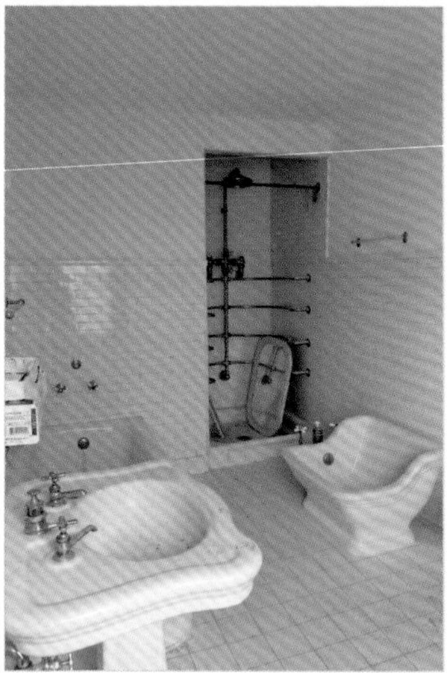

Popular upper-class luxuries included footbaths (at right) and elaborate showers. The mirror sitting in the shower should be hung over the pedestal sink. Here, the original floor tile has been replaced with a later style.

Another view of the room above, showing the pedestal sink, ceramic tub, and subway tile wainscoting.

Marble hexagonal tile was popular for decades and is still available in several sizes. Modern ceramic tile is also available in the hexagonal shape. Multicolored patterns in hex tile were not uncommon and are available today as reproductions. Specialty restoration tile comes with correct historic spacing—i.e., 1/16-inch grout lines—whereas commonly available hex floor tile typically has 1/8-inch grout lines. If the budget allows, go for the correct tile spacing.

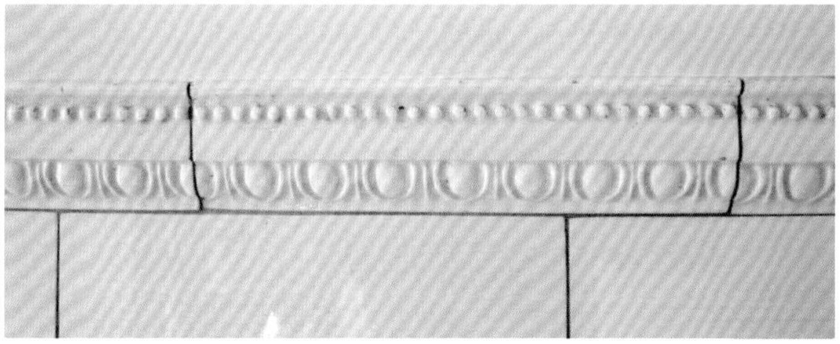

Cap tiles and other specially shaped pieces were used in many bathrooms, often in Classical motifs. Similar trim pieces are available today.

By the 1940s, square tile had largely replaced hexagonal and subway tile shapes in advertising and presumably in homes.

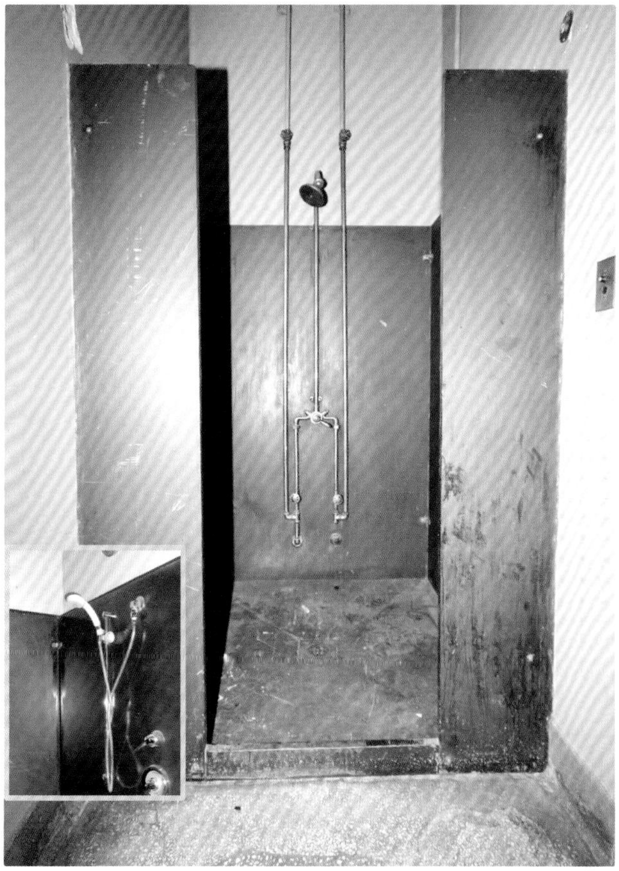

This early-twentieth-century shower is fabricated from large slabs of slate. During the rehab, no-longer-functional historic plumbing elements were disconnected but retained in place for their character, and a modern showerhead and control unit were mounted on one side of the shower (inset).

The bathroom in this 1950 ad features large square wall tiles and a striated sheet flooring material, possibly Marmoleum (which is being produced again). By this date, tubs, toilets, and sinks were available in colors other than white, including green, pink, and blue.

CRANE

This advertisement from around 1927 shows an early use of square tile on walls, extending all the way to the ceiling with a tile crown molding at the junction. Patterned borders for floor tile, as shown here, were widely used. *(Library of Congress, Prints & Photographs Division)*

Hardboard sheet material with a faux tile pattern came onto the postwar market and was used in many bathrooms. Extruded aluminum trim pieces, including edging and corner joints, finished the installation.

Black Magic...enchanting bathroom elegance !

Hardboard sheets were also available in marbleized patterns, as seen in this stylish bathroom ad from 1957.

In this c. 1960 ad, a Mondrian-like colored glass treatment is used to provide privacy inside a glass wall. Shag carpeting has made its appearance.

of 1890 or 1950. Prior to the wide adoption of chrome plating after World War II, polished nickel plating over brass was the principal finish for faucets, towel bars, and other plated bathroom fixtures. Nickel is silvery, like chrome, but has a slightly warmer tone. For utilitarian fixtures, polished brass was often used without plating. The very wealthy sometimes had fixtures plated in silver or gold or cast from solid bronze. By the late 1940s chrome was rapidly becoming the finish of choice, and it has remained so ever since.

Polished nickel is available again for higher-end fixtures and confers a lovely subtle difference from chrome, usually at a premium price. Brass fixtures are often available, either "old brass" of some sort or ultra-shiny polished brass. Avoid the latter—along with brushed nickel, which was not used historically, and gold plating—unless you are restoring a mansion with marble tubs and sinks. Chrome is an acceptable finish for pre-1940s houses if the budget does not allow polished nickel.

Example Project: A Compatible Bathroom

This combined bathroom/laundry room is located on the first floor of Whitten House. Although there are no bedrooms on the first floor, we installed a full bath in case it should ever be necessary to convert a first-floor room to a bedroom due to age or injury. The space was created by combining a walk-in closet beside and under the back stairs with a portion of a small adjoining room, the remainder of which was partitioned off to create a hallway outside the bathroom. Only 39 inches of wall between the closet and adjoining room was removed to combine the spaces, and the closet door, frame, and trim were relocated several feet and reinstalled in the new partition to serve the bathroom. The former door opening was covered with a removable MDF panel to provide access to the rear of the washer and dryer if necessary.

The design incorporates such common elements of late-nineteenth-century bathrooms as beadboard wainscoting, black-and-white hexagonal tile floor, and a pedestal sink. A traditional-looking widespread faucet set in chrome finish finishes the sink.

The modern, stacked washer and dryer occupy an alcove with a cabinet above. A 1940s cast iron bathtub was purchased through an online classified ad site, and subway tile covers the tub surround. Other fixtures are new American Standard pieces that reproduce early-twentieth-century designs. Because the room is located on the interior of the house without a window, an antique barn sash with frosted and colored glass was repurposed as a faux skylight with LED lighting above. The ceiling was dropped 6 inches to create the recess above the window sash, which also allowed for new plumbing and electrical runs without cutting into the historic plaster ceiling.

A built-in cabinet was sized to use doors that were found in the house, likely remaining from an early-twentieth-century kitchen that was removed with the former woodshed and barn in the 1960s. The cabinet provides storage and also contains the manifolds for the hot- and cold-water supply lines for the fixtures and washing machine. Antique wall sconces matching those installed in Whitten House c. 1898 were purchased on eBay and rewired. Modern recessed lighting was installed over the tub and the washer/dryer area.

The historic plaster was thin and roughly applied because the space had originally been a closet and back room. Consequently, the plaster buckled in several locations while structural repairs were made

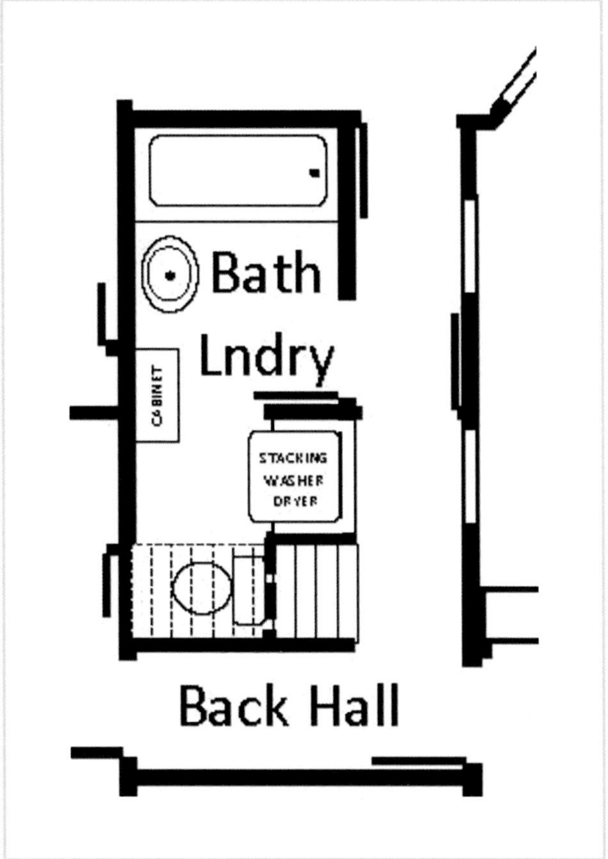

The plan for the downstairs bath/laundry room combined two small spaces while creating a hallway along the outside wall. A larger bathroom with a window and two doors would have been possible without the hallway, but the hallway provides better circulation to the rooms on the south end of the house. Tucking the toilet under the back stairs left limited headroom, but the tallest resident is 5 feet 7 inches, and there are other baths and lavs in the house for tall people to use.

A view past the former door opening to the closet under the back stairs, looking into the adjacent small room before the hallway was partitioned off (where the 2 x 4 is sitting). A portion of the floor has been opened to provide access for structural repairs in the crawl space.

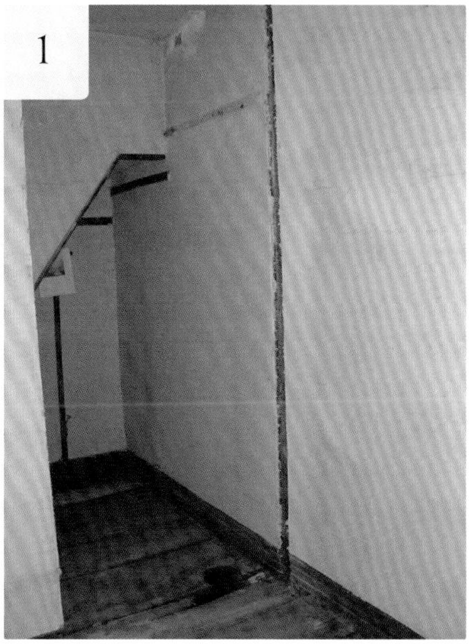

View from the small room into the closet after the removal of 39 inches of wall to combine the spaces.

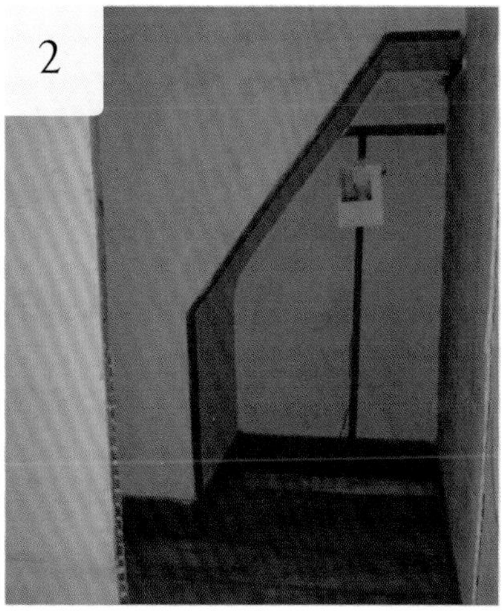

View into the closet. At left is the alcove where the stacking washer and dryer will go. The alcove under the stairs will accommodate the toilet.

In addition to replacing damaged subflooring with matching material, I removed several high spots with a power planer prior to installing shims to level the new tile floor.

A view into the small room showing the removed section of subfloor. The wide pine finish flooring removed from this room was reinstalled in the newly created hallway area and used for repairs elsewhere in the house.

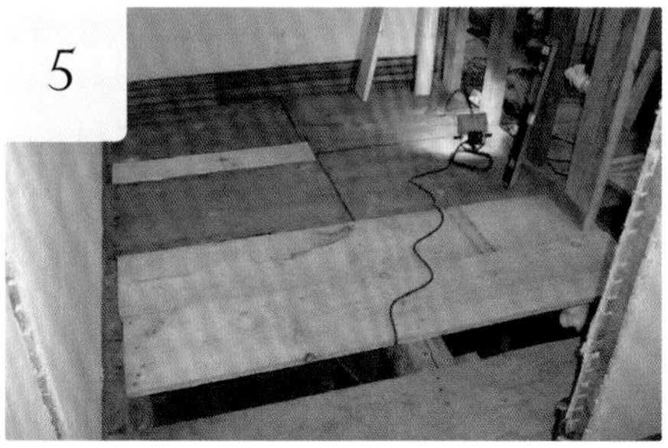

New rough-sawn pine subflooring was installed where the historic flooring had been damaged.

To level the new floor, a laser level was mounted to a tripod in the crawl space and its beam used to indicate cut lines on tapered shims cut from 2 x 4s. The shims were marked and cut one at a time, working from each end toward the level.

With shims installed, the red beam of the level can be seen just touching the tops of each, indicating that they are all on the same level.

In this view of the installed shims, additional shingle shims to adjust for unevenness of the subflooring boards can be seen. Once the shims were installed, ¾-inch marine plywood was screwed down and overlaid by 3/8-inch cement-fiber backing board for tile, which was installed with self-tapping flush-head screws on a 12-inch grid. This floor system combined with new support posts in the crawl space rendered the tile floor stable and unlikely to crack.

A view of the floor tiles extending from the small room into the former closet area.

This view shows the closet door relocated to the new partition to become the door to the bath/laundry room. In its original installation, the 1-inch-thick frame elements were nailed flush to the wall framing without shims or spacers, requiring the wall framing for the opening to be perfectly plumb and level. This installation method was reused and the original trim was reapplied to the hallway side of the relocated door. There had been no trim around the door inside the closet, so new trim matching the door and window trim elsewhere in the rear portions of the first floor was installed inside the bathroom after this photo was taken.

In this hallway photo taken after the tile floor and marble threshold were installed in the bathroom, the door at right is for the opening between the hallway and the small room beyond. This door had originally swung into that room and against the outside wall where a new bay window was planned; accordingly, the door casing and frame were removed and turned around so that the door would swing into the new hallway instead.

Within the former closet area, the flange for the toilet is visible in the alcove.

The 1940s cast iron tub—purchased from an online classified sales site—had to be slid on end into the bathroom on lengths of 1½-inch PVC pipe before it could be laid down for final positioning.

Here the tub and pedestal sink have been set in place. All plumbing drains and vents were installed before the floor was installed, so fixtures had to be selected and their locations determined in advance. Drains and venting can be challenging in a timber-framed house but can usually be done in a code-compliant manner with some thought and creativity. Drains must all pitch toward the house sewer outlet, but vents can travel in any direction before turning up to the roof.

In this photo, shims have been installed to plumb the wall behind the tub, and the ends of the tub have been framed. In addition to making the wall plumb and simplifying the installation of subway tile on the tub surround, the shims created a recess for radiant heat tubing behind the tile. The lowered ceiling has been framed in with a recess for the architectural salvage window sash "skylight" with colored glass. Antique electric sconces are in place but not yet wired, and construction has begun on the built-in cabinet in the foreground.

This view into the former closet space with the toilet, washer/dryer, and lighting installed shows hot- and cold-water supply lines in the foreground. The manifolds to turn them on and off will be hidden within the built-in cabinet.

Here the built-in cabinet framing is underway, and the back panel of beadboard has been installed to cover the water supply lines. The water comes in from above, and the lines to all fixtures run above and down within the walls to the fixtures. This keeps the bathroom plumbing out of the way of the radiant heat tubing and related aluminum plates to be installed under the floor.

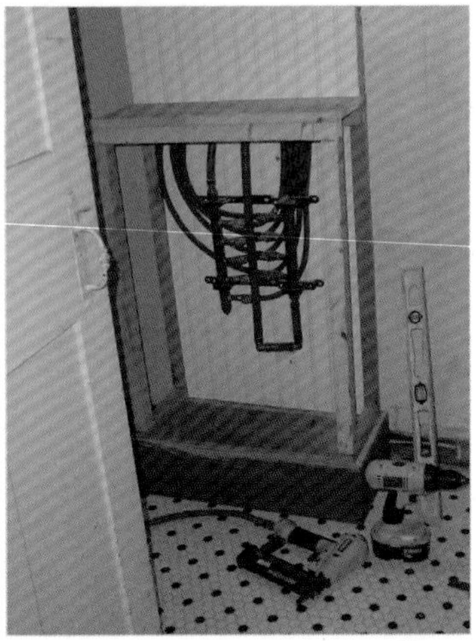

Here are the manifolds for shut-offs on the supply lines to each fixture. Note the salvaged baseboard from the room being reused on the base of the built-in cabinet. The original baseboard was reinstalled in the entire bathroom.

Here we see the reflective foil paper, tubing, and aluminum plates installed for radiant heat behind the tiled tub surround. A warm tile tub surround in winter is even better than a warm tile floor!

This detail view shows the original baseboard with a bead along the top edge and new beadboard wainscoting above. The baseboards were shimmed out ½ inch from their original location to allow them to project beyond the beadboard.

The original trim in the alcove under the stairs. The beaded detail on the original baseboard and in this trim provided a design direction for the new work in the room, which would use beadboard for wainscot and cabinetry with beaded edges on the cabinetwork.

This poplar stock for the new bathroom trim has been milled with a simple bead bit in a router mounted in a router stand. Having the stock milled by a millwork shop would have cost significantly more money. Because I owned the router and stand, the only expense for this custom trim beyond the material itself was the bit for the router, which cost under $30, and the bit can be used to run thousands more feet of beaded trim in the future if needed.

The built-in cabinet, shown here under construction, was designed around two paneled cabinet doors that had been stored in the house following the demolition of an early-twentieth-century kitchen when the original shed and barn were removed in the 1960s.

The cabinet with doors installed. I later replaced the beadboard panel above the doors (not visible in this view) with a metal grill, which allows ventilation in the cabinet in the moist environment of the bath/laundry room.

Here the stacked washer and dryer are enclosed with painted trim in their alcove, with a built-in cabinet above. Vertical boards on either side are hinged from behind and can be swung out of the way if the machines have to be pulled out. A section of plastic grid intended for use as a diffuser on a fluorescent ceiling fixture was used for the shelf in the cabinet. In combination with the metal grilles in the doors, this allows excess dryer heat to escape.

Here the baseboard has been shimmed out a half inch and reinstalled, and the new sheet beadboard has been installed atop it. Before the new 3/8-inch gypsum board and skim-coat plaster were applied to the upper walls, cellulose insulation was blown into the wall cavities (via holes drilled through the historic plaster) for sound insulation between the bathroom and surrounding rooms.

This view toward the alcove under the stairs shows the painted cabinetry and wainscoting. The paint color was selected to match the nineteenth-century trim color in the house. Note the wallpaper sample, which is a reproduction of the "Honeybee" pattern designed in 1881 by America's first female industrial designer, Candace Wheeler. The manufacturer of this reproduction paper offers it with an invisible acrylic coating for damp environments such as bathrooms. The new skim-coat plaster had not yet been applied to the upper walls and ceiling when this photo was taken.

Because the pedestal sink was already installed and in use, the beadboard sheet was scribed to fit around it.

The approach used on trim in the house's other secondary spaces was duplicated for the enclosure around the washer and dryer. Old new stock hinges (still in their original 1940s packaging) were purchased on eBay for the bathroom cabinets.

One of three antique electrical sconces acquired for the bath/laundry room on eBay. They match the surviving sconces in the house, likely installed c. 1898, when electricity first became available in the village. Shades matching the historic shades in the house were found separately. The rewired fixtures still have working pull-chain switches but are controlled with a reproduction push-button wall switch by the door. The reproduction wallpaper shown here will be used in the room in a different colorway. A small room like a bathroom can be the ideal space to splurge on an expensive reproduction wallpaper.

The architectural salvage barn sash used as a faux skylight in the bathroom, with lighting in the recess. Installation of minimal wood trim awaits plastering of the ceiling in this shot.

The cap for the wainscoting was also milled with router bits, except for the off-the-shelf ¾-inch cove, or scotia, molding at the bottom. The same bead bit discussed in the previous image was used for the bead band, and a nosing bit was used for the top piece. Note that this cap is "heavier" than most off-the-shelf wainscot caps carried by lumberyards and home improvement centers. Keeping new trim elements in scale and character with the historic trim in the house will help it to feel right.

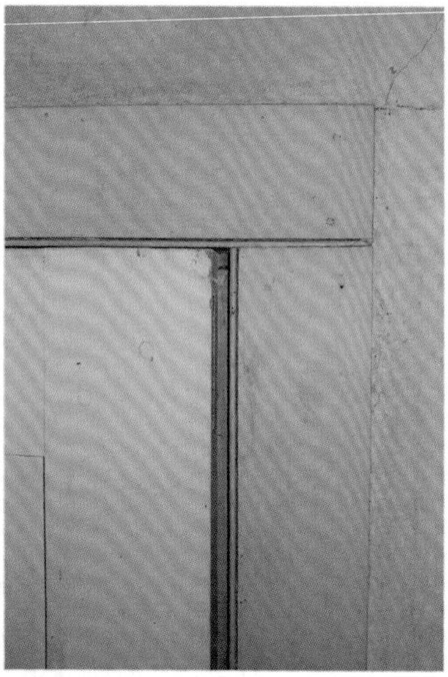

In Whitten House's secondary spaces, the beaded trim had not been mitered at the joints as it was in more important spaces. This closet door trim is in the room over the original kitchen.

A vintage glass-rod towel bar was installed in the room to continue the character of a historic bathroom.

The side of the door facing the walk-in closet had never been painted, since it opened inward into a dark closet. After being relocated for the new bathroom, it was left unpainted to be treated with mineral oil, which will bring out the warm color of the 190-year-old Maine white pine and provide some protection against water stains. The original thumb latch hardware retains the black paint it received in the factory in 1827. A late-nineteenth-century rim lock was installed above it, and the skeleton key hung by twine from an old square-cut nail. Although added in the twenty-first century, the lock looks as if it could have been installed in the late nineteenth century.

under the room. Returning the plaster to its original state would not have provided the desired quality of finish behind the planned reproduction wallpaper. Rather than removing the plaster, I used modern MDF sheet material for the beadboard wainscoting and added 3/8-inch gypsum blueboard (i.e., drywall) with skim-coat plaster above the wainscot. The historic baseboard was removed and reinstalled with the wainscot after the floor was leveled and tiled.

Additions

Adding on to a historic house is often an effective way to gain space for modern features without destroying historic spaces. The key to making this work is not to allow the addition to overwhelm or distract from the home's historic character. In the realm of historic tax credit rehabilitation projects, new additions need to be both *compatible with* and *differentiated from* the historic building. While seemingly contradictory, these two objectives can be met at the same time with a sensitive and thoughtful design. Many local historic preservation ordinances have similar requirements for additions.

Compatibility is the more easily understood of these objectives and can be achieved through scale, placement, materials, and fenestration (i.e., the placement, size, and patterning of window and door openings). An addition that is noticeably larger than the existing house is out of scale and should be attempted only if it will be minimally visible when viewing the home's primary elevations. No addition should ever be built onto the historic front façade of the house. Materials that were commonly used in your area when the house was built can be considered compatible, but a brick addition is unlikely to be compatible with a wood house even if brick was used historically in your community. The simplest solution is often to use the same materials that are used on the historic building. If the house has symmetrically placed rectangular window openings on publicly visible elevations, an addition with irregularly placed triangular windows is incompatible. You might, however, be able to vary the fenestration of an addition sited in the traditional location of a rear ell, woodshed, or carriage house, because these structures often had different (usually smaller) windows and doors than the main house.

Differentiation is a subtler matter. The point of differentiation is to avoid making something new look exactly like something old, the concern being that "fake history" undermines the educational value of historic districts. This is generally only a consideration for additions or alterations that will be visible from a public way in historic districts. While differentiation need not be a primary concern for homeowners outside designated historic districts, it is always worth considering. The addition you are building is going to become a part of the house's history, and making it somehow identifiable as twenty-first-century work will be helpful to people in the future. Differentiation can be as subtle as using consistent modern 3½-inch spacing on clapboards for an addition when the house has historic clapboards that start with 2-inch exposure at the sill and gradually increase to 3 inches by the first-story window sills, continuing at that exposure for the rest of the wall. Most observers will not consciously notice the difference but will register that there is one. Another approach is to use trim on the addition that is similar to but simpler than the trim on the historic house. Keep it in scale, but use less detail. This is easier to accomplish on a high-style historic house than on a vernacular house that has very simple trim to start with.

The large addition planned for Whitten House is based on evidence of a similar nineteenth-century addition that had long since been demolished. The intent is to rebuild a structure on the footprint of the original woodshed (part of which later became a kitchen) connecting the ell to a barn with garage and shop spaces on the first floor and a large family room in the loft above. A pantry will be tucked into the corner where the two buildings meet. The rebuilt woodshed section will house a larger modern kitchen, replacing the existing small kitchen in the 1960s addition, which will be demolished. The kitchen and family room will be connected by a staircase. Salvaged historic timber frames are to be reused for the two structures and will remain exposed. Both living spaces, kitchen and family room, will have cathedral ceilings.

These will clearly be contemporary spaces built using historic materials. The exteriors will be highly compatible with the historic house, using traditional materials in traditional ways, while the interiors will be distinctively differentiated by their contemporary character. Elements of the design are drawn from local examples, such as using clapboard siding on the most visible elevation of the barn with shingles on the other elevations. The design of the cupola on the barn is based on several similar local examples. Both structures will have multiple skylights placed on the north side of the roofs, where they will be less visible from the public way. Woodsheds traditionally had at least one door wide enough for a wagon, so an opening of similar size will be filled with glazed doors in the new kitchen. Hinged wooden doors will be mounted on the exterior and fixed open to reinforce the idea of a historic wagon door.

Example Project: A Bay Window Addition

The renovation plans for Whitten House called for two additions, one small and one large. The small addition, a two-story bay window, added only three feet to a room on each floor but made a large difference in those small rooms. The 8-foot by 8½-foot rooms are at the southwest corner of the house and were potentially the brightest rooms in the house for much of a winter's day, but with only one window in each room, and none on their west walls, that potential was unrealized. The bay window increased the size of the rooms by 25 percent and flooded them with light, making them feel much larger.

Because the house is in a local historic district, any addition visible from the public way would have had to be compatible and differentiated as defined in the district ordinance. The bay window, however, would not be visible from the street, allowing more flexibility in design. Bay windows became popular in the middle of the nineteenth century, particularly on Italianate houses of the 1850s and 1860s. In the Whitten House region, many earlier houses had bay windows added during that period or later, during the Colonial Revival era. The bay window on Whitten House was designed to appear as if it had been added in the later nineteenth century, and compatibility was given a much higher priority than differentiation for this small addition. The design process started with a photo survey of bay windows in the neighborhood. Most had been added to earlier houses during the second half of the nineteenth century, establishing a historic context for the design.

The new bay window was located so as not to interfere with the historic paneled corner pilaster, and the historic frieze and cornice at the top of the wall was replicated to run around the bay window. The

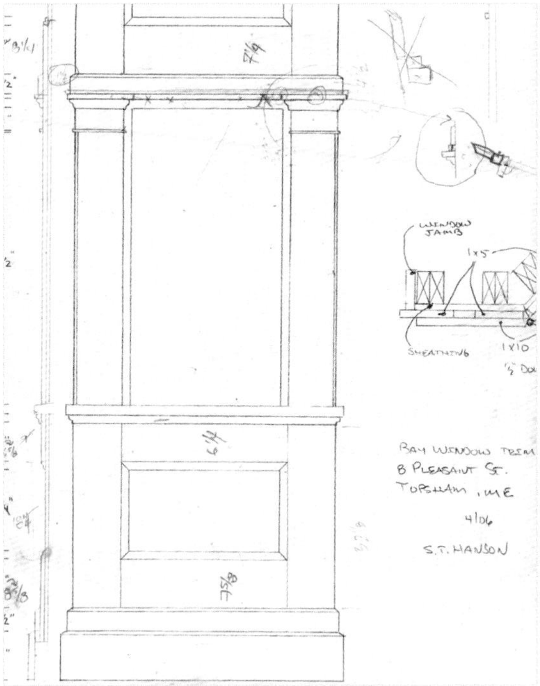

A portion of a trim detail drawing for the new bay window on Whitten House.

windows match the style and configuration of the historic windows in the house, though the widths are different to fit the bay. An architectural historian would recognize the design as reminiscent of Colonial Revival work from the end of the nineteenth century, trying to look Greek Revival but in a way that would not have been seen during the Greek Revival period (when bay windows were rare). To a casual viewer, the bay window blends into the house and does not look like an addition.

Because it was impossible to evaluate the existing framing before the walls were opened up, I left many structural design decisions to be worked out

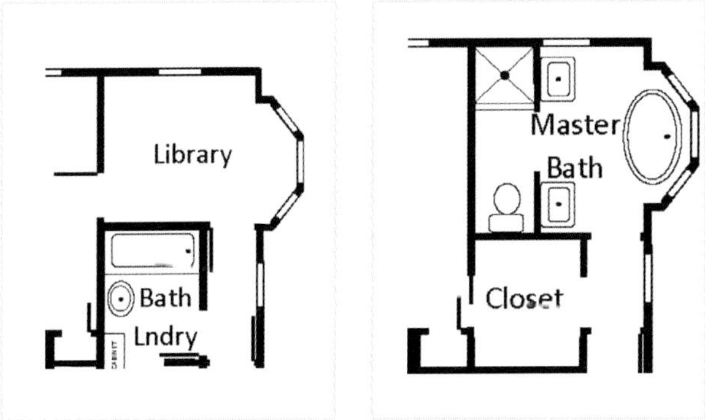

First- and second-floor plans with the new bay window.

during construction by the experienced carpenter doing the work.

The house's slab granite foundation sits on footing stones that are just below grade. This results in some movement of the house when frost penetrates the ground in winter. The bay window was therefore

View of the west (rear) elevation of Whitten House showing the expanse of wall without windows at right.

Historic bay windows photographed near Whitten House to provide design ideas.
Elements from several examples were combined in the Whitten House bay window.

Initial concept drawing for the bay window addition. As the design developed, it became taller and deeper.

Removing vinyl siding in preparation for the bay window addition.

Removing historic clapboards where the bay window will go.

Opening up the first-story wall.

A new built-up beam installed beneath a historic wall-plate beam to support the second-floor cantilever and future bathtub.

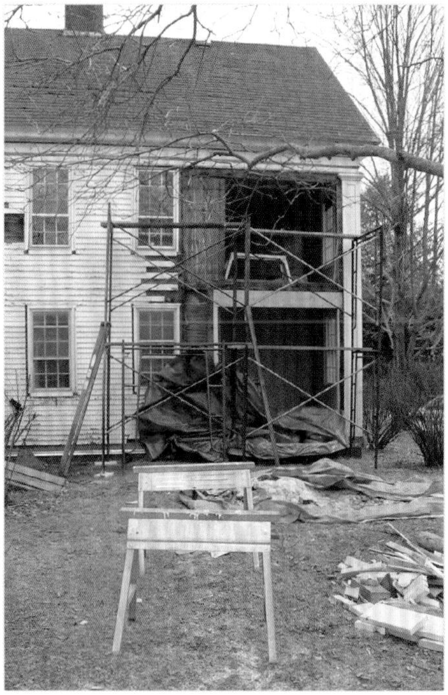

Both stories opened up.

Opening up the second-story wall.

cantilevered to allow it to move with the house. Placing it on a stable foundation with footings below the frost line (as would be required by modern code) would have created stress where the addition meets the house and likely resulted in cracked plaster and open trim joints over time.

Once the carpenter completed the framing, sheathing, and window installation, I did the time-consuming trim work myself, in part because I enjoy doing finish carpentry and in part to save money. If you are only going to do part of the work on a project, it is good if it can be the part with the most expensive labor!

First-floor cantilever beams installed. The beams were fabricated on site from 2 x 10s and ½-inch plywood glued and nailed together.

The first-story sills of the bay window installed.

Second-story cantilevers installed. The new beams cross the historic beams with a half-lap joint on both floors.

Detail of the attachment of the new beams at the interior end.

Framing up the bay window.

Bay window framed, sheathed, and wrapped.

Second-story interior view with windows installed.

Installing the new single-glazed, wood-sashed, double-hung windows matching the historic windows in the house.

Trimming out the bay window. The wide original frieze board that had been removed to make way for the bay window at the top of the wall was cut into three pieces for the tall panels under the first-story windows. I found nineteenth-century panel moldings at an architectural salvage shop.

The original moldings on the frieze at the top of the wall had been removed when the vinyl siding was installed. Around the corner, on the south elevation, the moldings terminated by returning against the house. Because the moldings were installed before the clapboard siding, the siding was coped to fit around the moldings. These profile cuts remained behind the vinyl, making it possible to exactly replicate the missing moldings on the bay window and the original frieze.

Here is the frieze with the replicated moldings installed. The vinyl siding pulled away from the house at right shows where the coped siding documented the molding profiles.

This detail shows the cut-back eave trim. The original wood gutter remains in place where it was roofed over when asphalt roofing replaced the original wood shingles.

The historic gutter was replicated in Spanish cedar for the bay window and for the remainder of the house, where it will be replaced and returned to use.

The nearly completed bay window addition.

This 1883 house by noted architect John Calvin Stevens was designed for his friend John H. Davis and built just down the street from Stevens's own home. This image shows it as it appeared in Scientific American's Builder's and Architect's Edition in 1893. (*Courtesy of Greater Portland Landmarks*)

By 2007, the Davis House had been added to several times and had a portion of its front yard removed to install a garage door into the basement level. Concrete retaining walls bore no relationship to the house. Adding to the awkwardness of the house, the porch had been replaced with unpainted pressure-treated lumber, and a pressure-treated 6 x 6 had replaced the boldly turned porch post. (*Courtesy of Barba and Wheelock, Architects, Portland, Maine*)

Dealing with Incompatible Existing Additions

Sometimes a historic house has already been added to in ways that are incompatible with the building's historic character. This is particularly unfortunate when the addition is on the front of the house or a highly visible side elevation. One of the more common incompatible additions is the placement of oversized dormers on the front elevation. Dormers have been used on houses in America for centuries and can be appropriate additions to increase light in attic living spaces, but they need to be in scale with the house to avoid looking like a cancerous growth on the roof. If the house you own or want to buy has an oversized dormer, consider reducing it in size or relocating it to the rear roof plane. This is not a small project, but it could go a long way toward restoring the historic character of the house.

The example in the accompanying images is an early Shingle Style house designed in 1883 by one of the creators of the style, John Calvin Stevens. Several twentieth-century additions to the side elevation had distracted from the original saltbox form. Adding to the distraction, the landscape terrace on which the house sits had been partially cut away for a driveway leading to a large basement-level garage door on the principal elevation of the house, and concrete retaining walls extended from the house to the street on either side of the driveway. This accumulation of changes had robbed the house of its original charm and left it awkward and unbalanced.

The solution managed to retain the added space within the house while making the additions feel a part of it by means of subtle alterations. A missing turned porch post was replicated by a local millwork shop using a historic rendering of the house published in *American Architect and Building News* in 1893 and a nearby matching example for dimensions. The concrete retaining walls beside the driveway were faced with half-brick, making them more compatible

These existing and proposed elevations show the relatively simple changes made to the exterior to incorporate the later additions more gracefully into the original house. *(Courtesy of Barba and Wheelock, Architects, Portland, Maine)*

The architect built a model to show how the proposed changes would look in three dimensions. *(Courtesy of Barba and Wheelock, Architects, Portland, Maine)*

This photo during construction shows the restored porch with turned post, which was made by a local millwork shop. The new granite steps relate to the materials of the house. *(Photo by the author, courtesy of the Portland Planning Department, Historic Preservation Program)*

No interior square footage was lost from the completed project, and the additions feel better integrated into the original house. The half-brick veneer on the concrete retaining walls of the driveway bear a relationship to the brick on the first story of the house. The house was painted with its original colors as described in the 1893 article.

with the historic materials of the house. The large garage door is still inappropriate to the principal façade of a Victorian era house, but with a new carriage-house door it doesn't distract from the rest of the house nearly as much as it did before the rehab project.

Enclosed Porches

Enclosed porches are another form of incompatible addition. The best approach to dealing with an awkwardly enclosed porch is simply to restore it to an open porch based on the documented historic design. Old photographs can help, and there is almost always some physical evidence remaining, if only marks in the paint. Often the same builder used the same details on more than one house in a region. If you have a photo that shows even a part of the detail, however indistinctly, look for similar work on another house nearby. You may be able to photograph and measure those details to replicate them

Enclosed porches add living space (at least for part of the year) but often significantly alter the historic character of a house. Porches should be returned to their historic appearance when possible. With luck, the turned posts or cased columns will remain within the infill walls of the enclosure, and paint marks documenting the heights and shapes of railings and other trim will still be visible.

exactly. If no evidence can be found for the original design, the solution is a compatible new design. Again, if you are in a designated local historic district, this design will have to meet local ordinance standards. Nearby houses of the same period can provide a design vocabulary from which to compose your new design. What types of posts are used? How are the railings shaped? Are the balusters flat or turned? Are brackets commonly used? What sort of skirting is used under the porch?

One often-overlooked aspect of a compatible design is making sure your new work is dimensionally proportionate to historic work. If the porch post brackets used historically in your area are 1½ inches thick and you use ¾-inch-thick brackets from a local home improvement store, they are not going to look right on the house. Creating or buying new design elements that are historically scaled and proportioned can be expensive and time-consuming. There are specialty mail-order millwork shops that reproduce historic trim elements, particularly from the Victorian era. Making the components yourself can save money. A friend of mine was quoted several thousand dollars by a millwork shop to replicate relatively simple sawn balusters for restoring a porch on his house. Instead, he bought the material for several hundred dollars and produced the balusters in his own shop in a weekend.

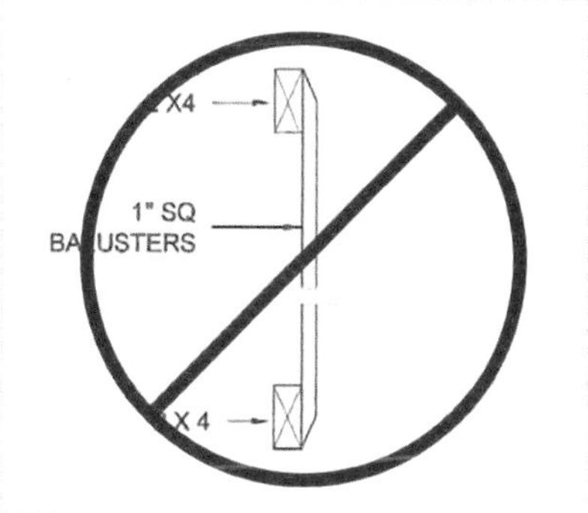

The columns on this historic entry porch are square posts cased in 1x boards with applied moldings creating the form of a traditional column, with a base, shaft, and capital. Deteriorated original columns should be replicated as closely as possible, but simplified versions of this type of column are easily fabricated with stock moldings if compatible replacements are needed for missing columns. It is advisable to study nearby historic examples for the proportions of the different elements.

It may make sense to simplify the design vocabulary of historic porch construction. For example, you might substitute 2-inch-square balusters for the turned balusters that would have been used in the late nineteenth century but place them in the same way—that is, spaced approximately two baluster widths apart between a shaped top rail and a supporting bottom rail. "California"-style railings with 1½-inch-square wood balusters screwed to the sides of the top and bottom rails, illustrated on the left, are never appropriate on a pre-1950s house. In the simplified compatible approach, cased square posts should have a base and a capital composed of simple moldings.

Columns and turned porch posts are available at the big-box retailers, some local lumberyards, and through specialty mail-order companies. Again, make sure their dimensions are similar to those on nearby historic examples. The city of Portland, Maine, offers an excellent guide for porch repairs on historic houses at www.portlandmaine.gov/DocumentCenter/View/2705/Guidelines-for-Porch-Repairs-and-Replacement. The booklet includes

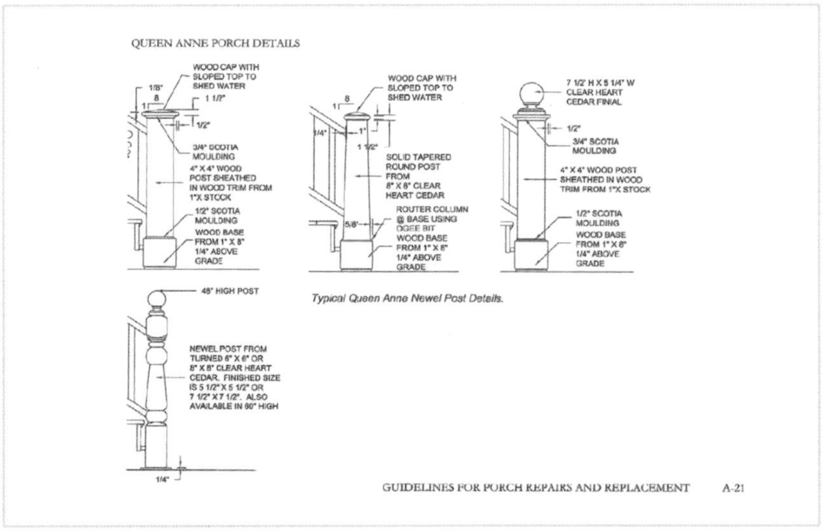

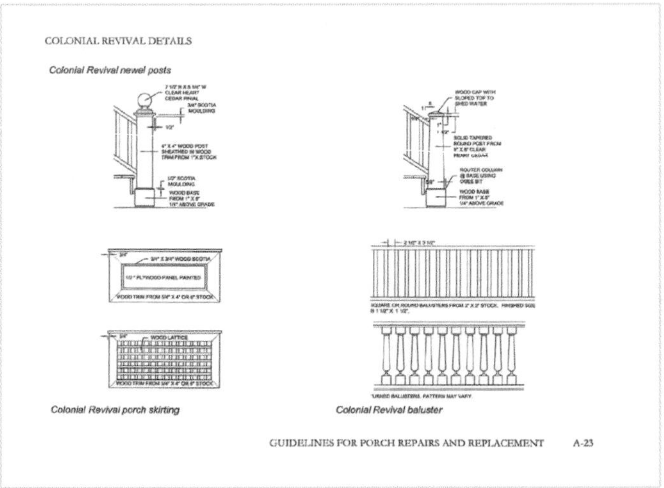

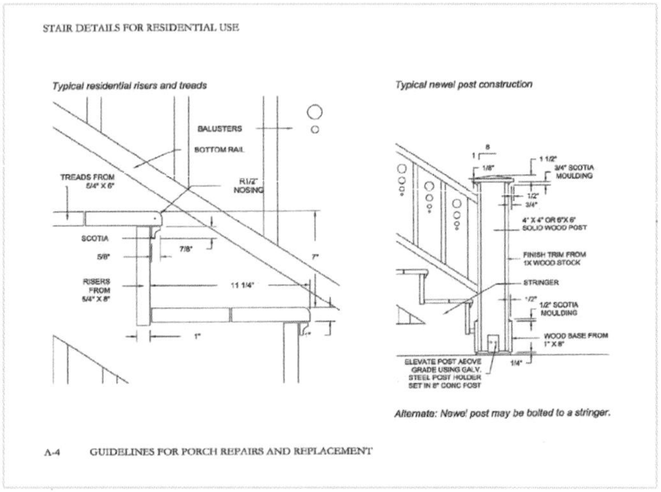

Example pages from the city of Portland's Guidelines for Porch Repairs and Replacement booklet, available at www.portlandmaine.gov/DocumentCenter/View/2705/Guidelines-for-Porch-Repairs-and-Replacement.

drawings of compatible porch design details for various styles of architecture.

Landscape Design for the Historic House

Although this is not a landscape design book, the setting of your historic house is important to its historic character. This is generally more of an issue for a village, suburban, or rural house surrounded by landscaping than for a city rowhouse with only a small paved areaway in front. But even that small areaway can have a negative impact on the home's historic character. Picture a Victorian-era brownstone rowhouse with an off-the-shelf 6-foot-tall wood stockade fence around the areaway. Not compatible.

For houses surrounded by landscaping, common areas of insensitivity are fencing, driveways, and modern approaches to plantings.

Dow Farm, 1769

Georgian Colonial Style with Additions

The Dow Farm house in Standish, Maine, is a one-and-a-half-story type known as a Cape Cod house, or simply a Cape. Its center chimney, typical of the type, allowed nearly every room to have a fireplace for heat. Two large rooms flank the chimney, and there is a small entry hall at the front of the house. A long room (the original kitchen) behind the chimney is flanked by small rooms in the two rear corners. A narrow, steep staircase provided access to two finished bedrooms in the attic; beneath that stairwell are stairs to the cellar. An ell was built onto the rear of the house about 1832 for a new kitchen and a staircase providing readier access to the two existing bedrooms and two new bedrooms upstairs. The original staircase was floored over, leaving the cellar stairs in their original location. In 1876 the house and ell were attached to a new barn via a woodshed and shop building. These progressive additions were typical of farmsteads in parts of New England during that period. A few incongruous Victorian-era alterations were made in the late nineteenth century, including replacement windows and a new dormer on the front elevation.

The property was purchased by the Dow family not long after the main house was built in 1769, and it remains in Dow family ownership to this day. It was actively farmed until the 1930s, when the last generation of full-time Dow occupants passed on. From that point until 1976, the house was used as a summer home by heirs living out of state.

By 1976, the buildings were suffering from deferred maintenance, and many of the fields were overgrown with trees. With the house vacant much of the year, break-ins were a problem, and family heirlooms had been removed for safekeeping. No updates had been made to the house since the early twentieth century. Exterior paint was flaking, windows were rattling, and the lawn and gardens were overgrown. Inside the house, plaster ceilings were falling, wallpaper was peeling, and a cast iron cookstove was the most modern appliance. A hand pump at the kitchen sink was the only indoor plumbing, with the privy between the shed and barn still serving its intended purpose. Sills and joists were rotting underneath all the buildings, and various critters had taken up residence.

In August 1976, Donald, a 23-year-old recent graduate of the nearby University of Southern Maine with a strong interest in historic preservation and restoration carpentry, noticed the Dow house from the road. When he spotted one of the out-of-state owners in the yard, he introduced himself and offered to move in as caretaker, doing needed repairs in exchange for rent. The owners agreed, and that began a relationship between Don and the Dow descendants that continues to this day and has resulted in the buildings being carefully restored and rehabilitated and the land returned to active farming on a small scale. In 1986, Don's now-husband Mike moved in and joined the effort. A major ice storm and a freak windstorm in the 1990s led to the clearing of acres of felled trees and the reestablishment of pastures. Don began planting an orchard of heritage apple varieties. Chickens, turkeys, ducks, sheep, pigs, horses, heritage breed cows, and a memorable cast of roommates and cats have been in residence through the years.

An early project was removing the fallen ceiling plaster in several rooms and replacing it with gypsum board. The house was repaired and restored little by little over the following years, usually with Don doing the work and the Dow descendants covering costs of materials. The Victorian-era two-over-two windows were replaced with historically correct nine-over-six single-glazed wood-sashed windows. The house was re-sided in new wood clapboards. Exterior walls were insulated. The first-floor framing was replaced with new timbers of the same dimensions. Both chimneys were rebuilt, and an 1887 cast iron brick-set range was installed for cooking and heating water. The awkward front dormer was removed. The house was rewired. An indoor bathroom was installed. The house was reroofed with wood shingles and later in metal. A compatible pantry was built off the kitchen. The floors were finished with decorative painting and the walls with period reproduction wallpapers. The house continues to be heated with wood- and coal-burning stoves, though central heating is in the planning stages—probably a radiant floor system that will be invisible in the historic spaces.

Beyond the main house, projects have included reconstructing and restoring the woodshed and shop to an earlier configuration; removing asbestos siding from the barn and replacing it with wood shingles; replacing sills in the barn; and recently installing a new metal roof and solar panel array on the barn.

In 1985, Don built a timber-framed Cape guesthouse for the Dows to use while visiting. After the passing of the brother and sister who first welcomed Don as caretaker, the brother's daughter Claudia and her husband Ken consolidated ownership of the buildings and surrounding land from various relatives and established legal life-tenancy for Don and Mike.

Upon retiring from careers in Washington, DC, Claudia and Ken began spending more time there, working with Mike to grow produce for the local farmer's market and establish a CSA (community-supported agriculture) program. Dow Farm has also become a major supplier of heritage apples to a local food co-op. Claudia and Ken now spend more than half the year in Maine, and in recent years the guesthouse was moved to a nearby field and expanded to accommodate them. The addition required removal of the screen porch, which was moved to the historic house and attached to the shed, where the most frequently used entrance to the house is located.

At a time when many Maine farms in the communities surrounding Portland were being sold off and subdivided into suburban house lots, the Dow heirs held onto their legacy despite living out of state and only being able to use it occasionally. And when vandalism and deferred maintenance threatened that legacy, Don arrived to change the trajectory of the farm's future. The unique relationship built between the Dow heirs and Don and Mike has resulted in a museum-quality restoration of the house and outbuildings, a return of the land to productive farming, and a determination on the part of the Dow heirs to ensure that the property will remain intact and protected for the future.

Four Decades of Work

1970s: Painted the exterior of the house; replaced fallen plaster ceiling in back parlor; set up woodworking shop in shed; cut back bushes; developed long-term work plan.

1980s: Replaced sill on front of house; re-topped kitchen chimney after lightning strike; replaced deteriorated windows with new wood windows; insulated walls, one side per year; re-clapboarded as needed (reusing some clapboards put on in the 1930s); installed a bathroom; removed dormer and re-roofed with wood shingles; built the guesthouse; rewired, replastered and painted more of the interior, one room at a time; took up many of the floors, reframed, then relaid the floors using the original boards or new scrubbed-pine floorboards.

1990s: Installed brick-set cooking range and rebuilt chimneys and fireplaces, upstairs and down; reconstructed the woodshed, including new granite foundation; installed pantry; repainted exteriors of all buildings; started installing reproduction wallpapers; logged a major clear-cut and reestablished former pastures after the Ice Storm of 1998 and a straight-line windstorm the following summer toppled acres of trees; planted a heritage apple orchard, a row of trees every year beginning about 1990, using original orchard map to plan the varieties, continuing into the 2000s.

2000s: Continued redecorating with appropriate paint colors and reproduction wallpapers in house; expanded farming activities; began work on the barn, including overhead mows and replacement cow tie-up sills, framing, and floors.

2010s: Reroofed the main house and outbuildings with metal roofing; moved guesthouse to hill; added porch salvaged from guesthouse to main house shed; installed new sills and did structural repairs to barn; resided barn in wood shingle; installed solar panels on barn roof; insulated attic of house; replaced 1980s bathroom with 1880s-inspired bath featuring custom milled beadboard walls and cherry cabinetry.

A tall case clock—one of several reproduction pieces Don built—stands in the parlor.

A brick-set range made by Cyrus Carpenter & Company (Boston) is used daily for cooking and heating water, which is stored in the nickel-plated tank on the left.

Views of the front façade and north side of the house around 1900 and today. The tall elms were lost to Dutch elm disease, but Don planted disease-resistant replacements in the 1980s. The chair seen near the corner of the house in the 1900 view remains in an upstairs bedroom today.

A desk and corner cupboard made by Don augment the front parlor's antiques. The floor stenciling was done in the 1970s.

The Dow Farm in the 1930s, shortly after the exterior was repaired and repainted.

The barn's hay loft. Built in the 1860s, the barn has received new sills, floor beams, and flooring.

An antique ogee clock from the 1830s keeps time in the kitchen.

Antique English transferware pottery is displayed with blown glass and creamware pieces in Don's front-parlor corner cupboard.

The barn, outhouse, shop, woodshed, ell, and house are visible beyond the orchard, which supplies heritage apples to a local food coop.

By the late 1950s, the towering elm trees in the front yard had succumbed to disease, and the paint applied in the 1930s was beginning to fail.

The eighteenth-century clapboards were in rough shape by the time this photo was taken about 1905.

The pantry was built by Don in the 1990s using wide floorboards salvaged from another building.

The floor of the kitchen (in the 1830s ell) was replaced in-kind with locally sawn pine in the 1980s and left unfinished, as it would have been historically.

These kerosene lamps on a painted pine dresser in the upstairs hall are ready for use in case of a power outage.

The back parlor was the original kitchen and retains the cooking hearth and brick oven (seen above).

Looking into the upstairs hall from the bedroom over the kitchen.

Elizabeth French Dow outside the recently replaced front door in about 1905. She and her husband, Herbert, were the last Dows to live fulltime at the farm.

The Dow family gathered on the front lawn, about 1895.

The orchard was planted in the 1990s based on a 1920s plan drawn by Herbert Dow.

The wood paneling in the first-floor master bedroom, papered in the nineteenth century, was uncovered and restored by Don. Mike did the checkerboard floor paint.

Herbert (shown with a grandson in the 1920s) was the last Dow to actively farm the property until his great grand-daughter Claudia revived farming on a small scale in recent years with her husband, Ken, and Don and Mike.

This portion of the upstairs hall, where the 1830s ell meets the 1760s house, has never been finished.

A view of the bedroom over the kitchen.

The farm as seen from the pasture across the road.

Inside Dow Farm during the Victorian era. The square grand piano is still stored at the farm.

One of the low-ceilinged bedrooms in the 1760s portion of the house.

The Work

Don installing wood shingles on the roof of the outhouse in 1988.

Eighteenth-century wood trim stripped of many coats of paint to expose the delicate Federal style molding profiles.

Removing the awkward Victorian-era dormer from the main house and reroofing with wood shingles in 1988.

Don and Claudia with newly delivered wood roof shingles, 1988.

Siding and flooring removed to expose deteriorated framing in a corner of the barn. 2017.

Installing new sills, floor framing, and flooring in the animal stalls in the barn, 2017.

Reroofing with wood shingles, 1988.

For its third repainting since the mid 1970s, the house was stripped to bare wood for a longer-lasting paint job. The project, here interrupted by the arrival of winter, took Mike several years in the early 1990s.

The deteriorated woodshed and shop were reconstructed to their mid-nineteenth-century form in the late 1980s, this time on a solid granite foundation.

Radius baseboard, baseboard cap, and corner moldings for the new bathroom, 2018.

Winter, late 1970s.

The freshly painted main house in 1977, after replacement of windows, doors, and some clapboards.

This pasture behind the barn was created after the Ice Storm of 1998 and a freak windstorm the following summer flattened acres of trees.

CHAPTER FOUR

Should You Do It Yourself, Hire a Contractor, or Both?

The author removing paint from a plaster ceiling.

Whether to hire a contractor, tackle the rehab yourself, or choose some combination of the two depends on your abilities, inclinations, and available time and resources. Good professional help—a general contractor who is experienced with historic rehab work—is expensive but will get the work done faster and possibly better. Avoid less expensive, less qualified help unless the workers will be under your near-constant supervision. One clueless human with a power tool can do a lot of damage quickly to a historic house.

If you are in no hurry to complete the project and you like the work and saving money, doing as much as possible yourself is probably the way to go.

Be Realistic

It is important not to overestimate your abilities and available time. No book can turn you into an experienced professional overnight. As with many things, experience and practice are the best way to become proficient at the trades and tasks associated with home rehab. If you are willing to learn new skills and can accept that the learning will take time, you can likely do much of what an experienced professional can do and maybe even do it equally well. You will probably waste some materials and have to do some of the work over to get it right—maybe several times—but so did the professional when he or she was learning the trade.

Most homeowners feel comfortable taking on some tasks but want a professional to tackle others. Building codes in some towns won't allow a homeowner to do certain tasks, usually electrical and/or plumbing work. A visit to your local code enforcement office can tell you what is or is not allowed. Many communities publish their codes online, and it can be helpful to review them before a visit to the office in order to identify where you need clarification or more information. Nuances of language or local interpretation can make a significant difference in how your project should be approached. For instance, some communities will allow a homeowner to run electrical wires but require a licensed electrician to make connections at the breaker panel. Other communities require that all electrical work, including running wires and installing fixtures, be done by a licensed professional.

If you're in reasonably good physical condition, you can undertake the following tasks with limited skills:

- demolishing non-load-bearing walls
- removing vinyl or aluminum siding
- cleaning up the worksite
- serving as "gofer"
- fishing wires through walls and ceilings
- installing insulation
- patching and skim-coating plaster
- prepping painted surfaces and painting
- removing paint and wallpaper
- wallpapering

If you possess basic carpentry skills and a willingness to learn, you can add the following tasks to the list:

- framing partitions
- making basic structural repairs
- installing pre-built cabinets
- repairing or installing doors
- repairing and weather-stripping windows
- installing windows
- repairing exterior trim and siding
- repairing or installing interior trim
- repairing or replacing porches
- installing flooring
- refinishing hardwood flooring
- tiling

If you're "handy," the specialized trade skills you can learn include the following:

- Doing custom cabinetry and fine finishwork
- plumbing
- electrical wiring
- HVAC work
- plastering
- decorative painting

Estimating Time and Cost: Double It!

One advantage of hiring professionals is that they can often accurately estimate the time and expense required for a job before it begins. Because it is possible or even likely that unseen problems will be encountered, a professional will usually add a contingency percentage to an estimate. Homeowners almost always wildly underestimate time and money; a good rule of thumb is to make your best estimates based on what you can see, then double them. You may still be underestimating, but you'll be closer.

If you can do some selective investigative exploration behind finished surfaces to determine the condition of structural elements, you'll have a better chance of making an accurate estimate.

It is important that your contractor know what parts of the project you intend to do, and it's important that you know how your involvement fits into the contractor's work schedule. Since painting comes at the end of the process, the contractor may not worry much about handing that task to you. But electrical work comes earlier, and if you fall behind in that, it will throw off the contractor's schedule for your job and others. Do not commit to work you cannot complete on time; if your contractor has to leave your job site because you're behind schedule, you may wait a long time for him or her to return.

Phasing a Project

If you lack the time or money to do a substantial rehabilitation all at once, it is possible to phase the project. Be sure to consider what effects this may have on your ability to get refinancing during the work. Lender requirements for mortgages and home equity loans can make it impossible to access these sources of money during a rehab. You may have difficulty getting a loan, for instance, if your house has exposed stud walls or subfloors, an incomplete kitchen, or a leaky roof.

One way to avoid this pitfall is not to tear something apart unless you will be able to complete the work on that area—including walls, ceiling, and floor—before you need to refinance.

Live in the Project or Move Out?

Living amid a big rehab project is hard. Depending on the extent of the work and local safety codes, it could even be illegal. On the other hand, if the alternative is renting or continuing to pay a mortgage on another house, living in the work area saves a lot of money—perhaps enough to justify the mess and inconvenience.

A substantial rehab is not an appropriate living situation for small children. Cosmetic work (paint and paper) and maybe a kitchen or bath rehab can probably be done with children in the house provided any lead dust from paint prep is contained. (See lead safety tips on page 658.) You may need to schedule some work to be done while children are at summer camp or visiting their grandparents. Whenever small children are around, you must *always* unplug power tools immediately after use and keep toxic materials out of reach. A good practice is to keep all tools and toxic materials in a room with a locking door that is off limits to kids. A little inconvenience is a small price to pay for keeping children safe.

Living in a House That Is Undergoing Rehab

Here are some ways to make a house undergoing rehab more livable:

1. If at all possible, keep at least one room intact, clean, and shut off from the mess of the work. You need a refuge where you can eat, sleep, and relax without breathing plaster dust.

2. Rough-sawn subflooring is difficult to sweep clean and is hell on socks and bare feet. If you have to live with it, cover it with inexpensive black "tarpaper," which comes in rolls and is easy to roll out and staple down. It's not pretty, but it's easy to sweep and will keep out drafts and dust that can enter through the cracks between boards.

3. Eating take-out gets expensive. At a minimum, you need a refrigerator, microwave oven, and sink with running water in your living space. A portable electric hot plate/burner can expand the cooking options considerably. A kitchen range is better. It is expensive to have a temporary 220-volt electric line installed for a range, and an old house may not even have 220-volt service at the start of a project. A propane gas range can run off a five-gallon tank with a pressure regulator (like an outdoor gas grill) and may be a better

Temporary kitchen facilities in Whitten House. A sink with hot and cold running water, a gas range connected to a five-gallon propane tank, a microwave oven, a coffeemaker, and a refrigerator were the essentials. Stacked antique shipping crates served as cupboard space. This was a large improvement over the period immediately after buying the house, when I had only an electric hot plate, microwave, coffeemaker, and minifridge.

temporary option. Used ones are available inexpensively through online and other classified ad sources. Check local codes to see if you need to have it installed by a licensed technician.

4. Install temporary bathroom facilities. You *can* get through a rehab project with a porta-potty in the yard, but you'll be much happier with a temporary hookup for a real toilet in the house—ideally with walls around it for privacy. PVC drain pipe and PEX (cross-linked polyethylene) tubing are inexpensive and easy to install. A one- or two-piece fiberglass shower stall can also be quickly installed. The walls can be temporary, too. Blue plastic tarps stapled into the ceiling framing will do.

Temporary shower hookup at Whitten House. Here an elevated base was built of 2 x 6 lumber and plywood to create space for a drain trap above the concrete floor in the 1960s addition. The copper supply and PVC drain lines were run along the surface of the wall. The shower control unit was mounted outside the fiberglass shower stall to avoid having to drill a hole in the stall. This setup remained in use five years, until the first new bath was fully functional. It was then disassembled, and the shower stall unit was sold on Craigslist for nearly what it cost new on sale. Today I would use PEX tubing for the supply lines, simplifying the installation.

Elevated base for temporary shower.

5. Hot water is a necessity. Rehab work is dirty, and the people doing it need the ability to clean up after a day's work. If you do not have 220-volt electrical service for an electric water heater, a gas-fired water heater can be quickly installed to run off a five-gallon propane tank with a pressure regulator (the type used for outdoor gas grills). If your hot water will ultimately be running off the heating boiler, pick up a used gas-fired water heater through online or other classified ad sources to get you through the rehab. You can sell it through the same source when you no longer need it.

6. You need to stay warm. If the house will be without a heating system for a time, and if that time is winter, you will have to install a safe temporary source of heat—electric, kerosene, or gas—in your living space(s). You will need to find one that does not emit toxic fumes. If you have a usable chimney, you could heat with a wood or coal stove. Whatever source you use, it needs to be installed and vented properly. Needless to say, you will want air conditioning in your living spaces during a rehab in a hot climate. This is easily accomplished with window-mounted units.

7. Install smoke and carbon monoxide detectors! They can drive you crazy in work areas, since the dust of the work can set them off regularly, but you *must* have them in your living space and any other spaces where work is not actively taking place.

8. Fire extinguishers are absolutely necessary. Have them both in work and living areas, and make sure they are charged.

9. Finish things! There is a temptation during a rehab project to jump to the next project when you near completion of a current one. Don't. That one piece of missing trim in the dining room, or those uninstalled cabinet doors in the kitchen, will prevent you from crossing that project or room off your list and calling it "done." It might be more efficient to paint several rooms at once, but if that means living with not-quite-finished spaces for months or years longer, it will be well worth the lost efficiency to have one room completed sooner. Psychologically, the more projects or rooms you can legitimately call "done," the happier you will be with the progress you are making. The people you live with will also be happier—much, much, happier.

A 110-volt five-gallon electric water heater provided hot water to the kitchen sink and shower. An on-off valve at the showerhead made it possible to get wet, turn off the water, soap up, turn on the water, and rinse off before the five gallons was used up. It was a happy day at Whitten House when the new natural gas boiler with endless hot water was installed.

10. Throw a party! There is nothing like inviting friends and family to see how the work is coming along to motivate the completion of projects underway. Depending on the stage of work you are at, you may have to close off some areas or install temporary railings or barricades for safety. Explaining your project to guests and describing what it will look like when completed will remind you why you are living through this particular hell, and positive reactions from your guests will help motivate you to keep going. There *will* be people who just don't see it or get it, but don't let their negativity and lack of vision get to you. And don't invite them to the next party.

Hiring a Professional

If you hire a contractor, hire one who is experienced in historic rehabilitation work. Most contractors are experienced only in new construction and have no understanding of traditional construction methods and materials. Sadly, many contractors who regularly work on old houses are not much better informed and know only how to rip out and replace. You need to find one who understands and appreciates traditional construction methods and materials—one who can work with your house, not against it.

In most communities there is at least one contractor who works on historic buildings regularly and is known for doing sensitive work. "Sensitive" does not include gut rehabs. Your local preservation organization or historical society may be able to steer you to a good contractor. If your town has a preservation ordinance, the municipal planning staff or a member of the review board can tell you which contractors have done good work on projects they have reviewed. If you are working with an architect with experience in historic rehab, he or she should be able to recommend contractors. Local or regional preservation organizations often publish directories and sponsor annual or semiannual old-house trade shows where contractors, craftspeople, and others involved in the restoration field exhibit their offerings and give lectures on what they do. Neighbors or friends who have done sensitive rehabilitation projects on their own houses may have recommendations or warnings. Listen to the warnings.

Interview potential contractors, ideally on site. Listen carefully to how they talk about the work. If their starting point is that everything should be torn out and replaced, you don't want them. If they notice the details of the house, talk about how to save plaster or trim or flooring, mention that they can get historic electrical fixtures rewired, etc., you may have found your contractor. Share what you love about the house and stress the importance of preserving as much original material as possible in the most important character-defining spaces. Are they listening when you talk? The contractor you hire is going to be part of your life for a period of months and will leave his or her mark on your house forever. Make sure you are getting into this relationship with someone with whom you can communicate. Always, always, always ask for references. You need to talk to past clients who have done similar projects, and ideally to see the work.

A good contract is essential to protect your interests as well as those of the contractor. If the contractor is unwilling to sign one, be wary of proceeding. In Maine, where I live, any construction project over $3,000 must have a contract by law, and the attorney general's website includes a section on construction contracts with a downloadable "Attorney General's Model Home Construction or Repair Contract." If your state does not provide such resources, this contract will give you a starting point. It may be prudent to have your attorney review a contract drafted by your contractor, especially if the estimated cost of the work is substantial and you are unfamiliar with such documents. Many standard home renovation guidebooks, including Michael Litchfield's *Renovation: A Complete Guide*, cover this topic in detail.

On this construction site, a propane-fired forced hot air furnace was installed in the middle of a room and vented through a window. Long, flexible ducts were used to conduct heat to various spaces as needed.

The annual holiday open house at Whitten House allows our family, friends, and neighbors to come by and see what progress has been made on the ongoing restoration project. It is good motivation to complete projects and provides a reason to take a break from the work to clean, decorate, and enjoy the house for a time—even though the restoration is far from finished.

A good job for a professional.

Proceed similarly with subcontractors for a job you are running yourself. A plumber or electrician who does not understand old houses can do enormous damage in a short time, as can an insulation contractor or a mason. Trade schools in the twenty-first century generally do not teach students to work on old buildings. You need to find contractors and subs who have learned how to do this—usually by working with older tradespeople who learned their trades when houses were built of local materials rather than assembled from parts fabricated all over the world.

Casa Roca, 1891

Nineteenth-Century Adobe

Casa Roca began as a small vernacular two-room house in Santa Fe, New Mexico. Constructed of adobe bricks on a rubble stone foundation, it was one of three small houses built over a period of years in a modest family compound. The adobe brick was covered in typical fashion with stucco on the exterior and plaster on the interior. Floors were brick, and the ceilings were exposed log vigas (beams) supporting sawn-board sheathing. Over time, between the 1920s and 1970s, the original structure was enlarged with additional rooms using the same traditional construction techniques. Rooms added on the north side of the house were cut into the hillside of the sloping site. The foundation was continued above grade to keep the adobe bricks above the soil, exposing several feet of rock wall on the interior (extending to roof height in the northeast corner). It appears that several of the rooms were created by enclosing open "portals," roofed outdoor living spaces that are a common feature of adobe houses.

When owners Ed and David purchased the house in 2005, it had experienced several rounds of less-than-sympathetic modern renovation, water infiltration through the roof and stone foundation, and structural issues in one wall and the roof. The area outside the house that would traditionally have been the patio was used for parking. The new owners were experienced in historic house rehabilitation on the East Coast but had no previous experience with adobe construction. With the assistance of local architect Gayla Bechtol, they developed a plan for rehabilitating the house that restored its historic character, met their needs for a winter home, and fulfilled the stringent requirements of the Santa Fe historic review board.

Many of the historic features of the house were hidden behind later finishes. The mortared flagstone floor of the living room was covered with ceramic tile, and the stone fireplace was coated with bright white Structo-Lite plaster. The historic vigas and board sheathing of the kitchen ceiling were hidden by gypsum board with a textured joint-compound surface. The one bathroom appeared tiny due to multiple layers of partitioning from past attempts to support and hide a structurally compromised wall. The structural issues were caused by the roots of a tree that had grown against the outside of the wall. The same roots had heaved the brick floor of the kitchen and bath in multiple places. Many original casement windows had been replaced with smaller aluminum sliding windows, often not the right size for the historic openings. Some openings had been partially infilled and others enlarged to accommodate the inappropriate windows. The main entry door had been replaced by one that would have been perfectly at home on a 1950s suburban Ranch house. The few surviving wood windows and doors were deteriorated beyond repair.

Ed's experience serving on the Portland, Maine, Historic Preservation Board and as a member of the advisory board of the National Trust for Historic Preservation was helpful in developing a plan that met the city's requirements, and David's careful preparation of their submission led to a positive experience with the review board. Due to the extent of work needed, the submission was in two parts, the first dealing with emergency stabilization work and the second with aesthetics and the details of a small addition. This approach allowed work to get underway quickly on the stabilization while giving the review board more time to deal with aesthetic concerns. Undertaking such an extensive rehabilitation project would be daunting under any circumstances, and Ed and David undertook it while living 2,300 miles away. During the 18-month project, they made 26 trips between Maine and New Mexico.

The plan for the rehabilitated house retained all historic walls and door and window openings with the exception of one door between the two bedrooms, which was infilled to create more privacy. A historic stone foundation and a fragment of an adobe wall on the west side of the house indicated that another room had once existed there. With this evidence, it was possible to get approval from the historic review board to reconstruct the room. This small addition created space for a second bathroom off the master bedroom. Historic features were retained, restored, or replaced in-kind when necessary. The only additions to the interior plan were a built-in storage wall between the kitchen and bathroom (see page 128) and a closet in each bedroom. These additions were kept below ceiling level to allow the original volume of the spaces to be understood. In the guestroom, the difference in thickness between the lower stone portion and upper adobe portion of the north wall was used to create built-in shelving and a covered recess for a pull-out television. A new kiva fireplace was added in the corner of the room, set 18 inches above floor level with an alcove for firewood storage below. Elevating the fireplace made it easily visible from the bed.

Adobe houses traditionally have few, small windows facing the street, which can leave interior spaces dark even in the bright Southwestern sun. Adding skylights allows more light into a house without altering its appearance from the street. At the architect's suggestion, several of the new skylights in Casa Roca were set against walls, allowing daylight to flow down the light-colored walls and be reflected into the rooms. The walls are finished in American Clay plaster—a natural-earth plaster manufactured in New Mexico—and tinted to colors found on the historic walls behind later finishes. The exterior walls received a new coat of natural cement-based stucco. A chicken-wire lath was used to prevent cracking in the stucco, with a heavier metal lath around doors and windows to protect against accidental impacts. The north wall received sprayed foam insulation before the stucco. The lath is nailed through the insulation into the adobe bricks with long nails. The parapets on the walls allowed the roofs to be covered invisibly with 12 inches of sprayed foam insulation, which helps to keep the house cool in summer and warm in winter (when the weather at Santa Fe's 7,000-foot elevation can become quite cold). The roof insulation is covered with a spray-on waterproof membrane. Over the bathrooms, the insulation and membrane slope to traditional canale drains, which carry water away from the exterior walls. The remainder of the foamed roof pitches to a curved sipper gutter on the east-side portal, which drains into a 3,000-gallon storage tank for irrigation of planting beds.

The existing roof had plywood sheathing, stained brown on the underside with mineral-roll ("tar paper") roofing and several inches of dirt, which was thought to be an insulator. Its weight was causing the plywood to deflect between the vigas. Additionally, much of the original roof framing of the house was deteriorated from water infiltration and had been undersized when new, requiring replacement with new, larger logs or timbers. This required removal of the entire roof during the rehab, during which a giant tarp protected the house from rain. The remaining historic vigas in the kitchen and master bedroom were hand-sanded by David to remove more than a century of soot and grime from them. One interior adobe partition had been carrying more roof weight than it was capable of supporting and had suffered consequent damage. A new 25-foot-long steel beam was inserted at the top of this wall to carry the load, and the wall was reconstructed in new adobe brick. The vulnerability of adobe bricks to rain was demonstrated when an unexpected storm struck on a night when the tarp had not been secured and the new adobe wall dissolved into a layer of clay mud and straw on the floor. The wall had to be rebuilt once the mud was cleaned up.

During the reconstruction of this wall, a bit of stone exposed by the old wall's removal suggested that the fireplace in the living room might be built of stone. When further exploration confirmed this, the stone was exposed by chipping and then grit-blasting to remove the Structo-Lite plaster.

Another thrilling discovery during demolition was the mortared flagstone floor of the living room, which had been hidden under modern 12-inch by 12-inch ceramic tiles. Ed and David immediately set about chipping away the tiles to expose more flagstone. All went well until they got about four feet from the south end of the room, where the stone suddenly stopped. Concrete filled the rest of the area. The flagstone had apparently been the floor of an outdoor portal that had been enclosed to create the room. Because the room was four feet longer than the portal, the unpaved area had been filled with concrete and the whole floor covered with tile. Ed and Dave's mason removed the concrete and filled in the missing section with matching flagstone.

The tree outside the bathroom wall had dislodged the wall and floor as its roots sought out and followed the cast iron drain pipe. Once the tree was removed, the wall was stabilized with a new sloped concrete buttress, which was covered with stucco matching the rest of the exterior. The freeform nature of a traditional adobe house allowed this solution to blend in naturally.

Radiant heat was installed in the new tiled bathroom floor. The brick floor in the kitchen was also taken up in order to run heating, sewer, and domestic water piping, then relaid.

For the bathroom addition off the master bedroom, a concrete foundation was poured where the remains of the missing walls had been discovered. Stones salvaged from the old foundation were laid up on the new concrete footing against the poured foundation to the previous foundation height. The walls of the addition were built up from that height with concrete blocks—a more practical choice for a bathroom than adobe—which were insulated and stuccoed like the rest of the house. As in the existing bathroom, radiant heat was installed in the tile floor of the new bathroom.

Ceramic tiles in both bathrooms were chosen to harmonize with the house. The floor of the existing bath was covered with a shaped Moroccan tile that is clearly related to the Spanish tiles that influenced tile work in

Mexico and the American Southwest. The 700-year duration of Islamic rule in parts of Spain ended the same year Columbus discovered America. The architectural influences of Islam were very much a part of Spanish culture when the conquistadors colonized Mexico and the Southwest. Tile was also used on the tub surround and wainscot. The tile chosen for the new master bath is compatible with the stone chimney enclosed within the new room.

To reflect the growth of the house over time, different roof support and sheathing styles were used in different rooms. The living room ceiling got new log vigas and angled, rough-split cedar latillas. The guest room got rough-sawn timber vigas with sawn-board sheathing, and the master bedroom got a sheathing of wood milled to evoke traditional latillas, which are small branches laid across the vigas. The original vigas and sheathing remain in the office, and the kitchen retains its original vigas with new narrow-sawn board sheathing.

Windows were returned to their original dimensions, and new custom wood casement windows were made for the house, copying the deteriorated surviving sash. The double doors from the living room to the upper portal survived but were in bad condition, so they were replicated as well. A similar door was made for the main entrance, replacing the suburban Ranch house door. Also reflecting the growth of the house, the stone chimney that served the historic fireplace in the master bedroom was left exposed in the new bathroom addition as part of the large walk-in shower. The shower's other, rounded walls were tiled in a compatible matte-finish ceramic tile. Because the tile was not available in the 2-inch by 2-inch size needed to follow the curve of the walls, larger tiles were cut up to make the size needed.

Such attention to detail sets this project apart from many. Other examples include antique radiators for the new heating system (with radiant floors in the bathrooms) and antique solid wood shutters, called alacenas, for a medicine cabinet and to hide the electrical panel and a television. Light fixtures are in character with the house without being overly decorative. Kitchen cabinets are of hickory, a wood that feels at home due to its rough character but has a smooth finish for ease of cleaning.

Outside the house, the driveway was shortened to allow space for a patio, which is accessed through a wooden gate in a new adobe wall. Flagstone steps lead to the main entrance just inside the gate. Flagstone paving and plantings create walkways and a seating area in the patio. A tiled hot tub is built into a stone surround against the wall. In the corner of the patio, a small brick niche containing a religious statue recalls the faith of the Hispanic original owners.

A view across Santa Fe from near Casa Roca captures the character of the adobe-built city 7,000 feet above sea level in the high desert.

A pre-restoration view of the house from what is now the patio.

Casa Roca from the patio showing the doors and windows replicated from the deteriorated originals. This lovely enclosed outdoor living space was used as a driveway prior to the restoration.

The portal over the main entry provides shade to the interior and a touch of drama.

A new closet in the master bedroom was topped below ceiling height to retain the full historic volume of the room.

The pre-restoration kitchen.

Hickory cabinets in the restored kitchen reflect a rustic adobe building character. David hand-sanded the original vigas (log beams) supporting the roof. An added skylight brightens the room.

This small room used as an office retains the vigas and board sheathing of its original ceiling. The textural qualities of the American clay plaster are apparent.

An antique adobe brick mold repurposed as a wall shelf.

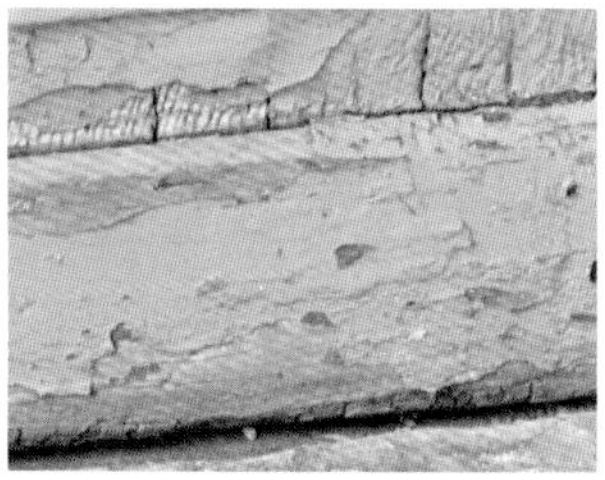

The earlier paint color peeking out from under the later turquoise on the surviving historic door was used for the restored exterior trim.

The stone chimney was left exposed in the shower of the new bathroom.

A new wall and gate reduce the driveway to two parking spaces, reclaiming the space beyond for the patio.

In the pre-restoration living room, ceramic tiles hide the flagstone floor and Structo-Lite plaster hides the stone fireplace.

The restored living room can feel cool and shaded in summer or warm and cozy in winter.

A pre-restoration view. The tree at left was causing significant structural problems for the end of the house, and its roots were dislodging the brick floor in the kitchen.

A street view toward Casa Roca (at left). The room nearest the road is the new bathroom, built on the footprint of an addition that was reduced to a fragment of wall and foundation when the project started.

The guest room before restoration was dark and had water seeping under the wall at the left.

One of the deteriorated casement windows that was replicated for the restoration.

The kiva fireplace added to the guest room is elevated to be seen from the bed. A long, narrow skylight against the wall conducts light into the room. Antique wood shutters, called alacenas, hide a television on a pull-out mount.

The bathroom after restoration. Moorish tile reflects Spanish colonial influences in the Southwest.

The Work

Looking into the living room while the roof was off.

Left: The end wall of the house after the encroaching tree was cut down. Removing the stump caused more disruption to the structure. *Right:* A concrete buttress was poured to stabilize the wall. Once coated in stucco (yet to happen here), it blended in with the house.

Outside the guest room and master bedroom, the foundation had to be extended downward with drainage installed to prevent water from seeping under the wall into the house. A trench was excavated, a reinforced concrete footing poured, and a block wall built on the footing up to the historic stone foundation.

The skylight opening over the kitchen in the new roof.

Electricians channeling through the plaster and drilling into the adobe to run wires.

Installing new log vigas for the replacement roof.

Historic flagstone was found under the ceramic tile in most of the living room, but a 4-foot-wide section at one end was filled with concrete. The mason was able to remove the concrete and lay matching flagstone to complete the floor.

Left: A fragment of adobe wall and stone foundation was evidence that another room had once existed here. *Right:* Based on that, the Santa Fe Historic Preservation Commission allowed the room to be reconstructed, providing space for a bath off the master bedroom.

Left: A new steel beam carries the roof load. *Right*: A new adobe wall had to be constructed beneath the steel beam after a rainstorm destroyed the historic wall.

Left: The existing portal at the main entrance was under-engineered and deteriorated. *Right:* A new portal frame under construction.

CHAPTER FIVE

Bureaucrats You'll Meet

Baldwin County Architectural and Preservation Review Board

Application for Certificate of Appropriateness

APPLICANT: ______________________________

**if applicant is not owner, an agent authorization form must be submitted*

MAILING ADDRESS: ______________________________

City: ________________ State: _______ Zip Code: ______________

Telephone: (_____) ________- ____________ Fax: (_____) ________- ____________

PARCEL ID NUMBER: 05-___ ___-___ ___-___ ___-___-___ ___ ___-___ ___ ___.___ ___ ___

HISTORIC DISTRICT: ( ) MONTROSE ( ) BATTLES WHARF/ POINT CLEAR ( ) MAGNOLIA SPRINGS

(CHECK IF APPLIES TO BUILDING OF THE PROPOSED WORK):

☐HAS A COUNTY HISTORIC MARKER ☐ON THE NATIONAL REGISTER OF HISTORIC PLACES

☐ON THE ALABAMA REGISTER OF LANDMARKS & HERITAGE

E-911 ADDRESS AT WORK SITE: ______________________________

WILL THERE BE A CHANGE INUSE AT THE SITE: ☐ YES ☐NO

IS A VARIANCE, CONDITIONAL/SPECIAL USE, OR REZONING REQUIRED FOR THE PROPERTY:

☐ YES ☐NO

DESCRIPTION OF PROPOSED WORK (CHECK AND DESCRIBE):

☐ NEW CONSTRUCTION ______________________________

☐ REPAIRS/RENOVATION ______________________________

☐ BUILDING RELOCATION ______________________________

☐ BUILDING DEMOLITION ______________________________

☐ PAINTING EXTERIOR ______________________________

☐ ROOFING ______________________________

☐ LANDSCAPING ______________________________

☐ OTHER ______________________________

COMMENTS: ______________________________

____

PAGE 1 OF 2

In most of the United States, rehabilitating a house will require some interaction with local officials. Depending on the ordinances adopted by your community, this could include the code-enforcement office, planning board, zoning board, and/or a historic preservation department, board, committee, or commission. If you are working with an architect or general contractor, he or she may handle these interactions and will likely be responsible for securing building permits and certificates of appropriateness.

In my current work as a historic preservation consultant, I often represent our rehabilitation projects to government bureaucrats. Formerly, as staff in the historic preservation program for the City of Portland, Maine, I worked with property owners to prepare materials for review by the city's historic preservation board, which reviews applications for certificates of appropriateness for exterior alterations on buildings within historic districts. I also worked regularly with the city's permitting and inspections department, since most historic rehab projects require a building permit as well as a certificate of appropriateness. Additionally, I have served on and been an applicant to the historic district commission in the town where I live.

In sum, I've seen the interactions between property owners and municipal officials from multiple perspectives, and I can report that few things look the same from all these points of view. It is helpful to bear in mind that the bureaucrats and volunteer board members you will encounter are charged with implementing policies they may not have written and have limited flexibility to interpret.

Applying for a Certificate of Appropriateness

If your house is in a local historic district and you are planning exterior work beyond regular maintenance, you probably need to apply for a certificate of appropriateness or similar permit from your town's historic preservation board, committee, commission, or planning staff. Who you apply to and what they are called depends on the preservation ordinance in your community. There are thousands of communities with historic preservation regulations in the U.S., and it is impossible to enumerate all the resultant variations, but the majority have many things in common. The following information is broadly applicable, but

you'll need to familiarize yourself with the specifics of your community's regulations.

If your project is complex, the local review board may want to schedule a workshop to become familiar with the project and offer feedback. Then you'll return for a public hearing in which the decision to approve, deny, or conditionally approve your proposal will be rendered. View the workshop as an opportunity to learn all you can from the board even as you are informing them about your project. The board will have reviewed many projects before yours and may be aware of solutions, materials, and other local resources that can help your project—resources you may not have considered or even known of.

A complete and well-organized application takes time and thought but will greatly simplify the review process. It will also please review board members, who struggle with far too many poorly prepared and incomplete applications.

Vocabulary Matters

As in most fields, historic preservationists use a specific vocabulary. Much of this language comes from the federal Historic Preservation Act of 1966 and the publications produced under it by the National Park Service and the National Trust for Historic Preservation. There is a good chance that the vocabulary used in your local historic preservation ordinance is rooted in these sources. A working knowledge of this terminology will help you communicate with your local preservation officials.

Some key terms to understand:

Compatibility is a term for the degree to which a proposed alteration or addition harmonizes with the size, scale, material, and possibly color of the historic building. Design elements such as the rhythm of window and door openings and the relative proportions of wall and window areas also affect compatibility.

Differentiation is desirable in an alteration or addition in order to distinguish it from the original house. The root of this term is in Standard 3 of the Secretary of the Interior's Standards for Rehabilitation, which reads: "Each property will be recognized as a physical record of its time, place, and use. Changes that create a false sense of historical development, such as adding conjectural features or elements from other historic properties, will not be undertaken." There is a wide range of opinion as to how much differentiation is appropriate. Too much differentiation renders an alteration or addition incompatible; too little makes it impossible to distinguish what is old from what is new. Learn how much differentiation your local historic preservation governing body has deemed appropriate in the recent past.

In-kind replacement of a building element that has deteriorated or been damaged beyond reasonable repair means replacing it with a matching new element made of the same material and possessing the same level of craftsmanship and quality. This may seem straightforward, but there are nuances when dealing with wood elements. Historic buildings were built with slow-growing old-growth wood that was far denser and more resinous than modern woods, which are grown quickly in managed forests. Today's woods are much softer than the same species 200 years ago; they are lighter, less dense, less resinous, and much more prone to rapid decay when exposed to the elements. Consequently, replacing historic white pine with modern white pine is not "in-kind" replacement. To match the rot-resistance of historic pine, it is necessary to use cedar, mahogany, or another wood with more natural resistance. Pressure-treated woods do not qualify as "in-kind" replacement materials for exposed surfaces because they hold paint poorly and are prone to excessive surface checking (cracks along the grain). Pressure-treated wood is, however, appropriate for structural work that will be covered by other materials.

Appropriateness depends on the review authority's interpretation of local historic preservation standards. A well-written ordinance will define key terms; always start by reviewing these. Sometimes the definition will be so vague as to provide little guidance, such as in this example from an actual ordinance: "*Certificate of appropriateness:* A certificate issued by the department evidencing approval of specific plans for alteration of a structure or new construction on a site in accordance with this article." Translation: Appropriate means meeting the standards of the ordinance. In theory, the review authority's interpretations of the standards will be consistent across many projects. Look at previously approved projects to get a sense of what is or is not considered appropriate.

Organize Your Material

Your application and supporting materials need to be clear and understandable to someone who has never met you or seen your house. Start with a description of the house and existing conditions. For example:

> 42 Elm Street is a two-story Queen Ann–style house with a round corner tower and a wraparound porch. The hipped roof has a projecting gable on the front, balancing the conical roof of the tower. The house sits back 20 feet from the sidewalk, in line with neighboring houses of the same period. The house has vinyl siding applied over the original clapboard and patterned shingles. Vinyl shutters have been installed beside many windows, most being of the wrong sizes for their respective windows. Original wood windows remain behind forty-year-old triple-track aluminum storm windows. Slate

roofing remains on the main roof and gable but has been replaced by asphalt on the tower roof. The porch is significantly deteriorated but retains nearly all its decorative wood elements including posts, balusters, and brackets.

This brief description sets the stage for outlining in broad terms the work you propose to do, as in the following:

> It is proposed to remove the vinyl siding and shutters to expose the historic siding and reinstall the historic shutters, which remain in the carriage house attic. The porch will be repaired and new asphalt shingles that better match the color and size of the slate roofing will be installed on the tower. A small addition is proposed for the rear (north) elevation of the house to create a mudroom and provide covered passage from the carriage house to the house. The addition will have limited visibility from the street.

Then the details of each part of the proposed work should be spelled out in their order of mention in the outline. Cross-reference attached drawings and photos and explain in detail how your proposed work meets the standards of the ordinance. For instance, you might describe an addition by saying:

> The proposed addition meets the standards for compatibility in its scale, placement, and form as well as its continuation of the fenestration pattern from the historic building. It also uses the same materials that are used on the historic building. It is differentiated from the historic building by the use of simplified trim profiles and the absence of Italianate brackets at the cornice. The placement of the addition at the northwest corner of the historic house ensures that it will not interrupt the visual rhythm of the historic buildings on the street and that the addition will have limited visibility from the street.

In sum, make a clear case, citing the standards that apply.

Cite National Park Service Preservation Briefs *or Local Design Guidelines*

The National Park Service's Technical Preservation Services department has published forty-eight *Preservation Briefs* on multiple aspects of historic preservation work. They can be found at www.nps.gov/tps/how-to-preserve/briefs.htm. A complete list of the *Briefs* is included in Chapter 20 of this book. Those most relevant to the rehabilitation of historic houses include the following:

2. **Repointing Mortar Joints** in Historic Masonry Buildings
3. **Improving Energy Efficiency** in Historic Buildings
4. **Roofing** for Historic Buildings
8. **Aluminum and Vinyl Siding** on Historic Buildings: The Appropriateness of Substitute Materials for Resurfacing Historic Wood Frame Buildings
9. The Repair of Historic **Wooden Windows**

10. Exterior **Paint Problems** on Historic Woodwork

14. New **Exterior Additions** to Historic Buildings: Preservation Concerns

16. The Use of **Substitute Materials** on Historic Building Exteriors

17. **Architectural Character**—Identifying the Visual Aspects of Historic Buildings as an Aid to Preserving their Character

19. The Repair and Replacement of Historic **Wooden Shingle Roofs**

20. The Preservation of Historic **Barns**

27. The Maintenance and Repair of Architectural **Cast Iron**

29. The Repair, Replacement, and Maintenance of Historic **Slate Roofs**

30. The Preservation and Repair of Historic **Clay Tile Roofs**

42. The Maintenance, Repair and Replacement of Historic **Cast Stone**

45. Preserving Historic **Wooden Porches**

Citing these *Briefs* while making the case for your proposed work adds credibility to your proposal. For instance, in making the case for the addition referenced above, you could add:

> NPS *Preservation Brief 14* states, "The new addition may include simplified architectural features that reflect, but do not duplicate, similar features on the historic building," which is what we have done with our proposed addition. The *Brief* also recommends that one should "base the size, rhythm and alignment of the new addition's window and door openings on those of the historic building," which we have also done.

Many communities with historic preservation ordinances publish design guidelines to help property owners meet the ordinance standards. Frequently these are drawn from the NPS *Preservation Briefs*. Quoting local guidelines is even more effective than citing the *Briefs*, as the local guidelines are almost always specifically referenced in the local ordinance.

The more effectively you can walk the review board through your proposed project using the accepted language of the historic preservation field, the greater the chances you will get approval for what you want to do—assuming you are sincerely attempting to meet the standards.

Include Drawings, Photos, and Other Supporting Information

Clear drawings and photographs are essential for communicating what you propose to do. Review boards are rightly hesitant about approving things they are not sure they understand. As described in Chapter 3 of this book, you can produce design drawings by hand, with a computer program, or by hiring a professional. The complexity of the project may well determine your approach.

Digital photography has made the task of supplying clear photos far easier and less expensive than it once was. You will want both overview and closeup photos of the building and the proposed work areas. Views from the public way are particularly important, as most ordinances are concerned only with what can be seen by the public. Historic photographs of the house can often be very useful and should be included if available. Often the easiest way to reproduce them is to take a digital photo of the historic photo and print copies.

If you are proposing to install new windows, it will be important to provide side-by-side drawings and photos of the existing windows and their proposed replacements, so the review board can judge how close to matching they are. If the historic windows are gone and you are returning to a documented earlier window design, use historic photos to show that the proposed new windows are a good match for what was formerly there.

Manufacturer's "cut sheets" should be included with the application to show what new materials look like and are made of. These can include windows, doors, light fixtures, shingles, trim elements, and so on.

Make Sure Your Application is Complete

Double check your application and supporting materials to be sure you've included everything necessary. (Some communities provide a checklist to help with this.) At the same time, make sure you are not burying the review board with information they don't need. Wallpaper samples and paint chips for the front parlor are not relevant to a review of exterior work.

Understand How the Review Process Works

Talk to municipal staff ahead of the meeting to learn how the review will be structured. In southern Maine, for example, a review generally starts with a municipal staff person introducing the agenda item to the board, as in: "Item three on your agenda is a review of proposed alterations at 42 Elm Street. The applicant is Joe Smith, who is here to present his application." The applicant then describes the proposed work and fields questions from board members.

Most boards then allow public comment. Having neighbors speak positively about your proposal is much better than having them speak against, so cultivate support ahead of time. After closing the public comment portion of the review, the board members will discuss the proposal among themselves. You and the public can listen but may not comment. Occasionally the chairperson will allow additional questions to be asked of the applicant if there is confusion about any issue.

Finally, the board will close their deliberations and the chair will call for a motion, which should include

the board's findings of fact relevant to the application, as in: "The proposed work meets the standards of the historic preservation ordinance." The motion will approve, deny, or conditionally approve the application, specifying the conditions in the latter instance, as in: "The porch stair treads must be painted or stained cedar and not plastic composite deck material as specified in the application." The motion, once seconded, is voted up or down.

Be Friendly! They Are Just People

When you are presenting your application, bear in mind that the board members are your fellow citizens, volunteers who give up one or two evenings a month to serve the community in this often-thankless duty. They have taken an oath to uphold the ordinance to the best of their ability, and the vast majority are trying to do just that. Be friendly, listen to their questions, and try to answer clearly and directly. If you don't know the answer to a question, say so. If board members appear troubled by some element of your proposed work, ask for suggested improvements. Architects, contractors, and people who have rehabbed their own homes often serve on these boards. Take advantage of their experience.

Building Permits

In most communities, a project involving more than cosmetic repairs is likely to require a building permit. Examples include additions (including porches), major structural changes, and new bathrooms or kitchens. Building permits are generally less about aesthetics than functional and safety considerations. The building code adopted by your community will determine what is and is not permissible.

Unlike historic preservation certificates of appropriateness, building permits are generally issued by municipal staff without any review board involvement. The application requirements for building permits vary widely, and it is impossible to generalize about such requirements.

Get to Know Your Local Inspectors

The best way to learn the requirements for a building permit application in your community is to ask the building inspections department. Obtain a copy of the application in advance (they are frequently available on the municipal website) and note anything that is unclear. Take it with you to the building inspections department.

Many communities assign inspectors by neighborhood or district. Try to meet with the inspector who will be reviewing the application for your house. Bring a few photos of the house and any plans or sketches of your proposed work to facilitate a conversation. Ask what aspects of the project will require review. Ask for a copy of a "good" application for a similar project that received its building permit. If you are planning an addition or change of use, ask about zoning requirements, particularly required setbacks from property lines. If you are making structural changes or the house needs major structural repairs, ask if the drawings need to be stamped by a licensed structural engineer.

An incomplete application will not be processed and may cause delays in your project. Double check that your application includes all the information and supporting materials required. Many communities now allow or even require electronic submission of permit applications. Make sure you pay the review/permit fee, or your application may sit in cyber limbo for some time before anyone notices it. Fees vary widely across the country.

Read the Code!

Building codes are public documents and usually available on the municipality's website. A baseline understanding of the code can be very helpful when dealing with code enforcement staff. Some codes allow exceptions to many standard requirements for historic buildings, but not all code enforcement staff are aware of this.

Codes generally contain language for determining when the extent of work on a project triggers "new construction" requirements for the whole project—usually when a threshold percentage of the property value is exceeded. Avoid this if possible, because any house built before the widespread adoption of modern building and life safety codes will have numerous features that are not "up to code." These can include the locations of electrical outlets and wall switches, the venting of plumbing fixtures, the rise and run of stairs, the heights of balustrades and handrails, and so on. Bringing all these items up to code is very expensive and can destroy character-defining features of the building. Knowing what the codes require can help you avoid triggering additional alterations simply to meet the code.

Breaking the work into distinct phases can help in this regard. For instance, if the threshold for new construction requirements is 50 percent of the building's value, and the interior and exterior work you plan to do is going to reach 60 percent of the value, you may be able to get approval for the interior work in one year and for the exterior work when the interior is completed, or vice versa. Also, any completed work in one phase will increase the value of the building, raising the bar for triggering new construction requirements in subsequent phases. Working with an attorney who is versed in your local codes and familiar with the local code enforcement staff may be worthwhile when navigating the approvals process.

Dealing with Unhelpful Bureaucrats

Many planning and building inspections staff people are willing to make suggestions for meeting code requirements, but some are not. It can be extremely frustrating to be told that you cannot do what you propose but not how to bring your plan into compliance. An "it's not my job to design your project" attitude can stem from a couple of causes. Perhaps the municipality's legal counsel is concerned about liability issues and has ordered staff not to offer design suggestions. Building inspections staff are seldom licensed architects or engineers and may be considered unqualified to offer advice. Another leading cause of unhelpfulness, sadly, is that people burn out on work that brings them into regular conflict with property owners. It is often not a lot of fun to be the enforcer of rules you had no role in adopting or enacting. Being regularly criticized for doing your job will wear on most people. Try to remember that you are dealing with real people. If a municipal staff person is resistant to offering help, ask to see an application for a project that faced a similar problem and resolved it successfully. This can provide your contact with a way to be helpful without being personally responsible for the suggestion.

On-Site Inspections

Your contact with the building inspections department does not end when you receive your permit. In most municipalities, the work must be inspected at defined points in the construction process: when interior framing is completed but not yet enclosed; when wiring and plumbing runs are roughed in; and when the work is completed. Depending on the scope of the project, you may need an occupancy permit before anyone can legally live in the house.

It is generally a good idea to schedule on-site inspections ahead of time. Inspectors are busy, and you cannot assume they will be available on short notice. Since you don't want to hold up the work for an inspection, make the appointment a week or so ahead of completing a phase of work.

Most rehab projects run into unexpected conditions. These can require changes to plans, and the changes may require approval. If you discover structural issues or other unanticipated work, ask the building inspector to come to the site and confirm that approval is needed. An inspector may be more willing to suggest appropriate solutions on-site. If the issue is straightforward, it may be possible to email the inspector photos of the problem area with a proposed solution and get approval by reply email. However the change is reviewed, be sure to get approval in writing.

Planning Boards and Other Bureaucracies

Planning boards review subdivisions and new construction projects, but not most rehab projects. Investigate what would trigger planning board review of a rehab in your community. One possibility is a site issue such as changing the number of dwelling units in the building (by adding an apartment or reverting an apartment building back to a single-family home) or seeking a variance from setback requirements to rebuild a porch that has fallen off the house. Planning board review may also be triggered by a change of use such as opening a bed-and-breakfast inn or operating a home-based business.

You should identify what zone your house is in, ideally before you buy it, and read the zoning code to understand what is and is not allowed in that zone. The zoning code will specify what changes require planning board review and where variances may be allowed under certain conditions. If there appears to be a serious conflict between your goals and the code requirements, it may be advisable to hire an attorney who is experienced in zoning issues. Sometimes a zone change is possible, particularly if the property abuts a zone where your proposed use is allowed.

Boards of Appeal

Nearly all communities have a board of appeals for zoning issues. If the planning board or historic preservation board will not grant permission to do what you want, yet you feel the code does allow it, your recourse is to appeal the decision to the board of appeals, preferably with the help of an attorney who is experienced in zoning issues.

Private-Sector Bureaucrats

Not all bureaucrats work for the government. Large utility companies, for example, can also be bureaucratic. The location of a meter on the exterior of a house can be a point of conflict, as can the means of bringing electrical power and cable TV into the building. Minimizing the visual impact of these modern necessities is seldom the goal of the utility's standard approach. If your house is in a local historic district, you may be able to get support for a nonstandard solution from municipal staff or the historic preservation board. Sometimes a utility will make exceptions for a recognized or designated historic building. Ask.

Dark Harbor House, 1896

Colonial Revival Style

Dark Harbor House was built as a summer cottage for Philadelphia banker George Philler and his wife, Ellen, in 1896. It was designed in the Neo-Georgian variation of the Colonial Revival style by Fred L. Savage of Bar Harbor, Maine, a noted designer of summer cottages. Large Colonial Revival houses of this type were built across the United States in urban, suburban, and resort locations. Located on an island three miles offshore, the Philler house commanded views of the ocean in three directions. The cottage was built as part of a summer colony development that included a large Shingle Style hotel as well as privately owned cottages. Around 1920, the hotel acquired Dark Harbor House and used it as an annex until the 1950s, when it was sold back into private hands shortly before the hotel was demolished. In 1978, the cottage was enlarged with an expansion of the service wing and converted to a bed-and-breakfast inn by new owners. It continued in that use seasonally until 2006, under several owners.

When the current owners, Bruce and Kerry, first saw the house, they were looking for a two-bedroom summer home on the island that would require minimal maintenance. Although it was not at all what they were seeking, they fell in love with Dark Harbor House, with its elegant Colonial Revival details and its thirteen bedrooms. They could see that the house was suffering from deferred maintenance and had experienced some unsympathetic alterations in its decades of commercial use, including exposed water sprinkler pipes throughout. But they mostly overlooked the failing foundation (due to water infiltration) and consequent structural issues.

Unaware of the seriousness of those underlying problems, they undertook a mostly cosmetic rehab in the first year they owned Dark Harbor House. This included replacing the four obviously rotted two-story columns on the east elevation, stripping the peeling exterior paint to bare wood, and repainting the entire house. Fiberglass replicas of the original wood columns seemed a practical option for Maine's harsh offshore winters. Like all the other building materials for the project, the 20-foot-long columns had to reach the island by ferry. Working from the original house plans (found at the Northeast Harbor Library), Bruce and Kerry had a missing wall rebuilt to restore the floor plan of the principal rooms on the first floor. Framed up plumb and square, the new wall highlighted the crookedness of the old work to such a degree that they had the carpenter angle the new door header to match the historic door openings, to disguise how crooked the house was.

Returning to the island the next summer, Bruce and Kerry found the expensive new paint peeling off much of the exterior. A new contractor refused to undertake more cosmetic work until the underlying structural and moisture issues were dealt with. Accepting the unpleasant truth, Bruce and Kerry told the contractor to do what needed to be done. This involved jacking the entire house off the foundation, slowly leveling it, and then setting it back down on a new foundation. The crumbling chimneys had to be rebuilt from the basement through the roof. Proper gutters and proper drainage around the foundation were installed to ensure that the basement and crawl spaces remained dry. This was essential to rid the house of excessive moisture, make it stable and dry enough to hold exterior paint, and avoid a repeat of the foundation issues in the future.

While the house was still up on cribbing, extensive work began inside, particularly in the wing added in the 1980s. Since that area had no historic significance, Bruce and Kerry felt free to make more extensive changes there. In the historic portion of the house, the wall that had been rebuilt the year before was rebuilt again, this time level and plumb. Plaster had been damaged by the slow settling of the house over time and would have suffered more damage from the leveling of the house during the rehab, so most of the historic plaster was removed and replaced with skim-coat plaster over fiberglass mesh on blueboard gypsum. While the spaces behind walls and above ceilings were accessible, all new plumbing and wiring runs were installed.

The kitchen, which had been reworked for commercial use when the house was an inn, was redesigned to function as a family kitchen with adjacent breakfast room. The floor framing in this area of the house was deteriorated and damaged from past plumbing and had to be replaced.

Finally, after significant foundation and structural work, the owners were able to focus on interior finishes, including the new kitchen and updated bathrooms. Custom beadboard cabinetry was used in the kitchen to be in character with the house. A master bed-and-bath suite was created in the upper floor of the 1980s wing, and the contractor had knives cut to match the historic molding profiles for the new work. The historic principal rooms in the first story and the bedrooms and bathrooms upstairs were left largely intact but restored. Paints and wallpapers were selected to be compatible with the historic character of the house, without attempting to exactly replicate period finishes. Padded fabric wallcovering was installed in the dining room to help address the odd acoustics of an oval-shaped room. A slate laundry sink was relocated from the basement to a mudroom at the rear of the house, where it is used for gardening and cleaning clams headed for the kitchen.

The extensive porches on the house were re-decked with Brazilian ipe wood, an extremely dense hardwood known for its beauty and durability in a harsh climate. New turned balusters and urns were replicated from deteriorated originals where needed. A new patio and gardens were created to extend the living spaces out toward the water and woods. A new timber-framed carriage house/garage was built on the wooded side of the property to store vehicles and property maintenance equipment.

This island home is an excellent example of the Colonial Revival style, gracefully combining Georgian, Federal, and Greek Revival elements. The many porches look toward the ocean.

A glimpse of the new kitchen. The lattice-pane windows are original.

A sweeping double stair brackets the entry hall. The door in the distance across the reception hall opens to a curving porch overlooking the ocean.

The porch continues around the corner to a Greek Revival portico on the east end of the house. Small balconies on the second story open from bedrooms.

A side view of the portico with its new fiberglass columns. A rooftop terrace opens off a bedroom in the attic.

Frederick Savage's original drawing for the driveway entrance façade. *(Courtesy Northeast Harbor Library)*

The entryway in 1994, while the house was being operated as an inn. *(Courtesy Sylvia Chanler Stumpfig)*

This antique tub received new fixtures.

An antique marble and nickel-plated sink was retained and replumbed for this bathroom.

An original bathroom after restoration.

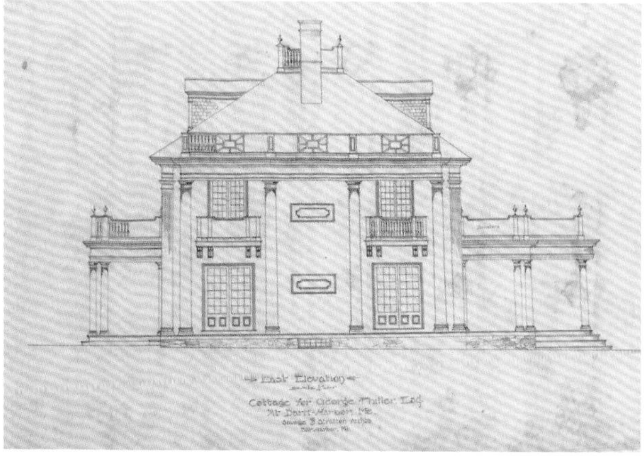

Savage's drawing for the east elevation. *(Courtesy Northeast Harbor Library)*

The pre-restoration parlor, showing the patched flooring where the original wall had been removed.

The parlor after restoration.

A pre-restoration view of a room in the 1980s addition.

One of the many bedrooms after restoration.

In this historic view, the Dark Harbor Hotel is seen in the distance. Dark Harbor was one of many Maine resorts where building lots for summer cottages were developed around a large hotel. The hotel no longer stands. *(Courtesy Maine Historic Preservation Commission)*

The oval dining room design allowed for open china cabinets in each corner.

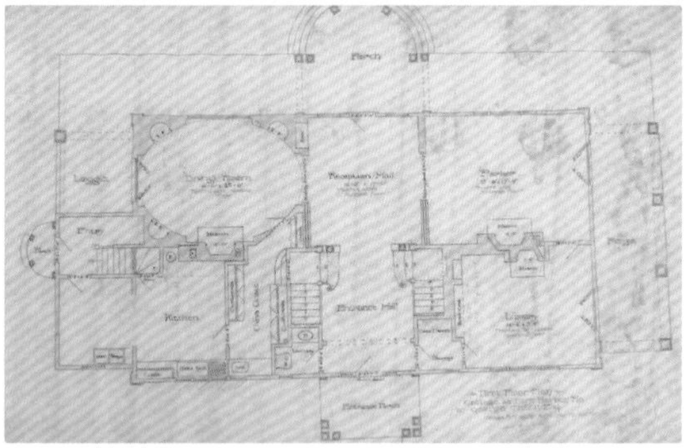

The original first-floor plan.
(Courtesy Northeast Harbor Library)

This early-twentieth-century postcard view shows how the portico and porch sit on a continuous terrace that wraps around the east and north sides of the house, creating extensive outdoor lounging space during good weather. The failing terrace channeled water into the basement for decades prior to restoration, causing significant foundation and structural problems. *(Courtesy Maine Historic Preservation Commission)*

The 1980s addition included this casual sitting room off the breakfast room, overlooking the gardens and ocean.

The view from the reception hall into the entry hall when Kerry and Bruce bought the house.

A detail of the double stair.

This attic bedroom opens onto the deck atop the portico on the east end of the house.

This bedroom opens onto one of the small balconies that project into the portico.

The new kitchen replaced the commercial kitchen installed when the house was operated as an inn. Reproduction period light fixtures are combined with recessed task lighting.

The restored Dark Harbor House at dusk. The portion of the house at right is the 1980s addition. The second-story porch opens off Bruce and Kerry's new master bedroom.

Pre-restoration views of two bedrooms.

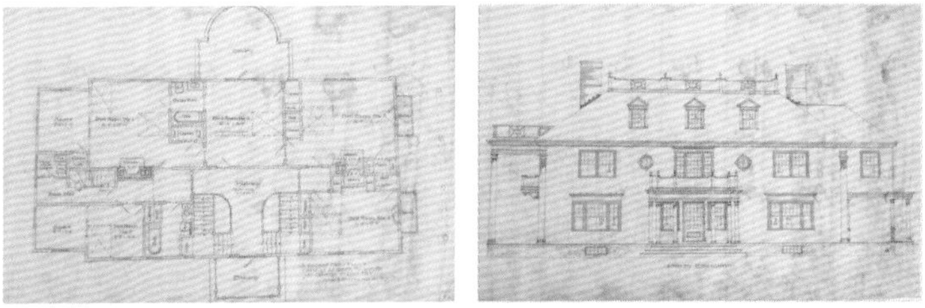

The original drawings for the second floor and the north elevation. *(Courtesy Northeast Harbor Library)*

An antique slate sink in the mudroom near the kitchen is used for gardening and cleaning the mud from freshly dug clams.

An early-twentieth-century view of the driveway front. *(Courtesy Maine Historic Preservation Commission)*

The breakfast room off the kitchen can be entered from the driveway when carrying in groceries.

Architect Frederick Savage's masterfully composed Colonial Revival driveway façade.

A view through the curved north-side porch across the new formal garden, toward the cove beyond the trees.

The restored library is entered though a passageway under the stairs or from the parlor and feels tucked away and secret.

The Work

Here the house is being stripped of paint in the first round of exterior restoration. Unfortunately, unaddressed moisture problems caused the first paint job to fail over one winter.

A view in the basement when the house was sitting on cribbing while the foundation was rebuilt.

The kitchen during restoration.

Looking from the 1980s addition through the dining room and reception room into the parlor. This space became the casual sitting room seen in a previous photo.

The back stairs were wrapped for protection during the restoration.

Demolition in the 1980s addition during restoration.

Temporary supports carry the portico roof while the rotted wood columns are replaced with new fiberglass columns. (Note the rot at the bottoms of the columns.)

A bucket loader was used to lift the new two-story fiberglass columns into place on the patio.

The view from the parlor toward the reception hall while the missing original wall was being reconstructed.

Looking through the space that became the new breakfast room.

PART TWO

Under the Surface

The next two chapters take us below and behind the finished surfaces of the existing house—to remove inappropriate materials, restore original finishes, add new features, and address structural issues.

This is a necessary phase of nearly every restoration project. Unfortunately, it is often the phase in which inexperienced owners (and contractors experienced only in new construction) get carried away and remove far more than is necessary or desirable for a sensitive restoration. Such excesses then require that new materials be purchased and installed to replace what was removed—often, materials of lesser quality than what has just been expensively hauled off in a dumpster. I advocate for removing only what *must* be removed to accomplish project goals, and then we explore potential structural issues and their solutions. We'll use an example project and several case studies to illustrate the twists and turns the work can take.

CHAPTER SIX

The Demo

A reasonably fit homeowner can generally undertake much of the demolition work for a rehabilitation project. It provides an excellent opportunity to learn how houses are put together, since demolition usually follows the construction process in reverse. Proceed slowly and methodically to ensure that salvageable historic materials are not destroyed and that vital structural elements are not inadvertently removed.

Remove no more than necessary. You will save time, money, and possibly your back if you think carefully about what does and does not need to be removed for the repairs and alterations you've planned. If you need access above a ceiling to install new plumbing, can you remove a channel of plaster instead of the entire ceiling? Patching plaster costs less than replacing it, even if the replacement is an inferior material like gypsum board. If water infiltration has damaged the plaster on one wall of a room beyond repair, remove only the plaster from that wall, not all the plaster in the room. Less is more in

A channel has been cut in the ceiling plaster to install new plumbing lines. The plaster will be patched invisibly when the work is complete. Note the staircase in the background that has been padded and cased in plywood to protect it.

demolition—more money, time, effort, and historic material saved.

Messages from the Past

It is not uncommon to find messages from past residents or craftsmen who worked on the house. Be alert to this possibility as you work. Working slowly and deliberately will allow you to notice things you will miss if you are ripping and tearing at a furious pace. Turn over scraps of paper that fall out of walls or ceilings. Look for inscriptions in the back of trim elements or even written in the plaster.

In this room, the plaster ceiling had to be removed to correct the undersized floor joists above, but the wall plaster was left in place.

Water infiltration through the foundation damaged the framing and plaster on the lower portion of the wall of this basement room. Only the damaged portion of the plaster was removed to allow repair of the rotted studs, and then new plaster was installed. A new drainage system outside the foundation solved the water infiltration issue.

Personal Safety

One of the first tools to purchase before starting any demolition work is a good respirator mask for every person who will be on-site. You need one that accepts different filters for different health threats.

For example, a filter that is appropriate for plaster dust will provide no protection against the volatile organic compounds in solvents. The inexpensive paper masks held in place by an elastic band provide very little protection from airborne particles and none from chemicals or toxins. They are no substitute for a good-quality respirator.

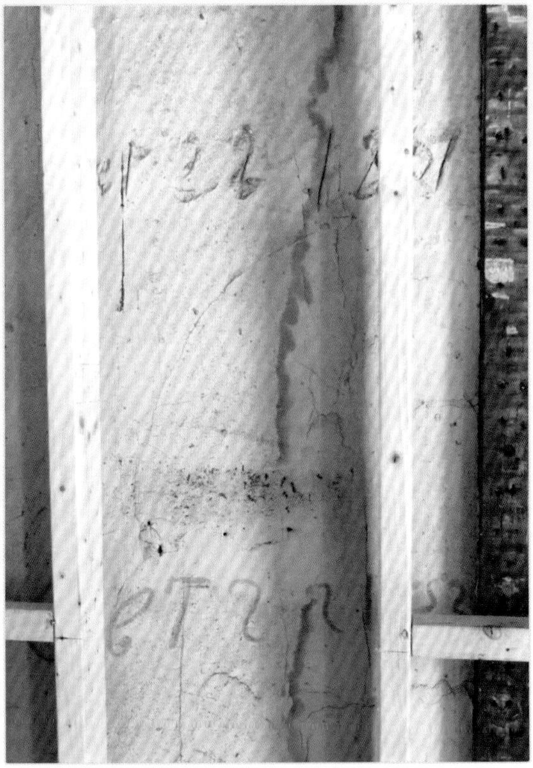

Two dates are inscribed in this plaster that was hidden behind a finish plaster surface. "Sep. 22 1851" was inscribed while the plaster was wet, and "Oct 22 1852" was scraped into the plaster after it had dried. Conclusion: The plasterer did the back plastering in 1851, then returned in 1852 to install the lath-and-plaster finished surface, hiding the inscriptions until the finish plaster was removed in 2016. Combined with deed research, these inscriptions confirmed the construction dates for the house.

Other important safety equipment includes a hard hat, eye protection, ear protection, heavy gloves, and work boots. If you will be handling potentially toxic materials, a Tyvek suit is a wise investment. Get a tetanus shot before starting the work and keep a first-aid kit on site. Remember, safety equipment can only protect you if you wear it. A hardhat sitting on a sawhorse while you drill a 1-inch hole in an overhead beam is not going to prevent a concussion when your heavy-duty right-angle drill hits a nail, binds up, and swings at your head. Trust my experience on this.

Structural Considerations

There are far too many do-it-yourselfer stories about removing a wall only to discover it was supporting the floor above. Learning to recognize the difference between a supporting, or load-bearing, wall and a non-structural partition is important before you knock out that last stud and watch the floor above drop ten inches closer to your head. Load-bearing walls can often be removed and the load supported in

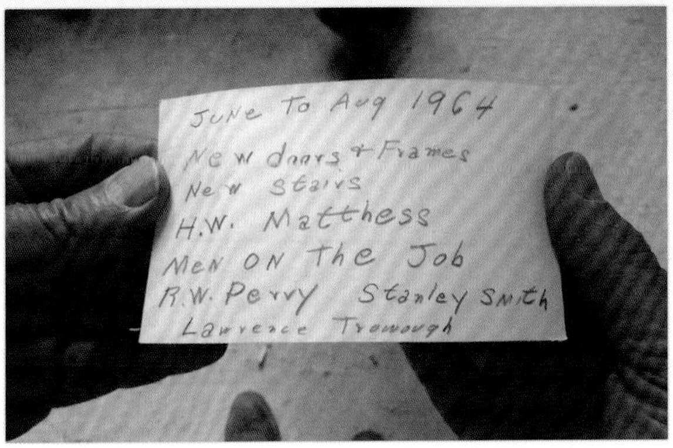

This scrap of paper was concealed in a wall in 1964 and uncovered during rehabilitation in 2010, providing a documented date for the alterations listed on the paper. *(Courtesy Sutherland Conservation & Consulting)*

A major structural beam spanned the width of this long room, with the floor joists for the second story meeting on the beam. The wood beam was undersized and sagged nearly four inches at the center after one hundred-plus years. It had to be removed to be replaced with a steel beam that would be cased in wood to maintain the appearance of the historic beam while providing sufficient support for the load above. Here, 2-inch by 6-inch stud walls have been built on either side of the beam, leaving room to work between them. Below the floor, cribbing has been installed to carry the load to the ground. With these temporary structures carrying the floor joists and the load above, the wood beam can be safely removed and replaced. See the *Beam Replacement Example Project* in Chapter 7 for more on this project.

an alternative way that opens up the space, but you need to work out how to do this beforehand, and you need to put temporary support in place before removing the wall.

Keep in mind that chimneys were often used to support structural elements of a wood house frame. Modern building codes do not allow this, but such codes did not exist when most historic houses were built. Do not remove a chimney without understanding what role it may be playing in the structural integrity of the building.

Load-bearing walls providing structural support to floors, walls, and roofs above can often be identified by looking for posts or foundation walls in the basement, cellar, or crawl space under the house. A load-bearing wall must have support all the way to the ground. In a multistory building, supporting walls will typically be aligned vertically above the foundation wall or posts under the house. Sometimes they will run all the way to the roof framing. In timber-framed buildings, there may be no support walls on the interior, but the frame may be supported by a chimney at the center of the house. The large timbers used in this type of framing (up to 12 inches by 14 inches) can span longer distances without additional support than more recent framing using 2-inch dimensional lumber. In masonry buildings, interior load-bearing walls may be masonry or wood-framed. In narrow row houses, the floor joists typically span from side wall to side wall. In wider row houses, the wall that runs parallel to the stairs, separating the hall from the other rooms, is often a load-bearing wall.

It is critical to understand the structure of your house before undertaking demolition work. The removal or alteration of a load-bearing wall must be carefully engineered, and adequate temporary support must be provided until the alternative support structure is installed.

Other Safety Concerns

Live electrical wires are an unseen but very real danger in demolition work. Wires snake through a house in walls, ceilings, and floors, all running from a circuit breaker panel or fuse box. If at all possible, the main power switch at the panel should be shut off during demolition work. This will kill the power to the entire house, eliminating the chances of cutting into any live wire except, conceivably, the main feed *to* the panel from the meter. It is essential to know where this wire enters the house and how it reaches the panel, because it will remain live even when the main power switch is turned off. You *must* avoid the main feed while working with reciprocating saws and other tools. If your demolition work requires the removal of the main feed, you will have to get the power company to shut off the power outside the house by removing the meter. Then a new line can be run to the panel or to a temporary panel during construction.

Shutting off the power to the entire house, of course, means working only during daylight hours with battery-operated tools or running tools off a generator. This is not always a realistic possibility, especially for a do-it-yourselfer working a day job. If shutting off the power is impossible, you will have to determine which circuits feed the portion of the house you are working in, and which lines pass through that area to reach other rooms. All of those circuits need to have their breaker switches turned off or their fuses removed. Run heavy-duty power cords from other rooms for lights and tools that need to be plugged in. Always check for power in a line with a non-contact AC voltage detector before cutting it, even if you're sure that line is shut off.

Less dangerous (except in combination with plugged-in electrical tools) but still a concern is the

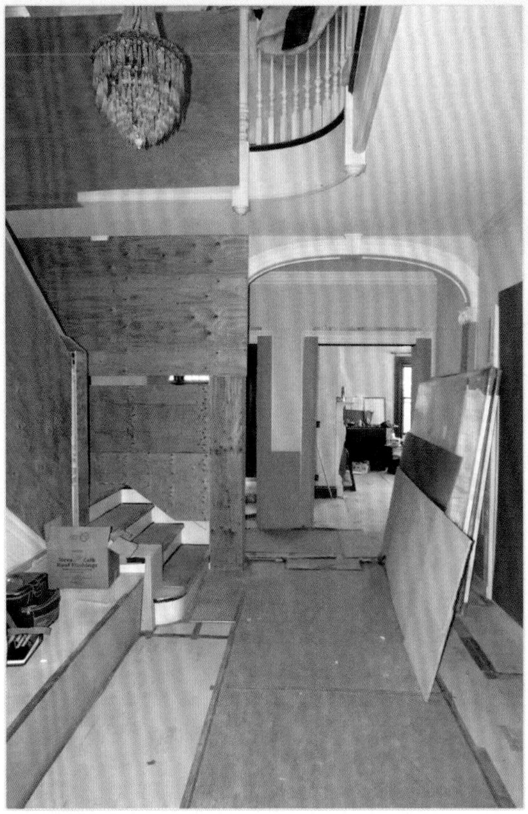

The floor in this formal Colonial Revival entryway has been covered with heavy paper and 3/8-inch plywood where the traffic will be concentrated. The fine woodwork of the staircase has been padded and enclosed in plywood casing to protect it during the project. Staircases often feature some of the most detailed trim in a house and are also the route up and down for workers and materials. Leaving the crystal chandelier in place, well out of the way of workers, seemed safer than storing it elsewhere. Enclosing it in a large plastic bag would have saved someone a tedious cleaning project when the work was completed.

services should be shut off at their main valves before demolition work begins. This is particularly important if you're preserving historic wallpapers, decorative painting, plaster ornament, or other features that might be destroyed or severely damaged by water.

Protecting Historic Features

Before starting demolition, cover floors and encase or pad trim elements. Remove historic light fixtures and store them in rooms outside the rehab area. Cover stained glass windows with plywood on one or both sides as necessary.

These historic plumbing fixtures will be reinstalled once new or expanded bathrooms are ready. In the meantime, they are stored on a padded floor in a room that will be largely untouched during the project. Note the bundled and labeled trim behind the fixtures, also awaiting reinstallation.

These floors have been covered with heavy paper to protect them during the rehab. Large pocket doors have been padded and stacked securely in this room, where little work will take place, and will be reinstalled in their original locations when the work is done.

water you can release by inadvertently cutting into a pressurized water pipe. This includes pipes running throughout the house to supply a hot-water heating system. A reciprocating saw can cut right through a copper water pipe without the user being aware of it until spraying water makes it obvious. (Older black iron water, steam, gas, and sewer pipes take a lot more effort to cut through.) The water and gas

Vermin

The spaces in houses that are removed from human traffic frequently provide shelter to creatures seeking warmth and food. During demolition, you may find evidence of long occupation by small rodents, bugs, birds, or bats. You may find their remains, or you may find them still living in the house. If the house has been unoccupied by humans for a time, you may even meet other, larger critters, such as skunks, racoons, opossums, porcupines, and feral cats.

Larger live creatures can usually be trapped humanely and removed to a more suitable location. Most communities have someone who specializes in this work if you do not want to undertake the task yourself. Pigeons can be particularly trying, as their homing instinct will keep them returning to where they were born. If their access into a building is closed off, they will gather on windowsills, the roof, or a nearby tree, and their offspring will imprint on your house just as strongly. Continually removing nests and eggs will eventually eliminate the problem, but that can take a long time. Calling an exterminator might be your best option. Smaller creatures like mice, chipmunks, and red squirrels tend to remove themselves when work on a house gets underway. Naturally, you will want to seal off their entry points to prevent their return.

Once the critters are gone, you may be left with piles of organic waste to remove. An attic full of bats or pigeons can produce an astounding amount of guano over a period of years. Even if it is old and dry, it is nasty and may contain things you do not want to breathe. A good respirator, gloves, and a long-sleeved shirt are essential for this particular cleanup.

Mold

Toxic mold has gotten a lot of attention in recent years, creating public awareness of the fact that certain molds can make humans very sick. Frequently lost in the message is the fact that most molds are nontoxic and pose no health threat. If you have to deal with mold in your house, it is important to determine what kind of mold it is so you can take appropriate precautions when removing it.

Removing Modern Materials

Often the first step in the rehab of a historic house is to remove modern materials that have been layered over historic finishes. The most common among these include drop ceilings, drywall, carpet, and vinyl flooring.

Removing drywall (also called sheetrock or gypsum board) is relatively straightforward. This material comes in 4-foot by 8-foot sheets, and the butting edges are hidden with drywall tape and joint compound. Nail- or screwheads are also hidden with joint compound. Removing the material in sheets creates far less mess than breaking it up into small fragments. The key is to identify the attachment points and remove the screws or nails holding it in place. A very strong magnet, such as one salvaged from a computer hard drive, can make this much easier.

This fireplace had been sealed up and hidden behind a modern wall finish for fifty years or more before being uncovered during a 2010 rehabilitation. *(Courtesy of Sutherland Conservation & Consulting)*

Slide it over the wall surface and see where it wants to stop. Once you locate a stud, slide the magnet vertically along the stud to find each screw or nail. Joint compound is a relatively soft material when dry, and it is possible to push a screw bit right through it to engage a screw that has been covered. Drywall nails, which largely went out of use in the 1980s, are a bit trickier. A "cat's paw" nail puller will do the job, but a 6-inch-wide putty knife should be placed between the back of the nail puller and the drywall to spread the force of the puller. Otherwise it will simply crush the drywall as pressure is applied. Even with the putty knife, it is critical to make sure the puller is aligned vertically on the stud so the pressure is transferred to the stud and not to a void between studs. It sounds complicated but is as simple as identifying where a nail is located with a magnet, using a hammer to engage the cat's paw with the nailhead, placing the putty knife blade flat between the puller and the wall, and pulling the nail. Once a sheet is free of nails or screws, use a utility knife to cut through the drywall tape and compound and remove the sheet. While somewhat slower than ripping and tearing, the time saved on cleanup will make the slower pace pay off.

Removal of a piece of wood trim installed about 1941 revealed the mid-nineteenth-century wallpaper and border that had been painted over for more than sixty years in the rest of the room.

Keep in mind that the removal of modern materials will often uncover previously unseen historic elements, from fireplaces to wallpapers. Working slowly and deliberately will prevent you from ripping out something of significance before you realize what you have discovered. Keep a camera handy to document discoveries that will have to be removed.

Ceilings

A variety of ceiling coverups have enjoyed periods of vogue over the past half-century. Pressed paper acoustical tile ceilings came into use in the middle of the twentieth century. Like pressed tin, they are typically attached to 1-inch by 3-inch strapping, generally with staples. These ceilings have little or no reuse value, and their removal need not be careful or slow. Once one tile is broken through and pulled down, it is easy to slide a flat bar along the strapping to break the tiles free. With the tiles off the strapping, a cat's paw can be used to pull the nails holding the strapping in place. If the intent is to save and repair the plaster above, the removal of the strapping should be done carefully.

Drop-grid ceilings, or dropped ceilings, use 2-foot by 2-foot or 2-foot by 4-foot pressed paper or fiberglass tiles set in a metal framework dropped below the level of the original ceiling. These became popular for inexpensive renovations in the 1960s, providing an easy way to run new wiring and plumbing without having to open up walls and ceilings. The metal frame hangs from wires attached to screws or nails in the original ceiling framing. With the tiles removed, it is easy to cut the wires and remove the grid.

Carpet

Carpet that is installed with tack strips around the perimeter of a room is relatively easy to remove. Work one corner loose with a pry bar and work along the walls, pulling it free of the thousands of little metal teeth holding it in place. Once the carpet is out of the way, use the pry bar to pull up the tack strips, which are usually held in place with thin wire nails. Wear heavy gloves for this work, because the tack strips are like sharks—angry, hungry sharks. You may want to turn the carpet over and leave it in place to protect the floor if you intend to remove plaster or other materials in the room.

Carpet that is glued down is more difficult to remove. You will have to work under one edge with a hand scraper to get it started and then use an ice scraper on a pole between the carpet and the floor to work it loose. The glue will mostly remain attached to the floor and will have to be dealt with separately.

Here I'm using a plastic dustpan to scoop cellulose insulation out of the bays between floor joists. It is nasty work, but a fan blowing out a window can help keep the dust under control. Shortly after this photo was taken, realizing that my paper dust mask was not nearly adequate for the task, I went downstairs for my respirator. *(Andrew Jones photo)*

Vinyl Flooring

Vinyl flooring is usually glued in place and has to be removed with a scraper, like the glued carpet described above. If the vinyl is glued to a thin (usually 3/8-inch) lauan plywood underlayment, it may be more efficient to remove both at once by prying one edge of the plywood free and working it loose with pry bars. Cut through the vinyl at the joints between sheets of plywood so you can remove one sheet with vinyl attached at a time. This type of underlayment is often nailed into historic flooring with 1½-inch ring-shank nails on a 12-inch grid. Removing it is no picnic, but once you find your method and rhythm, it goes fairly quickly. If you intend to save and refinish the floor under the lauan, use care in removal.

Insulation

Removing insulation is nasty. Goggles, respirator, gloves, and full body coverage are essential for dealing

After scooping the majority of the material into plastic trash bags, I sucked up the rest with a shop vac and emptied the vacuum canister into plastic bags as well.

with fiberglass, cellulose, rockwool, or any number of other materials that have been used for insulation over the past century.

Although particularly unpleasant to work with, fiberglass insulation is comparatively easy to remove, as it can be pulled from stud or joist bays intact and stuffed into large contractor's trash bags. Because it contains a large amount of air relative to its mass, it can be rolled or folded to reduce its bulk.

Loose-fill or dense-packed materials need to be scooped up. If they are in stud bays or between rafters, the interior or exterior wall, ceiling, or roof materials will have to be removed from one side to access them. They will fall to the floor if loose-fill or have to be knocked to the floor if dense-packed. If they are in floor bays, they will have to be scooped out by hand—dusty and unpleasant work. Large contractor's trash bags are good for scooping the material into. Since it is not very heavy (if dry), it is possible to pack more material in the bags by lifting them and dropping them to the floor periodically as you fill them. Gravity will pack the material more densely as the bag strikes the floor. A shop vac will easily pick up whatever material is left after scooping.

Removing Historic Trim and Flooring

Removing historic trim or flooring for reuse requires care and patience but can be done successfully. In general, the approach is to identify the means of attachment and the order in which elements were installed and then reverse the process, taking out the last piece installed first. You will often need to run a sharp blade along seams to break the seal of paint or a clear finish before attempting to pry elements apart.

Most historic trim and flooring is nailed in place. As a rule, it is best to carefully pry the piece loose

These three photos show a carpenter starting to remove the bottom piece from a false ceiling "beam." After gently tapping a thin pry bar between the bottom piece and one side, he applies enough pressure to start separating the pieces. Moving to the other side, he repeats the process and opens the seam a bit further. This continues along the length of the beam. Note the mouse nest exposed partway along the beam. Mouse droppings began falling from the cavity as soon as the seam was opened. Once the board was lowered enough for access, the nest was removed with a vacuum cleaner.

with the nails in place rather than attempting to pull the nails from the piece. Once the piece is loose, the nails can be pulled through from the back, unfinished surface, which will prevent the splintering around the nail hole that is likely to occur if the nail is forced back out the way it went in. This is particularly true of historic square nails.

There is an art to prying loose historic trim. Use a thin pry bar to start the piece at one end, applying pressure until you feel the nails at that end pulling free. Then move the bar along the piece, gently but firmly applying pressure at intervals. Place blocks behind the section you have already freed to keep it lifted and under tension. This will help you with the next set of nails you need to pull loose. As you get a feel for how much tension is required to release the nails, it will become faster and easier to move along a piece of wood. The occasional resistant nail—perhaps one that was driven into a knot or other denser material—may have to be sheared off. Try to position the notch in the blade of the pry bar on the nail shank under the board and strike the other end of the bar solidly with a hammer to cut through the nail. It may take several strikes. Once the trim piece is free from end to end, turn it over on the floor with nails pointing up. Grip each nail close to the board surface with a nail puller or pliers and pull it through the board. The resultant hole can easily be filled after the piece is reinstalled.

These pine subfloor boards are clearly numbered in order with a permanent marker before being removed. Where several shorter pieces equal a single full-length board, the sections are given a letter after the number, working from left to right.

Face-nailed wood flooring is removed in similar fashion. For wide floorboards, you will likely want a large J-hook pry bar. Be careful not to apply too much pressure to one edge of the board; you do not want to split it lengthwise. Lift one side a short distance, insert a block to hold that side, and then lift the other side. Work back and forth to lift the full width of the board evenly. Once the nails on one end are loose, work along the length of the board toward the other end. If the board is wide enough to be nailed in the center as well as at the sides (not uncommon with the 20- to

This piece of windowsill was damaged beyond repair, but the most intact piece of the sill was cut out and clearly labeled to guide replication with new material. Labeling the piece immediately after its removal will reduce the chances of its accidental disposal. *(Courtesy of Sutherland Conservation & Consulting)*

30-inch-wide pine floorboards found in New England), you will have to work the flat end of your pry bar under the board toward the center and then use a block under the bar to lever up and loosen the center nail.

Narrow hardwood flooring is typically tongue and groove, nailed through the tongues so the nails are hidden (see the illustration on page 508 in Chapter 14). Removing this flooring in reusable condition requires a lot of care and patience, but it can be done. The key is to remove the flooring without breaking off the tongues. The first row of boards is the most difficult to remove, and these often end up damaged beyond reuse. Once the edge is exposed, careful prying with the bar directly under the nail location can often accomplish the task. Another approach is to use the notch on the pry bar to cut through the nail under the board, or to use a rotary grinder with a cutting disc to cut the nail under the board. This will produce sparks and should only be done with a fire extinguisher at hand.

Early-nineteenth-century accordion lath was installed by nailing one edge of a board in place and then using a hatchet to split and stretch the board, creating gaps for plaster to key into. As the board was split, the hatchet was twisted to open up the gaps, and the board was nailed to the studs or joists to keep it there. In this view of lath on a ceiling, later sawn lath is visible at left, installed when the original chimney that passed through this ceiling was replaced by a smaller chimney in the late nineteenth century.

It is important to mark elements that need to return to their original locations or be replicated if deteriorated. This can be done with permanent marker on the piece itself or on a piece of tape applied to the piece. Wide floorboards should be numbered in order.

Removing Historic Plaster

Removing a plaster wall or ceiling is relatively straightforward and simple if you observe a certain order to minimize the mess. Using a shingle ripper or ice chipper—essentially a thick 6-inch blade mounted on a long wooden pole—work the blade through the plaster and slide it along the lath. This will separate the plaster from the "keys" that hold it in place. You can work from the top down or the bottom up. Just be sure that you're moving your blade parallel with the lath; otherwise it will go through the gaps and get caught up repeatedly.

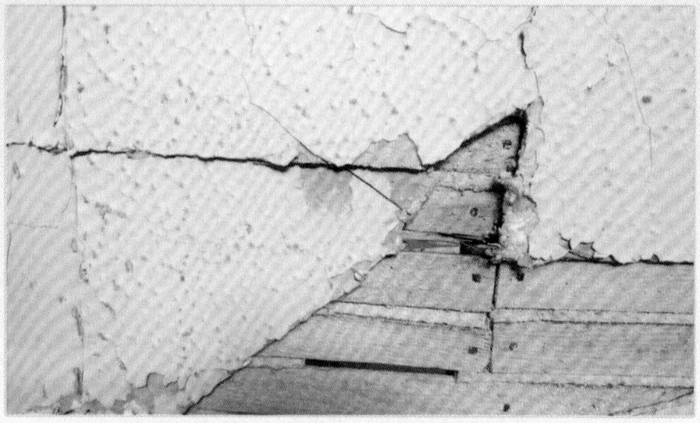

This is the sawn lath found in many tens of thousands of homes built in the nineteenth and early twentieth centuries.

Here, wire lath is seen between the metal corner bead and the wood window trim. Plaster generally keys tenaciously into this material.

Wear an appropriate respirator mask and goggles for this work. Historic plaster is typically made with lime and sand, with a binding fiber such as horsehair. The dust from the lime and sand is caustic and very unpleasant in the eyes or lungs.

If you are removing plaster in a room with a floor that is to be retained, protect the floor with heavy builder's paper, tarpaper, or even sheets of vinyl flooring taken up elsewhere in the house.

Once you've stripped the plaster from the lath, a wide snow shovel is a good tool for scooping it into bags or buckets for removal. Plaster is heavy; don't overfill the bag or bucket. A length of 12-inch-diameter Sonotube from a nearby window to a dumpster below will save a lot of effort in a big job.

Next remove the lath. You will want heavy gloves for handling this very dry, splintery, lime-coated material. Houses from before 1850 or so may have thicker riven or "accordion" lath. Because of the additional thickness, these boards are usually attached with larger nails and take more effort to remove.

The more common sawn wood lath is typically ¼ to ½ inch thick and installed with relatively small nails. If you can get behind the lath and strike it from behind, it will come off the studs easily. You can also use the hook end of a flat bar from the face of the lath to pull or pry it off.

Houses built after 1900 may have various types of wire lath, frequently with gypsum rather than lime plaster. Again, the best approach for removing the material is to run a blade between the plaster and the lath to separate them, clean up the plaster, and then detach the lath from the studs or ceiling joists. With some types of wire lath, the plaster is keyed tightly

This pressed paper lath board with gypsum plaster finish coat was installed in 1932. The paper board has deteriorated significantly due to moisture from a failed roof. The layer of gypsum plaster can clearly be seen at the left side of the broken area.

This gypsum lath board system is essentially the same one used today for skim-coat plaster on blueboard gypsum lath. Today's material comes in 4-foot by 8-foot or larger sheets, whereas some historic lath of this type came in much smaller modules.

The final type of plaster you may encounter in a house built after 1920 is skim-coat gypsum plaster on a gypsum board base. The base may be in narrow two-foot-long pieces or in larger sheets. Unlike the types of plaster discussed above, there is no way to separate the skim coat from the base, which is most often attached to the studs or joists with nails. It is possible to use a magnet to locate the nailheads, or you can cut or knock a hole into the plaster the full height of one stud bay and then work along the wall prying the gypsum board away from the studs from top to bottom. Move from stud to stud, taking off large sections at a time. The neatest way to start this is with a reciprocating sawcut on one side of a stud from ceiling to floor.

When the lath (of whatever type) is removed and cleaned up, chunks of plaster will remain in

through so many small openings that separating them is impossible and the plaster and lath have to be pulled off the studs or ceiling joists as a unit. A reciprocating saw with a metal cutting blade is handy for making vertical cuts in stud bays to expose an edge to work from and to cut the material into manageable pieces.

Although not common, pressed paper lath board was sometimes used as a base for skim-coating gypsum plaster. This system is lightweight and likely was less expensive than other available options. It provides a nice surface but has little resistance to puncturing if hit with any force, and it's readily flammable if fire gets into the stud bays. This may be why it never really caught on.

One section of the precast ornamental plaster on this beam had been removed prior to the beginning of this restoration project. All the remaining sections on the beams in the room were carefully removed to allow structural improvements.

Here, the cast plaster sections are being reinstalled. The plaster restorer made a mold from one section and cast additional sections to replace missing and damaged units. The completed work can be seen on page 650. *(Courtesy Sutherland Conservation & Consulting)*

the stud bays. These are the keys that once held the plaster to the lath, and that fell when you separated the plaster from the lath during removal. These need to be shoveled up and removed.

Removing Other Materials

Ornamental Plaster

Plaster ornament sometimes has to be removed and reinstalled after other work is completed. Try to avoid removing ornamental moldings that have been run in place. Removing these and returning them to their locations is very challenging if not impossible—you may need to replicate rather than reinstall them. Fortunately, much historic plaster ornament was cast in pieces in a factory and installed with screws. The joints and screws were then hidden with wet plaster. Once it dried and was sanded and painted, the patches disappeared. These elements can usually be separated, unscrewed, and removed with minimal damage for subsequent reinstallation.

Pressed Tin

The earliest popular way to cover plaster ceilings (and sometimes walls) rather than repair them was with ornamental pressed tin. Tin ceilings have come to be appreciated as historic in their own right and are generally preserved where they exist. If you have to take one down, remember that it's comprised of modular pieces with overlapped and nailed joints. Typically, the last part installed was a perimeter metal cornice. You will need to reverse the installation process, gently popping loose the nails holding the cornice in place after cutting along the joints between pieces with a utility knife to break the paint seal. A small cat's paw nail puller may work from the face side of the tin, or it may be necessary to work a very thin pry bar under the edge beside the nail and pop the nail by prying up on the tin around it. A pry bar has a notch on the blade's front edge. Work the bar under the tin until this notch is around and against the nail. Typically, these ceilings were installed against 1-inch by 3-inch wood strapping that was attached to the ceiling framing, either directly or through the original plaster. The

This barn is being disassembled to be moved to a new site. Every piece of the framing has had an orange metal disc nailed to it with a code number written on the disc in permanent marker. The numbers are recorded on a drawing of the frame before disassembly to provide a key for reassembly.

nails generally pull loose easily; if any are stubborn, a small grinding wheel, like a Dremel tool, can be used to remove the nailheads on the face side of the tin. This is slow work, but it allows the pressed tin to be removed in reusable condition.

Other Metals

Some metals can be challenging to remove with basic tools. Sheet materials that are nailed or screwed in place can be removed by locating their seams and removing the screws or nails. Soldered sheet material may have to be heated sufficiently with a torch to loosen seams. A fire extinguisher is as important for this job as the torch. Cast iron is brittle and can often be destroyed with a solid sledge hammer hit. This can work for old plumbing drain lines. Any decorative cast iron should be removed carefully and preserved for reuse on your house or another building. Cast iron elements are generally attached with screws or with bolts and nuts. Find the attachment points and remove or cut off the fasteners. A handheld grinder can be useful for this. If the elements are structural, use extreme caution and have shoring in place for whatever is above.

Framing

When removing studs from a non-supporting partition, it is generally best to minimize impacts to the structure. Studs can certainly be knocked loose with a sledgehammer, but there will be less vibration and damage to nearby finish surfaces if they're cut with a reciprocating saw about four inches above the floor and pulled loose by hand, using a pry bar to separate the stud from the nails at the top. Top and bottom plates can also be removed with a pry bar, a large J bar often proving ideal.

The removal of frame elements that are providing structural support requires planning and care. It is essential to provide an alternative means of support for the load; this is covered in detail in Chapter 7, Structural Concerns. As mentioned above, you should label elements that will be reinstalled. For complicated reassemblies, like a barn or house frame, every element needs to be marked and keyed to a set of drawings.

Masonry

Masonry demo typically requires a greater use of force. The material, mortar, and extent of the work will determine whether it can be done with a chisel

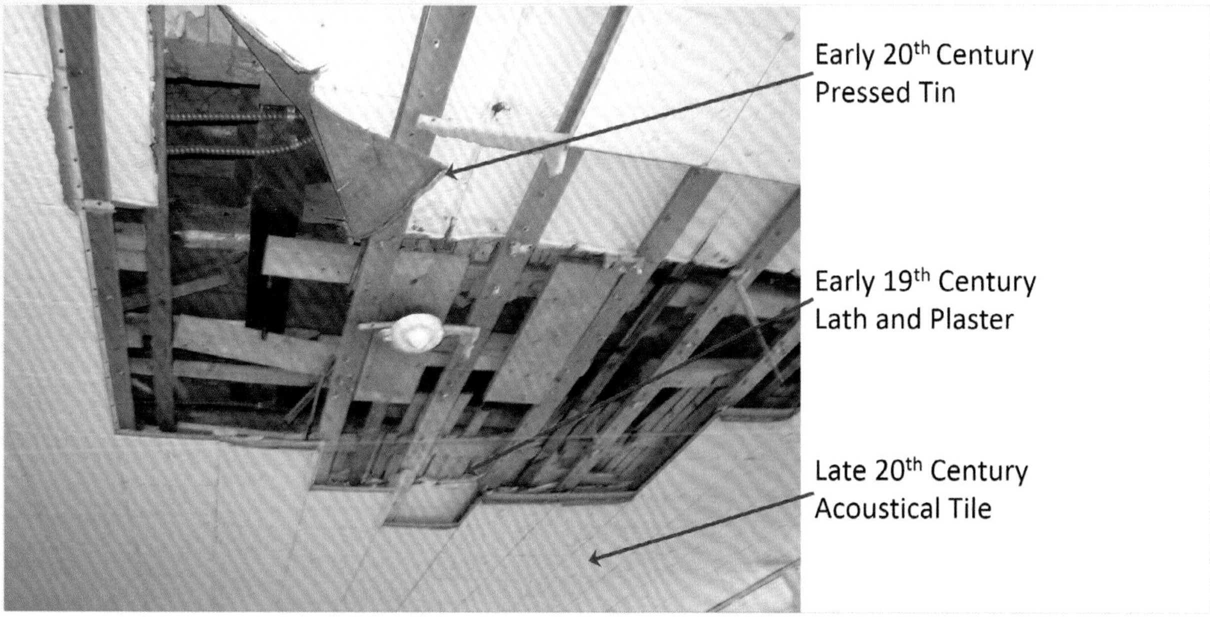

This nineteenth-century building had three layers of material on the ceiling: The original lath and plaster was covered with early-twentieth-century pressed tin and late-twentieth-century acoustical tile. The tin and tile layers were each installed with 1-inch by 3-inch wood strapping. The tin ceiling was contemporary with other surviving finishes through most of the building and was retained where it was in salvageable condition. *(Courtesy of Sutherland Conservation & Consulting)*

and mason's hammer or requires a pneumatic jackhammer. Demolishing masonry above waist level involves significant risk and should be undertaken with extreme caution. It may best be left to a professional. A masonry saw can be useful if a new door or window opening must be cut in a masonry wall.

Multiple layers from several periods

Sometimes the removal of one layer of added material will expose another. In cases like this, it is helpful to have determined your "period of significance" for the house as described in Chapter 2. If the middle layer is related to alterations that are still reflected in surrounding surfaces, it may be more appropriate to restore that layer than to return to the original surface.

Final Word: Stow the Sledgehammer

Despite what you've seen a hundred times on DIY television shows, *always* use the gentlest feasible method when removing materials or architectural elements from a house. It can be a lot of fun to swing a sledge and watch the splinters fly, and it looks dramatic on TV, but every percussive sledgehammer strike causes vibrations to radiate through nearby parts of the building. This can loosen plaster that you want to keep or disturb stable joints in nicely finished trim. With modern reciprocating saws and a selection of appropriately sized pry bars, there is seldom any need of a sledgehammer in demolition. Even brick and stone are demolished more effectively by striking between units with a broad mason's chisel, loosening several bricks or a stone at a time. More of the pieces will be reusable this way as well.

Weston House, 1957

Ranch Style

The Weston House exemplifies many elements of the postwar Ranch-style dwelling, including open living spaces on multiple levels, modern materials, recessed lighting, and an attached garage. Like many postwar homes, it only recently passed from its original owners to another generation, and its period character remains intact. The current owners, Nate and Adrian, have made thoughtful updates to address worn finishes (such as flooring) and elements (such as appliances) while preserving most original features.

The Westons were one of the early families to settle Augusta, Maine, and had been among the prominent families of the city for many generations when, in 1957, Pete and Hope Weston purchased a large, undeveloped property fifteen miles outside the city. In effect they joined a new, postwar wave of settlers who were departing city centers nationwide to populate suburbs carved from rural land. Most new suburbanites moved to developments of multiple houses on lots of an acre or less, but the Westons built their 4,000-square-foot house on a 65-acre lot with sweeping views of a nearby lake.

According to her published memoir, Hope Weston spent three months in 1956 sketching plans for the house on grocery bags before taking the plans to Ward Buzzell, a local architect. She wrote that the architect "basically took the layout we had wanted and then gave us many frills we never dreamed of." The plan features a large living/dining room on the main level, with a kitchen to one side and a guestroom and bath-and-master-bedroom suite to the other. A breezeway outside the kitchen separates the two-car garage from the house. An open stairway in the living room, just inside the front door, descends to an equally large recreation room. Taking advantage of the hillside site, the lower level is fully exposed on the south side with a large patio outside the rec room. Two bedrooms and a shared bath are located under the master bedroom suite. Service spaces are under the kitchen area. There are fireplaces in the living room, rec room, and one of the lower-level bedrooms. The house has two half baths and three full baths.

During the post-World War II period, America's automobiles essentially became family members and moved into the house. Earlier barns, carriage houses, and garages had been separated from the house altogether or set back and attached to a rear ell. The postwar garage was front and center, often appearing more important than the front entranceway. The Weston House was no exception, with a two-car garage sharing a continuous roof with the house, from which it was separated only by a narrow breezeway.

The house received press coverage when new, with the local newspaper writing, "Walls are softened with a textured paper. Natural finished wood trim contributes to the warmth of the room as does the wood sheathed ceiling. The built-in lighting of the cove and the recessed china closet are capable of producing a soft light for general illumination which may be complemented by recessed down lights for accent where needed. The planter on top of the rail surrounding the stairs to the lower level adds a decorative note and a softening of the whole." The recreation room on the lower level, with its pull-out stage for bands, stylish bar, and large area for dancing was described in detail. So was the passive solar orientation of the house, featuring large expanses of glass on the south elevation. Deeply overhanging eaves on the upper level and a sun screen on the lower level prevented the hot summer sun from penetrating the interior while allowing the low-angled winter sun to warm the house. Nearly all these features remain.

Mrs. Weston lived in the house for more than five decades before selling it to Nate and Adrian and their family. The new owners had rehabilitated several nineteenth-century houses but had no experience with postwar architecture. Although the mid-twentieth-century house was not what they had been looking for, they recognized that the Weston House was an important example of its style and might be ruined by buyers who did not appreciate its historic character. They decided to take it on.

Among the challenges they faced were an outdated and highly inefficient heating system; worn interior finishes; failing appliances; and deteriorated roofing and siding. The extensive gardens created by Mrs. Weston were overgrown and neglected, and the view of the lake had long since been hidden by tree growth.

Nate and Adrian first replaced the original kitchen appliances, with a carpenter carefully altering the original cabinetry to fit the new appliances. Modern countertops replaced the worn-out pink laminate, and the section of cabinetry with the most deteriorated finish was painted to complement new backsplash tile. Ceramic tile replaced the original kitchen's pink carpeting.

The off-white Berber carpet in the open living/dining space was worn well beyond its useful life, but new carpet could not be expected to hold up beneath Nate and Adrian's two large dogs. Also, the adhesive securing the living/dining ceiling's original sheets of dark paneling was failing, and the panels were sagging and pulling loose at their edges. To address these issues, the carpet was replaced with dark hardwood flooring—durable and easy to clean—and the paneled ceiling was replaced with painted white gypsum board. The room's original balance of light and dark surfaces was thus retained, though reversed. The grass cloth wall covering was worn and stained but largely intact. It was painted to retain its textural qualities while adding color to the spaces.

Deteriorated asbestos-vinyl tile on the floor of the large, lower-level recreation room demanded an immediate solution. To postpone the cost of asbestos abatement, the new owners left the tile in place and encapsulated it under a new laminate floor. This solution, though not permanent, bought time to address the home's numerous other pressing needs.

The three full baths and two lavatories, all with characteristic 1950s finishes and fixtures, were left intact, including their wonderful period wallpapers. Also intact is the full-wall collage of travel brochures, maps, and souvenirs from the Westons' world travels that their paperhanger created when the house was new. It was showing wear after more than 50 years, so Nate and Adrian gave it a coat of clear finish to protect it for another 50.

Built in a time of inexpensive petroleum, the Weston House was heated with a massive oil-fired commercial boiler, which the new owners replaced with a much more efficient and economical wood-pellet boiler. Heating with wood remains common in Maine, the most heavily wooded state east of the Mississippi, and wood pellets are a locally manufactured, cleaner-burning alternative to traditional cordwood. Modern, hopper-fed pellet boilers do not require the regular feeding of a traditional woodstove.

The septic system needed immediate replacement, but Adrian and Nate have retained the home's unusual low-voltage system for electrical switches. This postwar concept never caught on and is now, more than a half-century later, a complete mystery to most electricians. Replacement parts have proven impossible to locate, but so far—for the most part—the system has continued to function. Given its stylish momentary-touch switches and fancy cover plates, Adrian and Nate hope to keep it in operation indefinitely.

The exterior siding was partly brick and partly naturally finished vertical redwood planks; the latter had suffered from deferred maintenance. Returning the discolored redwood to its intended appearance would have required stripping the cloudy remains of the original clear finish, removing numerous stains and watermarks, and refinishing the wood, a task beyond the scope of available time and resources. Instead the redwood was painted in a color close to the original clear-coated wood, and the original finish was retained in the protected breezeway, where it survived in better condition. The paint can someday be stripped and refinished to match the breezeway siding if desired.

The deteriorated asphalt shingles on the 6,000-square-foot roof were replaced with matching new shingles. The taming of the overgrown surrounding landscape, bringing Mrs. Weston's gardens back to life, is ongoing.

Writing about the house, Adrian, a noted poet, spoke for many lovers of old houses, writing:

> But Hope's house is still beautiful and I love it because it is this outrageously oversized thing made by a woman in an American time that is long gone and will not—and probably should not—return. Hope's house is beautiful because it reminds me in its unwarranted optimism of Walt Whitman's *Leaves of Grass*. It's beautiful because it makes space for azaleas and dogwoods and deer and foxes and turkeys and any other wild thing I might name save a boar. And Hope's house is beautiful because, like too much in America right now, it was in ruins when we bought it—because it needed my husband and me to see and understand what it had been and could be. Hope's house was and is beautiful because it was and is an exquisite burden of a broken thing. Like each one of us and every school and town and city all the way up the ladder to all of America and the whole earth itself, Hope's house was and is beautiful because it is an impossible dream worth believing in and aching for and working on.

The built-in bar in the rec room. Materials of the postwar period include plywood, sheet paneling, acoustical ceiling tile, and decorative fiberglass used as a light filtering partition.

The stairway from the main-level living and dining room down to the rec room opens onto a stone patio with sliding glass doors. A pull-out stage for live bands is stored under the stairs.

Beyond the front door with its 1950s center doorknob, a built-in planter separates the living room from the stairwell.

This bathroom retains its original fixtures and finishes, including the metallic Pegasus wallpaper.

The deck off the kitchen overlooks the gardens, which were overgrown when Nate and Adrian purchased the house.

One of the stylish momentary-touch low-voltage light switches in the house.

A light-filtering fiberglass screen with a butterfly motif provides separation between the rec room and the door to the downstairs lavatory.

The angled mirrors of the vanity in the master bath reflect natural light from the ribbon window above.

The back-lit built-in adjustable shelving behind the bar.

The living and dining room as seen from the kitchen.
A built-in china cabinet is at left.

The pre-restoration living room featured a dark paneled ceiling and a 60-year-old off-white carpet with its most worn sections covered by area rugs.

Given the worn-out carpet and the sagging ceiling paneling, the decision to reverse dark and light materials in the room was inspired.

The wood paneled ceiling remains in the master bedroom.

Nate's home office on the lower level has a raised fireplace on the back side of the rec room fireplace.

The hallway outside the downstairs bedrooms features a collage of maps, travel brochures, tickets, and other ephemera of the original owners' globe-trotting lives.

The kitchen with new flooring, appliances, and countertops. Cabinets were painted as necessary.

Two photos of the pre-rehab kitchen. Though intact, the appliances, countertops, and pink carpeting from 1957 were at the ends of their useful lives.

An uncluttered view of the worn carpet and sagging paneled ceiling of the pre-rehab living and dining area.

This hand-colored black-and-white aerial photo of the house was taken shortly after it was completed in 1957.

The Work

Here the new hardwood floor has been installed, and the ceiling paneling is being removed.

Taping the seams of the new gypsum board.

The new gypsum-board ceiling is being installed.

Priming the new ceiling.

Painting the worn and faded grass-cloth wall-covering after the ceiling was completed.

CHAPTER SEVEN

Structural Concerns

If your house has structural issues or you are planning structural alterations, it is essential to understand how the house was built. In general, early wood-framed houses were built with heavy, widely spaced timber framing, and later houses were built with much lighter (typically 2-inch-thick), more closely spaced framing.

In masonry houses, the wood framing for floors, walls, and roofs is typically supported by the brick or stone walls. Early masonry houses, like their wood counterparts, will likely have larger, more widely spaced framing than later houses. More recent masonry houses may actually be wood-framed with only a brick or stone veneer on the exterior.

Whether wood or masonry, early houses may have their structural timbers partially supported by chimneys.

This nineteenth-century story-and-a-half kitchen ell frame is typical of timber-frame construction of the later eighteenth and first half of the nineteenth centuries. Large vertical posts join with horizontal bottom and top plates using pegged mortise-and-tenon joints. Large crossbeams tie the side walls together at the posts. Angle beams at the junctions of major timbers make the frame more rigid and stable. Smaller but still substantial vertical wall studs are set into mortises in the plates. At left in the near wall, a door opening is framed with a header running between the first stud and the corner post (obscured in this view). Window openings are framed in a similar manner, but with a sill added.

Historic Framing Methods

A cellar, basement, or crawl space is often the best place to get a view of the framing that supports the house. In houses built before 1850 or so, you will likely see large square- or rectangular-sectioned sills and other major beams that show adze marks from hand-hewing or the parallel saw marks of a vertical (not circular) saw. These timbers will be tied together with traditional joinery, most commonly mortise-and-tenon joints. Smaller floor joists will be set into notches cut into the tops of the sills.

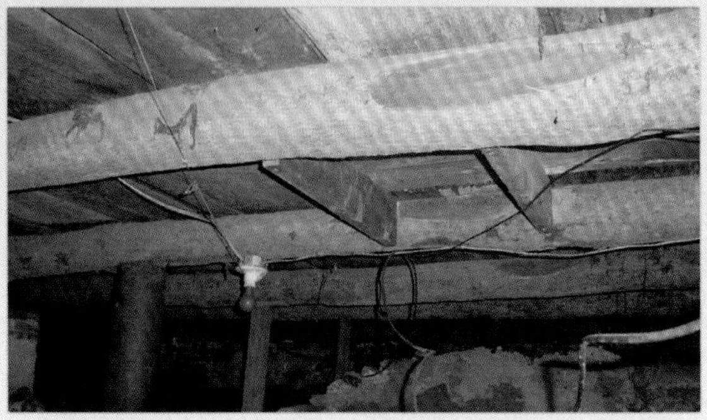

This early-nineteenth-century floor framing combines squared timbers with logs flattened on one side and set into notches in the beams. This was a typical New England solution for floor joists in the period, since log joists saved considerable effort. More substantial houses were likely to be built entirely with squared timbers.

You may find floor joists that are round logs flattened on the top surface. Keep in mind that joists may have been replaced, particularly if the cellar or crawl space is damp. I live in a neighborhood built over a layer of marine clay, which prevents water from soaking down into the soil. Nearly every old house in the neighborhood has a wet cellar and has had at least

In the cellar of a substantial house built about 1810, the 2-inch-thick floor framing is clearly a later replacement, as such framing was not in use when the house was built. The sills surrounding the cellar show mortises and pockets for the original timber framing.

some original heavy floor framing replaced with 2x dimensional lumber.

An unfinished attic with exposed roof framing is another place to view the structural system. The roof framing in a house built before about 1850 would likely have been heavy (8-inch by 8-inch or larger) timber rafters set 10 or more feet apart, supporting smaller horizontal purlins (4-inch by 6-inch or larger) that carry sheathing boards installed vertically. This technology shifted around 1850, when roofs began to be framed with smaller but still substantial rafters set closer together—for instance, 4-inch by 7-inch rafters set 30 inches apart—and horizontal sheathing boards. These rafters might be attached to their opposite members at the ridge with traditional joinery or with large nails. Later common roof framing, still used today, uses even smaller rafters (typically 2 x 10s or 2 x 12s) placed more closely (usually on 16-inch centers today).

Technological developments in the mid-nineteenth century increased the speed and efficiency of sawing lumber and led to the development of framing techniques that used much lighter lumber, typically 2 inches thick. Balloon framing with 2 x 4 studs running uninterrupted from the sill to the eaves was the first widely used version of this light framing. Studs were typically spaced between 16 and 24 inches on

The nineteenth-century kitchen ell pictured above illustrates an early timber-framed roof construction, in which large rafter beams set 8 feet apart support smaller horizontal purlins to which vertical sheathing boards are nailed. The intersections of the major rafters with the wall top plates and with their opposite mates at the peak require pegged mortise-and-tenon joints. This framing, common before the middle of the nineteenth century, was eventually supplanted by newer roof framing techniques, sooner in some regions of the country than others.

center. Floor joists were typically 2 x 10s with similar spacing. Upper-story floors were "hung" from the studs on 1-inch-thick ledger boards that were either set into or nailed to the studs. All joints were nailed. Rafters were also 2 inches thick and spaced 24 inches or less apart, and they sat on and were nailed to doubled 2 x 4 top plates attached to the tops of the studs.

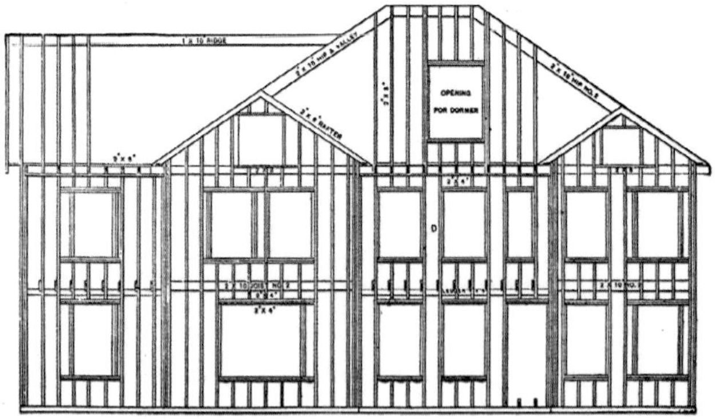

Fig. 146.

This illustration shows balloon framing for a house. The 2 x 4 wall studs run continuously from a bottom plate to the roof. Upper-story floor framing sits on ledgers set into or nailed to the studs. *(From Light and Heavy Timber Framing Made Easy, by Fred T. Hodgson, Chicago, 1909)*

Balloon framing eliminated heavy timbers and their complicated joinery, allowing house frames to be erected by smaller crews of less skilled workers. It also lent itself to the complex forms of many Victorian-era houses, with their projecting bay windows, round towers, multiple intersecting gables, etc. The transition from timber to light framing occurred quickly in much of the United States, advanced by steam-powered sawmills and expanding railroads. By the end of the nineteenth century, traditional timber framing was mostly limited to barns, although it enjoyed a bit of a revival during the Arts and Crafts period, when exposed timber framing with complicated joinery was occasionally used (most often for ceilings).

This mid-nineteenth-century kitchen ell shows the transitional period of roof framing, with horizontal sheathing boards nailed to smaller rafters placed 3 feet apart. In this approach, rafters are often simply nailed to the top plates of the walls and to each other at the peak, a much less labor-intensive system than mortise-and-tenon joinery.

Platform framing was a refinement of balloon framing that eliminated the very long studs running from sill to eaves. With platform framing, each story was built in succession, with the floor of the next story sitting atop the wall studs of the one below. This further simplified the construction process and coincidentally eliminated the uninterrupted stud bays that could allow a fire to quickly spread from basement to attic. Platform framing is still widely used in residential construction today.

Regardless of the framing, there are likely to be interior walls or posts that are supporting floors

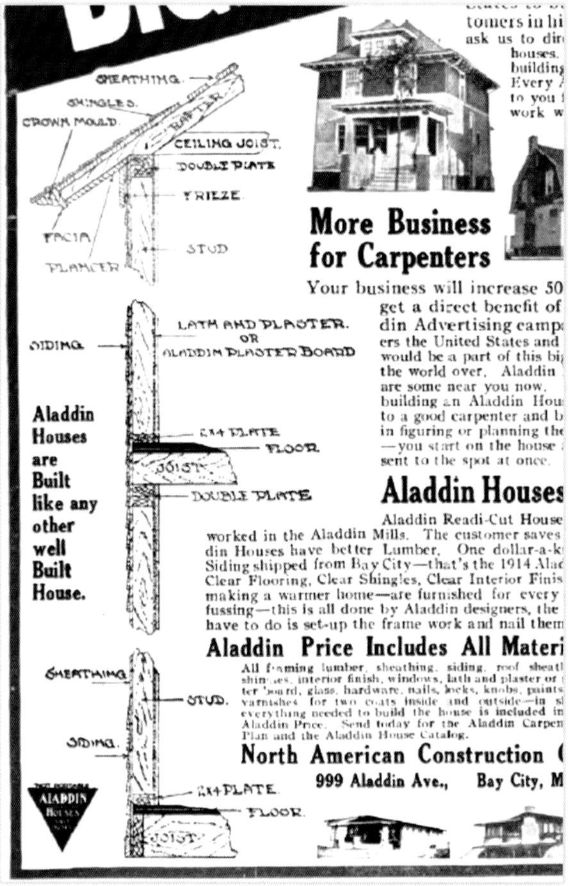

This framing illustration from a 1914 advertisement for Aladdin kit homes clearly shows how platform, or Western, framing separates each floor with framing that sits atop the walls of the floor below. In addition to using shorter, more easily transported and handled 2 x 4s, the method also created a fire break in the wall cavity. This system for framing houses remains in common use more than a century later.

and walls above. These should be identifiable in the basement or crawl space by the presence of a line of piers or posts or a masonry wall inside the foundation perimeter. There is usually a large beam (often assembled from multiple 2x planks nailed together after 1900) sitting on the piers or posts beneath the bearing wall. Interior load-bearing walls shorten the spans of horizontal joists, allowing joists to be smaller than would otherwise be necessary.

Common Structural Issues

Structural issues can be the result of damage that has occurred over time or, sometimes, flaws in the original construction. The three greatest threats to the structural integrity of a house are water, insects, and plumbers, with ill-conceived past renovations running a close fourth. Some issues are obvious and require solutions that can be undertaken by a competent do-it-yourselfer. Others may require the services of a licensed structural engineer and professional contractor to solve.

Look for evidence of structural issues before buying a house. Minor issues are to be expected in any house that has stood for a century or more and can usually be addressed without a serious financial impact. Structural issues, however, can add tens of thousands of dollars and months of additional time to your project. In an ideal world you could rely on the home inspector the mortgage company insisted you hire to identify structural issues, but in my experience these inspectors are focused primarily on surface issues that might derail the closing. Many are more likely to report peeling paint than the water infiltration from failed flashing that is causing the paint to peel. If you have the opportunity to select the home inspector, do your research and find one who understands old houses.

Signs of potential structural concerns include cracked foundations; cracked plaster (especially in masonry buildings); sloping or bouncy floors; obviously out-of-plumb walls; evidence of long-term moisture on floors, walls, or ceilings; evidence of

Hidden behind a trim board that was undamaged on the surface, carpenter ants destroyed the structural integrity of this porch sill beam supporting a two-story column. The beam will have to be replaced.

wood-boring insects; or wood framing members that can easily be penetrated by an awl to a depth of more than a half-inch. Nearly all old houses have some cracked plaster and floors that are out of level or show a bit of bounce; these are not necessarily signs of foundation or framing problems, but they should prompt you to look more closely to ensure that there are no deeper problems in play.

The structural steel reinforcement installed in the basement of the Hench House in the 1940s helped the house survive an extended period of neglect and water infiltration in the late twentieth and early twenty-first centuries.

It is not uncommon to find evidence of old structural issues from problems that no longer persist. If doorframes are clearly racked and have sloping top trim, but the doors were trimmed to fit the openings long ago and function well, it is possible that the house settled a century ago and has been stable since. If removing wallpaper exposes extensive cracks in the plaster that were patched long ago, but the cracks haven't reopened and the patches are still well adhered, it is unlikely the cause of the cracking is still a concern. A building that has stood through multiple generations may have had significant repairs made already. The Hench House (featured on the cover of this book and shown beginning on page 103) is an example: Although it had experienced decades of neglect and significant water infiltration before its dramatic restoration by the current owners, its structural components survived in relatively good shape due to major structural repairs made in the in the 1940s. The steel beams and posts installed at that time were undamaged by the later neglect.

Virtually any structural problem *can* be fixed; the question is whether it *should* be. The outcome for an exceptional house may justify the effort even if the resultant cost is higher than you would like. A run-of-the-mill old house, however, might not be worth the effort unless its selling price accurately reflects the necessary expense. If you love the house and the property, a larger investment may feel perfectly justified.

Foundations

A solid, firm foundation is essential for a stable house. Historic foundations are typically stone below grade and may be stone or brick above grade. Brick was rarely used below grade because it will absorb water from the soil and eventually crumble. Twentieth-century foundations are more likely to be concrete, either poured or block. Many historic houses were built over a crawl space and supported on timber, brick, or stone piers sitting on stone footings. In the Northeast, granite slabs on edge were often used above grade, standing on a rubble stone foundation or on flat footing stones. Southern houses were often built on tall brick or wooden piers on stone footings, with lattice around the perimeter to allow airflow under the house. In areas with wet or boggy soil, masonry houses were sometimes built with wood pilings below the footings.

While many historic foundations remain in excellent condition more than a century after construction, others were insufficient when new or have been compromised since. These will generally need to be repaired or even replaced.

When a foundation is replaced, the local building code will likely require the new foundation to meet current requirements, meaning that replacement in-kind may not be an option. Because much of a foundation is below ground and invisible, a change in material and design should present no aesthetic problem as long as the new foundation allows an above-grade facing that matches the historic foundation, whether brick, slab stone, ashlar block, rubble stone, wood lattice between brick piers, or something else. Even slab granite can be sliced into 2- or 3-inch-thick facing stones by a stone yard.

A foundation can sometimes be replaced piece by piece in short sections, with the house sitting on the foundation throughout. Often, however, the entire

When the eighteenth-century rubble stone and granite slab foundation of this building was replaced with a new concrete foundation, the facing 2 inches of the original granite capstones was sliced off and installed in a recess at the top of the new concrete wall. The cornerstones were cut to create L-shaped facing stones. There is no exterior evidence that the foundation has changed.

Inside the new foundation shown on the previous page, the crawl space is now insulated and dry, with a concrete floor, walls, and piers. The eighteenth-century floor framing above will last indefinitely in these conditions.

This view shows some of the wood cribbing and steel beams used to lift Dark Harbor House during foundation replacement work. The house is featured beginning on page 211. *(Photo courtesy the homeowners)*

This Shingle Style cottage has a fieldstone base and piers forming its front porch. (See the Beam Replacement Example Project, page 270, for more about this cottage.) Much of the stonework sat directly on ledge that was at or close to the soil surface, but some sections of the wall were built with a shallow stone footing supported only by soil. After 121 years, the wall was showing clear signs of movement, and the homeowner elected to address the underlying issue rather than repointing the stones and hiding the problem for another several decades. Here, the area in front of the wall has been excavated to expose where new concrete footings are needed and where the wall is sitting on ledge and needs no additional support. (Note the ledge under the feet of the mason.) New reinforced concrete footings were poured with rebar inserts to tie the wall to the underlying footings. A new footing drain running to daylight was installed along the footings before they were buried.

This early-twentieth-century board-form concrete foundation has been destroyed by insufficient drainage, which caused subsidence of the soil along the concrete, which further increased the amount of water against the foundation and ultimately undermined the footings. The owner's attempt to hide the problem with wood lattice did nothing to help. The foundation needs to be replaced.

Insufficient drainage around this mid-eighteenth-century house was causing runoff from the gutters to enter the basement, threatening the structure. A French drain failed to halt the problem, so a dehumidifier had to be operated continuously (and expensively) in the basement. A new, more involved solution meant installing catch basins for the downspouts and a drainage system running downhill to daylight some distance from the house. While this trenching work was underway, a trench was dug to the street to allow electric and telephone lines to run underground to the house, an aesthetic improvement that also eliminated a route used by squirrels to enter the eaves.

house must be lifted on steel beams and cribbing. Three of the featured houses in this book underwent this process, one more than seventy-five years ago and the others during recent restorations. Any foundation replacement should involve a structural engineer and/or a foundation contractor with experience in this work. To lift a house off its foundation and support it without significant damage to plaster and other interior finishes requires skill, experience, and patience.

This section of an 8-inch by 8-inch post shows how insidiously insect damage (powderpost beetles, in this case) can advance without showing any evidence on the surface of the timber. Water infiltration through poor flashing where a roof met a wall allowed this to occur. Wet wood is much more vulnerable to insect damage than dry wood.

Repairing or reinforcing a foundation rather than replacing it may provide more leeway on code issues. This can sometimes be accomplished by pouring a new concrete footing and foundation wall inside or outside the historic wall (below grade) using forms to contain the wet concrete during pouring and curing. Alternatively, a reinforced concrete wall can be installed using shotcrete (pressurized concrete sprayed from a nozzle) without the need for forms. Either approach will permanently cover a historic foundation wall.

Much damage is the result of insufficient drainage away from a foundation. Whatever approach is taken with repair or replacement, proper drainage must be a part of the solution to prevent future foundation issues and to minimize moisture levels under the house. Rot and mold do not thrive in dry spaces.

Sills

Wood sills atop the foundation support the first-floor joists and the exterior walls. Their location makes sills particularly vulnerable to damage from water and insects (which rarely attack dry wood). Sections of sill under doors are particularly prone to deteriorate due to water dripping from the eaves and splashing on the door stoop. Even covered porches can facilitate sill rot if the porch deck is not sloped away from the house to drain rainwater and snowmelt away from the sill. Anywhere the grade of the soil is as high as or higher than a nearby foundation invites a rotted sill, as does shrubbery that is too close to the house for airflow to keep the wall dry. Gutter downspouts that do not direct water away from the house may also lead to sill rot.

Other Damage

Sills are the most common but by no means the only areas where water and insect damage occur. Roof and plumbing leaks frequently cause wet and rotting

Aluminum panning and vinyl siding were hiding significant moisture and insect damage on the corner of this nineteenth-century house. Pulling back the panning at the base of the paneled corner board revealed the damage.

structural members. The dripping of condensation from toilet tanks can rot the flooring around a toilet. Toilet drain seals, bathtub drains, and tub surrounds with poor caulking are other sites of potential structural problems.

Insufficient ventilation under a house can allow moisture from the soil or from water seeping into the basement or crawl space to raise humidity levels high enough to rot floor framing. Soil that is allowed to come in contact with wood framing members in a crawl space is equally a danger. One of the problems with vinyl and aluminum siding over trim is that it can hold in moisture and hide insects doing structural damage. Probing for soft wood with a sharp awl behind vinyl or aluminum siding along the sills may tell you whether it is prudent to remove some of the cover-up material to explore more fully.

Damage from Plumbing, Heating, and Electrical Installations

Plumbers and heating contractors do an enormous amount of unintentional structural damage to houses. Joists and beams are obstructions to their work, often located exactly where a drain line or duct "should" go for the most efficient installation or operation. Nearly everyone who has worked on historic houses has at least one horror story about structural problems caused by indiscriminate drilling, cutting, and hacking of joists and beams to accommodate pipes and ducts. In the crawl space under my own house, the heating contractor installing the furnace and ductwork for a forced hot-air system in the 1960s notched floor joists to make room for the plenum atop the furnace and pushed dirt out of the way of his ducts,

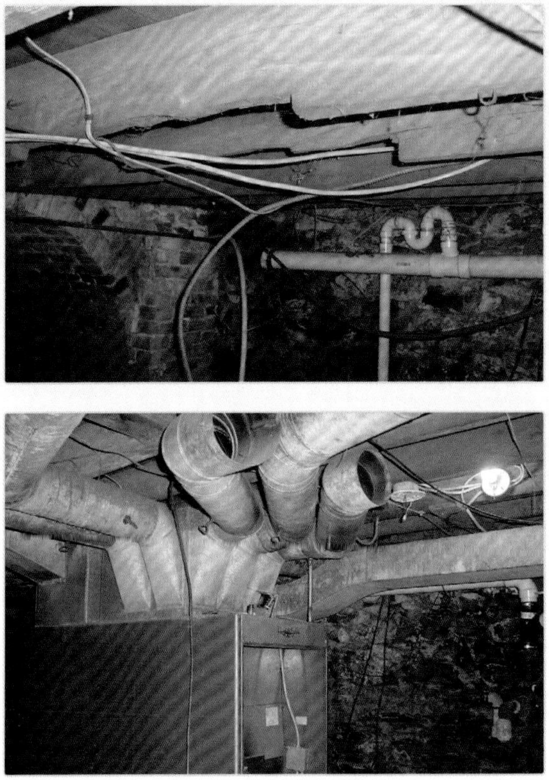

These joists were notched to make room for the plenum of a forced hot-air furnace installed in the 1960s, weakening the joists. The furnace had been removed before the upper photo was taken.

piling it to within inches of wood flooring beams. By the time I came along fifty years later, substantial structural repairs were needed as a result of the long proximity of damp dirt to the wood framing.

Electricians seldom cause as much damage, having smaller and more flexible lines to work with, but are not immune to unthinking installations. I know of one case where a number of tenons (pegs) were knocked out of timber framing joints so the electrician could run his wires though the pre-drilled holes rather than taking the time to drill his own holes through the beams. Needless to say, this was a bad idea.

Damage from Previous Renovations

As mentioned frequently in this book, few houses exist today exactly as they were built. Changes made over time—especially if extensive and thoughtless—may impact the structural integrity of a house. A structural issue caused by a later alteration is addressed in the "Replacing a Beam Example Project" below. Chapter 3 of this book discusses the importance of understanding how your house came to assume its present condition. That analysis can help you work out how structural issues originated and how they might be resolved.

Structural issues are more likely to result from subtractive than additive changes. Taking out a wall can cause the floor above to lose needed support, while adding an interior wall rarely weakens a structure. An addition to the side of a building will seldom cause issues in the original structure (unless the joint is not waterproofed well). Additions on top of existing structures, however, can create structural issues if the original building is not strong enough to support the additional weight or if the added weight is not supported all the way down to the ground.

Do It Yourself or Hire a Professional?

Many do-it-yourself rehabbers have successfully made structural repairs on their houses, while many others have opted to bring in professionals for structural work. Assess your competence, comfort, patience, and time for this type of work and, as always, be realistic. If you opt to tackle the work yourself, it may be prudent to have a structural engineer review your plan and confirm that it is safe and will accomplish your goals.

The work can be dangerous, particularly if you will need to lift or support an enormous weight. I learned this lesson early in my career when, working with my boss, we jacked a support beam under a house in order to replace a deteriorated wood post with a new basement bearing wall. The boss in this case was the property owner, who was acting as his own general contractor and had little more experience than I did. In retrospect, we did almost everything wrong. Once the jack was in place we removed the deteriorated post and from that moment on were entirely dependent on the single bottle jack we were using under a 4 x 4 to lift an 8-inch by 10-inch beam beneath a two-story bearing wall. We had placed no shoring on either side of the jack to prevent the beam from dropping back down if the jack or 4 x 4 failed. The jack was sitting on the concrete floor without any sort of jacking plate to spread its load. We were not wearing hardhats or any other protective gear, and we were working by the insufficient light of a single droplight. As we jacked the beam to remove its sag, it became clear that the 4 x 4 was under extraordinary stress. Nevertheless, we decided to jack it just a little higher. I do not recall the exact cause—perhaps the concrete floor gave way on one side of the jack—but I recall vividly that one second I was standing near the jack and the next second I was lying on the concrete floor ten feet away, in darkness, with water spraying everywhere. When the 4 x 4 kicked out, it had knocked me across the basement and taken out a copper water line. The droplight hit the concrete and went out. The boss was not hit but was nearly as stunned as I was. Remarkably, I was not badly hurt and quickly realized that the beam carrying the bearing wall above us was now unsupported. I scrambled to find a post to jam under it while the boss scrambled to find the shutoff for the water line in the dark. My bruises faded eventually, but the enormous respect I gained for the forces created by jacking under a house has never left me. We were extremely fortunate not to be killed or seriously injured.

The single most important advice I can give for jacking in or under a house is to go slowly—very, very slowly. When something needs to move any distance, use multiple jacks over the length of the member to be raised, shore surrounding structural members, and plan to spend weeks or months raising the jacks in small increments. Movement of 1/8 inch per week is not too slow. This is even more important when there is relatively intact plaster and trim above the jacking, and it is critically important when there are historic decorative finishes such as period wallpaper or decorative painting above. The building very likely took a long time to settle into its current position. Framing joints and finishes accommodated that change as well as they did because it happened slowly. They will not just snap back into their original places if the settlement is reversed quickly. Things will come apart. Go slowly.

Basic Structural Repairs

The following case studies address five basic and common structural repairs that can often be undertaken by the owner of an old house: jacking to remove sag from floors and provide a solid footing to prevent future sagging; sistering joists to stiffen and strengthen a floor; replacing a sill; pulling together a separated timber sill corner joint; and strengthening under-engineered roof framing.

CASE STUDY 1

Strengthening Floor Framing in a Crawl Space

1

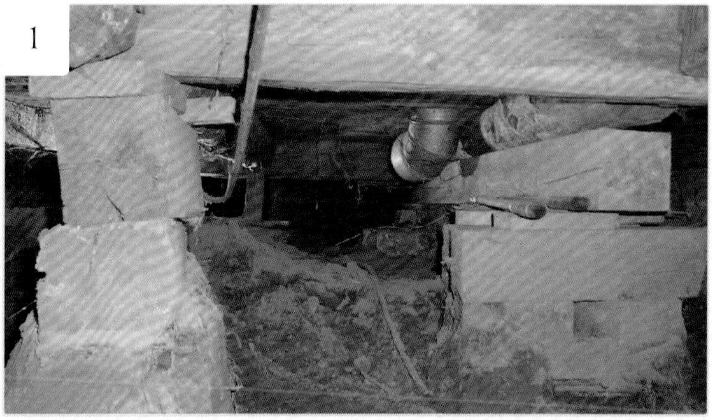

In this crawl space under an early nineteenth-century house, dirt had been shifted around when ductwork for a hot-air system was installed in the 1960s. The resultant close proximity of soil to floor framing in a damp space accelerated deterioration of the structure. Over the following decades, efforts were made to strengthen the bouncy floors with wood blocking stacked on the damp soil—a "solution" with a built-in failure mode. This photo was taken while the space was being excavated to a uniform 28-inch clearance. All the ductwork and blocking had to be removed to make way for more appropriate long-term solutions.

Case Study 1 continued on next page

CASE STUDY 1 (continued)

2

3

Once excavated, numerous jacks were set up on precast concrete blocks throughout the area with out-of-level and bouncy floors, and the floors were slowly jacked closer to level. All the first-story doors—which had been removed and stored in the attic in the 1960s—were returned to their respective openings (as determined by matching hinges with the mortises cut for them in 1827), and the jacking continued until all the doors closed properly. At this point the sagging floors had been returned to their 1960s position.

4

5

Once the floors were jacked to the desired height, eighteen new 6 x 6 pressure-treated posts were installed on precast concrete footings. The bottoms of the posts were wrapped in Bituthene membrane roofing to ensure that no moisture could wick into the wood. Subsequently, the soil was covered by a heavy polyethylene vapor barrier to reduce the moisture rising from the soil into the space.

CASE STUDY 2
Sistering Joists

1

The 2 x 10 joists supporting this floor were 1940s replacements for the original timber framing, which had likely deteriorated in this damp cellar. The replacement joists were inadequate to the span and had been further weakened by thoughtless installation of a heating system.

2

This joist was notched for more than half its height to accommodate the fill pipe for an oil tank installed in the 1960s.

3

New 2 x 10 pressure-treated joists were sistered to the existing joists. Here a new and old joist are clamped together to hold them in place to be screwed together. In a dry environment it would not be necessary to use pressure-treated members, but it was not clear that this cellar could be made sufficiently dry.

4

In this view of the sistered joists, one of the metal joist hangers used on the ends of the paired joists is visible in the foreground.

5

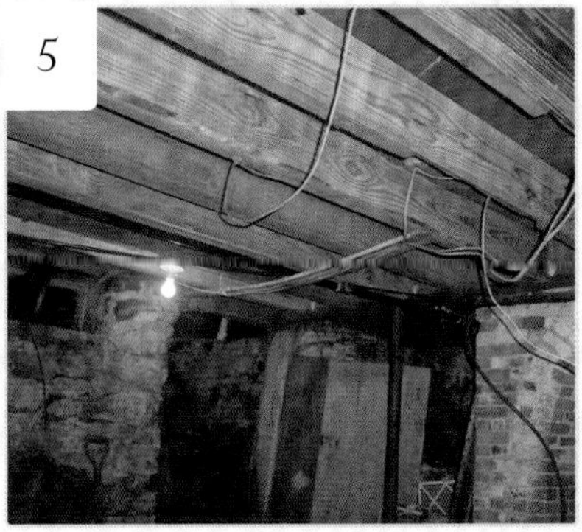

A view of the new pressure-treated joists that will add strength to the floor many decades to come.

CASE STUDY 3
Replacing a Sill, First Scenario

Removing the deteriorated front steps and landing from this early-nineteenth-century house exposed a rotted sill under the front door. This is a common location for sill rot on houses without gutters, as water dripping from the eaves splashes on the landing and wets the sill. The absence of a granite capstone under the door is somewhat atypical; it may have been intended to allow ventilation under the house through lattice in the sides of the landing.

Here the sill is being cut back to solid wood. A jack has been installed to provide support under a load-bearing wall that meets the front wall just behind the door surround.

A second jack topped by a steel plate and wood blocking to spread the load is supporting the intersecting beam at left. The rot extended under the tenon of this beam, so the sill was cut back to allow the new length of sill to support the joint. Note the narrowness of the granite capstone slab at the opening, which created some concern about the replacement sill piece having sufficient support.

Here the rotted sill has been cut back to solid wood at both ends, where lap joints have been cut for the replacement sill.

Case Study 3 continued on next page

CASE STUDY 3 (continued)

5

The replacement length of sill was cut from a pressure-treated timber to prevent a future recurrence of the problem. The new wood was attached to the old with long screws. To address concerns about the new sill rotating inward over time due to the narrowness of the granite capstone slab it sits on, 6 x 6 pressure-treated posts were installed on precast concrete footings at either side of the opening. A piece of sheathing salvaged from a project on the other side of the house was used on the face of the sill to match the thickness of the original sheathing on either side.

CASE STUDY 4

Replacing a Sill, Second Scenario

1

In this house a combination of water dripping from intersecting roofs and insufficient clearance above the soil grade resulted in a long section of significantly rotted sill. Hidden by vinyl siding for more than 30 years, the problem went undetected until the damage was extensive.

2

Rather than installing a network of posts, brackets, and jacks to support the wall during the work, the sill was replaced in two sections that were short enough to remain unsupported for the brief time necessary to cut and install the replacement lengths. In this photo the first piece of rotted sill is being removed. Although much of the wood remained solid, the portion resting on the narrow granite capstone had rotted to such an extent that it no longer provided sufficient support for the wall above. Had the foundation been thicker, it might have been possible to cut back the rotted portion of sill to solid wood and scab on new material.

Case Study 4 continued on next page

CASE STUDY 4 (continued)

3

Here a replacement length of sill has been made by combining a piece of 2 x 6 with a 6 x 6 timber to match the dimensions of the original sill. A single full-sized timber would have been preferable but was not locally available. The shape of the piece—dictated by how the rot had affected the original sill—enabled the retention of as much original material as possible while creating a joint that could be solidly connected with long screws.

4

The first length of new sill is in place. The lap joint at right will allow a solid connection to the second piece. Note how the 2 x 6 applied to the back side of the new sill extends past the lap in the 6 x 6 but stops short of its end. This will allow multiple points to screw through the 2 x 6s and tie the joint with the next new piece of sill.

5

The next rotted section of sill has been removed. The tenon of an intersecting floor beam can be seen to the right of the cutaway section. Although the portion of the sill over the granite foundation was rotted, the bulk of the sill timber was solid and still supported the tenon. To avoid disturbing the joint, a new support post was installed inside the crawl space, visible left of center, reducing the weight resting on the tenon. This enabled only the rotted depth of this portion of the sill to be replaced, using structural screws to scab the new and old material together.

6

With both lengths of replacement sill installed and screwed to each other and to the ends of the remaining original sill, a pressure-treated 1 x 6 was installed as new sheathing. Despite the drainage improvements planned for this area, it is likely to remain prone to moisture, and pressure-treated material will better protect the new sill from future damage.

Case Study 4 continued on next page

CASE STUDY 4 (continued)

7

House wrap was slid behind the remaining clapboards before reinstallation of the clapboards that had been removed for the sill work.

CASE STUDY 5
Strengthening a Corner Sill Joint

To pull together and secure a corner sill joint, this steel reinforcement plate was fabricated by welding a length of threaded rod to a steel plate drilled for screw holes. The joint had been pulled apart when firefighters used a pry bar to force open the door above. By prying sideways to separate the deadbolt from the jamb, they saved the 1827 door but destroyed the structural integrity of its frame and separated this corner of the sill.

A steel plate and nut allowed the sill joint to be pulled back together and secured without having to take apart more of the historic wall to repair the joint. A new oak sill for the door will cover the plate and rod, and a recess in the exterior sheathing will hide the plate and nut.

CASE STUDY 6
Strengthening Roof Framing

This early-nineteenth-century roof was framed with 3-inch by 7-inch rafters spaced approximately 30 inches apart. Knee walls reduced the span and carried some of the roof load to the floor joists below. In the 1980s, engineers determined that this framing was insufficient to handle the increased snow load that would result from insulating the floor between the attic and the story below. Their solution was to remove the plaster and lath from the sloped and flat ceilings and then remove the original knee walls in order to install new 2 x 10 rafters between the original rafters. Once the new rafters were in place, new knee walls were framed up. Then cellulose insulation was blown into the floor joist bays through holes drilled in the floor. (In 2010, the roof itself was insulated so that these attic rooms could again be finished and used.)

Example Project: Beam Replacement

This project addressed a noticeably sagging beam in the primary living room of an 1890s Shingle Style cottage. The cottage had been well built, and the renovations that expanded it in 1910 were designed by a respected architect. It remains in the family that undertook those renovations more than a century ago, and the current owners recognized that some corrective work was needed to make sure the cottage would stand another 125 years.

The first step was to analyze the cause of the sagging beam and devise a solution. The owner (who is an architect), the structural engineer, and I (as preservation consultant) thought initially that the beam had been undersized from the beginning and simply could not carry its load without bending over time. Measurements showed that the 19-foot-long beam had deflected more than 3 inches at the midway point. Stress cracks suggested that it needed to be replaced to prevent further sagging and possible failure.

A sheet metal plate about the size of an electrical box had been nailed to the bottom of the beam at its midway point and painted to match. We suspected that an early electrical light fixture had hung there, with wires run through a hole drilled through the beam, perhaps contributing to the beam's weakness. When the metal plate was removed, however, it became clear that the 1910 alterations had removed a key structural element. Lying inside the small recess above the plate was a large flanged iron nut, unattached to anything, and above it was a ¾-inch hole that had once accommodated the threaded iron tie rod that the nut was attached to.

Photos of the cottage before (c. 1905) and after (c. 1910) the expansion. *(Courtesy Neil Kittridge)*

Reconstructing the evidence, we concluded that during the 1910 alterations, when a portion of the roof had been raised to make a large covered porch on the second floor, a window had been cut into the wall above the beam to look out through the porch toward the ocean. This window was placed where the iron tie rod had run up through the wall to the peak of the roof framing, creating a truss to stiffen the beam below. Unfortunately, when the rod was removed to install the window, no alternative support for the beam had been provided. The combination of the missing tie rod and new static and live loads in the added porch caused the beam to sag over time.

Our solution to the problem—designed 107 years after the removal of the tie rod—was based on the premise that it would be unrealistic to return the floor framing to where it was in 1896. The mere attempt would likely cause serious damage to the wall and ceiling plaster and the window and door

The sagging beam in the cottage, with a yellow line to highlight the amount of deflection.

One end of the beam, showing stress cracking from the strain of deflection.

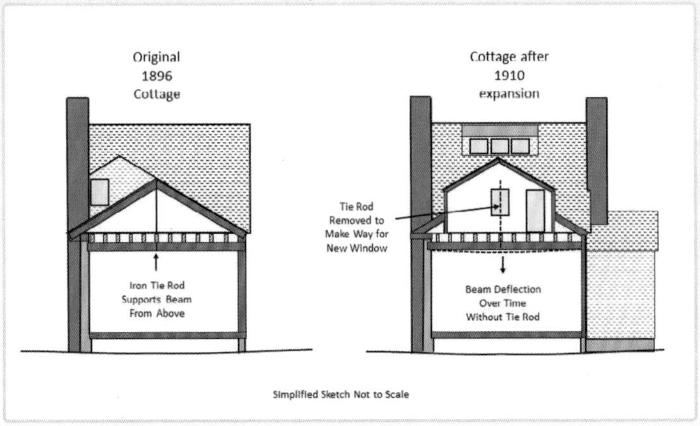

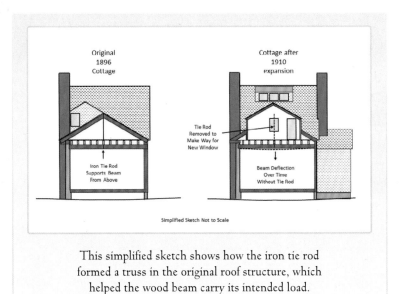

This simplified sketch shows how the iron tie rod formed a truss in the original roof structure, which helped the wood beam carry its intended load.

trim of the room above the beam, and it might even have negative consequences for the roof valleys above that. We decided instead to jack up the second-story floor joists and wall as far as they would go without damaging the wall and the roof above, and then to install a new steel beam to prevent any future sagging. The steel beam would be encased in wood to maintain the appearance of the old wood beam. New rectangular tubular steel posts would be set into the walls to carry the load on the steel beam to the sills.

To accomplish this, the joists that rested on the sagging beam were supported on

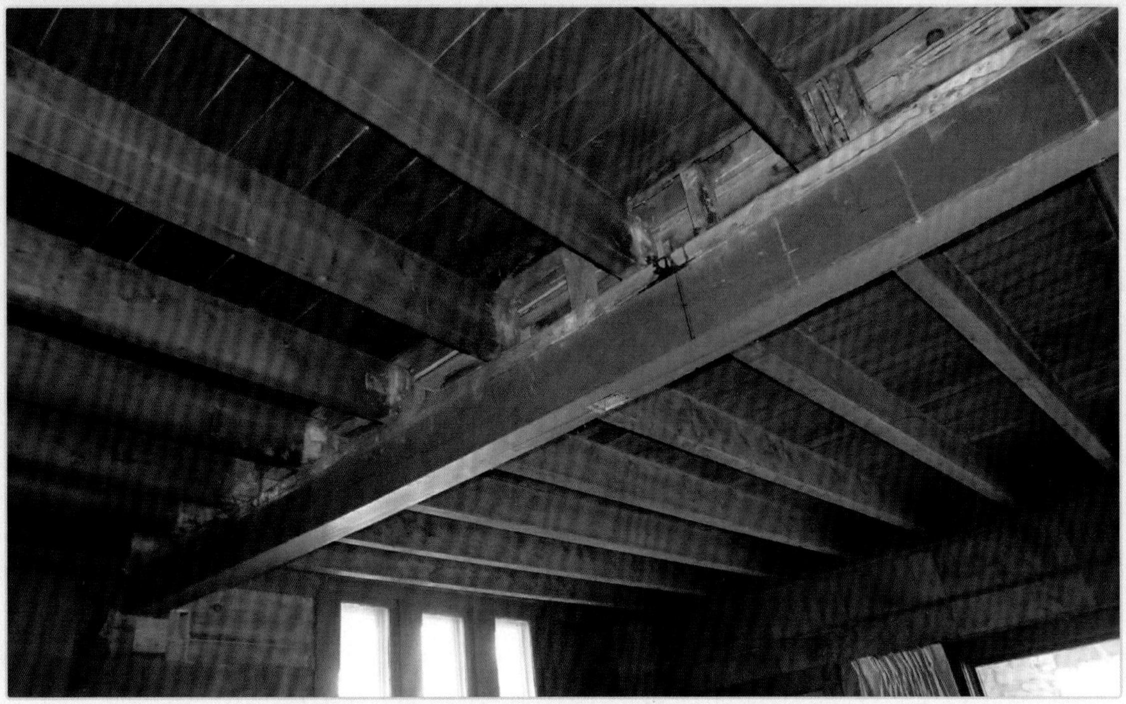

This view of the original beam shows the uncovered recess at center where the iron tie rod had been connected by a flanged nut.

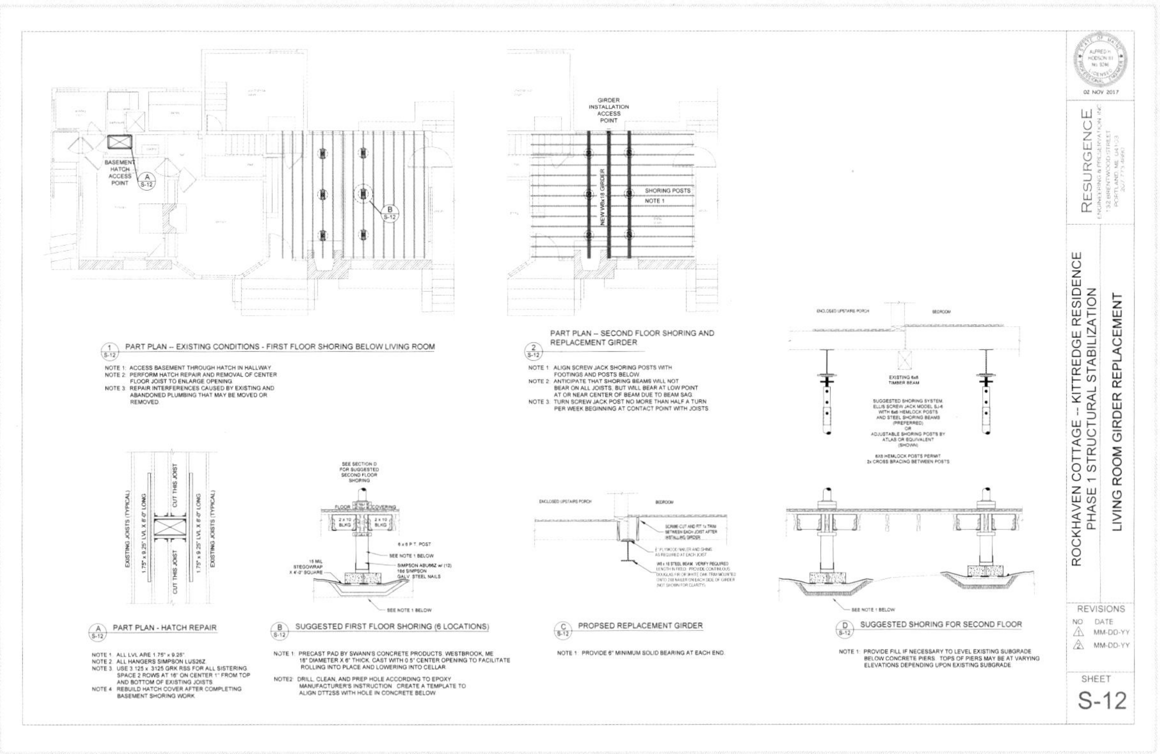

A structural engineer with experience on historic buildings was hired to assess the problem and design a solution. Stamped plans from a licensed engineer are often required for a permit to do structural work.

temporary shoring walls set 28 inches back from the beam on either side. These shoring walls were in turn supported by temporary beams and posts in the crawl space under the floor. Once this support was in place, the beam was removed and replaced with the new steel beam, which had to be brought in through a window on one side of the room and threaded through a hole cut in the opposite outside wall, then slid back into its recess in the wall with the window. The steel beam was placed on several jacks and posts set on steel beams supported by the shoring under the floor. The beam was leveled and lifted by the jacks until it came in contact with the lowest of the floor joists above. Over a period of months, the beam was slowly raised in small increments to remove much of the sag from the floor above.

To prevent movement if the jacks were to fail, ¼- and ½-inch-thick steel shims were placed atop the steel posts in the wall as the beam was raised. The wall and roof above were, of course, rising as well. As the beam was slowly raised, more of the joists came to rest on it and were lifted off their temporary shoring walls. A doubled top plate on each shoring wall allowed the joists to rise while the walls remained anchored to the floor. The second-floor walls, ceiling, floor, and trim were carefully monitored for damage as the lifting continued through several months. When the beam was as high as it could go without causing serious

Bringing the new steel beam in through the window, lifting it into position, and sliding it out through a hole in the opposite wall and then back into its recess in the first wall required a crew of six. *(Courtesy J.B. Leslie Construction, Inc.)*

The steel beam is in place between shoring walls constructed of 2 x 6s. Steel beams run across the space between the temporary walls to support another beam parallel with the beam above. Two screw jacks sit on the lower beam, with 6 x 6 wood posts rising to support the new beam to be jacked. Wood blocking between the shoring walls ensures that the jacking posts cannot move laterally.

A closer view of a screw jack. Four screws in the sides of the U plate atop the jack attach it firmly to the 6 x 6 post that extends up to the steel beam.

At the tops of the jacking posts, lengths of 2 x 6 screwed to the sides of the posts keep them aligned directly under the beam. Note the magnetic level attached to the underside of the beam (upper left) to check that the beam remains level throughout the jacking.

In this view of the steel beam during jacking, note that the joists above sit primarily on a ledger atop the steel beam, just as they sat on the wood beam being replaced. This structural detail is atypical for a location like this but was accommodated by the new beam design.

A tubular steel post inserted into the wall to carry the new steel beam. In addition to the fireplace and chimney immediately adjacent to the post, a fieldstone veneer on the exterior of the wall made clearances tight. A bit of masonry work was required to make the post fit. Below the floor, on both sides of the room, new concrete footings were installed to support the posts. The board sheathing on the interior of the room was originally covered with burlap, which was restored as part of this project.

A close-up of the steel shims installed as the beam was jacked. Thin shims were placed first, then replaced with thicker shims as the beam moved 1/8 inch higher per week over three months. Once the jacking was complete, a steel bracket extending down in front of the shims was bolted to the beam to prevent them from moving outward in the future. Welding them in place would have been an option if not for the fire risk.

Arrows show carriage-bolt locations in this view of the doubled top plate on one of the shoring walls. Holes were drilled to allow the upper 2 x 4 to move up freely as the joists were jacked, while the bolts ensured that the wall remained plumb. The upper 2 x 4 was screwed to the joists to help stabilize them and keep the wall plumb though the jacking process. When this photo was taken, the jacking had moved the beam up nearly 1½ inches, as shown by the space between the 2 x 4s.

Encased with wood, the new steel beam mimics the wood beam it replaced. The face boards have been scribed to match the remaining curve of the floor above but are straight along the bottom of the beam, rendering the remaining sag unnoticeable. The filler boards between the joists and the applied ledger board under the joists replicate original details.

damage to finishes above, the shims on the new steel posts in the wall cavities were secured in place with a bolted bracket, and the jacks were removed.

Even then, not all the joists were sitting on the steel beam (since it did not return all the way to the 1896 position of the wood beam), so blocking was installed between the steel beam and those joists to support them. Then the temporary shoring walls were removed along with the temporary shoring in the crawl space. Finally, wood blocking was installed in the recesses of the I-beam, and a three-sided wood box was fabricated to cover the sides and bottom of the beam. The top edges of the sides of this casing were scribed to meet the line of the joist bottoms, hiding the blocking atop the beam. With the more visible bottom surface of the beam level and straight from wall to wall, the remaining sag in the joists above is not readily noticeable.

Red Farm (Rowell-Cummings House), 1796

Federal Style with Twentieth-Century Colonial Revival Restoration

Red Farm is a wonderful example of an eighteenth-century house that experienced a Colonial Revival "restoration" in the 1930s and 1940s and is now maintained in its Colonial Revival form with recent mechanical updates and other rehabilitation work. It has been under institutional ownership since the 1970s and continues to be used as a seasonal residence by the Skowhegan School of Painting and Sculpture.

The house was built in 1796 for farmer David Rowell, an early settler to the upper Kennebec Valley of Maine. It remained in the Rowell family through several generations and experienced a variety of Victorian-era alterations, including the construction of a kitchen ell, installation of two-over-two windows in place of the original nine-over-six windows, and the addition of tin ceilings in portions of the interior. Coincidentally, David Rowell's daughter Myra married William Whitten of Topsham in 1830 and spent the rest of her life in Whitten House, which is now owned by the author and is used as an example in several places in this book.

The farm was purchased by Willard and Helen Cummings of Skowhegan in the 1920s. Mr. Cummings operated woolen mills there, and the family had a summer camp on a lake near the farm. Mrs. Cummings had an interest in historic buildings and owned several in the area, including a meetinghouse. In the late 1920s, Mrs. Cummings and the Cummings children lived in Paris, France, where several of the children—including son Willard (Bill), a painter, and daughter Muriel (Petie), a pianist—were training under noted artists. This life came to a sudden end with the stock market crash of October 1929. Mrs. Cummings and the children returned from Paris, and the family moved into the vacant and neglected former Rowell farm to raise chickens. Mrs. Cummings ran the farm while Mr. Cummings struggled to keep the mills out of bankruptcy.

It was at this time, starting in 1930, that the first phase of the "restoration" took place. Mrs. Cummings was a cultured woman who likely read the popular magazines of the time, which included photo spreads and articles on the restoration of historic houses. The Colonial Revival movement had started with the United States International Exhibition, the first official World's Fair, held in Philadelphia to celebrate the American Centennial in 1876. An exhibit of a "Colonial Kitchen" with a massive hearth and antique furniture and cookware was among the most popular exhibits with the nearly ten million visitors to the fair. In the decades that followed, America began to recognize and appreciate its own history, including the architecture of the colonial period. The restoration of Colonial Williamsburg in the 1920s and 1930s contributed greatly to an interest in the architecture of the colonial period. For many homeowners, what was lacking in academic research of the period was made up for by a vision of "olde tyme" character. Houses from the late- and even post-colonial period (1760–1800) were often "restored" with features common in the early colonial period (1620–1700). Wildly inaccurate though they sometimes were, these "restorations" are now valued for their documentation of how the early Colonial Revival movement viewed history.

According to the recollections of her children, Mrs. Cummings had the tin ceilings removed, multiple layers of wallpaper stripped, and the painted floors scraped bare. She purchased eighteenth-century windows from a nearby house to replace the two-over-two replacement windows in the house. In addition to the tin ceilings, she had the original plaster ceiling removed in the original kitchen of the house, which was by then a living room, exposing the hand-hewn beams of the frame—a very desirable feature for a Colonial Revival restoration, although plastered ceilings had been far more desirable in the 1790s. Mrs. Cummings also had a bathroom built in the nineteenth-century kitchen ell and had a compatible modern kitchen built using old boards. A basement-level fieldstone garage with a roof terrace was built on one end of the house, tucked into the hill the house sits on. It is said that the long dining table in the house came into being when Mrs. Cummings was at the grain store buying feed for her five thousand chickens and noted that the seven-foot-long grain-bin lid was a single board; she purchased it and had the table made from it. It was during this time that the white house was painted red and acquired the name Red Farm.

As the Depression began to ease for the Cummings family and prosperity returned to the woolen industry, Mr. Cummings purchased a mill 30 miles from the farm. With the children married or in college, Mr. and Mrs. Cummings left Red Farm and moved closer to the new mill. The farm sat vacant for several years, with occasional summer visits from college-age Cummings children and their friends.

During World War II, the eldest son, Bill, by then a noted portrait painter, got to know several other young artists also serving in the Army Art Unit. This group was created to document army life during the war. Following the war, Bill and three of these friends, Henry Varnum Poor, Charles Cutler, and Sidney Simon, decided to found an art school where the teachers were active artists, not art professors. Cummings later said, "We wanted a school for artists run by artists. We wanted to give the students a place where they could work and not face commercial problems." The vacant farm property became home to the new Skowhegan School of Painting and Sculpture in 1946. Barns and chicken houses were converted to studio and lecture space, and Red Farm became the summer home of the school's director, Bill Cummings.

Bill Cummings had an eye for antiques and American folk art, and during his twenty-nine summers at Red Farm he continued the "restoration" begun by his mother and filled the house with beautiful objects and fine art. Among his most notable changes was the installation of two bathrooms in the main house and rebuilding the chimney, which included replacing the original cooking hearth and brick oven in the living room with a massive open fireplace. Such fireplaces were a common motif of the Colonial Revival movement. The house became the social center and heart of the school, with weekly dinners served to all the students, faculty, and visitors.

Since Bill Cummings's death in 1975, Red Farm has served as a seasonal home for directors of the school and visiting artists. Much of the Cummings collection of art and folk art was left to the Colby College Museum of Art in Waterville, Maine, but to the degree possible, the character created by Bill has been maintained. In recent years, the deteriorated clapboard siding has been replaced, damaged plaster repaired, and mechanical systems updated. A few unfortunate changes made after 1975 are being undone as funding is available, with the goal of returning Red Farm to its appearance during Bill Cummings's occupancy, while it continues to be a living and useful part of the school's annual summer programs. The Skowhegan School of Painting and Sculpture continues as one of America's oldest and most prestigious summer art schools. Alumni include noted American artists Alex Katz, Robert Indiana, Lee Bontecou, and Ellsworth Kelly.

Red Farm as seen from the driveway in the 1970s. *(Courtesy Skowhegan School of Painting and Sculpture)*

One of the second-story bedrooms.

The living room is the original kitchen. The ceiling beams were exposed when the eighteenth-century plaster was removed in 1930.

At right can be seen the doors to the fieldstone garage, which is built into the hillside and opens into the basement of the house.

The 1930s garage addition originally had a deck on its roof, as seen here in a 1950s photo.

This late-nineteenth-century view shows members of the second and third generations of the Rowell family. The shed visible at left was moved across the pasture in 1941 to become Bill Cummings' studio. *(Courtesy Skowhegan School of Painting and Sculpture)*

This 1940s view shows the back of the house when it still had a wood shingle roof. *(Courtesy Skowhegan School of Painting and Sculpture)*

The wide boards in the wall of the green bedroom were probably salvaged from another building and installed as part of the 1930s work by Helen Cummings.

The founders of the Skowhegan School of Painting and Sculpture—Henry Varnum Poor, Charles Cutler, Sidney Simon, and Willard (Bill) Cummings—outside a chicken coop that was converted to an artist studio. *(Courtesy Skowhegan School of Painting and Sculpture)*

Students and faculty gathered around the dining table, about 1960. *(Courtesy Skowhegan School of Painting and Sculpture)*

The parlor retains its eighteenth-century plaster ceiling.

Red Farm as seen from the footbridge at the pond inlet.

The house in the 1940s, before the large barn on the left was lost to fire. *(Courtesy Skowhegan School of Painting and Sculpture)*

The dining table in the living room. The top was a feed store grain-bin cover purchased by Mrs. Cummings while picking up chicken feed.

The recent rehab of the downstairs bathroom retained the historic fixtures.

Red Farm in the mid-1950s before Dutch elm disease killed many of the towering elms planted by the Rowells in the nineteenth century. *(Photo by Paul Cordes, courtesy Skowhegan School of Painting and Sculpture)*

Bill Cummings seen through a living room window at Red Farm in the 1950s. *(Courtesy Skowhegan School of Painting and Sculpture)*

In the kitchen, an early cast iron cook stove and antique cookware and implements coexist with a mid-twentieth-century sink cabinet.

A 1960 gathering on the brick patio built by Helen or Bill Cummings in the Rowells' former kitchen yard. *(Courtesy Skowhegan School of Painting and Sculpture)*

Rowell family members outside the kitchen door in the late nineteenth century. *(Courtesy Skowhegan School of Painting and Sculpture)*

Actress Betty Davis (front center) performed in summer stock theater nearby for a number of years and regularly visited Bill Cummings (front right) at Red Farm. Helen Cummings is behind Ms. Davis in this photo from about 1970. *(Courtesy Skowhegan School of Painting and Sculpture)*

The blue bedroom features artwork by Bill Cummings and others.

The north side of the house, shown here, nestles into the landscape for protection from winter winds, while the south, or front, side (photo above) is exposed to the warmth of the sun.

PART THREE

Systems

8. Heating and Cooling

9. Insulation and Ventilation

10. Plumbing and Electrical

Whether your house needs only minor updates to wiring and plumbing or a complete replacement of its electrical, plumbing, and/or HVAC (heating, ventilation, and air conditioning) systems will depend on the age and condition of the existing systems, the scope of your rehabilitation project, and your budget.

The majority of houses built before 1890 were electrified between 1890 and 1920 (later in many rural areas), and that early wiring was often updated in the 1940s, 1950s, or 1960s. Central heating systems were increasingly installed in the early twentieth century—following electrification—and were updated in the middle of the century. Plumbing followed a similar pattern. Such houses are now coming due for another reinvestment in systems. Likewise, houses built since the 1950s seem to be coming due for system updates as they turn fifty or sixty years old.

Expectations for household systems have evolved through time. In the seventeenth century, a safe well in the dooryard and a house that (barely) kept us from freezing to death in the winter—cold, dark rooms notwithstanding—were the fulfillment of our dreams. By the mid-nineteenth century, we expected more consistent warmth and wanted oil or gas lighting to hold the darkness at bay. We may even have had a handpump to deliver water into the kitchen sink. In the early twentieth century, if we lived in town, we could expect electric lights, automatic furnaces or boilers, plumbed bathrooms with pressurized water, and even a few electrical outlets for the new appliances coming on the market. By mid-century we could expect those conveniences in the country, too, and our appliances included a refrigerator, a washing machine, and maybe even a television. Today we expect our heating and cooling to be consistent and efficient, and we want "smart" lighting, more bathrooms with more fixtures, and lots and lots of outlets for our ever-expanding arrays of appliances and devices.

If you are undertaking a hundred-year rehabilitation of a house—which seems to be the approximate interval at which houses built before 1950 require major reinvestments in foundations, framing, and/or finishes—it only makes sense to bring the systems up to current expectations if possible. It is impossible to anticipate what new technologies and conveniences will be coming in future decades, but if the past is any guide, the pace of change in household systems will continue to increase and will be matched by our expectations.

It is said that a child born today will experience more technological change in the first twenty years of her or his life than has been seen in the past 150 years. There is no way to anticipate the impacts of such rapid change on houses, but we can bet that those impacts, for the most part, will need to build upon existing systems. The more fully a historic house satisfies current expectations for comfort and convenience while preserving its character-defining features, the more likely it will continue to be used, maintained, and technologically updated in the future. A useful and comfortable building stands a much better chance of being preserved.

The Built-in Efficiencies of Historic Houses

Heating, cooling, and lighting systems that were in vogue a decade ago are largely outdated now, and it is likely that many current trends will be obsolete a decade from now. So why not incorporate historic approaches that not only still work but can also accommodate evolving systems without loss of character-defining features?

We often hear that energy was cheap prior to the 1970s, but in fact it was really cheap only from about 1945 to 1973. The investment required to maintain heat and light prior to the 1940s was significant in most areas of the country, and as a result, houses back then were generally built to be efficient. Decades of increasingly cheap electrical power and heating fuels during the mid-twentieth century caused many of these inherent advantages to fall from use and be forgotten before the cost of energy began to climb again in the 1970s. The potential for energy efficiency remains inherent in many old houses and can be reactivated now that the true cost of cheap power is becoming clear in the age of global climate change. Combined with properly used modern insulation materials, heating and cooling systems, and alternative energy sources, these legacy efficiencies can dramatically reduce operational costs and increase comfort for homeowners while contributing to a green lifestyle.

America's early English, French, and Dutch settlers brought tools and technologies from their homelands and built replicas of the buildings they had left behind. In the Southwest, the Spanish adopted and adapted the Native American adobe construction technology. Adaptions of traditional European house plans to accommodate the New World's varied climate soon followed. New England homes, for example, began to be built with their cooking and heating chimney masses at the center of the house, minimizing the heat lost in winter, while Southern houses came to have fireplaces on exterior walls and kitchens in separate buildings to keep heat out of the main house.

Houses were frequently sited to optimize passive solar and prevailing wind effects. New England houses were often built low to the ground and sheltered from cold winter winds by surrounding terrain, while Southern houses were built on elevated sites or tall foundations to catch as much

These solid paneled shutters in a 1796 house slide into recesses in the wall. They provide privacy when closed and significantly improve the energy efficiency of the single-glazed double-hung windows. Such shutters are sometimes found painted in place, long hidden in the wall and forgotten.

summer breeze as possible. Making use of such inherent advantages can contribute to the efficiency of your house today.

Other common strategies for controlling heat and ventilation before the development of electrically powered systems included the deployment of shade trees, window shades, paneled interior window shutters, and functional louvered window shutters. Deciduous trees planted on the south side of a house created shade to control solar gain through south-facing windows in summer, while their bare branches allowed the winter sun to provide passive solar heating. A row of evergreen trees on the north side of a house can screen it from cold winter winds. Window shades can be closed on sunny summer days to keep heat out or opened on winter days to allow solar heat in, then closed on winter nights to keep heat from escaping. Louvered window shutters are generally considered an aesthetic accent today, but historically had an important ventilation function. Closed shutters over an open window will keep out the sun's heat while allowing air to flow through the house.

Double-hung windows allow both the lower and upper sashes to be partially opened at the same time, encouraging cooler air to enter at the bottom and hot air near the ceiling to escape through the top.

An understanding of how long-ago occupants achieved comfort in your house can guide the changes you make to achieve comfort today. Modern expectations may be higher, but they can cost less when we reuse our ancestors' strategies.

CHAPTER EIGHT

Heating and Cooling

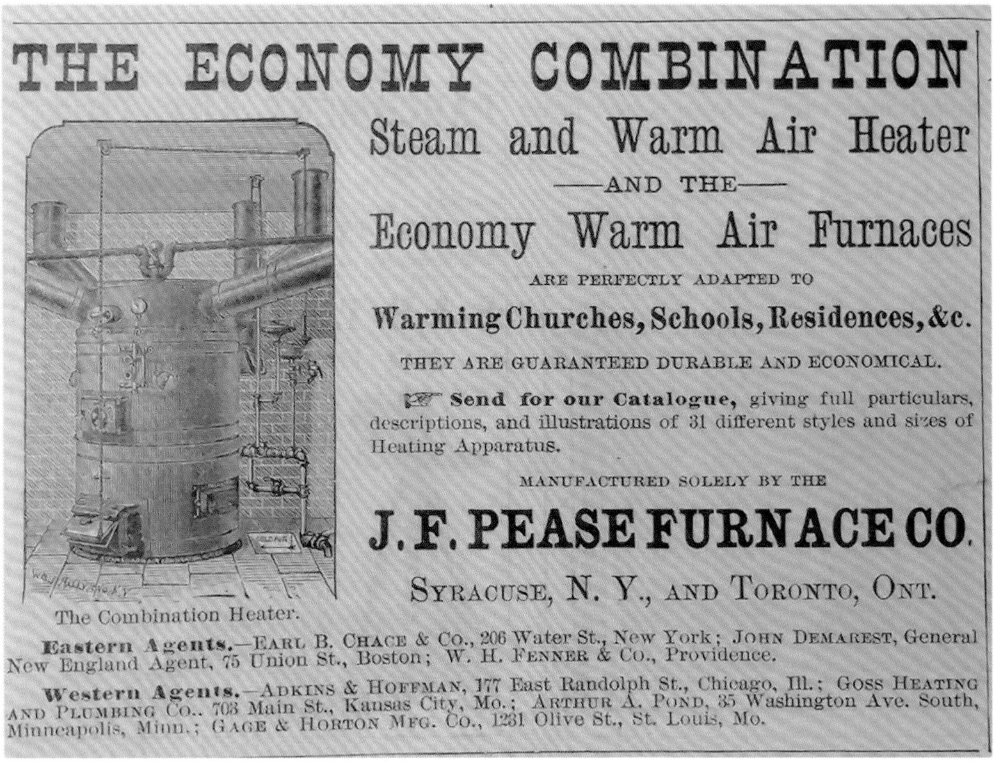

There have been significant advances in the efficiency of HVAC (heating, ventilation, and air conditioning) systems over the past several decades. If your house has a system that is more than twenty or thirty years old, it is worthwhile to consider replacing the heat source if not the entire system. The older your existing system, the more quickly a new system will pay for itself with savings from increased efficiency. You may also wish to consider whether the time has come to introduce air conditioning to a house that has not previously had it. Thanks to our changing climate, people in many parts of the U.S. who traditionally got by with fans in summer are increasingly installing mechanical cooling. At the same time, parts of the country that historically had minimal need for winter heating are enduring more and longer cold spells.

Climate scientists predict that these changes will continue and even accelerate in the future.

Heating

In much of the U.S. before 1850, houses were typically heated by open fireplaces. Many of these were replaced with more efficient cast iron woodstoves in the middle of the century, and central heating followed in the late 1800s. The most common systems installed over the following century were steam with ornamental cast iron radiators; forced hot air with floor registers or grates; and hot water with ornamental cast iron radiators or (later) cast iron or sheet-metal baseboard radiators.

Where components of historic heating systems contribute to the historic character of a house, it is best to retain and reuse them if possible. Cast iron radiators may be 50 or 100 years newer than the house in which they stand, but they will seem "in character" to most observers. These floor-standing radiators (occasionally they are mounted to a wall or ceiling) can serve as well with an efficient modern boiler as with the inefficient asbestos-covered beast they were once paired with.

Old cast iron floor grates and registers also have historic character. (Floor grates or grilles—the terms are used interchangeably—cannot control the flow of air through them, whereas registers have louvers below the grille or grate to direct or shut off the flow.) Cast iron floor registers that once delivered heat from a coal-fired furnace can as readily deliver heat and cooling from a state-of-the-art natural gas– or wood pellet–fired unit or a heat pump and will not detract from a character-defining space.

This is not true of the newer sheet-metal grates and registers, however, and it is certainly not true of sheet-metal baseboard radiators. These often feel too modern for old houses and will detract from the historic character of pre-1940 rooms. Eliminate these from character-defining spaces if feasible, and disguise them (as discussed below) if not.

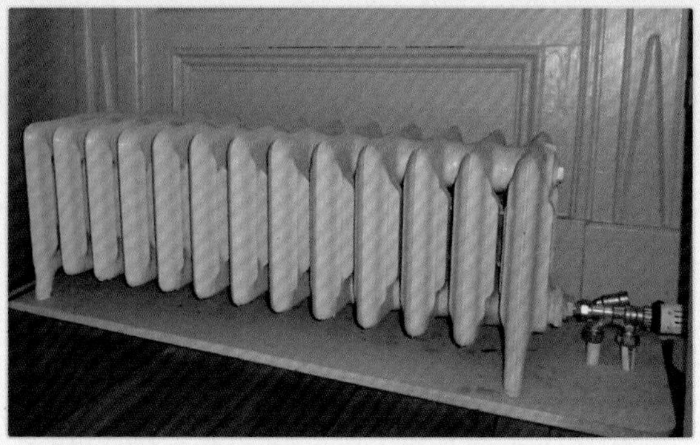

This historic radiator has been retrofitted to function with a modern high-efficiency boiler and remains in the location it has occupied for nearly a century.

This cast iron floor register and wall-mounted cold air return were installed to serve the original 1850s wood-fired hot air heating system in this house. Since then, they have served with a coal-fired furnace, an oil-fired furnace, and now a modern high-efficiency gas-fired furnace, the last two being forced hot air with a blower. The painted canvas floor cloth dates from the mid-nineteenth century.

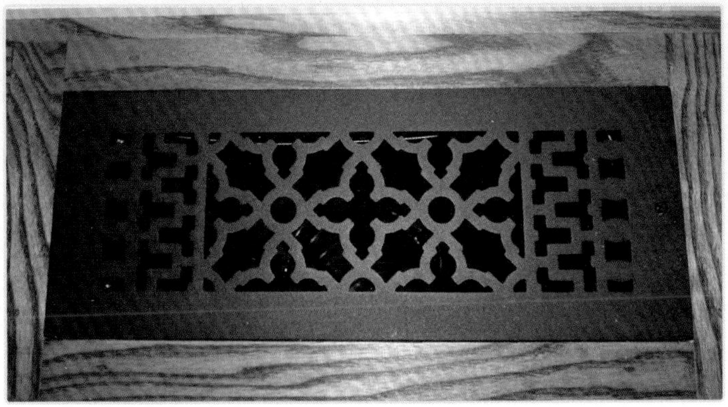

A modern reproduction cast iron floor grate installed to serve a new heat-pump-supplied HVAC system in a historic house.

The different fuels impose different and variable financial and environmental costs, which are impossible to predict far into the future with any degree of confidence. Global economic forces and other dynamics can dramatically affect the prices of fuels, as can changes in discovered reserves or extraction methods. Even as technology is advancing rapidly for improving the efficiency of traditional fuels, new possibilities are on the horizon. For many of us, environmental considerations also factor into the equation. We'd like to stay warm while minimizing the stress we add to the environment, and we'd like to be able to afford it.

Do your research and understand the options before deciding on a course of action. If you know the options and the vocabulary to discuss them, you'll be more likely to know when an HVAC contractor is pushing you toward a solution that suits his interests

If some or all of the cast iron radiators or registers have been removed from your house and you want to replace them with appropriate period pieces, look in local salvage shops or place a "wanted" ad in online classifieds or marketplace sites or in the local newspaper. There are companies that specialize in restoring historic radiators, and several companies reproduce cast iron floor registers and grilles.

Your HVAC contractor might insist that old components be replaced even if in serviceable condition. If so, you may want to look for another contractor.

Fuel Sources

The fuels commonly used for heating in the United States today include oil, gas (natural and LPG or propane), electricity, cordwood, wood pellet, and solar energy. Coal remains in very limited use for domestic heating. Most of these fuels can be used with more than one type of heating system. Most systems use either water (liquid or steam) or air to move heat from the source to living spaces. Cast iron heating stoves fueled by wood, coal, or gas heat spaces directly with radiant heat.

For more than two centuries of European presence in America, a stack of dry split cordwood beside a fireplace was the primary source of heat in homes. It still is used (in more efficient airtight stoves and furnaces) in some areas today.

more than yours. If the job will be substantial, get multiple bids and check references. Ideally, the contractor will give you contact information for customers who had similar work done in a similar house at least several years ago. Knowing how that work has held up over time and how well the contractor responded to any problems or annual maintenance is valuable.

An HVAC contractor installed new gas-fired boilers in several houses in my neighborhood shortly after natural gas was installed on the block. He had previously done maintenance on the old oil-fired systems in several of the homes without any issues. He was personable, seemed reasonably priced, and was recommended from one neighbor to the next. None of us checked past references, and several of us followed his recommendation for a gas boiler to purchase. Several years later, after he was out of business, serious problems developed in several of the installed systems and other HVAC contractors pointed out issues with the original work. This was an expensive lesson for me (who should have known better) and several others. Replacing an eight-year-old boiler because it was installed incorrectly is painful, especially if the contractor is no longer in business and cannot be held responsible.

As always, maintain the historic character of your house to the greatest degree possible when updating your current heating system or installing a new one. The faint hiss of air escaping from a radiator valve, followed by the sound of expanding cast iron as entering steam warms the radiator from within, has been associated with warmth and comfort in many homes for a century or more. It's a subtle part of the historic character that attracts us to these houses in the first place. We retain a sensory memory of these sounds and expect to feel heat radiating from the cast iron shortly after we hear them. In a cold world, little things like this can be powerful and profound if we think to keep them.

Venting Modern Boilers and Furnaces

Because modern boilers and furnaces are extremely efficient, they often can be vented with a two- or three-inch PVC pipe rather than a masonry chimney. If the boiler or furnace is installed in a basement, the vent pipe often exits through the sill and extends several feet up the side of the house to elevate the exhaust outlet above the expected snow depth. A vent like this can be a distraction on a primary elevation

This gas heater vent in a brick wall has been made much less apparent with a coat of paint to match the brick. A shiny stainless-steel vent is a highly noticeable distraction from the historic character of a building.

of a house; it should be located as inconspicuously as possible and painted to match the siding or masonry material behind it. Gas-fired wall-mounted heaters like the Rinnai direct-vent wall furnaces require metal vent caps on the exterior of the building. These too should be painted to reduce their visibility.

Heating Systems

Steam Heat

Steam heat came into wide use at the end of the nineteenth century and is still functioning in many buildings in the twenty-first century, although the installation of new residential steam-heating systems largely ceased in the mid-twentieth century. It is not uncommon to find a historic house with steam radiators supplied by a large cast iron boiler that originally burned coal and was later converted to oil or gas. These old systems typically produce a lot of heat but use fuel inefficiently and are difficult to regulate from one room to another.

Steam systems must allow condensed water to flow back to the boiler from the radiators, either by pitching a one-pipe system (so that condensate flows down to the boiler in the bottom of the pipe while steam flows upward to the radiator) or by adding a second pipe to drain the condensate. A vent on the radiator allows air to escape as steam enters. In a two-pipe system, a trap at the outlet prevents steam from entering the return pipe. If the house has settled, or alterations have altered the pitch in one or more pipes, or steam traps have failed, steam and cold water can come in contact, causing water hammer, the knocking and banging associated with poorly maintained steam heating systems. The cast iron radiators are often decorative and may be a character-defining feature of the house, but ornate detail cast into the iron is often lost under many layers of chipping or peeling paint. Stripped and returned to their original finish—which was often "radiator gold" metallic paint—they can contribute positively to the historic character of a house.

These systems were often installed when fuel was less expensive and houses were poorly insulated and not tightly sealed. Their big boilers blasted enough heat to overcome the inefficiencies. With the rapid increase in fuel costs in the 1970s, many houses were insulated and tightened up with storm windows and caulking to save heating costs. This often left the boiler significantly oversized for the heat load of the house, decreasing efficiency and undercutting the intended effect of insulating and air sealing. Where such a boiler is still in use in a house that is already well insulated or about to become so, serious consideration should be given to replacing it with a modern, high-efficiency boiler of the appropriate size. An energy audit should provide you with the information you need to determine what is needed and what the likely payback time will be.

A skilled HVAC technician who is familiar with historic steam heating systems will be able to replace clogged air vents and failed steam traps, correct pipe pitch, and resolve other issues to allow relatively even heating without clangs and bangs in the night. One of the advantages of a steam system is its simplicity and the limited number of moving parts. A modern thermostat can provide electronic or even internet-connected control of the temperature settings at different times of day, helping to increase efficiency. It remains true, however, that steam systems are not as easy to zone as hot water or forced hot air systems.

Encapsulation of Asbestos-Covered Furnaces and Boilers

Concerns about the efficiency of heating systems are not new. It was long ago recognized that insulating boilers, furnaces, and their associated pipes and ductwork improved their efficiency. However, the material of choice for insulating these systems in the past was asbestos, which is now recognized as a serious health threat when disturbed. The danger of asbestos is from breathing in its microscopic fibers, potentially leading to respiratory disease. Undisturbed, asbestos is not a threat.

Most jurisdictions require that asbestos be removed by trained abatement specialists to ensure that the material is not released into the surrounding environment. This can add significant cost to the removal and replacement of a large old boiler or furnace. Removal is unnecessary, however, when encapsulation can be used instead to eliminate any risk of the hazardous material being disturbed. Leaving an asbestos-covered boiler or furnace in place with an enclosure around it can save thousands of dollars in abatement costs. Gypsum board on wood or metal framing can be used for this enclosure, although concrete block walls will better ensure that the enclosure is never breached.

Encapsulation can also be used for pipes that have been insulated with asbestos. Indeed, this is often the best approach when these pipes will remain in use. The two primary methods are coating with an encapsulant paint or wrapping with a material that is applied wet. Either method can be undertaken by a homeowner, although some states require that you take a training course first. If the pipes are obsolete, having them removed by an abatement professional may be worthwhile to resolve the problem once and for all. This will eliminate any possibility of a future buyer or lender objecting to the presence of asbestos in the house, even if it is encapsulated and poses no threat.

Hot Water Heat

Historic hot water, or hydronic, heating systems have some things in common with steam systems. They often use cast iron radiators and massive old boilers that have been converted from coal to oil or gas. More recent hydronic systems generally use baseboard

This wall-mounted antique cast iron radiator was purchased in New England and shipped to Santa Fe for use in an historic adobe house. The new hydronic heating system uses a combination of antique radiators and invisible radiant heat in tile floors.

radiators, which are usually less decorative than earlier floor-standing radiators. As mentioned above, the ubiquitous white sheet-metal baseboard radiator is usually at odds with the historic character of a house, but some cast iron baseboard radiators are more attractive and appropriate. These can most often be found in houses built between the 1920s and 1950s. Sometimes a house that once had ornate cast iron radiators has had them replaced by unattractive baseboard radiators. Telltale holes in a hardwood floor often identify former radiator locations. It is possible to find antique cast iron radiators to restore to former locations or even to add in new locations.

Everything written above about steam boilers is true of hydronic boilers as well. Hydronic systems require more moving parts than steam, however—particularly mechanical pumps to move the hot water through the pipe loop(s) of the system. Water needs to flow from the boiler to each radiator in turn, then back to the boiler to be reheated. Because the water cools as it moves through the system, there is a limit to how many radiators each loop can serve. A small house may get by with a single loop, but larger houses generally need more than one. Multiple loops provide the opportunity for introducing multiple thermostats for a zoned system. Each loop will have its own circulator pump or zone valve, allowing them to operate separately or all at once, depending on the room temperatures at the various thermostats.

As with steam systems, a knowledgeable HVAC technician can often tune up a system by replacing radiator controls where needed and balancing the heat supply between rooms. Replacement of an inefficient old boiler with a new high-efficiency boiler can significantly reduce the cost of operating the system while retaining the character-defining historic radiators in living spaces. Once again, an energy audit is the best starting point for determining the most sensible and cost-effective solution. As with steam heat, modern thermostats can generally be used.

If your hydronic system was installed or substantially updated in the past fifty years, it probably has baseboard radiators installed on the interior side of all or most of the outside walls of the house. These radiators came into postwar vogue when a clean, sleek, modern look was all the rage. Unfortunately, the baseboard units seldom remained clean or sleek for long. Airflow through their fins filled them with lint, pet hair, and dust, and furniture and shoes left dents and scratches in the thin sheet metal. Endcaps have a habit of coming loose and disappearing, leaving the sharp edges of the sheet metal exposed. After a few decades of service, baseboard radiators are often the worst-looking element in a room, their ugliness highlighted by scratches in a dirty white finish. And a baseboard radiator within three feet of a toilet used by men will be rusty.

In case it is not abundantly clear by now, I am not a fan of sheet-metal baseboard radiators. I do not much like them in contemporary houses and *really* do not like them in historic houses. They are out of character, look cheap, and interfere with the placement of furniture and the length of window treatments. If you buy a house that has them and intend to continue using the hot water heating system, it is possible to replace baseboard radiators with antique cast iron radiators, at least in the most significant rooms.

Typically, a loop of copper tubing runs around the perimeter of the house inside the exterior wall, dropping down into the basement or crawl space to get past exterior doors, fireplaces, kitchen cabinets, and other obstructions. When reverting from baseboard to cast iron floor-standing radiators, this loop of tubing will need to be moved under the floor except where it connects to a radiator. In many cases today, it makes sense to replace the copper with cross-linked polyethylene tubing (such as PEX), now the industry standard for this application. Upper stories can be a bit more complicated to switch over because access to the underside of the floor is difficult. It can be done, however, with strategically located access holes that are cut through the plaster ceilings below and patched after the installation. If the electrical system is also being updated, both can be done from the same access holes.

This sheet-metal baseboard radiator has been painted to match the baseboard behind it, reducing the incongruous effect of this radiator in a historic room. This is especially important when the trim color is dark. A bright white radiator would be far too obvious.

If floor-standing radiators aren't an option, you can now buy low-profile baseboard radiators that are less distracting than the units they replace. If you must keep the existing baseboard radiators, at least replace any damaged or missing elements, do what you can to minimize their presence, and paint the entire assembly (except the foil fins) to blend with the surface behind it. If the unit is mounted on a varnished wood baseboard, it can be painted with a faux woodgrain finish.

Sometimes a baseboard radiator is taller than the baseboard it is attached to, leaving a catch-all void behind it for dust and debris. In such a case, remove the historic cap molding from the baseboard behind the radiator; install a filler piece of wood to raise the baseboard flush with the top of the radiator; and reinstall the cap atop the filler piece. Alternatively, you may find that the baseboard was removed and the radiator mounted directly to the wall surface. In that case, paint the radiator to match the room's remaining baseboard and, if possible, install a cap molding on the radiator that matches the surrounding baseboard molding. Resist any impulse to paint baseboard radiators metallic gold because historic cast iron radiators were often painted that way. They will not look "Victorian."

Forced Hot Air

Forced hot air systems evolved from pre-electric hot air systems that relied on convection, the natural tendency for heated air to rise, to move hot air from a furnace to rooms above. Early versions of such systems can be found in the ruins of Roman buildings. Central heating largely disappeared from the West following the collapse of the Roman Empire, making the Dark Ages even colder, and did not reemerge in any significant way until the nineteenth century.

American inventors of the late nineteenth century embraced central heating with a passion, and thousands of patents were issued. Meanwhile, many British scientists embraced theories about the dangers of overheated air in enclosed spaces and largely convinced the public there that open fires were the only safe way to heat a house. As a result, central heating did not really catch on in Britain until after World War II.

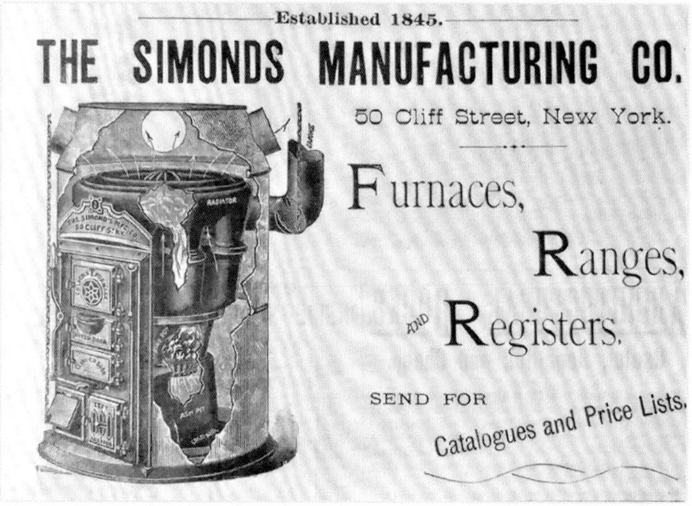

1888 advertisement for a coal-fired furnace encased in a sheet metal plenum with multiple ducts to conduct heat to the rooms above. Furnaces of this type were also enclosed in brick plenums.

Most hot air systems involved some sort of enclosure, or plenum, around and above a wood- or coal-fired furnace (essentially a utilitarian cast iron stove), with ducts running from the plenum to floor grates or wall grilles to distribute the captured heat. Early ducts were often wood, with sheet metal coming into almost universal use later. Control of the heat distribution was usually accomplished with dampers at the plenum or at the outlet grates. Because the fire had to be maintained by hand, someone had to go to the cellar at least several times a day to feed the fire. Beyond the human effort to control the fire and dampers, the systems were passive, using no energy except natural convection. Mechanical dampers to open or close ducts, controlled by thermostats in the rooms above, came along later.

The introduction of domestic electricity made possible the use of electric blowers to push heated air through ducts—even long horizontal runs—much more effectively. An electric thermostat could control the blower, but humans still had to manage the fire. Automatic-feed systems for coal came into use for larger furnaces, typically in commercial buildings and homes of the wealthy. Many older people still can recall their parents going to the basement several times a day to shovel coal onto the furnace grates. From the 1940s on, these systems were increasingly operated with oil burners, which were sometimes simply installed in the old door opening used to add coal to the firebox. New oil-burning furnaces came on the market then too, of course. When natural gas came into use a bit later, it too could be retrofitted with the conversion of an existing furnace or the installation of a new one. Furnaces converted to oil or gas (and boilers for steam or hot water systems) were usually less efficient than new ones.

An advantage of modern forced hot air systems is the ability to humidify the air passing through the system. Winter air can be very dry in cold climates, and dry skin, eyes, sinuses, and throats—which are more susceptible to infection—result from time spent in dry buildings. What's more, very dry air is no better for antiques or art than for people and pets. Many

This new HVAC equipment has been installed in an attic to serve the second story of the house, eliminating the need for vertical ducts through the first story. The previously uninsulated attic has been insulated and air sealed with gypsum board to make it conditioned space as well. Heating/cooling supply ducts are all insulated for greater efficiency. Cold-air supply ducts are not insulated.

people run portable humidifiers in winter, but these units must be kept filled with water and provide little benefit beyond the room the unit sits in. Building a humidifier into the heating system is an attractive option in a cold climate. The humidified air will be distributed throughout the house without the need to monitor water levels and fill reservoirs.

If you buy a house with a forced hot air heating system in place, improving its efficiency is likely to be more cost-effective than replacing it. For example, a new high-efficiency furnace providing hot air to the existing ductwork may pay for itself quickly with fuel savings, depending on how inefficient the old furnace was. If the existing system was installed in the second half of the twentieth century, it may have sheet-metal grilles and registers that are out of character with the historic rooms of the house. These can be replaced with antique or reproduction cast iron, brass, or even bronze grilles and registers. Pick ones that are appropriate to the character of the house. (Bronze would be overkill in a modest cottage.)

If your house has forced hot air on the first floor and no heat on the second (not uncommon in historic houses in New England; for example), consider where the ducts to serve the second story (and the third, if necessary) can pass through the floor below without impacting character-defining features or taking up all the closet space. In a two-story house with an attic, it may be more practical to install a second heating system in the attic to serve the second floor. This can often be done with a high-efficiency electric heat pump to avoid running fuel lines to the attic.

An existing forced hot air system can make the installation of central air conditioning much less expensive, since the same system of ducts can be used for both. For a new system, it may make more sense to use heat pumps or a high-velocity small-duct system that can be installed with less impact to historic spaces (see the "Cooling" section below).

Forced hot air systems do have downsides. They tend to deliver wide swings in room temperature as the blower cuts in and out, and they blow dust around the house even when filters are changed regularly. The ductwork can carry sound from room to room or even floor to floor (this was useful for eavesdropping on my parent's conversations as a kid). With older systems, noisy blowers can be another issue.

If you buy a house with a forced hot air system, you may decide to replace it with another type of system altogether. Whitten House had a fifty-year-old forced hot air system serving only the first floor when I bought it, with an enormous oil-fired furnace on its last legs in the wet cellar under the kitchen ell. Water marks twelve inches up the sides of the furnace offered eloquent evidence of past sump pump failures leading to flooding of the furnace. The blower bearings were going, and complained with constant rattles and occasional squeals when the blower ran. In my first several winters in the house, I spent a small fortune on oil and was cold much of the time.

Because the furnace needed to be replaced and nearly all the ductwork needed to be removed for structural repairs under the house, I saw an opportunity to replace it with a completely different system. I chose to replace the forced hot air system with natural gas–fired hydronic radiant heat, installing the boiler on the first floor, out of the perpetually wet cellar.

Radiant Heat

Radiant heat works by heating people and objects rather than the air around them as other home heating systems do. It is invisible but can be felt as heat radiating toward you. Hot air rises but *heat* is omnidirectional, radiating in every direction equally unless blocked by insulation or directed by reflection from a metallic surface.

Traditional radiators use radiant heat to warm the air immediately around them, and their many fins or openings allow air to flow through them to maximize this warming. The heated air rises and moves across the ceiling toward the colder side of the room, where it is cooled and sinks to the floor to be drawn back to the radiator and warmed again. This convective cycle is what warms the objects and occupants in the room.

With radiant heat, the "radiator" is an entire floor, or wall, or ceiling, which heats the objects and occupants of the room directly, without relying on air convection. The U.S. Department of Energy website (energy.gov) has a good page explaining the basics of radiant heat at https://energy.gov/energysaver/home-heating-systems/radiant-heating.

Hydronic radiant heating has become increasingly popular for new construction in recent decades, particularly since the introduction of cross-linked polyethylene tubing, often generically called PEX, a leading brand. This flexible plastic tubing has excellent properties for transferring heat, is moderately priced, and can be connected with crimp or friction-fit fittings. Radiant systems turn surfaces like floors, walls, or ceilings into large radiators by pumping hot water from a boiler through coils of tubing attached to the backs of the surfaces.

These systems are invisible. Also, since radiant heat does not require the movement of air to heat spaces, it does not cause dust and other airborne particulates to circulate around a room and cover every horizontal surface as forced hot air and baseboard radiator systems do. The great challenge of radiant heating systems in historic houses is the difficulty of installing them behind existing surfaces, which are often character defining. If you have the necessary access (to the undersides of wood floors or the back sides of plaster walls or ceilings), the installation is very feasible for a do-it-yourselfer. (See "Example Project: A Radiant Heating System in a Historic House," page 314.)

Radiant ceilings have at least one advantage over radiant floors: They will never be covered by carpet, rugs, or heavy upholstered furniture that could interfere with heat radiation. If a ceiling system can be installed from above without disrupting historic ceiling plaster, this can be a great approach. It would likely work as well for a pressed metal ceiling installed over plaster. A radiant system installed above a new gypsum board ceiling could use a manufactured hydronic panel system for the installation and might be acceptable in a historic house if the plaster ceilings have to be replaced anyway. This work would be done from below; in this case, I recommend applying a skim-coat plaster over the gypsum to achieve a more appropriate look for the finished ceiling. (See Chapter 15, "Interior Walls," for more discussion of plaster versus gypsum board.)

One key to a functional hydronic radiant heat system under wood floors or behind plaster is to create an airspace for the tubing (and the aluminum heat distribution plates that hold the tubing in place) with a backing of foil-faced paper and sufficient insulation behind the foil paper to ensure that heat is radiated in the direction you want it to go. Radiating heat into a basement or crawl space is generally not the objective. The insulation may not be necessary in an interior wall if radiating at least some heat toward both sides of the wall is acceptable. The same is true of ceilings between floors with finished living spaces.

I can affirm from personal experience that switching from cold floors to floors with radiant heat is a wonderful experience. When I moved into Whitten House, the uninsulated floor over an unheated, uninsulated crawl space was so cold in the winter that water splashed on the floor while watering houseplants would freeze there. (The plants did not thrive.) I herniated a disc in my back while installing radiant tubing in the crawl space, but it was worth it! (My mistake was forgoing lower back support while lying

Jake is a big fan of the radiant-heated floors in Whitten House.

in the crawl space and reaching up repeatedly to staple hundreds of aluminum plates to the subflooring; you have to pay attention to what your body is telling you!) The cats are fond of the warm floors too.

Speaking of cats, litter boxes on radiant-heated floors can be an issue. The odor of cat urine can be pungent in the best of circumstances, and heating it does not help. Either leave an area without tubing for the litter box to sit on or elevate the litter box on ½-inch strips of wood to allow airflow beneath. I read half a dozen books on radiant heat before installing mine, and not one offered this warning.

If you will be installing ceramic tile flooring in a cold-winter climate, radiant heat under the tile will make the space much more comfortable in the cold months, especially in bathrooms. Bare feet on cold tile feel wonderful in August but not so good in February. Warm tile makes a huge difference. Radiant tubing can also be run in tiled walls, which is a great idea for tiled tub surrounds. And radiant tubing in a tiled wall is easily connected to a wall-mounted towel rack, adding the luxury of heated towels at no added expense beyond the cost of the rack. Some people run radiant tubing under stone countertops in kitchens to eliminate the uncomfortable cold feeling of the stone in winter.

Electric Heat

Baseboard Radiators. Electric baseboard heating came into wide use in the middle of the twentieth century, at a time when knowledgeable experts were declaring that nuclear energy was going to make electricity so cheap it would not have to be metered. Needless to say, that did not happen. Instead, the hidden costs of nuclear power generation became apparent, the construction of these plants stopped, and the price of oil, a leading fuel for electrical generation plants, skyrocketed. Electric baseboard heat became very expensive to operate. I was growing up in northern New Hampshire during those years. I remember an elementary school teacher commenting on how odd it was to see new houses without chimneys, since electric heat did not require them. Only a few years later, the parents of friends living in new homes heated by electricity were setting their thermostats at 52 degrees and still worrying about the electric bill. People heating with oil also faced skyrocketing heating costs at that time, but at least they had chimneys, and a great many woodstoves were connected to those chimneys during that time, our house among them. Many houses later had their baseboard electric removed and hydronic baseboard installed in its place.

Nevertheless, where electricity is less expensive and the heating needs moderate, baseboard electric heat remains a reasonable option. As with hydronic baseboard heat, the standard sheet-metal electric units generally clash with the character of a historic room. Since these units run hotter than hydronic units, proximity to upholstered furniture and window treatments is a serious concern, further limiting furniture placement and drapery lengths. If these units will remain in use in your house, do what you can to make them blend in; see the discussion in the "Hot Water Heat" section above.

Electric wall unit heaters can be useful for supplemental heating in bathrooms or other confined spaces. These units can be recessed into the wall framing or surface-mounted. A modern flat, metal radiant heat panel can be used with minimal visual impact in these situations, particularly if painted to blend in.

Electric Radiant Heat under Tile. Electric radiant heat mats for installation under tile floors are useful for adding heat in a bathroom or other space with a ceramic tile floor. In cold climates, where tile floors can be cold even in well-heated rooms, a radiant mat under the tile can make an enormous difference in comfort. Although less often used this way, they can also be installed in tiled tub surrounds to create a surface that radiates heat toward your naked body—a nice feeling on a cold winter morning!

This type of heat is not widely used as a primary heat source in cold climates. In the Northeast, electricity is expensive and other fuel sources are generally more affordable. In areas with more moderate heating needs and a tradition of tile flooring, such as the Southwest, this type of electric heat might be all you need to take the chill off cold weather.

Electric Radiant Heat in the Ceiling. Electric radiant heat is being installed in gypsum board ceilings as the primary heat source in some new construction. Lacking personal experience with its use in a retrofit, however, I hesitate to recommend it for a historic house. I fear that it might have a limited lifespan and prove difficult to replace.

I favor modern conveniences that have no visual impact on historic rooms, and I've installed hydronic radiant heating in my own historic house. I knew I could install it without disturbing or covering historic surfaces in primary spaces; I was confident that the system could operate for many decades undisturbed; and I knew that if and when a better system comes along, it can be removed with no lasting impact to historic surfaces.

An electric radiant system above a gypsum board ceiling might be acceptable in a historic house if the plaster ceilings were going to be replaced anyway. I would recommend applying a skim-coat plaster over the gypsum to achieve a more appropriate look for the finished ceiling. The caveat is that the ceiling will have to come down and be replaced when the electric heating units fail at some future point. If you are planning to install expensive reproduction ceiling papers or to have decorative painting reproduced on the ceiling, this could be a painful experience.

This is a typical wall-mounted head for a mini-split heat pump system. A sensitive restoration project will require careful placement to minimize the impact on historic character.

Heat Pumps

Heat pumps extract heat from the air and transfer it via a refrigerant fluid to another location. Even cold air contains calories of heat that a heat pump can extract to warm the air in a house. The pump can also work in reverse, extracting heat from interior air and transferring it outside to cool the house. Although electrically powered, heat pumps do not *generate* heat as traditional electric radiators do, and are consequently much less expensive to operate. In use for decades in mild climates, heat pumps have only recently entered wide use in cold-winter regions. In part this is due to the improved efficiency of the units, and in part it's due to their air-conditioning ability, which is increasingly appreciated as summers become warmer everywhere.

Installing a heat pump can be significantly less expensive than a central air-conditioning system and does not require the installation of ductwork throughout the house. "Mini-split" units have a wall- or floor-mounted "head" connected by coolant lines to an outdoor compressor. A fan within the head pulls air from the room; the air is heated or cooled in a heat exchanger (a coil of copper tubing surrounded by metal fins) in the head, then blown back into the space. There is a second heat exchanger outdoors.

Refrigerant circulates in a closed loop between the indoor and outdoor coils.

When you are heating a room, the refrigerant extracts heat from the outside air and evaporates into a gas. The gas then condenses back to a liquid in the indoor portion, releasing its stored heat to the room. When the direction of flow is reversed for air conditioning, the refrigerant extracts heat from the room and transfers it outdoors.

The most significant downside of a heat pump in a historic house is the appearance of the wall-mounted units. These are universally made of stamped metal or molded plastic colored white, off-white, or gray. They need to be mounted six feet or more off the floor and simply do not blend into most historic rooms, and they are often aligned with doorways in order to blow air into several rooms, making them even more prominent.

Floor-standing units are also available and are similar in size and placement to cast iron radiators, making them less visible than wall-mounted units. However, their placement renders them somewhat less effective at warming and cooling the space.

Wall-mounted heads can be painted to match the surface they are mounted to, somewhat reducing their visibility (check to make sure this will not void the warranty). They can be built into cabinetry and screened with wood or metal lattice, though this will reduce their airflow and efficiency. Careful placement can help reduce visibility. For instance, mounting a unit over the primary entrance to a room will hide it from those walking into or passing by the room. If the room's seating faces away from the door, toward a fireplace or other focal point, the unit will not be particularly noticeable even when the room is in use. Hopefully manufacturers will soon recognize that there is a market for more compatible head covers in place of the current stamped metal and molded plastic covers.

This floor-standing unit for a mini-split system is similar in size and placement to a cast iron radiator and may have less visual impact than a wall-mounted head. For this unit (which actually is fastened to the wall), the insulated coolant tubing and electrical lines are contained in a three-sided wood box matching the historic baseboard.

Another option is a concealed ceiling unit. This can be recessed into the ceiling framing, leaving only narrow slot vents or grilles visible. The installation requires removal of plaster and lath (or flooring above), and future replacement would require a further disturbance, but this can be an attractive option if there is an unfinished attic above the space to be conditioned. The market for heat pump/HVAC systems is expanding, and additional options are likely to become available. Explore all the options before committing to one.

Another issue with a mini-split in a historic house is that installers prefer to run the insulated refrigerant tubing and electrical line along exterior walls, penetrating the wall where the head is mounted. These exterior runs are typically covered with a formed plastic shield or conduit that is quite noticeable and incompatible with a clapboard, shingle, brick, or stone wall even when painted to blend in. This installation approach also ensures that the head will be mounted on the interior face of an exterior wall, rather than in a less visible location on an interior wall. If at all possible, the supply conduit should be run through a basement, attic, or wall or ceiling cavity. Closets and mechanical chases provide other opportunities to hide these elements. The conduit should never be run on the exterior of a principal elevation.

This mini-split compressor is mounted on the front façade of a modest Greek Revival house, close to the sidewalk. Adding insult to injury, the white plastic line shield runs up from the compressor to the interior location of the wall-mounted head. Avoid this.

Exterior compressors should, of course, be located in an inconspicuous location near the house. Your installer will probably want to place it in the most convenient location for his work and may imply that his preferred spot is the only possible one. Do not accept this if the location will make the compressor highly visible. If your house is in a local historic district, you may need a certificate of appropriateness for the compressor; the same may be true for any exterior line shields that can be seen from the street. The threat of a local historic review board proceeding can sometimes cause a contractor to realize that his preferred location is not the only possible one after all.

Heating Controls

Historically, most central heating systems were controlled by a single thermostat using mercury to sense temperature. The desired air temperature was set with a dial, and the thermostat would turn the system on and off as needed to maintain that temperature.

Modern electronics have dramatically altered this century-old standard. Programmable thermostats have become the norm, and internet-connected controls are rapidly catching on. Any recommendations I might make would soon become dated, but I'll venture a few general comments.

Energy efficiency gets a lot of attention in this time of rising sea levels and extreme weather events. The overwhelming scientific consensus is that human effects on the climate, primarily from the burning of fossil fuels, are to blame. The simple act of rehabilitating and living in an already-built house is a positive contribution to solving the problem. Far fewer materials and much less energy are consumed to rehab an existing house than to build a new one. A rehab also makes use of the energy already embodied in the structure—the materials, labor, and power that went into creating it. As the saying goes, "The greenest building is the one that is already built."

The rehabilitation of a house can and typically does greatly improve its energy performance. Electronic controls for the efficient use of the system will further benefit your wallet and the environment. The challenge in a historic preservation project is that modern HVAC controls are almost universally made of formed plastic and are not aesthetically compatible with historic finishes. They are also increasingly large, to accommodate touch screens. As long as these units need to be in living spaces to monitor the temperature, they will remain an incongruous element that distracts from historic character. To control the new radiant heat system in Whitten House, I rejected modern control units in favor of new-in-the-box metal thermostats from the 1930s, purchased on eBay. I was able to locate five matching units to handle the five heating zones in the house. Since radiant systems are best operated at a continuous temperature, eliminating any need for a programmable control, these work as well as a modern control could and feel far more compatible with the character of the spaces. As with heat-pump units, it would be wonderful if manufacturers would begin to offer less glaringly contemporary options. Alternatively, cast-metal cover units with perforated side panels and a hinged front might be used to hide modern plastic units without impeding their function. Perhaps one of the reproduction hardware manufacturers could take this on.

Fireplaces

Although fireplaces were the only source of heat in most houses before the mid-nineteenth century, few people use them as a primary heat source today. Nevertheless, the symbolism of a fireplace—as epitomized in the phrase "hearth and home"—has never waned. Fireplaces have never ceased to be a desired amenity and continued to be built in the principal rooms of many homes long after their replacement as the primary heat source by the cast iron stove and later by a central heating system. Some of the most iconic Modernist houses in America, such as Frank Lloyd Wright's Fallingwater and Robie House, have their living spaces built around a massive chimney and fireplace.

Many old houses have fireplaces that have not been used in many years. They may have been sealed up with brick, plaster, or sheet metal, with a thimble for a stove pipe. Often these fireplaces can be returned to usable condition without a complete reconstruction of the chimney. If the house is old enough, it may have had its fireplaces altered early on to improve efficiency. Seventeenth- and eighteenth-century fireboxes were typically rectangular in plan,

with straight-up vertical sides and back. The flue opened into the firebox without a smoke shelf to prevent downdrafts from blowing smoke into the room.

Technological advancements in the late eighteenth and early nineteenth centuries led to much more efficient fireplaces with a shallower box, slanted sides, and a back that was tilted forward to support a smoke shelf at the base of the flue. Often called a Rumford fireplace, this configuration throws much more heat into the room and sends much less heat up the chimney. Many eighteenth-century fireplaces were rebuilt into the Rumford design, leaving a space behind the new back wall of the firebox. Because fireplaces today are used more for ambiance than heating, the earlier design can be more desirable (with a smoke shelf and damper added at the flue). An experienced restoration mason can determine whether a chimney was altered in the past and often can reconstruct it in its original form if desired.

This late-nineteenth-century photo shows a cast iron insert installed in the fireplace of an 1820s house, probably about twenty years after the house was built.

Another approach to improving the efficiency of fireplaces before the widespread adoption of closed cast iron stoves was the cast iron fireplace insert, also known as the Franklin stove. This consisted of two curving iron sides and a top, sometimes with an iron floor or frame, and was usually installed so it sat out on the brick or stone hearth, projecting in front of the old fireplace opening by a foot or more. The remainder of the old firebox was usually bricked in around the insert, and a new brick back wall was laid for the insert if it did not have a cast iron back. The insert was open in the front and frequently featured ornamental brass balls or finials on the front corners of the top. By pulling the fire forward on the hearth and taking advantage of the radiant properties of cast iron, a Franklin stove could heat a room faster, better, and with less fuel than a masonry fireplace.

Hearths

Fireproof hearths are an essential part of a functioning fireplace. Made of brick, stone, or tile, they project into the room in front of the fireplace, typically flush with the flooring. In historic kitchens, the hearth was used to cook on before the introduction of cast iron cooking stoves. Hearths in the Victorian, Arts & Crafts, and Colonial Revival eras were frequently made of ceramic tile and were typically accompanied by a tiled firebox surround inside the mantel. Depending on the period and style of the house, the tiles could be hand-painted, transfer-printed, encaustic, terracotta, or Art Tiles—often with relief patterns. Hearths are a character-defining feature of a room, and it is important to preserve original materials and

An unvented gas log unit has been installed in this cast iron Franklin stove (essentially a freestanding cast iron fireplace insert). The homeowner carefully drilled a hole in the cast iron for the gas pipe, starting with a very small bit size and enlarging the hole by increasing the bit size in consecutive passes.

to use appropriate replacements for missing materials. A fireplace that retains high-gloss glazed Art Tiles on its surround should not have missing hearth tiles replaced with 6-inch by 6-inch white bathroom tile or by rough eighteenth-century hand-molded brick. Do the research to understand what is appropriate and select accordingly.

Cast Iron Coal Grate Inserts

A bit later in the nineteenth century, cast iron inserts with coal grates came into use, primarily in urban areas where coal was readily available. These inserts were often highly ornamental, and some incorporated a ceramic tile surround. Because these appeared when the Italianate and Second Empire styles were in vogue, they were frequently made to fit the round-arched opening of the marble mantels commonly used in these houses. (Ornamental cast iron ventilation grilles designed to fit in false fireplaces can resemble coal grate inserts in appearance but not in function; see the photo of Farnsworth House mantel on pages 308-309.)

Historic coal grate inserts can sometimes be found in usable condition, although the flue may need to be lined to meet code requirements (see below). It is also not uncommon to find part of a coal grate insert remaining, usually the cast iron frame, with other parts missing, usually the coal grate itself. Check the attic, basement, shed, or barn to see if the missing pieces are in storage; if not, you may be able to find replacements at a local architectural salvage yard.

Reproduction inserts that burn natural gas or propane (LPG) are available as an alternative to burning coal. The gas units can be connected to a thermostat and used as a secondary or even primary heat source for the room.

Gas Conversions

Historic coal- or wood-burning fireplaces can often be converted to gas with a vented or unvented configuration. Unvented gas inserts, gas logs, or gas "coal" baskets can be used if the flue is not usable, either due to its condition or because it vents another fireplace or a furnace. There are requirements to ensure an adequate air supply without depleting the oxygen in the room to dangerous levels, and all modern units have built-in CO_2 sensors and other safety features to turn off the flame if conditions warrant. Burning gas in an unvented unit can increase the humidity of a

This modern vented gas-fired unit resembling a nineteenth-century coal grate is installed in an 1850s fireplace opening with a new slate surround. As a vented unit, the fire is contained behind glass. A stainless steel flue liner runs up the chimney.

room, which can be a detriment in a humid climate or an advantage where the winter air is dry.

Flue Liners

Depending on the age of a chimney and whether or not it was repaired or rebuilt in the past, the flues might have a fired clay liner, a modern stainless-steel liner, a parged cement coating, or no liner at all. In general, modern codes require a chimney flue to be lined if it is to be used. A reasonable code officer may agree that an intact parged lining is equivalent to other more common modern flue liners. A restoration mason can drop a camera down the chimney on a line to inspect the condition of the parging.

Most insurance companies will not be happy about an unlined chimney being used. Options for installing a liner in a chimney that lacks one include—besides clay tile and stainless steel—a grouted liner formed around an inflated tube in the chimney.

Flues and Building Codes

It used to be common for a single flue to serve several fireplaces or stoves, but modern code does not allow this. You may have to select which fireplace or stove thimble on a flue to maintain as usable when installing a new liner. If the existing flue is large enough, it may be possible to subdivide it with two stainless steel sleeves. If only one fireplace can remain in use for wood or coal fires, it may be possible to install an unvented gas log in a second fireplace that previously shared the flue.

It was regular practice in the part of the country I live in for the original brick oven flue (in a house old enough to have one) to be reused for a furnace in the basement when central heating was installed. Cast iron cookstoves had made brick ovens obsolete, and the oven flue could be repurposed without affecting the use of the adjacent fireplace. It has been long recognized that a flue should not be used for both an oil or gas appliance and a wood fire. If a furnace or boiler is vented into the bottom of a chimney arch in the cellar/basement and the oven door is sealed shut or bricked in, there is a good chance the oven flue was repurposed.

This can be reversed if you want the oven to be usable again, but an alternative vent for the boiler or furnace will have to be found.

Many modern heating appliances are so efficient that their exhaust is primarily water vapor and they can be safely vented with PVC pipe. Replacing an old furnace or boiler with one of these units can free up the chimney flue for its original use.

If a chimney needs to be rebuilt anyway, installing a clay tile liner is the way to go. Many restorers have chimneys rebuilt in concrete block with historic brick used for visible surfaces of fireplaces and above the roof. This approach is substantially less expensive than an all-brick chimney, and if the chimney is large enough (and the mason small enough), it can sometimes be accomplished without disturbing the plaster or paneled walls around the chimney. Clay lining can also be installed by opening up one side of a chimney (which requires removing the plaster or wood paneling covering that side), mortaring in the flue tiles, and rebuilding the chimney wall as the tiles rise. If a chimney has more than one flue, you'll want the side you open up to grant access to all of them.

A typical modern cap for a stainless-steel flue liner. Although required by code, these are distracting and alter the character of a historic chimney. They should at least be painted flat black.

A traditional stone chimney cap set on brick piers will keep rain and snow from entering the flues.

Stainless-steel liners are popular for chimneys that are to be used for wood or gas stoves, gas fireplace inserts, or oil burner/boiler flues. A downside to these relatively easy-to-install liners is the code requirement that they be terminated with a metal cap above the top of the chimney. All too often, the

This stainless-steel chimney cap is even more distracting and out of character than the flue liner cap shown in the preceding photo. If painted flat black it would be less eye catching.

installer leaves a shiny steel cap sticking up nearly a foot above the brick, clashing with the character of the house and drawing attention to itself by glinting in the sun. There are several ways to disguise these caps, but perhaps the best is to install a bluestone or similar chimney cap above the metal flue cap. Chimney caps are a traditional means to keep rain from entering a chimney. Brick piers 9 to 12 inches tall are mortared in place at each corner of the chimney, and the cap is set atop the piers. A stainless flue cap is far less obvious in the shadowy recess under a stone chimney cap, and painting it black will make it even less visible. Full-sized black-painted metal chimney caps can achieve a similar result, although they were not nearly as common historically. Another alternative is a lower-profile flue cap in copper or another metal that will patina over time and look compatible with the period and materials of the house. If no better solution is affordable or your local code enforcement officer insists on a freestanding tall metal flue cap, you can at least paint it flat black to reduce its prominence.

A grouted flue liner can be a good solution if the flue shape allows it. This is installed by placing an inflatable tube or bladder down the chimney, centering it in the opening to leave space on all sides, and pumping cement grout around the bladder. Once the grout cures, the bladder is deflated and removed, leaving a solidly grout-lined round or oval flue. The

A traditional chimney cap on taller brick piers can be used to conceal a stainless-steel flue liner cap. If painted flat black, the flue liner cap will be virtually invisible under the stone cap.

result is similar to a historic parged chimney but with a thicker liner. Some historic chimneys carry three or four long, narrow, rectangular flues side by side, usually with a single wythe of brick between them. Each flue serves a single fireplace, or perhaps a fireplace on one floor and stove thimble on another. The round or oval inflatable bladders used for installing grouted liners will not fit in narrow flues and cannot be used with flues of this design.

Dampers

A damper to close a fireplace flue did not become a priority until central heating relegated the fireplace to occasional use. Before then, it was common to use "fireplace boards" to seal up the opening of a fireplace when not in use. These were often decorated with paint or wallpaper.

If your fireplace is old enough, it may have been constructed without a metal damper. A damper may well have been retrofitted at some later time if the fireplace remained in occasional use, but not if it was sealed up long ago. Several options are available for installing a damper in a fireplace that does not have one. The classic cast iron damper in a frame can be installed in the throat of the fireplace by a mason or a skilled homeowner. This will require some level of deconstruction and rebuilding of the firebox. How extensive and invasive this work will be depends on the original design of the fireplace and flue.

There are modern dampers that mount on the top of the chimney and operate with a sprung lid controlled by a cable that hangs down the flue. Restoration masons I have talked to, however, express concerns about the possibility of these dampers trapping moisture in the chimney and causing long-term damage. Modern lifestyles—daily showers, cooking, humidifiers, etc.—put a lot of moisture into the air of a house, and modern energy efficiency practices insulate and seal the exterior envelope of a house to such an extent that moisture can't easily escape. It is possible that a good deal of that moisture could end up in a top-dampered chimney flue, whereas a flue with a damper at its base can easily vent moisture out the top. (The latter will, however, also allow rain and snow *in* at the top, along with an occasional bird or squirrel.)

The simple solution I chose for the undampered fireplace flues at Whitten House was to have custom sheet-metal dampers made. These are flat, with a handle mounted on the bottom, and slide between 1-inch angle iron and the masonry neck of the flue.

A nineteenth-century fireplace board with period wallpaper and border. These were used to close undampered fireplaces in summer and when idle in winter.

The angle iron pieces are screwed into the mortar joints. The damper slides out and is set into the adjacent chimney closet while the fireplace is in use, to be returned to its place once the ashes are cold.

I recommend working with an experienced restoration mason to determine the best approach to installing dampers in undampered fireplaces.

False Fireplaces

The presence of a mantel does not always mean the room had a fireplace. Many fireplaces were bricked up after cast iron stoves came into use, leaving a thimble into the former firebox or into the flue above the mantel so a stove could be set on the hearth. It's also true, however, that many mantels were installed in front of chimneys without fireboxes, intended for stoves from the beginning.

A common clue to whether a mantel surrounds a closed-up fireplace is the presence of a brick, stone, or tile hearth in front of it. Many mantels installed without fireplaces never had a hearth; instead, the wood flooring runs to the wall. Cast iron stoves commonly stood on wood floors without any combustion protection other than the stove's elevation on its legs, though sometimes an iron sheet would be placed under the stove. (Insulated heat mats are required for stoves installed over wood floors today.) Further confusing the issue, sometimes a brick or tile hearth was removed and wood flooring patched in when the fireplace was bricked up. This can be detected if there is an obvious patch in the flooring or if the underside is visible in a basement or crawl space. Putting a light into the thimble hole is often a quick way to see whether it opens into a narrow flue or a larger firebox.

Going one step further, decorative mantels continued to be installed after the advent of central heating. Though no longer paired with fireplaces or woodstoves, they continued to embody the symbolism of hearth and home in marble or carved oak. Such an installation might contain cast iron frames and grilles that resemble decorative coal-burning fireplace grates

This impressive cast iron and faux marble (reverse painted on glass) mantel appears at first glance to surround a coal-burning fireplace, which would be consistent with the mid-nineteenth-century Italianate house's age and style. But close examination, as shown in the two following photos, makes clear that this is a ventilation grate, not a fireplace. A 2-inch-deep recess behind the grille front leads up to a small opening into a chimney flue. Before the availability of electric blowers, this passive ventilation system allowed warm air to rise into the chimney, which drew freshly heated air into the room from the basement furnace through a floor grate.

but, in fact, delivered heat from a furnace duct. In other cases, the grille opened into a ventilation flue, allowing airflow up the chimney as part of a passive heating system before electric blowers were available for forced hot-air systems. A slow, steady, draft of warmed air moving up the chimney would draw fresh hot air into the room from a floor grate ducted from a furnace below.

If you discover a false fireplace where you were hoping for real one, a good restoration mason may be able to construct one. This is not a small or simple task and will likely involve a good deal of demolition to create space for the masonry and required clearances from wood framing. Any existing chimney will have to be reconstructed from the basement up. If the false fireplace has no chimney behind it, a new chimney will have to be constructed.

If this level of work is beyond your scope or budget, it may be possible to install a modern gas fireplace behind the historic mantel without the need for a masonry chimney. There are zero-clearance units that are allowed by code to contact wood framing members and can be vented through a small-diameter PVC pipe or unvented altogether. Other units require a larger vent pipe and/or a stainless-steel vent.

A closer look at the "firebox door" reveals the absence of any way to remove or open it to light and maintain a fire.

Moving in even closer confirms that there is in fact no firebox, as the brick wall is set only 2 inches behind the cast iron front.

A false fireplace mantel is often an important character-defining feature in a room. Its preservation should be a high priority whether it can be made functional or not.

Stoves

Cast iron wood-burning stoves largely replaced open fireplaces for heating after the mid-nineteenth century and are still widely used as primary heat sources in some parts of the country, typically where cordwood is abundant and more affordable than oil or gas. Modern woodstoves are more efficient and less polluting than historic stoves, but many historic stoves remain in use. A historic room requires either an antique stove or a modern stove of traditional design. Historic cast iron stoves were often highly decorative, reflecting popular architectural styles of the period in which they were made. Cast iron was the plastic of the nineteenth century. Once the technology was developed, the Victorians cast virtually every form and pattern imaginable.

There are companies across the country that specialize in restoring and selling antique cast iron stoves for heating and cooking. Some of these retrofit stoves with catalytic converters to reduce pollution or install gas burners in place of historic wood or coal grates. Cook stoves are often converted to gas or electric, preserving the

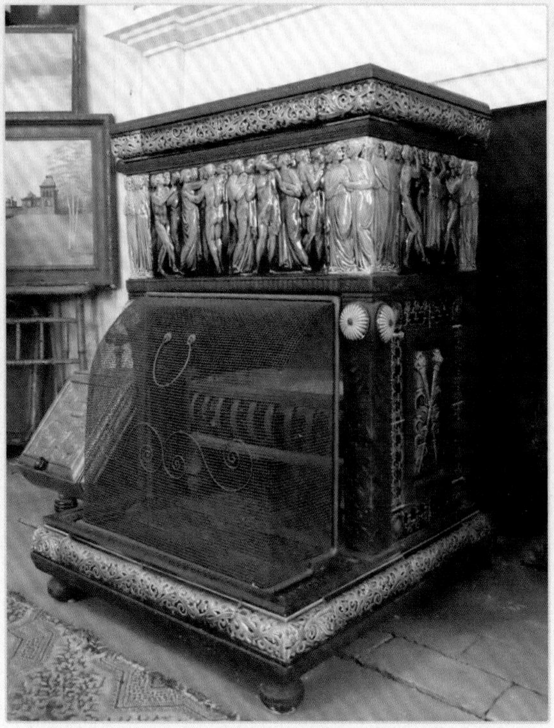

During the nineteenth century, heating stoves were often highly decorative and intended to provide aesthetic pleasure as well as warmth. This stove, fabricated with a cast iron frame and ceramic tiles, was made by the Low Art Tile Works of Chelsea, Massachusetts, in the 1880s.

appearance of a historic wood- or coal-burning range with the convenience of a modern range.

Coal-burning stoves were used in much of the country for heating and cooking from the late nineteenth well into the twentieth century. A few people still heat with coal, and a handful still cook with it (see for example the Dow Farm, page 175). Coal has become a dirty word (literally?) due to its well-documented role in global warming and related climate change, but small-scale localized use of coal for domestic heating and cooking contributes very little to atmospheric pollution—certainly no more than heating with cordwood or an old, inefficient oil-fired system. The environmental damage comes from large-scale industrial use and the generation of electricity in massive coal-burning power plants with

This cast iron stove with Gothic Revival detailing may have been used in a location much like this when new in the mid-nineteenth century. The absence of a masonry hearth indicated that the mantel never surrounded a fireplace but was intended for a stove. With a heat-resistant floor mat (now required by code), this stove could be used for heating with wood or coal or could be converted to gas.

The stove on the previous page was made with a removable front panel so it could be used as an open fireplace-type stove or as a closed "airtight" stove.

tall stacks that send the smoke high into the atmosphere. One modern power plant in Georgia burns 1,288 tons of coal per hour at full operation. Dow Farm burns about three tons a year. The power plant burns soft, bituminous coal, which produces more pollutants. Dow Farm uses hard coal, which burns much more cleanly. This is not to a pitch for widespread adoption of domestic coal heating in the twenty-first century—just a suggestion not to conflate the pollution spewing from massive coal-fired plants with your neighbor's tiny bit of carbon from coal.

Gas Conversion

As with fireplaces, it is possible to install gas burners in historic cast iron stoves. This can be an effective way to enjoy supplemental heat from an ornamental antique stove without having to deal with wood or coal and their resulting ash. An open-front stove can be fitted with a vent-free gas log or coal basket. A closed stove should be vented through a chimney. Air-control dampers on the stove will likely have to be fixed open to ensure that the fire gets enough oxygen to burn properly. Many antique stove restoration shops specialize in this work, and your insurance company might be more comfortable with having the conversion and installation done by a professional.

Gas-fired Heating Stoves

Modern gas-fired heating stoves are widely available, many in more-or-less traditional forms. Although generally less ornate than antique stoves, a modern stove that resembles a historic form can be quite compatible

This modern gas stove resembles a traditional woodstove and is compatible in a historic context. Colored enamels on cast iron stoves came into use in the early twentieth century, most often on cookstoves. This stove is connected to a thermostat powered by a thermocouple and the stove's pilot light. It is the heat source for the room.

with a historic setting. Depending on the size of the unit relative to the space it occupies, it can be the primary or secondary heat source. Many of these are available with thermopiles to control a wall-mounted thermostat. The thermopile is heated by the pilot flame, generating a small amount of electricity to operate the thermostat. A significant advantage of these is that loss of electrical power in a storm does not affect the operability of the thermostat and stove. You will stay warm even through an extended power outage. Like gas logs and fireplace inserts, these are available in vented or vent-free models.

Place these stoves where a wood or coal stove might historically have stood. (Look for a stove thimble in a chimney.) The building code may require a heat mat over wood flooring, but back and side clearances are generally minimal for a new gas stove—far less than for a woodstove. A simple enameled metal heat mat or a slab of slate, marble, or another traditional fireproof material can be used. Avoid building up a brick hearth on top of the floor. Historic hearths were nearly always flush with the floor and built in front of a fireplace. When stoves came into use, sheet metal was generally used to protect floors, not brick. A mantel shelf may be mounted over a gas heating stove, just as mantels were often mounted above woodstoves in the nineteenth century.

Example Project: A Radiant Heating System in a Historic House

While increasingly common in new construction, radiant heating systems are not often installed in historic houses. And yet, for houses built before the advent of central heating systems, radiant heat can prove an ideal way to enjoy modern comfort and heating efficiency without impacting historic character. It is also possible to install such a system with minimal impact to the fabric of a historic house. Unlike the ductwork required for forced hot-air systems, the tubing used in radiant heating systems is small and flexible, making it possible to run it between floors or in walls.

The house in this project is Whitten House, my own house, which appears in several places in this book. It was built in the first half of the nineteenth century and has a timber frame, with large beams spaced at relatively wide intervals. It is a two-story house with an attic that was originally finished but had lost its historic plaster walls and ceilings. A story-and-a-half ell extends from the rear of the main house, and a one-story modern addition extends from the ell. The main house has only a crawl space beneath it; the ell has a low cellar; and the modern addition is on a concrete slab at grade. A forced hot-air system was installed in the first floor of the house in the 1960s, with a large oil-fired furnace in the cellar under the ell.

The fifty-year-old forced hot-air furnace that I replaced in this project is shown being prepared for removal. Note the water marks on the side of the furnace, the result of periodic flooding in the cellar. The new gas-fired boiler will be located on the first floor, well above any future flooding.

This oil tank was in the cellar with the old furnace, a typical arrangement in this part of the country. In this damp environment, there was a real chance of the tank rusting through and leaking after fifty years. I was not sorry to see it go.

The existing system was approaching fifty years old when I replaced it. It was inefficient and heated the house poorly.

Designing the System

The installation of a natural gas line in the street on which the house sits presented the opportunity to switch to a cleaner and more affordable fuel source. With the existing heating system on its last legs and the cost of oil increasing rapidly, the decision was easy to make.

The simplest and cheapest solution would have been to install a new gas-fired furnace to feed the existing forced hot-air system, reusing the ductwork and extending it to the second story. The metal floor registers from the 1960s did not add to the character of the house, but they were not a major distraction either. Cast iron registers could have been installed to improve their appearance. However, large-duct forced hot-air systems come with blower noises and blowing dust. They tend to heat rooms unevenly, and creating separate heating zones is complicated. Additionally, all the metal ducts in the unheated crawl space were uninsulated, a major source of heat loss, and would have had to be insulated or replaced with new insulated ductwork. In sum, the cheapest solution was not the best in this instance.

Another option was to connect a new gas-fired boiler with metal baseboard radiators in all the rooms. Convection radiators of this type heat more evenly than forced hot-air systems and are quiet and reasonably efficient. Because they move heat through small-diameter tubing, they're easier to install than hot-air ducts, which must be hidden in closets and chases. The biggest downside of baseboard radiators is that they are not aesthetically compatible with the character of historic rooms, even when painted to match the baseboard they're mounted to. Cast iron baseboard radiators are more attractive but also significantly more expensive than the standard sheet-metal versions, and either type forces furniture away from walls and prohibits window drapes from reaching the floor, as they often did historically. Of lesser concern, baseboard radiators set up

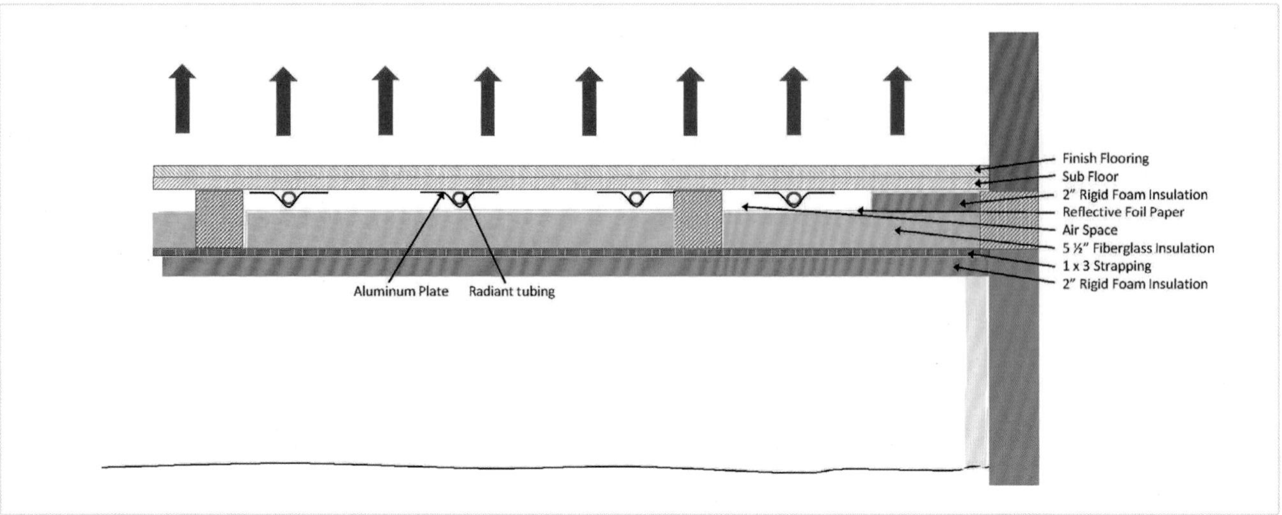

This drawing shows the first-floor installation. Aluminum plates help to spread the heat from the tubing under the floor. The air space, foil-faced paper, and insulation are critical, directing the heat up into the living space.

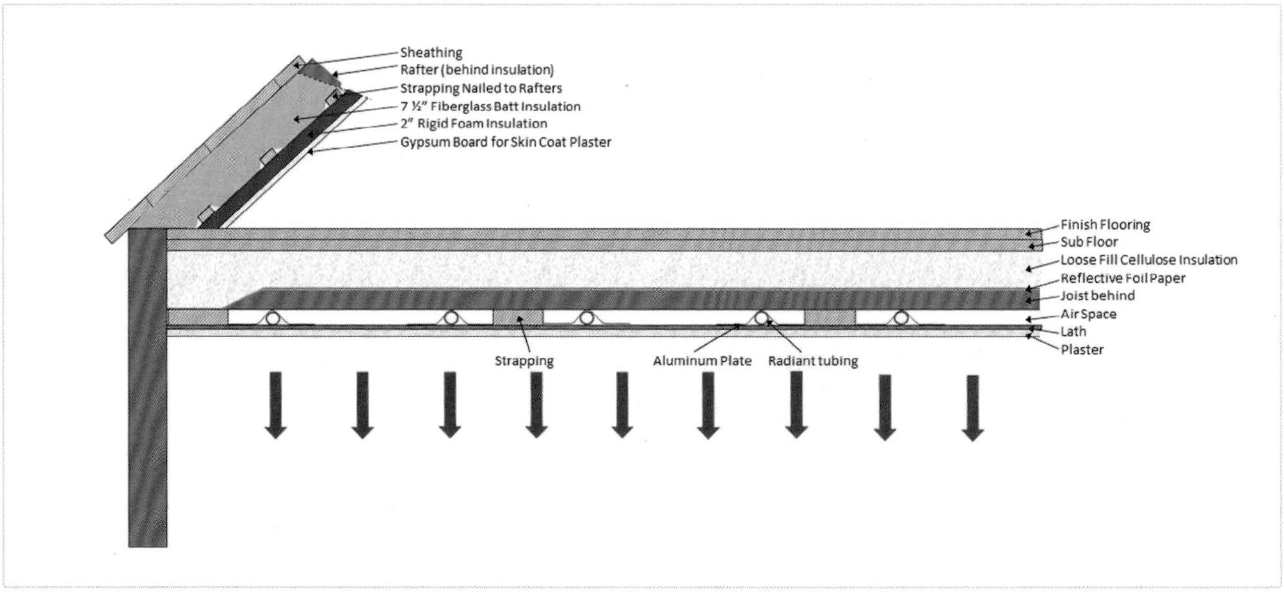

This drawing illustrates the upper-floor installation. Here, the air space, foil-faced paper, and insulation are configured to direct the heat down into the primary living space. I used less insulation above the radiant tubing and air space to allow some heat up into the finished attic rooms. Had the attic been unfinished, I would have increased the insulation substantially.

an air-circulation cell in which cool air returns to the radiator across the floor, creating an ankle-height draft and moving dust around.

The third option was to couple a new gas-fired boiler with radiant heat tubing under the floors and above the ceilings, an efficient system that would not be visible at all in the historic rooms and would eliminate the air currents of a convection system. With a radiant system, there are no limitations on furniture placement or drapery lengths. It is the most complex system to install and, if done by professionals, probably the most expensive. However, a reasonably handy homeowner can do the labor-intensive installation of the tubing and heat-distribution plates, leaving the boiler installation and hookups to a professional. A braver homeowner could do the whole system.

A radiant system works by circulating hot water through loops of plastic tubing that transfer heat to the adjoining surfaces. In new construction, this is often done by encasing the tubing in concrete, turning the entire slab into a heat-radiating surface. In existing construction, aluminum plates are used to

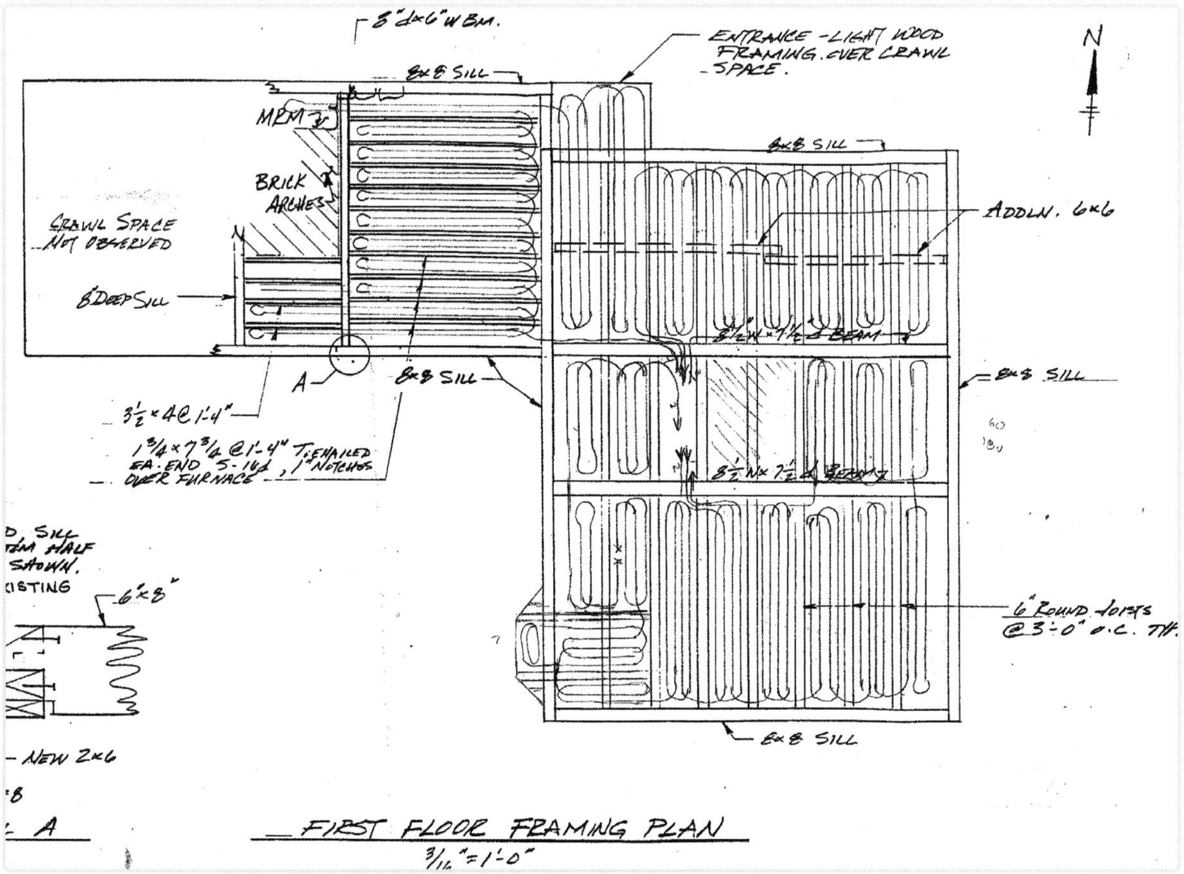

After penciling in the layout for radiant tubing under the first floor on a photocopy of an existing drawing of the framing, I made copies to work with in the crawl space under the house.

I used this homemade tubing spooler to uncoil the PEX tubing in the crawl space under the house. It was constructed from 2x wood, plywood, and a Lazy Susan turntable. The skids allowed me to move it relatively easily on the dirt.

Once the tubing was tacked roughly in place with U-shaped clips, I stapled up the aluminum plates to hold it in position.

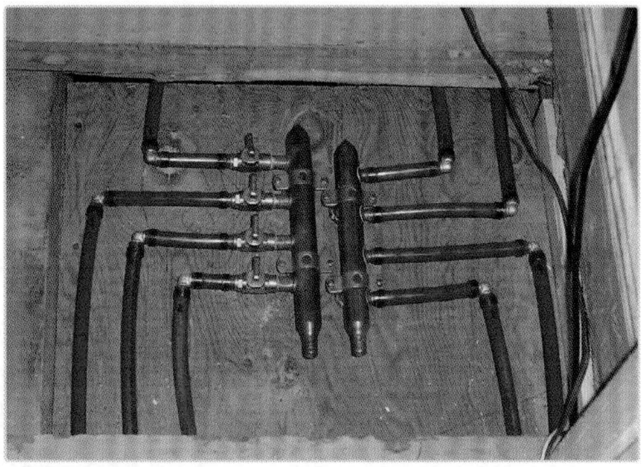

The manifolds for the four loops on the first floor were located under a removable section of flooring in a closet near the center of the house. An accessible basement (versus the largely inaccessible crawl space under Whitten House) would have eliminated the need for access through the floor.

Here a section of tubing has been run in the framing bays, passing through 1-inch holes drilled in the beams with a heavy-duty right-angle drill, and tacked in place. To create continuous 300-foot runs of tubing, I had to thread it all the way through the series of holes to the manifold location and then pull it through from the spooler to form the loops in each bay.

There was no access from below to install tubing in a new bay window that is cantilevered from the floor framing. Instead I installed the tubing from above and then sealed the underside of the bay with pressure-treated plywood. I filled the bays with spray foam insulation to just below the tubing, then installed foil-backed paper atop the insulation to direct the heat up into the room. A photo like this is helpful for avoiding the tubing when it is time to nail down the flooring.

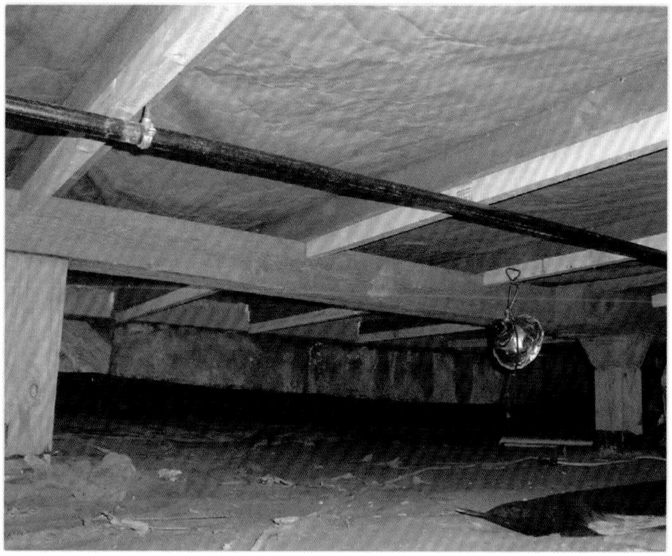

Here the foil-faced paper has been stapled up below the radiant tubing, leaving a 2-inch air space below the floor. Note the 2x material screwed onto the smaller floor beams to allow for the necessary depth of the fiberglass insulation that will be installed below the paper.

In this view, 1 x 3 strapping has been screwed to the framing to help support the fiberglass insulation and the rigid foam insulation that will be below it. Installation of the fiberglass is underway in the distance.

Here the 2-inch rigid foam insulation is being installed. It is screwed to the 1 x 3 strapping with screw washers to prevent the screws from pushing through the foam insulation.

A close-up view of the insulation assembly. Note the rigid insulation against the exterior sill at the left and the spray foam insulation used to seal the edges and irregular spaces. The damaged beam is the result of long-term moisture issues in the crawl space. Fortunately, the old-growth wood was dense and resinous enough to retain its structural integrity in spite of ½ inch of rot near the surface.

In this nearly completed section, the rigid insulation seams need to be sealed with foil tape. Spray foam insulation will be applied around the partially visible black iron gas pipe. Note the heavy plastic that has been glued to the foundation, the first piece of a moisture barrier that will cover all the dirt in the crawl space.

Here the foil tape has been applied to the seams of the rigid insulation, and a spray foam seal has been applied between the insulation and the stone base of the center chimney. All that remains is to install the plastic moisture barrier over the dirt.

transfer heat from the tubing to the wood subfloor, usually with a narrow air space for holding heat as well. Foil-faced paper faces toward the room, helping to direct the heat toward the living space. The wood or tile floor becomes the radiating surface. Unlike convection systems, which heat the air, radiant heat heats the objects in the space, including people.

Installation Under the House

I drew the radiant tubing layout in pencil, creating loops of approximately 300 feet each; these connected at supply and return manifolds that would be located under a removable section of floor in a closet next to the center chimney. My goal was an even coverage of the floor area with a minimum of holes drilled through framing timbers. I kept the tubing 12 inches from exterior walls to allow for insulation between the heated air space and the wood sills; this would prevent thermal transfer through the sills to the exterior. The boiler would be located next to the chimney in the ell, so I ran larger-diameter supply and return lines from that location to the manifolds. (I did not run a loop of tubing in the framing bay that carried the supply and return lines.) The framing bay dimensions varied in different parts of the building, so I adjusted the tubing layout to fit the circumstances. The floor framing in the ell had been replaced in the past with 2 x 10s and required a different layout than under the main house, which retained its original heavy framing.

Insulation

Without a heated air space, foil-faced paper to direct the heat, and insulation to force the heat in the desired direction, it is likely the tubing and aluminum plates would not be able to heat the floor sufficiently to condition the space above.

The modern natural gas–fired boiler occupies a small closet and vents through the PVC pipes at the top (one is the air intake; the other, the vent). Only one of the three heating zones has been connected to its zone valve. The other two need to be run to their zone valves (center) and the return manifold (right).

I found enough matching new-in-the-box metal Minneapolis Honeywell thermostats from the 1930s for all of the radiant zones. There is something very cool about opening a sealed 80-year-old box and taking out a brand-new yet vintage item.

Installation in the Second-floor Ceilings

To heat the second story, I elected to lift the attic floor and install the radiant tubing above the plaster ceilings. Because the knee walls had already been removed from the attic, it was not difficult to lift the finish floorboards and the subflooring below them, exposing the framing bays beneath. The cellulose insulation that had been blown into these bays in the 1980s had to be removed, a nasty process involving a plastic dust pan to scoop with and contractor's trash bags to receive the insulation. A respirator and eye protection were essential.

The Boiler

I selected a wall-hung combination heat-and-domestic-hot-water boiler to run the new system. With a 95 percent efficiency rating, the boiler could be vented with 2-inch PVC pipe and would fit into a first-floor closet, getting it out of the wet basement, where the old furnace had been flooded more than once. Unlike the old oil-fired furnace with its roaring burner, rattling blower, and occasionally squealing bearings, the new boiler can barely be heard from the adjacent room when it is running. The on-demand endless hot water in the shower makes my husband very happy.

Compatible Thermostats

Finding a modern thermostat that is aesthetically compatible with a historic house is no small task. In fact, it is nearly impossible. A months-long search prior to the beginning of this project failed to turn up any modern thermostats that are not obviously plastic. Even very expensive wireless digital thermostats that can be controlled remotely look cheap and flimsy. There is a great opportunity for one of the reproduction hardware companies to produce a metal cover for modern plastic thermostats.

My eventual solution was to purchase "new old stock" metal thermostats from the 1930s. These were in their original packaging and had never been installed. While the 1930s arrived a century after the house was built, the thermostats were nevertheless made of materials in use in the nineteenth century and are of a style that might have been used in the house had a central heating system been installed in the 1930s. They feel far more compatible with the character of the house than any plastic thermostat could. I found five matching thermostats on eBay for the five zones planned in the house.

Cooling

Whole-House Fans

Where days tend to be hot and nights to be cool and not overly humid, a whole-house fan can help maintain comfortable temperatures in a house. This is a large fan installed in the ceiling beneath the attic to draw hot air up from the floors below and blow it through a louvered opening into the attic above, where vents allow it to exit. A whole-house fan is used at night, after the outside air has cooled to a temperature below the interior. Open windows on the first floor allow cooler air in to replace what the fan pushes out. A properly sized fan can often exchange all the air in a house in a matter of hours. In the morning, the windows should be closed before the outside air gets hot again. The interior air warms through the day, then the process is repeated when evening comes.

The size of the fan depends on the size of the house. The manufacturers specify how much cubic footage their various models can handle. Ideally, a fan should be capable of one complete air exchange in two hours. Manufacturers will also indicate the attic vent area needed for a given size of fan. There is no point in trying to pull more air from the house than can be vented from the attic. Vents can be any combination of open windows, louvered openings, ridge vents, and/or eave vents. Insect screening will reduce air flow and needs to be factored into the calculated vent area.

Whole-house fans are sometimes used in combination with central or window-unit air conditioning, using the fan to vent the hot air from the house in the evening before closing the windows and turning on the A/C. This is effective if the outside air is not excessively humid. If it is, pulling that humid air into the house will simply cause the air conditioning to work harder, as it cools air by removing moisture from it. In general, whole-house fans are less effective at increasing comfort in high humidity, since humid air slows evaporation from the skin and consequently feels warmer.

Because whole-house fans require a large louvered opening in the ceiling between the conditioned living space and the attic, insulating the opening during the winter is essential to prevent heat loss. A common solution is a custom-made insulated box that sits over the fan unit in the attic seasonally. It should be weather stripped around the base and latched firmly to seal the box to the attic floor. The box can be hinged on one side to make it easier for a single

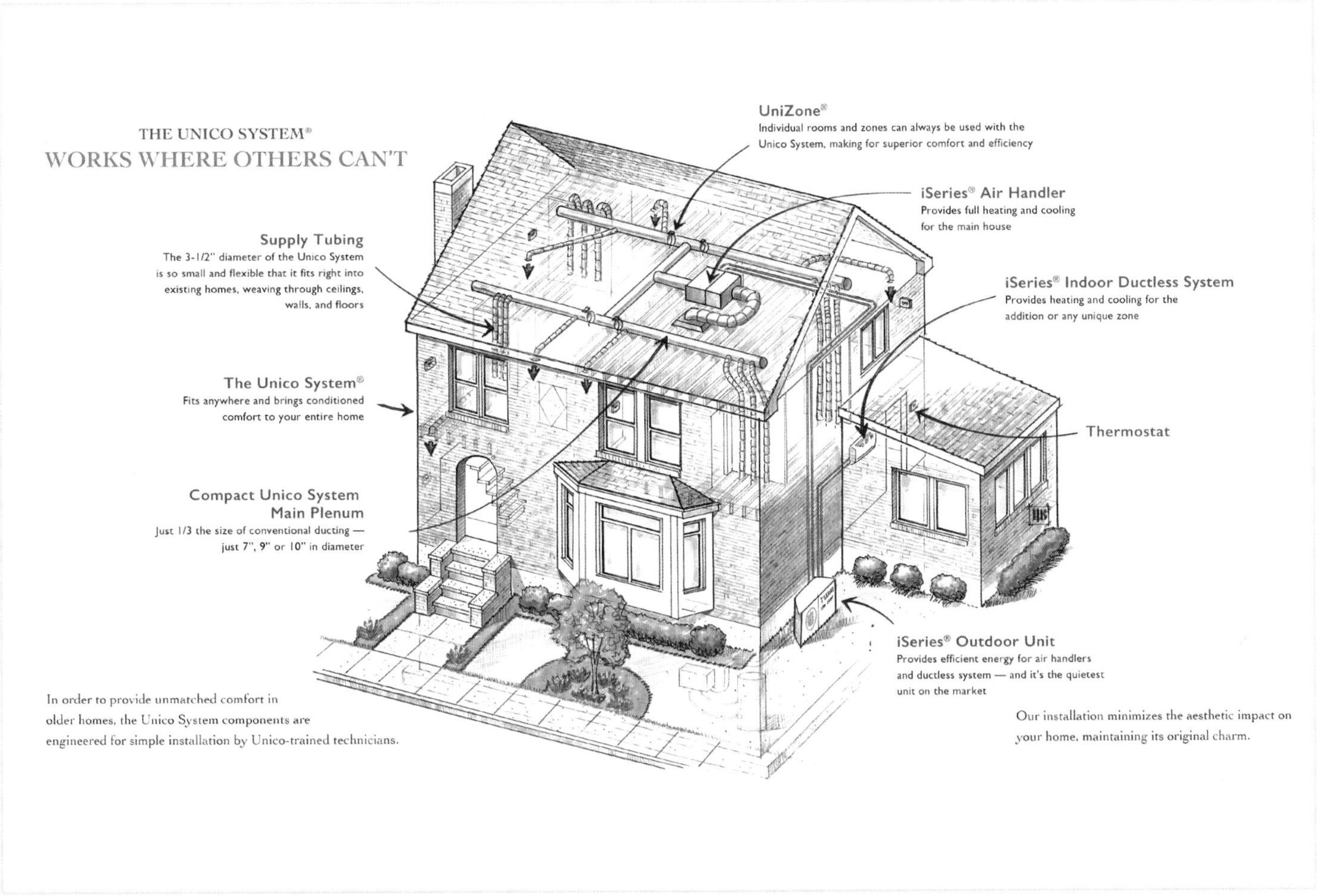
THE UNICO SYSTEM®
WORKS WHERE OTHERS CAN'T
UniZone®
Individual rooms and zones can always be used with the Unico System, making for superior comfort and efficiency
iSeries® Air Handler
Provides full heating and cooling for the main house
Supply Tubing
The 3-1/2" diameter of the Unico System is so small and flexible that it fits right into existing homes, weaving through ceilings, walls, and floors
iSeries® Indoor Ductless System
Provides heating and cooling for the addition or any unique zone
The Unico System®
Fits anywhere and brings conditioned comfort to your entire home
Thermostat
Compact Unico System Main Plenum
Just 1/3 the size of conventional ducting — just 7", 9" or 10" in diameter
iSeries® Outdoor Unit
Provides efficient energy for air handlers and ductless system — and it's the quietest unit on the market
In order to provide unmatched comfort in older homes, the Unico System components are engineered for simple installation by Unico-trained technicians.
Our installation minimizes the aesthetic impact on your home, maintaining its original charm.

High-velocity, small-tube systems like the Unico illustrated here can have much less visual impact on a historic house than a traditional large-duct system. *(Courtesy The Unico System)*

person to open and close it. Some fan units are made with insulated louvers on the ceiling side of the unit, eliminating the need for a box above.

Central Air Conditioning

In some areas of the country, air conditioning has been considered necessary for decades, and this attitude is spreading into areas where it was considered an unnecessary indulgence not long ago. The large ducts of a traditional central air system—like those needed for forced hot air—are challenging to install in a historic house. Sometimes ducts can be hidden in basements, closets, and attics, but often they intrude into living spaces. With careful detailing to continue the baseboard and any wainscoting or crown molding around a duct enclosure, the visual impact can be minimized, but it can't be eliminated.

An alternative to the traditional large-duct central A/C system is one that uses higher velocity to move air through smaller tubular ducts. These are popular in historic buildings because the tubing can be snaked through walls and ceilings and because small,

The outlets for a small-tube system are less noticeable than traditional wall or ceiling grilles. Here, a floor-mounted outlet is set into reproduction nineteenth-century carpet (to the right of the trunk) in the President Lincoln Cottage in Washington, DC. Similar outlets can be placed in a ceiling or wall and painted to blend in. *(Courtesy of The Unico System)*

less visible duct heads are used in place of the large louvered openings of a traditional system. The Unico system is probably the best-known small-duct system and is widely advertised in publications devoted to historic houses and home renovation. A major complaint about these systems when they first came on the market was that the high-velocity airflow was loud, but Unico claims to have solved this problem.

With any cooling or heating system, it is essential to insulate all ductwork that passes through unconditioned space. Cold ductwork passing through warm, humid air in an attic or basement will cause water to condense on it and then drip onto surfaces where it does not belong. Hot ducts passing through the same spaces will radiate heat where it is not needed, reducing the efficiency of the heating system.

Heat Pumps

Heat pumps are discussed above in the Heating section of this chapter. One of their attractive features is that they can cool as well as heat; they have become increasingly popular in areas where air conditioning used to be uncommon. The advantages and aesthetic challenges of these units are discussed in the Heating section.

Green Technologies

After a century of relative stability, new technology in recent decades has begun to significantly alter the way homes are heated and cooled. For the previous century, some form of fossil (coal, oil, or gas) or bio (wood) fuel was burned in a central unit from which heat was distributed throughout the house by hot water or hot air. All these sources consumed fuel and exhausted the combustion byproducts on-site. For half that century, electric heat attempted to impact the market, bringing power to the house from a distant source (fossil or hydro) and producing heat at various locations in the house, but it had little impact in areas with significant heating requirements.

Now, heat from the sun is being harvested on-site to heat water or produce electric power, and heat is also being extracted from the ground on-site. These efforts, underway since the 1970s, are achieving the necessary critical mass to become far more affordable and available. On a smaller scale, wood is being processed in centralized facilities to remove most of its moisture and reconstitute it into easily shipped and stored pellets that burn cleanly and can be used in automatic-feed hopper systems.

Solar Power and Hydronic Heating Systems

Through the developmental decades for the solar energy industry, historic preservationists and solar power advocates were often on different pages. Preservationists were prone to see solar collectors as unsightly distractions from the historic character of a building, while solar experts insisted that collectors be very large and positioned at the most efficient solar-collection angles possible. Consequently, solar collectors were largely kept out of local historic districts due to their visual incompatibility.

In recent years, both camps have begun to reassess. The goal of sustainability, a common objective now, has encouraged compromise. Solar installers will still inform customers that an ideal placement requires a framework standing off the roof at an angle but will often gladly install panels directly on an existing roof that is less than perfectly suited for maximum solar gain. Solar panels themselves have become thinner and are now available with an overall matte-black finish, making them far less distracting than the shiny blue-black panels of the past. When placed on roof planes facing away from primary views of the building and kept back from roof edges, solar panels are now often approved by local historic district commissions.

Even less visible solar options are now becoming available and will likely increase significantly in the coming years. These include the solar-electric roof shingles introduced by Tesla in 2017, which can effectively mimic slate or other traditional roofing materials, and solar-hydro collection systems that can be installed under standard metal or asphalt roofing. The large domestic batteries also introduced by Tesla in 2017 may signal a new era in storage capacity for electricity generated on-site. It remains to be seen whether products of this type will deliver on their possibilities.

Geothermal Heating

Geothermal heating extracts heat from the ground and uses it to heat buildings. It can also be used in reverse, to remove excess heat from buildings and store it in the ground. This approach has great potential if the high cost of installation can be reduced. Its greatest environmental advantage is that it burns no fuel to produce heat and uses a relatively small about of electricity to operate. Its greatest historic-preservation advantage is its invisibility. All the work of extracting or exhausting heat for the building

happens underground. Once installed, virtually silent pumps circulate liquid through coils or lengths of tubing buried below the frost line, and a heat exchanger transfers heat in or out of the house as desired. The cost of operation is the cost of the electricity needed to run the pumps, heat exchanger, and whatever moves the heat through the house (more pumps or blowers/air handlers). If that electricity is being generated on-site from a solar-collection installation, the system can operate for very little cost.

However, the cost of installing the tubing underground—either in long trenches four feet or more in depth or in deep drilled wells—can be significant. Since geothermal is one of the most environmentally friendly HVAC systems in existence, tax credits to help offset the installation cost would seem a worthwhile investment by states and the federal government. It makes more sense to invest in strategies that will slow climate change than to continually reinvest in communities devastated by natural disasters while making no effort to prevent those same disasters from happening again.

If electricity can be generated and stored on-site to power heat pumps, geothermal pumps, or other heating and cooling systems (with backup from the commercial electric grid) *and have no significant visual impact on historic houses*, it will bring the environmental sustainability and historic-preservation communities together in a way that was unimaginable at the turn of the twenty-first century. Many additional options for generating or conserving energy at a domestic scale will likely be introduced in the coming decades.

Smith House, 1877

Italianate style

The C. Sydney and Emily Smith House, in a coastal New England village, has many of the defining features of the Italianate style, including a low-hipped roof with cupola or belvedere, eave brackets, elaborate window hoods, paneled corner boards, bay windows, and double doors with etched glass. Houses like this were built across the United States in the mid-nineteenth century, when the Romantic movement inspired architecture that evoked foreign lands or times past. The original owner of this house was the treasurer of a local bank; his wife was the daughter of a sea captain. The Smiths' only child, Rita C. Smith, lived her entire life in the house apart from her college years. She became a local schoolteacher, world traveler, and talented photographer. The house passed through several owners after her death in 1962, remaining relatively intact.

When the current owners, Valerie and Jeff, bought the house, it was suffering from long-deferred maintenance. Poor drainage and plantings too close to the foundation had resulted in rotten sills. The rear shed was severely deteriorated, and the slate roof was failing. Coming from southern California, where they had owned modern houses, the new owners had to climb a steep learning curve in order to restore a large, historic New England house.

First the house was jacked up six inches to level it and replace rotted sills. To reduce future moisture infiltration at the base of the house, plantings were removed and the soil around the foundation was properly graded.

Roof repairs were needed to stop infiltration from above, but a new slate roof proved too costly. Instead, Valerie and Jeff selected an asphalt shingle that echoes the color and shape of slate shingles. The roof framing had never been as substantial as it should have been to support the slate, and reinforcement was necessary in the attic even for the new asphalt roofing.

Historic photos guided the restoration of the front porch to its original design, including its elaborate square columns that had previously been replaced. The house's original windows were repaired and retained, with new storm windows added. A photo from the early twentieth century showed the side porch extended into an uncovered deck with a balustrade of turned balusters that had been repurposed from a wooden fence along the street (shown in earlier photos). The missing deck extension was recreated, and the long-gone balusters were replicated in a new balustrade.

One of the most noticeable changes was restoring period colors to the exterior, which had been painted entirely white for many decades. Historic photos clearly show the polychrome paint scheme that would be expected on an 1877 house, and those photos and surviving evidence on the building guided the return to such a scheme. The result dramatically altered the presence of the house on a street lined with historic white houses. Not everyone in town was thrilled by this change, but the house now appears as it was intended to, and as it did for many years when new.

A stone patio was built outside the back door to create outdoor living space off the kitchen. The nineteenth-century gazebo on the property was moved back to its original location (as documented in historic photos of the house) and restored.

An attached, severely deteriorated shed ell was rebuilt and repurposed, with a bathroom upstairs and a guest-room downstairs. Since this formerly utilitarian space would now be primary living space, new trim was milled to match other trim in the house. The closet in the new guestroom is on the footprint of the original two-story outhouse, which featured two holes up and two down.

In the original back entry, the soapstone laundry sink was moved across the room, and a wall between the entry and the original icebox was removed to open up the space. Cabinets from the original scullery were moved to the laundry area, and a large built-in icebox was relocated within the space. The grain-painted doors in the back rooms of the house were preserved in place.

Cabinets in the new kitchen were selected to be compatible with the historic pantry cabinets, which remained in place. The historic wood floor was retained. Throughout the house, hardwood floors (including an elaborate quartered oak and cherry parquet floor in the front hall) were cleaned and oiled to bring out the natural color of the wood and protect them from water. The floors had never had a shiny varnish finish, and Valerie and Jeff wanted to preserve the more subtle glow of oiled floors.

The home's historic plaster was repaired and retained wherever possible. New gypsum board was used only for walls and ceilings in the spaces that were substantially rehabilitated, primarily in the kitchen and shed ell. Many layers of wallpaper were removed, and the walls were skim-coated to repair plaster damage, then painted for a more contemporary treatment than wallpaper could provide. A fragment of historic wallpaper uncovered in the dining room was framed to hang on the wall there, and the new wall color was taken from this pattern. The chosen colors reflect the Victorian aesthetic but are used without much of the pattern also common then. Prints of historic photos of the house and the Smith family are framed and hung in other rooms, and the Smiths' presence is intimately preserved in a section of wall plaster upstairs with marks and notes recording Rita Smith's heights from 1889 to 1906, with height marks for "Mamma" and "Papa" as well.

Valerie laboriously scraped away layers of paint from the parlor ceiling to uncover the original decorative painting, which she left unrestored, offering visible evidence of its history and survival. She also partially uncovered the decorative paint on the sitting room ceiling, making parts of the design visible through the later flat-painted surface, but decided to leave the remainder of that job to a future owner. The cast plaster ceiling medallion in that room was given a polychrome paint treatment using the colors in the exposed decorative painting.

Historic light fixtures were rewired and retained where they existed, including a fabulous brass five-globe gasolier mounted on the entry-hall newel post. Gas lighting had been available in the area for several decades when the house was built, and it originally had gas fixtures throughout, some of which had later been converted to electricity. Valerie and Jeff retained the pushbutton electrical switches throughout the house.

They placed gas inserts in two fireplaces, one of which received a new marble surround. In the other, the still-intact historic tiles on the surround and hearth were preserved. Over the mantel in the parlor, where a large mirror would traditionally have been located, a flat-screen television was installed with a wood frame made from molding that copied the room's original door and window trim. This placement and framing help the TV blend into the space. The intact historic bathroom on the second floor, probably installed in the 1890s, was retained with its original fixtures; several modern bathrooms were added in secondary spaces. A maid's bedroom was repurposed as a walk-in closet with the original built-in cabinetry.

Primary heating is provided by a high-efficiency oil-fired hot-air system and several gas-fired Monitor heaters in secondary spaces. Use of the Monitor heaters avoided the need to run ductwork up through significant rooms.

Of the transition from one-story informal modern living in California to the Smith House, Valerie says, "We didn't know how we were going to live in a house like this. It is large, formal, and has many more rooms than our previous houses. But we discovered that this floorplan works really well for our lives." She notes the convenience of being able to use a room for a particular task or project and leave it for another time just by closing the door, which is hard to do in an open-concept house. Jeff and Valerie have found the house to be great for entertaining. They host a lot of parties and open the house to the public for fund-raising events.

The restored façade of the Smith House. The walkway and metal fence are compatible new landscape elements.

The house in the late nineteenth century. Although black and white, this photo guided the restoration of the original paint scheme. *(Courtesy Thomaston Historical Society)*

The renovated laundry room with relocated original cabinets and slate laundry sink.

The nineteenth-century bathroom was preserved during the restoration, and several modern bathrooms were added for daily use.

This bright sitting room retains its marble mantel, gilded window cornices, and the early-twentieth-century electric fixture that replaced the original gas lighting.

The second-floor alcove over the front door, showing an original light fixture that was rewired and reinstalled.

Italianate detailing on the pediment of the projecting entry bay, with the belvedere and chimneys behind.

This grain-painted built-in icebox in the shed was relocated when the deteriorated shed was reconfigured.

C. Sydney Smith in his parlor. The photo was probably taken by his daughter, Rita. *(Courtesy Thomaston Historical Society)*

Etched glass in the front door.

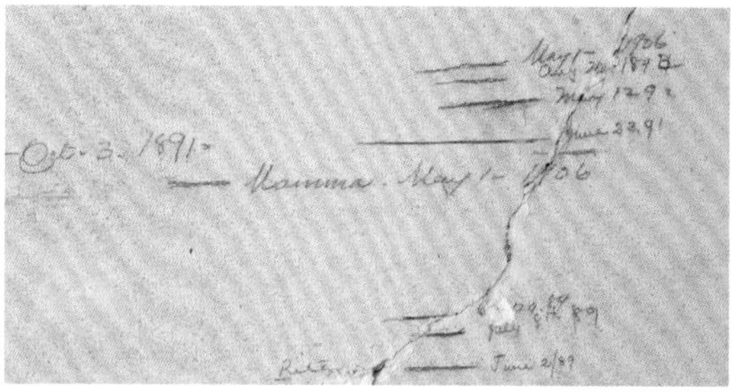

Valerie and Jeff preserved these pencil markings on the plaster wall of the upstairs back hall, tracking Rita Smith's growth from 1889 to 1906.

The house around 1970 during its "white elephant" period.

This pre-restoration view of the shed's second floor shows the upper level of two-story privy.

The Smiths' nineteenth-century bathroom was state of the art when new.

The kitchen before restoration.

The restored dining room. The wall color matches one of the colors in a piece of wallpaper found in the room (visible, framed, over the mantel).

The kitchen was the only room in the main house that had to have its plaster replaced.

This plaster, damaged by settling, was repaired after the house was jacked up and the sills and foundation repaired.

This early-twentieth-century barrel top addressed to Rita Smith was found in the house.

From the new kitchen, the historic pantry cabinetry is visible through the door at right.

The front hall with its parquet floor and original newel post light fixture. The wall at right is covered with historic photos of the house and Smith family copied from the Rita Smith negative collection at the local historical society.

The restored belvedere.

The new paint color in this restored parlor harmonizes with the historic hearth and fireplace surround tile work.

The pre-restoration pantry retained its original cabinetry, but the lower cabinets had been painted at some point.

Wooden toys in the upstairs back hall.

The dining room before restoration.

The pantry after the lower cabinets were stripped of white paint and restored to their original appearance.

The Smiths' gazebo in the early twentieth century.

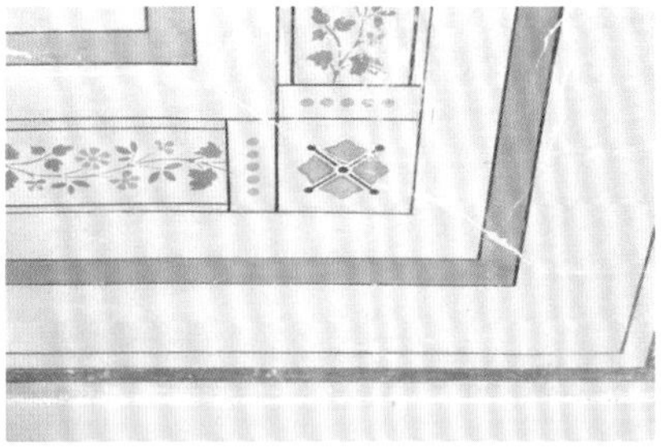

A corner of the paint-decorated parlor ceiling laboriously uncovered by Valerie.

C. Sidney and Emily Smith on the front steps of their stylish new home, about 1880. This photo was valuable for reconstructing the missing porch columns during the restoration.

The Smith House when Valerie and Jeff purchased it in 2007.

A pre-restoration bedroom.

The same bedroom after restoration.

The parlor in which Valerie uncovered the original paint-decorated ceiling. A large flat-screen television is framed with wood that matches the historic trim.

The Italianate double front doors.

The Work

Here the structural and sill work has been completed and the restoration of the front entry portico is underway.

The portico detailing, most of which had been lost during an earlier replacement, was recreated using historic photographs for guidance.

The second story of the severely deteriorated shed during restoration.

The gazebo had been relocated at some point. Valerie and Jeff had it returned to its original location.

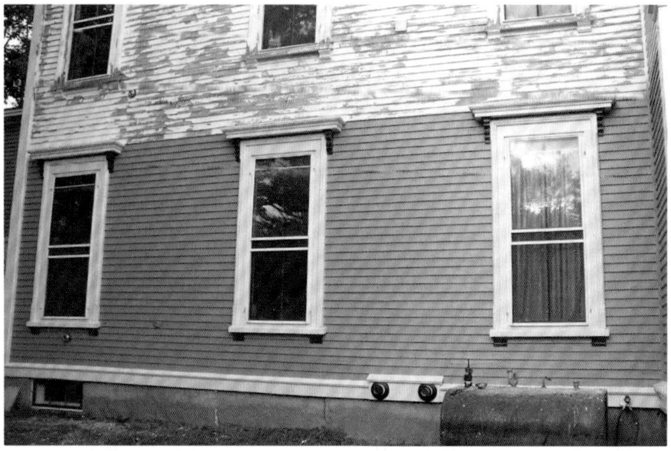

Here the house has been scraped and painting is underway. After decades of being all white, the return to original colors was startling for some neighbors.

The gutted kitchen during restoration.

Painted decoration on a ceiling emerging from later paint layers.

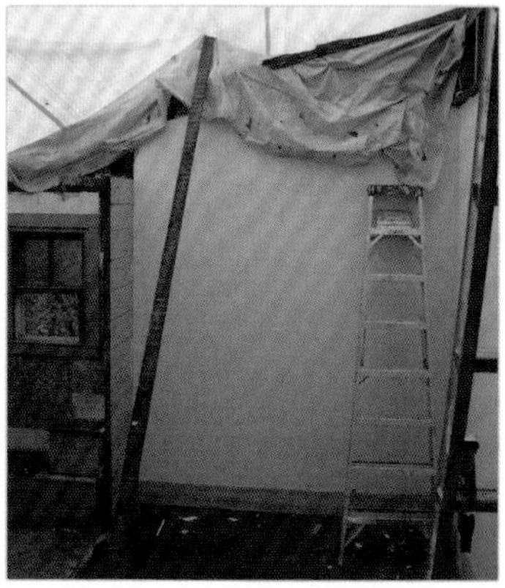

A tarp covers the shed after removal of the rotted roof.

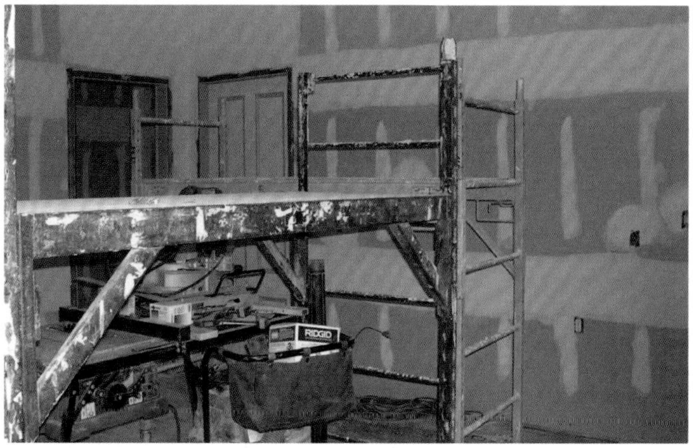

New gypsum board being taped in the kitchen.

The paint-grained door remains in place in the gutted kitchen.

CHAPTER NINE

Insulation and Ventilation

An industry devoted to reducing heat loss from buildings has grown up over the past generation as energy efficiency has become a primary goal. Much of the research and most of the advances in this field have focused on new construction, and manufacturers and contractors often assume that the same approaches and materials are equally useful for the renovation of existing buildings. Many even believe that an old building should be insulated and air-sealed to the same standards as a new building—and that anything less is an environmental sin—despite the fact that there *are* separate efficiency standards for existing buildings. To a contractor or architect who is "green in the extreme," many of the character-defining features of a historic house are impediments to achieving the zero-emission, carbon-neutral, LEEDS Platinum sustainability badge of honor they have committed their careers to. While the work they do is important, these are not the folks you want working on energy-efficiency improvements for your historic house.

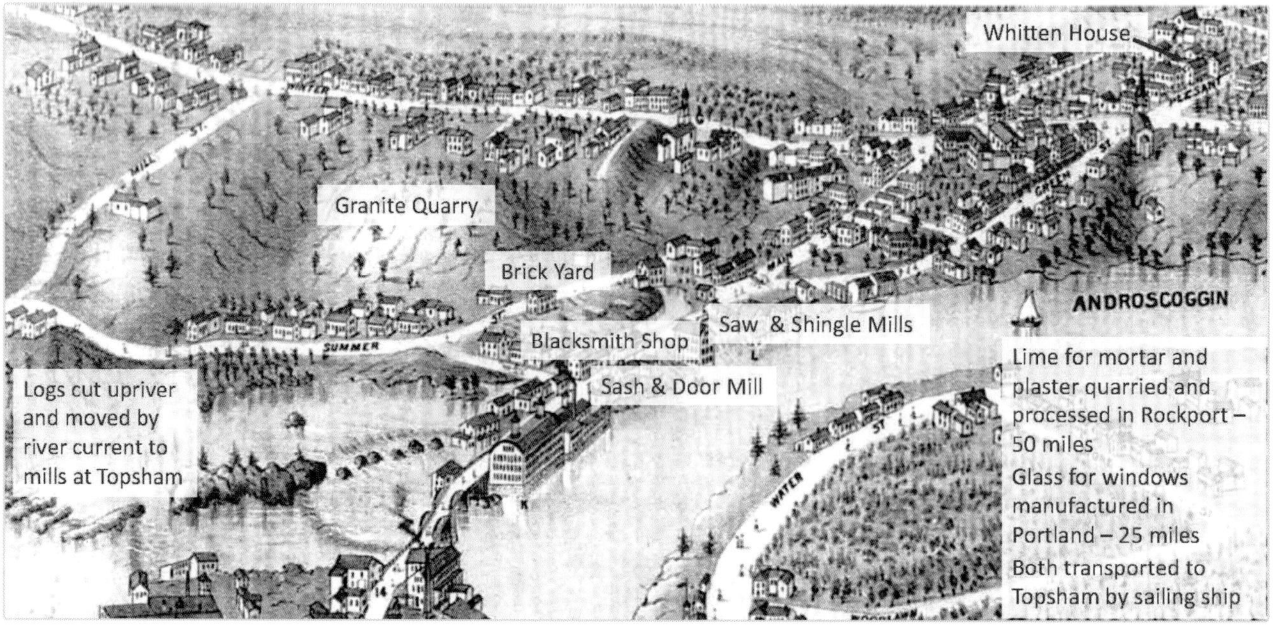

Whitten House provides a good example of the inherent sustainability of historic houses. The origins of its building materials are identified in this drawing. All the wood was cut from native forests upstream from Topsham and moved by the river's current to the sawmill, shingle mill, and sash and door mill, all located within a half-mile of the building site and powered by water. The foundation granite was quarried near the mills, and the chimney bricks were made in the same vicinity. Glass was manufactured in Portland, 25 miles away. Lime for mortar and plaster was quarried and processed in Rockport, 50 miles away. The local blacksmith had to import his iron, likely from the Mid-Atlantic region, and the door latches were probably made in Portland or Boston and purchased from a local merchant. All transportation was by sailing ship. Compared with the building materials for a comparable modern house, the carbon footprint of Whitten House's construction was tiny.

If your goal is to live in a home that will have as small a daily carbon footprint as is humanly possible, it is probably best to build a new house. Renovating an old house to achieve that goal will likely be an exercise in frustration. For a larger picture, however, spend some time thinking about what went into building your old house. Was the lumber harvested locally, with no transportation-related carbon footprint? Was it sawn in a local water-powered sawmill, with no carbon footprint? Were the bricks made of local clay and fired nearby with locally harvested wood? Was the lime used for the mortar and plaster in the house brought to the community by sailboat, steamboat, train, or wagon? How many petroleum products (plastics, Styrofoam, synthetic carpeting, asphalt roofing) were used in the construction of the house? Now think about building a comparable-sized modern house, the building materials for which might be transported from around the world.

It doesn't require an advanced mathematical degree to conclude that, on a sustainability scale, most historic houses start off with a significant advantage. In this context, is it necessary for the historic house to perform as efficiently as a new house to match its sustainability? The old house starts the carbon-footprint race two laps ahead; even if it doesn't run as well, it may still finish ahead.

Enough venting about the sustainability police. Let's look at how to improve the energy performance of a historic house in ways that will not destroy its historic character and materials.

The good news is that you do not have to gut the house to vastly improve its energy performance. What you should do first is hire a qualified independent energy auditor to inspect your house, tell you where the problems are, and help you prioritize the solutions. Tax credits have been available to homeowners who complete certain energy efficiency improvements; some states offer low-interest loans to finance the work; and some communities and states make grants to low-income homeowners for this purpose. Check to see what may be available in your state and community. As of 2019, nearly all such programs require an energy audit.

When looking for an energy auditor, get recommendations from local preservation organizations; municipal officials who staff the historic preservation commission, board, or committee; and/or local contractors who do restoration work. You want an auditor who understands how historic houses are constructed and ranks the preservation of historic material and character as a priority. I recommend against having the audit done by the contractor who will be doing the work unless you have a high level of trust for that contractor.

A good energy audit should include:

- Thermal imaging of the exterior of the house to identify where and to what degree heat transfer between outside and inside is happening. This will show where insulation and air sealing are needed.
- A blower door test to measure air infiltration.
- Identification of moisture problems and recommendations to address them, since tightening the house envelope will aggravate any such issue unless you take remedial action.
- Documentation of existing HVAC equipment, energy costs, and efficiencies.
- Documentation of domestic hot water production (if separate from the HVAC systems) and efficiency.
- Documentation of the types of lighting and appliances in the house and calculations of their efficiency.
- Recommended priorities for insulation and air-sealing efforts.
- Recommended improvements to the HVAC system and calculations of their efficiency benefits.
- Recommended lighting and appliance changes and calculations of their efficiency benefits.
- Approximate costs of the recommended improvements and the savings to be realized from improved efficiency, including estimated payback periods.

Recommended priorities are house-specific and might vary wildly for two superficially identical houses depending on their condition and the efforts of previous owners. Most previously unimproved houses get the most bang for the buck from improving the insulation of the "cap" (the top of the house) and sealing cracks all around the house. Hot air rises, and the insulation cap is what keeps it from rising right out of the house. Cracks occur from the sills to the soffits; wherever two pieces of building material come together, there is potential for a crack that will allow cold air to seep in. Cracks around door and window openings are frequent offenders.

Close behind cap insulation and crack sealing is the installation of storm windows in a house with single-glazed glass. A properly installed storm window turns a single-glazed window into an insulated window, the performance of which can be improved even further by tightening up the historic window. (See the discussion of this topic in Chapter 13, Exterior Trim and Windows.)

Window shades and/or drapes can also increase the energy efficiency of windows. Insulated cellular shades can have an R-value of almost 5, but even a traditional roller shade creates a dead air space in front of the glass, providing added R-value.

Insulation

As mentioned above, the most important insulation is generally at the cap of the house. This is true whether you are trying to retain heat in Vermont or keep it out in Arizona. Next on the insulation priority list is usually the base of the house—either the foundation (or at least the part of it that is above ground) or, if the house is on piers, the floor itself. Heat gain or loss through walls is typically a comparatively small percentage of a house's total energy loss. Air sealing will often do more than insulation to improve the performance of a wall assembly.

Dew on a lawn is life-giving, lovely, and will be absorbed by the soil or evaporated by the sun, but dew droplets inside a wall assembly can cause significant problems. Moisture-laden vapor can pass through many building materials, but liquid water, not so much. The dew point—the temperature at which air of a given humidity becomes saturated with moisture—must therefore be considered when insulating walls and roofs. When saturated air contacts a cool surface, water will condense onto that surface. If moisture cannot be kept out of the wall or roof, that condensation surface could be inside a wall. One objective of insulation is to prevent that.

Cellulose insulation is a popular option for historic houses. Made primarily from recycled newspaper, it can be blown through small holes into stud bays, between floor joists, and into other voids and open areas like an attic. This insulation had been under an attic floor for more than thirty years when the floorboards were lifted to expose it.

Insulation Types

The primary types of insulation in the U.S. are fiberglass batt, blown fiberglass, blown cellulose, blown

Insulation Comparison Chart for Use in a Relatively Intact Historic House

Scale 1 to 5 – Poor to Good

The scoring below is based on a hypothetical wood-framed historic house with its interior and exterior finishes relatively intact and restorable

The scoring would come out differently for a house that has already been gutted or which has finishes beyond saving

Insulation Type	Disturbance to Install	Reversibility	Vapor Permeability	Air Sealing Ability	R-Value per Inch	Cost to Enviro.	Score
Dense Pack Cellulose	5	5	5	5	4	5	29
Std. Pack Cellulose	5	5	5	4	4	5	28
Blown Fiberglass	5	5	5	4	3	5	27
Blown Mineral Wool	5	5	5	4	2	3	24
Mineral Wool Batts	1	5	5	4	3	5	23
Fiberglass Batts	1	5	5	3	2	5	21
Open Cell Foam	1	1	3	5	3	1	14
Closed Cell Foam	1	1	1	5	5	1	14
Rigid Polyurethane Foam	1	1	1	2	5	1	11

Disturbance to Install: The more historic fabric to be disturbed/removed to install product the lower the score.
Reversibility: Products that can be removed easily at a later date without permanent damage to historic fabric have a higher score.
Vapor Permeability: Products that allow moisture to pass through are generally better for historic fabric and score higher.
Air Sealing Ability: Products that more fully fill available spaces without leaving gaps for air infiltration score higher.
R-Value per Inch: Higher R-Value equals more comfort and energy efficiency and score higher.
Cost to Environment: Some products are made from natural or recycled materials and score higher, others are made of petroleum products with processes that may be polluting and score lower.

Insulation chart

and batt rockwool, rigid foam, open-cell spray foam, and closed-cell spray foam. Other insulation materials, including natural-fiber materials like cotton, wool, straw, and hemp, are much less commonly used. The most reliable information can be found in independent sources such as government or university websites, not in the advertising of manufacturers. The U.S. Department of Energy website (www.energy.gov) is a good place to start. The full URL as of this writing is https://www.energy.gov/energysaver/weatherize/insulation/types-insulation. Rely on independent sources to select an insulation for your house, then use the manufacturer's marketing to decide which brand to use. The best choices are not necessarily the same for a historic building as for new construction. The chart above provides some guidance.

Much of what follows pertains to wood-frame buildings. We'll turn to masonry buildings below.

Blown-in Insulation

Blown-in insulation will generally have the least impact on historic finishes. Blown in through small access holes in the exterior sheathing or interior plaster, it does a good job of slowing the transfer of heat through the wall or ceiling and minimizing air filtration if all voids are filled. When working from the exterior, do not drill through the clapboards; instead, remove clapboards selectively to drill through the sheathing, then plug the holes with cork and reinstall the clapboards in the same locations. Alternatively, or in addition if areas of the exterior are not accessible, holes can be drilled through interior plaster if there are no historic wallpapers or painted finishes to protect. Holes in plaster are relatively easy to patch, and this approach may be preferable if there is back plastering in the house, as it will leave this air-sealing layer intact. Blown insulation can also be used for sound insulation between rooms.

Blown insulation is also useful in attics, where it can be installed in the bays between floor joists. If there are no floorboards, it can extend above the joists to create a cap with no thermal bridging through the joists. This is the most important insulation in a house, as hot air rises and most heat loss is through the roof. Seal all penetrations between the living space and the attic (plumbing, electrical, HVAC, etc.) before insulating.

The three main types of blown-in insulation are cellulose, fiberglass, and mineral wool. Cellulose is made from recycled newspaper that has been treated for fire and pest resistance. Fiberglass insulation is manufactured specifically for this use and is not recyclable. Fiberglass will melt at very high temperatures (it is glass). Mineral wool (also called rock wool) is made from waste byproducts from the manufacture of other products and is water- and fire-resistant.

Batt Insulation

Batt insulation cannot be installed without complete removal of interior finishes, making it less than ideal for restoration work if the plaster doesn't need to be removed for structural work. Removing intact plaster only to insulate a wall is not recommended. Not all batts are created equal. Mineral wool is denser than fiberglass and can be fitted more tightly between studs, reducing air infiltration. It resists water and will not be ruined if it gets wet temporarily, and it is made from industrial waste that would otherwise go into a landfill.

Rigid Insulation

The several types of rigid foam insulation have a high R-value per inch of thickness but require full access for installation. This insulation is impervious to water and should not be used where moisture must be allowed to escape. It's a good choice where the goal is to keep moisture out, such as under floor framing above a damp basement or crawl space. The key to making this work is to seal the seams between sheets to create a seamless impervious surface. Rigid foam can be a better choice than spray foam where you need maximum insulation with minimal space; even if sealed around the edges with spray foam from a can, it can be removed with relative ease. Most building codes require this material to be covered with a fire-rated material such as gypsum board or intumescent paint in habitable spaces. This includes usable basements and attics but not unusable crawl spaces.

To Foam or Not to Foam?

People may be surprised at the ranking for spray foam in the chart above. Why is such an excellent insulator, one that expands to fill every crack and void, not scored the highest? Spray foam has one great flaw in a preservation context—the difficulty of removing it when removal becomes necessary. Anyone who has worked on old houses for any length of time knows that wood will rot, pipes will leak, insects will consume the cellulose in wood, etc. Sprayed-in-place

This adobe house roof has 12 inches of spray-foam insulation on the roof covered by a spray-on membrane coating to make it waterproof. A poly barrier separates the foam from the roof sheathing, making future removal of the foam possible. The parapet, typical of adobe construction, completely hides the additional thickness of the roof from the ground. This is an appropriate use of spray-foam insulation.

Open-cell spray foam insulation was chosen for this new bay window addition for several reasons. The amount of framing in the window left relatively little area to insulate, making it more important to maximize the R-value. Also, there were a number of small voids between framing members, including nearly inaccessible Vs at the outside corners of the bay. Expanding foam was virtually the only insulation material that could fully and effectively fill these spaces. Finally, because this was new construction, it was possible to install a complete vapor barrier on the interior to keep moisture out of the insulation material. Though not yet installed when this photo was taken, ½-inch rigid foam was placed over the studs and foam insulation to create a thermal break between the many studs and the interior plaster wall.

insulation is nearly impossible to remove completely, whereas all the other insulation types in the chart are relatively easy to remove if you—or another homeowner fifty years from now—need access to something for repair. Furthermore, the jury is still out as to whether spray foam insulation can contribute to rotting of wood by preventing moisture from passing through it. A complete and thorough interior vapor barrier and exterior rainscreen siding can be used in new construction but are rarely possible with restoration. If moisture gets into wall assemblies of an old house through cracks in a finish wall surface, trim, floors, attic, or basement and cannot escape, it will be destructive.

There is general agreement that closed-cell foam is of greater concern in this regard than open-cell, which has a fair degree of vapor permeability. Both materials are flammable and create toxic smoke when burned.

Where cooling is of more concern than heating, outfacing reflective foil is frequently used in wall and roof assemblies to reduce the amount of heat radiating through the assembly. Foil can also be used with insulation, facing in, to help keep heat inside a building; standard fiberglass batt insulation is available with a foil-faced paper backing. Foil has no effective R-value but does affect the transfer of radiant heat.

The expected life of much new construction is measured in decades. In our throwaway society, even buildings are treated like consumer products. Historic houses were built to last indefinitely and some have stood for 350 years. Our work on these buildings should not shorten their remaining lives from centuries to decades. First do no harm. And bear in mind the distinct possibility that far superior insulation materials might be developed in the future. Installing these new materials would require removal of existing insulation—a near impossibility if you insulate with spray foams. (One exception: Spray foam applied over a drainage mat on a basement wall allows water penetrating the wall to flow down to a drainage channel and around the perimeter of the space to a drain or sump pump. The mat and foam should not extend up to the wood sill. They can be removed together if access to the wall is necessary.)

Sheets of 2-inch rigid foam insulation have been installed between the 2 x 10 rafters, covering the original 3 x 7 rafters. The resultant 7-inch-deep voids behind the rigid foam are to be filled with blown-in cellulose insulation. Strapping (1 x 3) has been installed across the rafters on 16-inch centers for the installation of ½-inch gypsum blueboard as a base for skim-coat plaster, restoring the original appearance of the room. The ¾-inch air space created by the strapping between the insulation and the blueboard adds further R-value.

This view shows the attic of Whitten House and the 2 x 10 rafters installed in the 1980s between the original 3 x 7 rafters to strengthen the roof frame. The original knee walls and ceiling plaster were removed so that the 2 x 10s could be added. New knee wall framing was then installed to help transfer the roof load to the floor joists, but the plaster was not replaced.

Here the strapping has been continued across the flat portion of the ceiling. There is no rigid foam on this portion of the ceiling because, even without it, there is only 6 feet 7 inches of floor-to-ceiling height. Blown-in cellulose insulation will fill the void above the ceiling.

This is the area shown in the first photo of this sequence, now with the gypsum blueboard installed and taped. PVC pipes near the floor are connected to flush vents in the eaves to vent the three bathrooms in the house with the help of blowers to be mounted in the crawl space outside the attic bedroom.

The hole in the ceiling used by the contractor to blow in the cellulose insulation can be seen—now patched—at this end of the attic. A worker carrying the hose from the insulation blower climbed into the triangular void over the ceiling through the hole and crawled to the far end of the house. Once there, he sequentially slid the hose down each angled rafter bay to the eaves, filling the bays one at a time by pulling the hose back toward him. Filling the bay on one side and then the corresponding bay on the other side, he worked his way back to the entry hole. As he finished a section of bays, he would use the hose to fill the triangular void above the ceiling where he had been working as well. Another worker with a thermal imaging camera monitored progress from inside the attic, ensuring that all areas were filled. Once all the bays were filled, he climbed out of the hole, and a piece of gypsum board was screwed in place to fill it. Then a small hole was drilled through this patch so that the remaining area above the ceiling could be filled. Finally, this hole was patched with setting plaster.

Combining Insulation Materials

It is not necessary to use only a single type of insulation in a wall, ceiling, or roof assembly. When insulating the cap of Whitten House, I used a layered combination of densely packed cellulose and 2-inch rigid polyurethane where the historic plaster had been removed by the previous owner, maximizing the R-value of the assembly without reducing the headroom more than necessary. This combination filled the entire cavity between the interior surface and the roof sheathing, necessitating some other means to ventilate the roof. To accomplish this, I installed 2 x 4 sleepers vertically from peak to eaves on the exterior of the sheathing, then added a new layer of sheathing and roofing above. Airflow through the resultant 1½-inch space between the original sheathing and the new roof was accomplished with a ridge vent and a ventilated drip edge at the eaves.

If interior wall finishes are gone or must be removed, it may be possible to cover the studs with a thin rigid insulation after filling the bays with another insulation type. This will block thermal bridging through the studs. If you are replacing a previous plaster-on-lath finish with skim-coat plaster on blueboard, the ½-inch-thick rigid foam under the

Taking advantage of the thick walls of masonry buildings, builders often installed hinged paneled shutters in the recesses, which could be closed to cover the window for privacy and increased energy efficiency. These may be single-leaf or bifold, with the second leaf hidden behind the first when open. It is not uncommon to find these shutters long unused, sealed in place with many layers of paint as shown here. The only clue that this paneled recess has shutters is provided by the painted-over hinges that can be detected upon close examination. Returning these shutters to functioning condition (and using them) would improve the efficiency of the window unit.

blueboard may bring the new wall close to its original thickness or only slightly over it.

Masonry Buildings

Masonry buildings—particularly historic ones—pose special concerns when it comes to insulation. There is disagreement about how (or even whether) to insulate historic masonry walls. Some say that these walls have always needed to pass heat from the interior to the colder exterior in order to stay dry and solid. Without this heat, they say, the dew point temperature will occur in the middle of the wall assembly, causing moisture to condense and destroy the wall from within. Others point out that many historic masonry buildings were barely heated (by current standards) when new and others (such as abandoned mill buildings) have stood vacant and unheated for decades without apparent problems from condensation within their walls. As long as the wall exterior has appropriate soft mortar pointing and is not coated with an applied sealer, they say, it will breathe and dry out just fine when it gets wet.

In my work as a preservation consultant on historic tax-credit rehab projects, this issue has come up many times, particularly on projects combining historic tax credits with low-income housing

One of the shutter hinges can be seen here (circled in red) under many layers of paint.

This early-twentieth-century garage was converted to living space. To insulate the buff brick walls, rigid foam insulation was installed against the brick. A new 2 x 3 stud wall constructed inside the foam accommodated electrical work. The splayed brick window lintels were left exposed, showing off the only interesting architectural feature of the original space. New wood framing was installed for the floor, over a vapor barrier that covers the historic concrete floor. This allowed space for plumbing and electrical as well as insulation.

tax credits. In my state, there are stringent energy-efficiency requirements for the use of low-income housing tax credits, and these include R-values for wall assemblies that cannot be met without insulating the walls. These projects also must meet the U.S. Secretary of the Interior's *Standards for the Rehabilitation of Historic Buildings* as interpreted by the Technical Preservation Services (TPS) division of the National Park Service. This oversight responsibility has required the TPS to determine whether historic masonry walls can be insulated to meet the standards without doing long-term damage to the historic fabric of the wall. They have concluded that it can be done and have approved the insulation of historic masonry walls for a number of projects I have worked on.

This is most often done by removing the historic interior finishes, constructing a new stud wall set ½ to 1 inch inside the face of the masonry, and filling between and behind the new studs with spray foam insulation. Electrical and plumbing are done before the insulation is applied. A new gypsum board interior finish is applied to the studs, and salvaged historic trim is reapplied in its original locations. Window and door jambs are extended to accommodate the increased thickness of the wall assembly. These low-income housing projects do not have the budgets to allow skim-coat plaster on the new wall surfaces, but I recommend it for a private home.

In one project—converting an early-twentieth-century garage to living space—rigid foam insulation was installed against the historic brick with a new 2½-inch-deep wood stud wall abutting the foam, putting all the insulation outside the interior wall assembly.

We cannot know how new products perform over the long term until a long term has passed. Until there is strong evidence that spray foam insulation will have long-term benefits without long-term negative consequences for historic buildings, I recommend that it be limited to new additions and other locations where it can be installed with a complete vapor barrier and cannot cause long-term harm to historic materials. Examples of such uses can be found in this book, including the bay window addition in Chapter

3, the new insulated roof added on top of the historic roof in the Ridlon Log Camp (page 529), and the new roof on Casa Roca (page 193).

A similar approach can be taken with plank-on-frame and other historic wood construction types that make walls hard to insulate. Again, most heat loss is through the roof and floor, not the walls. I would never recommend destroying good, intact interior finishes to insulate walls. Focus on the cap and base, install storm windows, seal every possible point of entry for cold air, and upgrade to more efficient HVAC equipment instead. Your money will be better spent.

Almost there. These domestic water supply pipes have been insulated to protect them from freezing and to prevent heat from escaping the hot water line, but an uninsulated elbow leaves this pipe vulnerable to freezing there. The location in front of a window makes it particularly vulnerable to a draft, although the presence of a storm window mitigates this risk. This photo was taken shortly before the large old oil-fired furnace in this basement was replaced by a high-efficiency gas-fired furnace. This conversion likely reduced the amount of heat escaping into the basement from the heating unit, lowering the temperature in the space during winter. The risk of frozen pipes increases with such a change, making it more important to insulate pipes completely.

Insulating Ducts and Pipes

While insulating the envelope of your house, think about impeding the unwanted transfer of heat from ducts and pipes that pass through unconditioned spaces like attics and basements. In cold weather, heating ducts and hot water pipes radiate heat to the cold air around them, leaving less heat where it is wanted. This reduces the efficiency of the system and increases costs of operation. In hot weather, humid air will condense on air-conditioning ducts and cold-water pipes, again reducing efficiency of the A/C system and dripping water where it does not belong. Well-insulated ducts and pipes solve these problems and should be a priority. Be sure to seal around all pipes and ducts with caulking or spray foam crack sealer where they pass from conditioned to unconditioned spaces.

Also consider the possible need to insulate pipes in unconditioned spaces to prevent freezing in winter. Many houses have had monstrous cast iron beasts radiating heat in their basements since indoor plumbing was installed. When these inefficient furnaces and boilers are replaced with new high-efficiency units, sometimes hung on a closet wall upstairs, previously warm basements can become quite cold. If an unmortared stone foundation allows much air infiltration, pipes near the foundation wall (where water lines generally enter) can freeze. Pipes in unconditioned spaces in cold-weather regions should be insulated, ideally with electric heat tape under the insulation. Heat tape is available with built-in thermostats to operate below a set temperature (usually 40 degrees). The small amount of electricity used by these tapes is well worth the cost if even one frozen and burst water line can be avoided.

Air Sealing

Along with insulation in the right locations, air sealing is often the most important thing you can do to improve the energy efficiency of your house. Few people realize how quickly narrow cracks in a house envelope become the equivalent of significant openings in the walls. A 1/16-inch gap around a standard door creates a void equivalent to a 12-inch by 12-inch hole in the wall—yes, a one-foot-square hole allowing your expensively heated or cooled air out into the world. Cracks and gaps can appear all over a house—around doors and door trim, windows and window trim, corner boards, sills, eaves, between clapboards, around basement bulkheads, around utility penetrations of the walls or foundation, and on and on. And that's just the exterior. Inside, the joints between floors and exterior walls, plaster and trim, and walls and ceilings are possible culprits, along with every penetration between heated and unheated spaces in the attic and basement. This includes electrical

Back plaster applied to the interior face of wall sheathing between the studs. In this house built in 1850, the back plastering continued up to the roof in the gable ends of the attic, where it remains exposed. In the lower stories it was hidden from view by the installation of lath and plaster on the inner surface of the studs. Back plaster does a good job of air sealing the exterior as long as it remains intact.

In this historic log camp, quarter-round molding was installed in the joints of the logs (at right) to help seal the cracks between logs against air (and mosquito) infiltration.

The joints between logs on the exterior of the Ridlon Camp were sealed with acrylic Perma-Chink when the camp was converted from seasonal to year-round use during a recent restoration. Chinking material of this type adheres to the wood and remains elastic, moving with the wood as it expands and contracts with changes in the weather. Because the exterior had previously been painted, the chinking material was simply painted with the logs when the building was repainted. Had the logs not already been painted, it would have been advisable to match the log color as closely as possible with the sealant.

wires, ventilation fans, plumbing, ducts, chimneys, and so on. Wherever two building components come together there is the potential for air infiltration. On masonry houses, the focus of air sealing should be around windows, doors, utility penetrations, and at the roofline, assuming the pointing is in good repair.

The more such voids you can seal with caulking, weather stripping, or spray foam sealant, the less energy your house will waste and the more comfortable you will be. There are numerous sources of information on air sealing available from basic renovation books and many websites, including the energy efficiency sites maintained by a number of states. They will provide guidance as to which sealant to use where. Use this opportunity to inject fire-stop sealant around all penetrations into the attic, basement, utility chases, or framing voids. This is generally required in new construction but not typically used for residential rehab. It can do a lot to prevent a fire from spreading.

If your house was built before about 1860, you may have back-plastered walls, which are very effective at air sealing the exterior sheathing. Back plaster was applied to the inside of the exterior wall sheathing, between the framing members, before the interior lath and plaster was installed. Because it was applied as wet plaster, it fills the stud bays completely, often lapping up onto the sides of the studs. If you find this treatment in your walls, leave it there.

Some houses have more cracks than others, but few can compete with log cabins. Residential log construction in the United States has a long history that includes early homes as well as nineteenth- and twentieth-century recreational cabins and camps. Log homes are usually chinked between the logs and sometimes clapboarded on the exterior, and they may also be plastered on the interior. All of these measures help with air sealing. Recreational log structures nearly always have the logs exposed on the interior and exterior as part of their character. Most of these were intended to be used seasonally, usually the summer season, and were not designed or equipped for year-round occupation. Fireplaces or wood stoves provided whatever heat was needed for cool summer nights or spring and autumn conditions. Some log camps had quarter-round molding installed between logs—on the interior, exterior, or both—to partially seal the joints.

When a recreational camp or cabin is rehabilitated for year-round use, sealing between the logs can be challenging. Products such as Perma-Chink, a flexible acrylic latex sealant that resembles mortar, can be used but will alter the appearance of a log wall that was not chinked historically. Perma-Chink is available in a number of colors, including brown and tan; come as close as possible to the color of the logs if there was no chinking historically. Alternatively, the entire structure can be painted to hide the sealant, as was done on Ridlon Camp, the featured house on page 529. This is an easier choice if the building has already been painted or stained.

A Note on Weather Stripping

There are many types of weather stripping available at a variety of price points. In general, the more expensive ones are the longest lasting. A window or door properly fitted with new interlocking bronze weather stripping will stay tight a very long time. That said, sometimes the budget requires a less expensive solution, and many of these are effective if not as long lasting. (See example in Chapter 13.) Find products that are within your budget and will do the job well and with as little visual impact as possible. The money saved on energy will help pay for replacing less-durable materials when the time comes, hopefully with longer-lasting substitutes. Simply choosing colors that match or blend into the base material can make a big difference. If you are using common aluminum-strip vinyl-bulb weather stripping around a door, and your door trim is painted a color other than white or brown (the two colors to choose from in this weather stripping), paint the aluminum part to match the trim after it is installed. If your door and trim are a dark color, choose weather strip with a brown vinyl bulb, because white would stand out like the proverbial sore thumb. A little care can greatly reduce the visual impact of weather stripping. See Chapter 13, Exterior Trim and Windows.

How Much Sealing is Enough?

The goal in new construction today is usually to seal a house so tightly that mechanical ventilation is needed to remove moisture and bring fresh air in the house. Short of extreme measures, as in the "Deep-Energy Retrofit in a Victorian Cottage" project on page 355, most old houses cannot be sealed tightly enough to require mechanical ventilation. A blower door test is the best way to determine how well sealed your house becomes after insulating and air sealing. It will be interesting to compare the new test results with the ones you got from your energy audit before the work began.

Ventilation

Unless thoroughly sealed, most historic houses exchange significant quantities of air with the exterior. This is an obstacle for energy efficiency but allows moisture to escape from the interior. Warm air carries more moisture than cold air, and heat is drawn to cold. These two properties of physics remove a lot of moisture from houses that are not tight.

When a house is tightened, trapped moisture can damage the house and cause health issues for its occupants. Toxic mold, not a notable problem when

Frost coating the inside of a storm window suggest that a lot of moist warm air is escaping around a poorly sealed primary window. Chapter 13, Exterior Trim and Windows, discusses how to tighten historic windows.

most houses could breathe freely, has become one in recent decades as more and more houses are built or rehabbed to be much more airtight. Moisture issues, particularly from infiltration through the roof or wetness under the house, must be addressed before or while any insulation and air-sealing work is done. Moisture produced within the house can usually be controlled with bathroom and kitchen vent fans *if* they are used consistently.

Part Four of this book, The Exterior Envelope, begins with the premise that water is the enemy and focuses on keeping water from penetrating the building's envelope. Here we will focus on getting rid of excess moisture that does penetrate the envelope or is produced in the house. A family taking daily showers and cooking meals produces a great deal of airborne moisture, more than is good for a reasonably airtight house. The first line of defense against these is to install and use bathroom and range hood vents.

Many people dislike the sound of an electrical ceiling vent fan running overhead while they're getting dressed in the morning or taking a bath, and I count myself among them. You can purchase expensive fans that are much quieter though not completely silent. Another alternative is to locate the fan in the attic, some distance from the bathroom, and exhaust the air through a vent in the eaves. Eave vent exhausts are nearly invisible and a much better option than the standard plastic or aluminum vent hood sticking out from the siding. I have used this approach with 3-inch PVC pipe for ducting. Moisture not expelled from the pipe when the fan is turned off will not affect the PVC as it would metal. The fan units (one per bathroom) hang from the rafters on bungee cords and are connected to the PVC pipe with flexible rubber couplings. Virtually no sound or vibration is transmitted along the pipe. The ceiling vent inlets in the bathrooms are covered with antique cast iron register grilles. Bathroom vent fans can operate on their own switch or be connected to a light switch so that they always operate when the light is on. Alternatively, motion sensors can be set for the fan to run for a set time after the motion stops. Whichever method you chose, make sure you and your family members use it.

Similarly, whenever possible, install a vented range hood in the kitchen. Recirculating hoods filter the air and remove grease and odors, but they do not vent moisture.

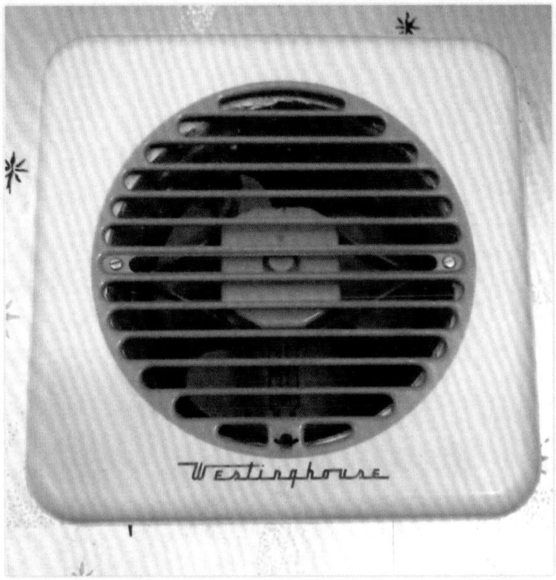

This Westinghouse bathroom vent in the Weston House (see page 243) is original equipment from 1957. Vintage ventilation equipment of this type can be found in more recent historic houses and should be maintained and used as it was intended. One of the wonderful things about electrical equipment of the 1970s and earlier is that it was usually made to be repairable. Parts may be challenging to find, but the chances are good that you can find one or two local electrical repair shops. If not, you may have to send your appliance farther away for service. The internet will guide you to a source.

Another frequent route by which moisture enters a house is through a basement or crawl space. This can be particularly true in old houses, which may have foundations of rubble stone or rough blocks of cut stone. Such foundations are porous even when mortared and will admit water that drains near the foundation. (See Chapter 12 for ways to reduce this.) If you're very unlucky, you'll find a natural spring in your cellar or basement that cannot be stopped. Whatever the cause, water that enters the underside of a house must be removed or it will migrate into the house as damp air.

It is not uncommon to find basements and crawl spaces under historic houses with dirt floors. Although a dirt floor may appear relatively dry, moisture passing up through this soil can add a great deal of unwanted water vapor to the air under the house. Such spaces should be ventilated if possible, and the soil must be covered by a heavy plastic vapor barrier with its seams sealed to create a continuous impervious surface that is affixed to the foundation on all sides. This can also be effective over a damp concrete or brick floor.

Modern foundations are constructed with footing drains to keep water outside the basement. In an old house, a drain channel around the inside of the foundation wall leading to a sump pit or, better, a pipe drain to daylight can be effective. If the cellar floor is concrete or brick, a strip will have to be removed around the perimeter. The interior drain

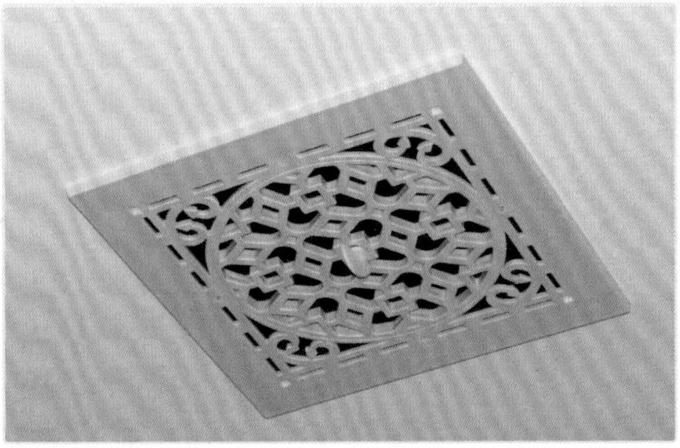

Bathroom vents do not need to be cheap-looking plastic. This modern bathroom ceiling vent has a new cast aluminum grille covering the opening. Whether new aluminum or salvaged antique cast iron, such features enhance the historic character of a space.

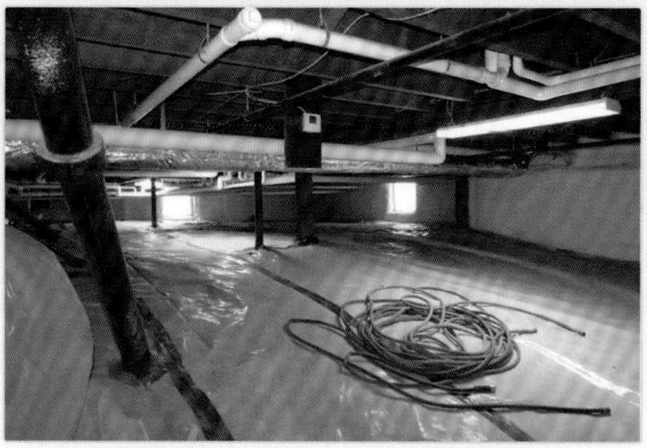

This crawl space under a recently rehabbed historic building has had its dirt surface covered with a heavy poly vapor barrier that is sealed at all seams and along the foundation.

can be a trough formed of concrete or a perforated pipe wrapped in filter cloth and covered with gravel. In either case, it needs to pitch toward the outlet. In most cases, a drain like this, with a vapor barrier on the floor if needed, will solve the problem.

Sometimes a dehumidifier has to be added to complete the job. If so, it needs to be connected to a covered sump pit or drain to daylight. Pulling moisture from the air and dumping it back on the floor will not accomplish the goal!

Minor moisture seepage through a concrete foundation can often be halted with a waterproofing agent on the interior. (Numerous products are made for this application.) The potential problem with this approach is that it traps water in the concrete. I have seen instances in which these coatings bulged out from the concrete with pockets of water behind, slowly expanding and separating the coating from the concrete. Eventually, these burst open. Trapping water in the concrete may not be the best long-term solution if no effort is made to reduce the water reaching the outside of the foundation wall.

It is possible to excavate around the foundation and apply waterproofing to the exterior, ideally with a footing drain installed at the same time. For rough stone foundations, heavy poly sheeting can be used as the waterproofing, with all seams tightly sealed together. The poly can be attached to the sill during installation and then cut off close to grade level after the trench is backfilled. Leave several inches above grade and seal the flap to the foundation with an appropriate adhesive. If the basement is heated to any degree (for instance by the furnace), applying rigid foam or spray foam insulation over the poly before backfilling will reduce heat loss through the foundation. While not a small project, exterior waterproofing has a greater likelihood of success than an interior coating.

There are contractors who specialize in drying out basements. If you go this route, follow the same guidelines suggested in Chapter 4 for hiring

A dehumidifier is installed in this crawl space under a nineteenth-century home, and a small pump drains the unit to a covered sump pit in the taller cellar area under the house. The poly vapor barrier covering the soil should have its edges sealed to the foundation stone.

any contractor. Just because a contractor is working under a franchised national brand doesn't mean their work is good or the national brand will stand behind it. Check references and read the fine print before signing any contract.

In a cellar with a natural spring or other underground water course, more drastic measures may be needed. This situation can cause significant damage to a house. I know of one case where the spring was left to run for generations in spite of evidence that it was causing problems in the house. By the time a new owner pulled away the crumbling plaster and rotted flooring, virtually all the timber framing inside the brick shell of the house was rotted beyond salvage. The entire interior had to be gutted and rebuilt after the engineer and contractor solved the water problem. The solution generally involves the creation of a closed channel to remove the water to an exterior drain, because it cannot be stopped at the source. It will just find another way in. Professionals may be required.

Before and after views of the Victorian worker's cottage that was the subject of this deep-energy retrofit project.

Example Project: A Deep-Energy Retrofit in a Victorian Cottage

Contradicting much of what I have written in this chapter, this 1887 Queen Anne–style worker's cottage underwent a "deep-energy retrofit" involving measures I usually object to—with my involvement and support. I helped make the case for this approach as a demonstration project, to see if super-insulating and air-sealing a historic house can be done in a way that does not destroy its historic character. The only way to know what will or will not harm a historic house over time is to test the approaches on real buildings that can be monitored into the future. This house was

The cottage before rehabilitation.

rehabilitated to become part of a community center, where it is one of three neighboring identical cottages with a large, modern facility behind them. Because the buildings are public, the building envelope engineers and architect who designed the project can monitor it going forward to a degree that would be difficult with a private home.

The cottage had been vacant for several decades and was in very poor condition. There was significant water infiltration through the failed roof, and more than a few four-legged inhabitants had moved in. The second story had been gutted at some point. Yet in spite of the neglect, or perhaps in part because of it, the first story was remarkably intact with few changes since it had been built by a progressive industrialist in the late nineteenth century to house millworkers. He built several groups of similar cottages in the vicinity from several designs. This cottage was one of a group of four that have survived into the twenty-first century.

The local organization that owned the two neighboring cottages and the attached modern community center secured a federal grant to rehab this cottage as a demonstration project with the goal of making it as energy efficient as possible—a "deep-energy retrofit." While no legal definition of a deep-energy retrofit has been established by any standard-setting agency, the term generally refers to measures that reduce a home's energy use by 50 to 90 percent below that of a code-minimum new house. A good source for additional information on this topic and other practical approaches to energy efficiency improvements in historic houses is *The Energy-Smart House* by the folks at *Fine Homebuilding* magazine (Taunton Press).

Because federal funding was involved and the cottage row was eligible for the National Register of Historic Places, the proposed work had to be reviewed

The subject cottage is in the foreground. The two previously rehabbed cottages are attached to the modern community center, which is out of sight to the right.

by the state historic preservation office to determine whether it would have an adverse impact on the building or the eligible district. My firm was brought in to assist in this review process, and we worked with the design team to find a balance that would achieve their energy-efficiency goals while preserving the historic architectural integrity of the cottage.

The drawings illustrate the wall and roof assemblies that were ultimately used. For much of the building, insulation was added both outside and inside the wall sheathing. The outside insulation was 2 inches of foil-faced polyisocyanurate insulating sheathing, which was applied over an air-barrier membrane that was fully adhered to the existing board sheathing. The inside insulation was cellulose blown into the stud bays. Salvaged historic and matching new materials were used for the exterior surfaces, and the clapboards and shingles were installed with a rainscreen gap. Because the façade had more architectural detail, including a bay window and porch roof, and was the most character-defining feature of the exterior, the insulation on that elevation was confined to the interior.

The roof received 4 inches of new insulation above the sheathing (two layers of 2-inch foil-faced polyisocyanurate insulating sheathing) plus cellulose in the rafter bays. The eaves and gable ends were detailed carefully to disguise the added height, and the chimney was rebuilt and heightened to maintain its historic relationship with the roof. Spray foam and cellulose insulation were applied in the basement wall assembly, and a new insulated basement floor slab was installed. Triple-glazed windows (matching the historic configuration) replaced the badly deteriorated original windows. The interior received an impermeable vapor barrier to seal the walls completely before

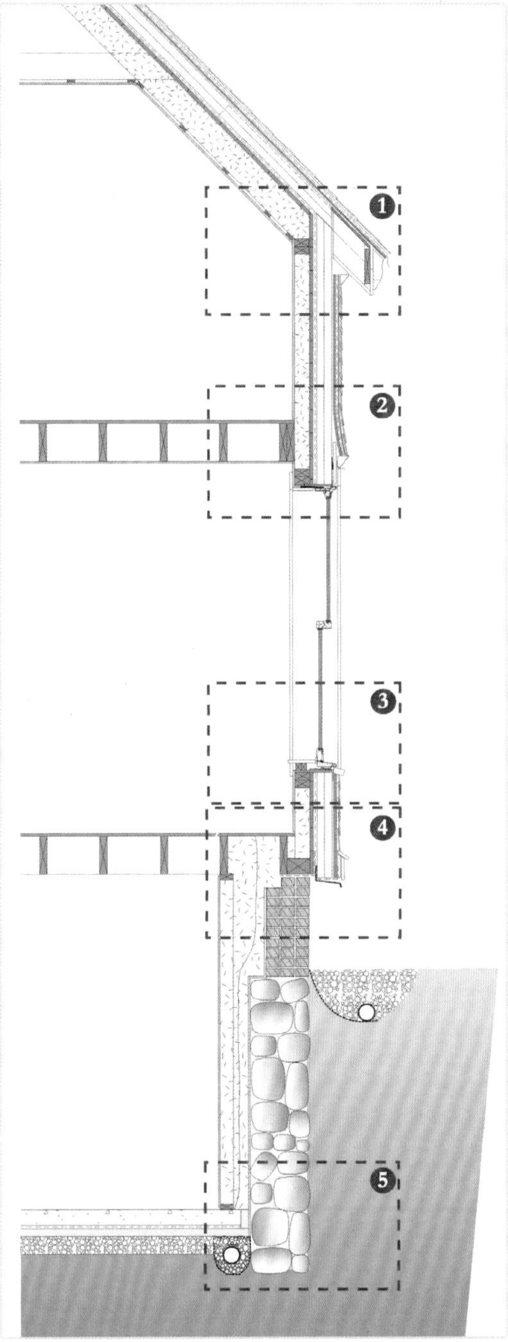

These drawings by Building Science Corporation (www.buildingscience.com) detail the approach taken to achieve a very high level of insulation and air sealing in the cottage.

Drawings continued on the next two pages

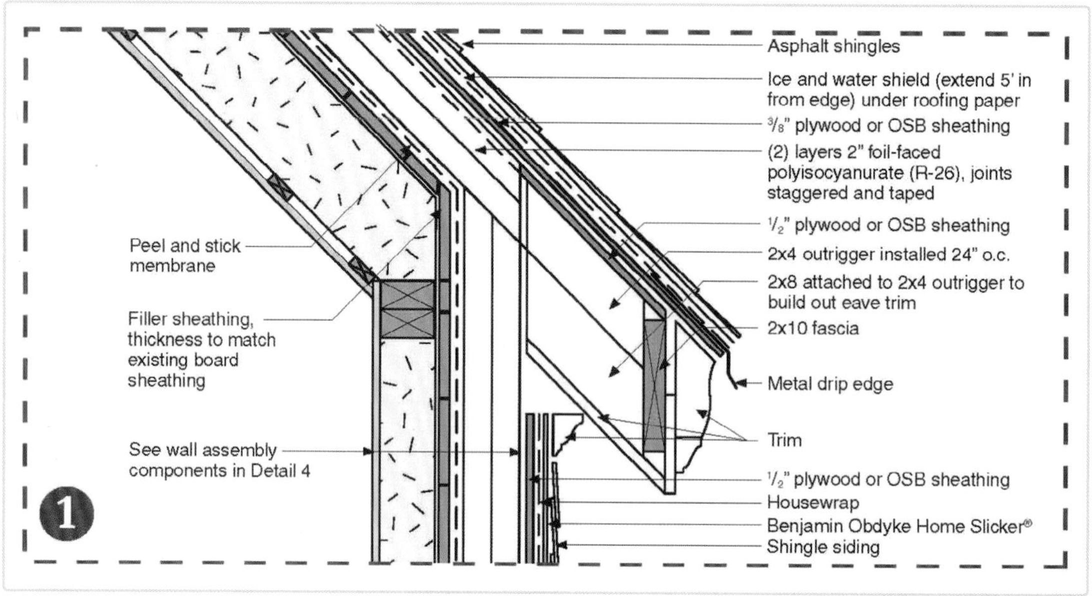

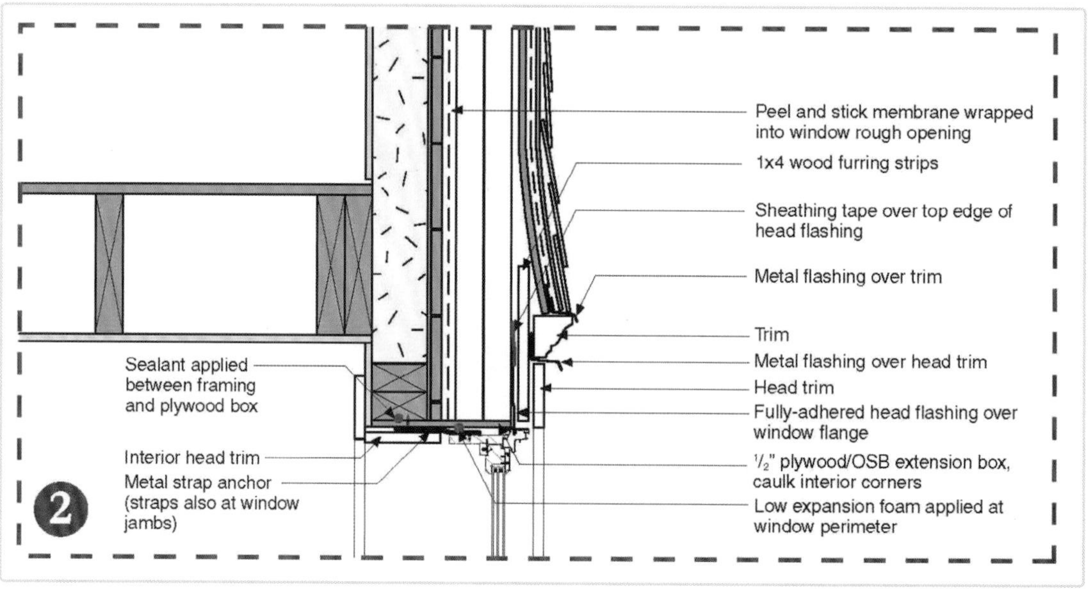

new gypsum wallboard was installed. Historic trim elements were stripped of lead paint and reinstalled. The sills and basement were insulated, and water infiltration to the basement was reduced or eliminated. Every possible avenue for air infiltration was sealed from the interior and exterior.

All this insulation and sealing necessitated mechanical ventilation to bring fresh air in and exhaust stale air out. The chosen solution was a superefficient heating and cooling system using a hybrid gas furnace and air-source heat pump with an integral heat recovery ventilator (HRV). An HRV transfers heat from the stale

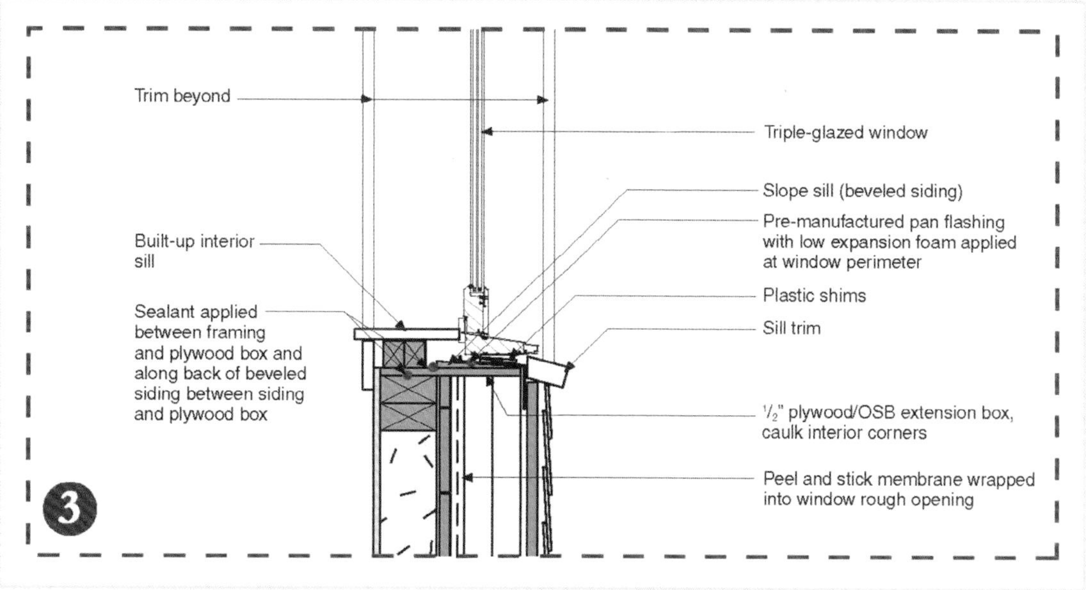

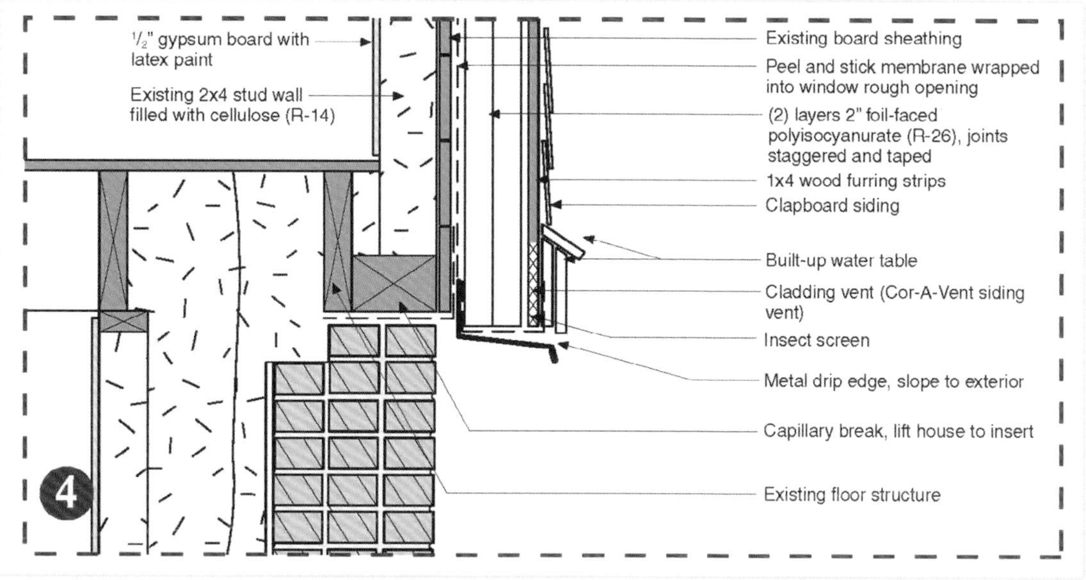

air being mechanically expelled to the fresh air entering the house, reducing both the heat wasted and the need to warm incoming air with the gas furnace.

The need for fresh air in a tight house has received a lot of attention in recent years. The concern is not new. In their 1869 book *The American Woman's Home, Or, Principles of Domestic Science*, sisters Catharine Beecher and Harriet Beecher Stowe wrote:

> Whenever a family-room is heated by an open fire, it is duly ventilated, as the impure air is constantly passing off through the

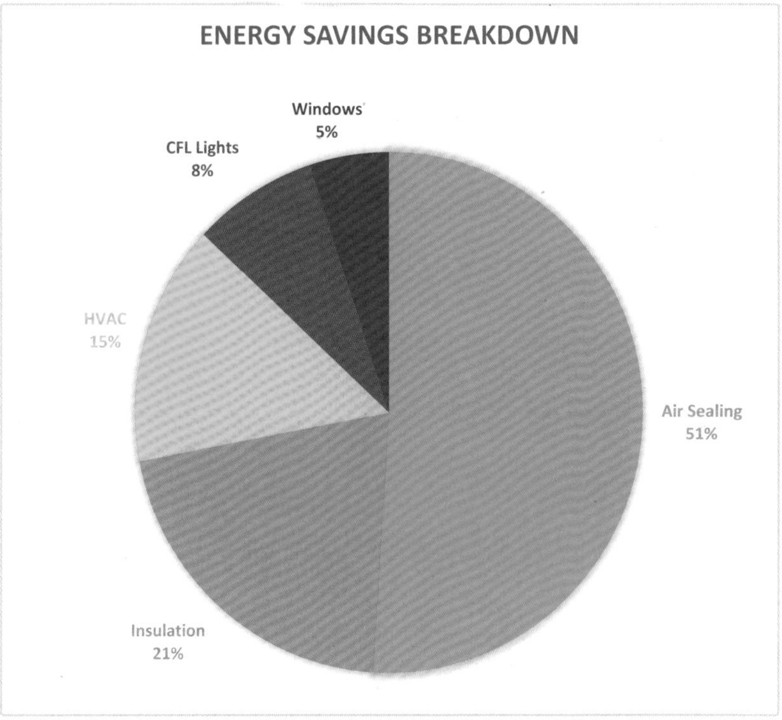

This pie chart shows a breakdown of energy reductions from the improvements made to the cottage. Because the building was vacant and uninhabitable at the beginning of the project and no utility records were available for comparison, reductions were measured against a code-compliant building in the same climate zone.

> chimney, while, to supply the vacated space, the pure air presses in through the cracks of doors, windows, and floors. No such supply is gained for rooms warmed by stoves. And yet, from mistaken motives of economy, as well as from ignorance of the resulting evils, multitudes of householders are thus destroying health and shortening life, especially in regard to women and children who spend most of their time within doors.

The famous sisters lost their argument against cast iron stoves (and later furnaces) largely because the uninsulated houses of the period weren't airtight enough to cause a sick house—until the insulation and air-sealing products of the late twentieth century came along. The twenty-first century has provided a solution to this problem in the form of HRVs.

The final component of the deep-energy retrofit for the Victorian worker's cottage was the installation of energy-efficient lighting.

Tests indicate that the cottage is now nearly as efficient as a well-built new house. It experiences 1.47 air changes every hour, versus a goal of 0.60 air changes per hour for a superefficient new home. The average existing house experiences three air changes per hour. An analysis by Building Science Corporation attributes 51% of the efficiency gains to air sealing, 21% to insulation, 15% to the high-efficiency HVAC system, 8% to the change to CFL lighting, and 5% to the new windows.

You read that right: Only 5% of the reduction resulted from the installation of expensive new triple-glazed replacement windows. This topic is addressed in Chapter 13.

The rehabilitation of the cottage cost approximately $300,000 in 2015 (not including the cost of building-science professionals, historic preservation consultants, lawyers, etc.), of which $30,000 was attributable to the deep-energy retrofit. The project managers estimate resultant gas and electricity savings of up to $4,645 per year, resulting in a payback period of less than seven years.

The parlor before rehab.

The parlor after rehab.

The previously gutted attic before rehab.

The new office space in the attic after rehab.

An eave detail before rehab.

An eave detail after rehab. The 4 inches of additional insulation and new sheathing on the roof are completely hidden by the carefully detailed eaves.

A detail view of the front gable end before rehab.

This detail view of the front gable during construction shows the several additional shaped shingles added to each course to disguise the added thickness of the roof. Once painted, the change became imperceptible to anyone not looking for it.

A detail view of a corner of the cottage at the foundation before rehab.

A detail view of a corner of the cottage at the foundation after rehab. The several additional inches of wall thickness over the foundation are the most obvious evidence of the deep-energy retrofit.

The front porch roof before rehab. The porch itself had rotted away.

The front porch after rehab.

A section of historic clapboard and trim before rehab. There was not a lot of lead paint left on the exterior to be concerned about.

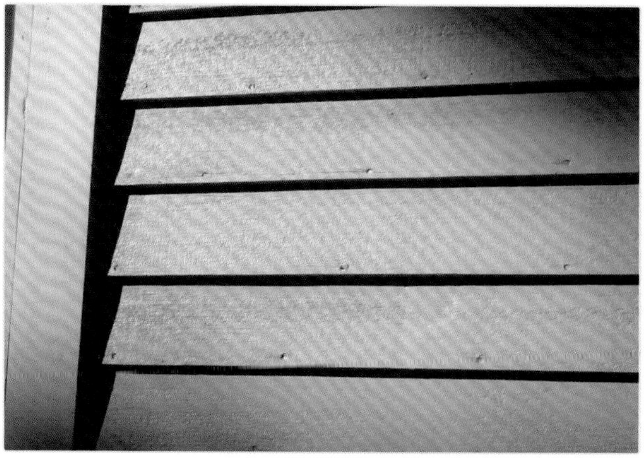

Historic clapboards and trim were removed from the house, stripped of paint where necessary, repaired, and reinstalled on the new wall surface several inches out from their original locations.

The ornamental Queen Anne chimney before rehab.

The rebuilt chimney. Original bricks were reused for reconstructing the chimney in its original form but several inches taller to retain its historic vertical clearance above the new, thicker roof.

The deep-energy-retrofit cottage is in the foreground, nearly completed, with two of the three matching cottages in the distance. Can you tell the difference?

Note that the $30,000 cost is *on top of* the costs of bringing the house up to code in a standard rehab. Not many private homeowners would undertake a deep-energy retrofit of a historic house due to its cost and possibly longer payback period, but this demonstration project shows what can be done. It will be instructive to see how the work holds up over time.

The historic review of this project was interesting. Despite several rounds of refinements to the original design—during which the additional wall thickness was reduced from 4 to 2 inches and the additional roof depth from 8 to 4 inches, the state historic preservation commission ruled that the work would have an adverse impact on the historic character of the district and required a mitigation for that loss. The agreed mitigation was to engage an architectural historian to produce an in-depth report on the history of the worker cottages in question, which I researched and wrote.

In the commission staff's opinion, the rehabbed cottage would lose sufficient integrity (historic character) as to no longer qualify for the National Register as a contributing building to the district. The accompanying before-and-after photos show the cottage in relationship to the (nearly) identical cottages next door. The other two cottages—which are attached to the community center—had much of their exterior materials replaced in-kind when they were rehabbed. Technically, they have less integrity of materials than the subject cottage but are nevertheless considered as "contributing" to the district. I will let you decide for yourself whether the deep-energy-retrofit cottage does or does not have equivalent post-rehab architectural integrity.

CHAPTER TEN

Plumbing and Electrical

There are many good books and websites on household plumbing and wiring. This chapter assumes a basic understanding and focuses on the plumbing and wiring circumstances encountered in historic houses. Many jurisdictions will allow a homeowner to do plumbing or electrical work, but it must be done to code. There may be required inspections before new plumbing or wiring is sealed into walls and ceilings. Know your local rules.

Plumbing

The first step in determining what to do with your plumbing is, of course, to figure out what's there. The two primary parts of a plumbing system are the supply pipes and the DWV (drain, waste, and vent) pipes. Supply pipes are pressurized to deliver water to fixtures. Typically, hot and cold supply lines run in parallel. DWV pipes carry away the water and waste from the fixtures after use and provide venting to keep sewer gas out of living spaces.

You can start your assessment by tracing each line from its entry or exit through the basement floor or wall, a crawl space, or a slab floor. There is usually a cleanout T or Y in the DWV pipe near its exit. DWV and supply pipes may or may not run close together. If the house was plumbed long after it was built, you may find vertical DWV pipes (soil stacks) enclosed by simple wood chases (casing) in the corners of primary rooms. A DWV may be fully exposed (usually with many coats of paint on it) in a secondary space, and I have seen DWV pipes wallpapered to match the walls behind them. Once you locate the DWV pipe(s), trace its route to the "stink pipe" vent through the roof. Don't be surprised if the vent pipe terminates in an

In this plumbing arrangement of about 1900, the large, vertical, cast iron pipe is the "soil stack" waste pipe. The iron pipe emptying into it from the left is from the kitchen sink in the adjoining room; this is joined by a painted lead drain from the bathroom sink. The iron waste line entering through the foundation wall is for a gutter downspout outside the house. They all flow down through the soil stack to an outlet from the basement, which empties to a private septic system or a public sewer line. If the system is old enough and hasn't been altered since installation, it may drain directly to the nearest stream or lake, which is definitely not to code! The two smaller galvanized iron pipes that come down through the floor and terminate in 90-degree elbows are former hot and cold supply pipes, since removed.

What is atypical about this configuration is that the soil stack is capped by a toilet flange and outlet on the floor immediately above. This arrangement—never common and now illegal—required the fitting of a vent outlet at the toilet base, just beyond the trap, to allow air in, maintain neutral air pressure, and vent sewer gases. That's the purpose of the 2-inch galvanized-metal vent pipe that drops down through the floor next to the soil stack and runs to the chimney (out of view) in the basement. A modern DWV system requires the soil stack to continue up through the house to a vent through the roof. Toilet drains enter it through T or Y fittings, just as other drains do, and every fixture has its own vent to the roof. Venting through chimneys is not allowed.

unfinished attic. This is not allowed today (it can fill the attic with poisonous gases) but was sometimes done in the past to avoid a pipe penetration through the roof. (Plumbers do not do roofing!)

DWV pipes are most often cast iron with heavy hub fittings that are packed with oakum or filled with lead solder. Earlier lead DWV pipes survive in some houses; they do not generally pose a health hazard (unlike lead pipes for incoming water) but are not good for the downstream environment and should be replaced. Updated DWV systems are more likely to be PVC pipe or a combination of PVC and cast iron. Fired clay DWV pipes are rare in a house though once commonly used underground.

The supply pipe (much smaller than the DWV pipe) will run from its entrance to a tee, where it splits. From there, one line continues as the cold-water supply and the other runs to the water heater inlet. From the water heater outlet, this line continues as the hot water supply. Cold- and hot-water lines typically run in parallel to the kitchen and bathrooms. They may be hidden in walls and ceilings, enclosed in chases with the DWV pipes, or fully exposed along walls or ceilings in secondary spaces. Supply lines for toilets and outdoor hose bibs will usually be cold without a parallel hot line. The most common materials for supply lines installed before about 2000 are threaded iron pipe (often galvanized) and soldered copper pipe. Unfortunately, lead supply pipes are still in use in some houses; if you find these, they must be replaced as soon as possible. It is usually easy to determine if your pipes are iron, copper, or lead by scraping the surface with a sharp blade. If the metal is painted, scrape right through the paint. Iron will be silvery and shiny, copper will have the color of a new penny, and lead will be a dull gray-silver and is noticeably softer than the other two metals. Tracing the route of your pipes will provide you with the map you will need to make any changes or repairs and will give you an opportunity to inspect for leaks or for corrosion that could lead to leaks.

At the business ends of the DWV and supply pipes are the fixtures they serve: sinks, tubs, toilets, washing machines, dishwashers, etc. Each should be inspected and its condition noted. It is not unusual for hundred-year-old kitchen and bath fixtures to remain in use, often in good condition. This is particularly true of ceramic fixtures, which are impervious to leaks if not cracked or broken. Do not assume a stained porcelain sink with a dripping faucet needs to be replaced. Stains can often be removed, and most old faucets were designed to be repaired (unlike many modern ones). Many old faucets can remain in use indefinitely after a tune-up to replace washers and any other parts that have worn out.

Most communities still have at least one surviving hardware store with knowledgeable plumbing department staff who can identify and supply what is needed. It is generally best to remove the faucet, valve, or other fixture and take it with you to the store. While you're there, buy a reserve supply of rubber washers and other parts that will have to be replaced periodically in the future; independent hardware stores have been closing at an alarming rate for decades, and there is no guarantee your local store will be there next year. Nor is there any guarantee a big-box home-improvement store will carry what you need, particularly if it is not something many people have a need for. The big boxes succeed on volume, not on small parts for obscure faucets, and there is more profit in selling you a $50 replacement faucet than a 30-cent rubber washer.

If the locally owned hardware stores in your community have already disappeared, you may be able to find what you need at an architectural salvage shop that deals in antique fixtures. As a last resort, send your faucets to a mail-order plumbing fixture restoration shop to be repaired and returned. Antique fixtures contribute to the historic character of a house. When fixtures are beyond repair, it is important to find replacements (salvage or new) that are aesthetically compatible.

Once you have assessed the existing plumbing, consider what can remain and what needs to be updated or replaced to address deficiencies or accommodate planned changes to the house, such as adding a bathroom or a laundry room. Plumbing that does not have to be replaced can often be left as is even if it does not meet code (except lead supply pipes in most states), but once you remove it, whatever replaces it will have to satisfy the code enforcement officer.

Laying Out New DWV Pipes

Many old houses were built without indoor plumbing, but few remain that way. The difficulty of snaking pipes into existing walls, floors, and ceilings is nothing new. Any experienced house restorer can tell horror stories about the structural damage done by long-ago plumbers. In the past, many a plumber was too often willing to cut a 4-inch by 4-inch notch in a 6 x 8 beam to let through a 3½-inch cast iron drainpipe. When we update or expand the plumbing in a historic house, we should repair the sins of the past as we encounter them, and we should definitely not repeat them. It complicates the task of routing pipes—DWV pipes in particular—but it must be done.

Drains are the principal difficulty because they are large and need to be sufficiently pitched toward the sewer pipe outlet to effect a thorough removal of water and waste by gravity alone. (Supply pipes are usually much smaller and are pressurized, which means that incoming water will flow up, over, and down to where it is needed.) Laying out a drainage system can be challenging in a balloon- or platform-framed house, and even more so in a timber-framed structure.

Main drains requiring a close-to-horizontal run pose the greatest challenge, especially in upper stories. If such a drain is pitched at less than ¼ inch per foot it will clog easily; if it's pitched at more than 3 inches per foot, it will drain too quickly to carry away the solids. Thus, a drainpipe needs at least 6¼ inches of pitch over a 25-foot run; a 3½-inch pipe will exceed the depth of a 2 x 10 floor joist (9½ inches) in a run of that length.

If your house is timber-framed, a drain will run into a large timber, 8 x 8 or larger, at the joist ends. Going through the timber will weaken it more than is acceptable unless, perhaps, there is a load-bearing wall immediately beneath. Dropping down to run under the beam will usually put the pipe in the living space below, and most people do not want to listen to the waste of a toilet swooshing down an overhead drainpipe while sitting in a parlor or dining room. And what if the joists run in the other direction and your drainpipe needs to cross them at right angles? Simply put, it cannot. The necessary notches or holes would so weaken the joists that they could not do their job unless their span was very short, say three feet.

The two most common solutions in such circumstances are to build up the portion of a floor where the drain needs to run horizontally, putting all or part of the bathroom a step above the surrounding floorplan, or to drop lines down through walls, closets, or chases to get below the first floor, where you can pitch the pipe and run it below the first-floor joists as needed. The latter approach requires each upper-story bathroom to have its own vertical drain if they are any distance apart; the drains can usually be tied together with the required pitch in the basement or crawl space. This is usually the preferred approach, but it is not always possible; sometimes a step-up bathroom is necessary.

When adding onto an existing DWV system, it is usually possible to cut out a section of cast iron drainpipe and insert a heavy rubber T or Y fitting to receive a new PVC drain from the added work. Sometimes, the cast iron looks fine but is thin from age (due to rusting on the interior) and will collapse when cut. In such cases, the prudent action is to remove the cast iron back to the house outlet and replace it all with PVC. It's more work, but it's far better than having a drainpipe fail inside the walls of your newly restored house a year later.

Venting

The "V" in DWV piping is venting, and every fixture needs it. Another challenge for installing new plumbing in old houses is that modern code requires every fixture to be vented through the roof. This was rarely done historically; instead, most houses had a single vent stack extending from the vertical soil pipe up through the roof, and all the fixtures on all floors drained into the soil pipe. If sink and tub drains glug-glugged a bit when draining because of their distance from the soil stack, it was still a vast improvement over carrying dirty water outside to dump it, as had been the norm for thousands of years previously.

Before and after views of the same house. In the first photo the new white PVC vent pipes stand out from the dark roof and are incongruous with the historic character of the eighteenth-century building. In the second photo, the pipes and their sheet-metal flashing have been painted flat black to blend with the roofing.

Vented fixtures drain better and eliminate the possibility of negative air pressure sucking the water out of the trap under a fixture, allowing sewer gas to flow into living spaces. But laying out drainage and venting systems has become more complicated. As mentioned above, work that is not being changed is usually not required to meet current code, and this includes venting every fixture. You need a vented system for the safety of your household, but if a single vent serving multiple fixtures is working well enough and does

A close-up of one of the painted vent pipes. Five dollars in materials and 20 minutes of time have made the attention-grabbing modern requirement virtually disappear.

not need to be replaced, you will save a lot of effort (and money if you are paying a licensed plumber) by leaving it alone.

In a lot of new construction, the vents from every fixture run up through the roof and extend at least a foot above the surface, higher in areas prone to heavy snow. For a house with multiple bathrooms, this can result in a little forest of 1½- or 2-inch white PVC pipes bristling across the roof—not a character-enhancing look on a historic house, where a single 2½- or 3½-inch black cast iron vent pipe has been the norm for over a century. My preferred solution for the many vents required for modern baths with multiple fixtures is to tie the small vents together in the attic and run the combined vent through the roof with a single 3-inch PVC pipe painted flat black. Horizontal vent runs need a bit of pitch back toward the vertical pipes in case rainwater enters through the main vent. Some code enforcement offices may insist on separate vents through the roof, but if you are in a local historic district, the review committee, board, or commission may back you up. If there is absolutely no way to get around multiple vent stacks, route them to inconspicuous locations on the back side of the roof, behind chimneys or ornamental cresting or similar, and paint them flat black.

Supply Lines

As mentioned, old supply lines are typically threaded iron or soldered rigid copper tubing. Sometimes the iron pipe is galvanized for rust protection. Some houses retain lead supply lines, which must be replaced. Because supply lines are continuously filled with pressurized water, even a small leak can release a lot of water. A fully ruptured line can quickly cause damage in the tens of thousands of dollars. A careful inspection of the supply lines in your house is essential. If their condition is at all questionable, replace them while you are rehabbing the house. Having to go back and tear out completed work to access leaking pipes in walls or ceilings is just painful.

If you are adding bathrooms and plan to tap existing supply lines, be sure they have adequate capacity to supply multiple fixtures at the same time. Do not assume an 80- or 100-year-old pipe retains the carrying capacity it had when new. Internal corrosion or mineral buildup may have reduced its capacity significantly. Test it. You do not want to find yourself standing naked under an anemic trickle of water in your beautiful new subway tile shower with radiant-heated walls and a thousand dollars' worth of nickel-plated hardware. A house that has existed for a century with a single bathroom and a kitchen sink but is now going to have three bathrooms, a laundry room, a dishwasher, and a kitchen with two sinks will probably need a larger waterline from the street or a larger line and more powerful pump from a private well.

A salvaged antique toilet slate has been installed under the high tank toilet in this new Victorian-style bathroom, nearing completion in the Dow Farm (see page 175). Toilet slates protect wood floors from the damage caused by dripping condensation. The marble sink top and backsplash and the sink and faucets are antique. The cherry base cabinet and wall unit were custom made for the room by the resident carpenter. The beadboard walls and ceiling will be painted.

You might find the floor around a toilet rotting away, or there might be evidence that the flooring in this area has already been replaced. The culprit may not be a leaky toilet drain or splashed water from the toilet, but condensed water dripping off the toilet tank in hot, humid weather. Like moisture dripping from a glass of cold iced tea on a humid day, cold water inside a toilet tank cools the porcelain, and moisture condenses on the outside surface from the surrounding warm air. This problem has existed for as long as toilets have had porcelain or metal tanks. Many early toilets were installed on slate or marble slabs (toilet slates) with slightly raised rims to keep the condensed water from contacting the wood flooring around it. These went out of use as ceramic tile and linoleum floor covers entered bathrooms, but the tiles or linoleum were often less effective at protecting the flooring beneath.

Toilet slates can sometimes be found at architectural salvage shops and are still a wonderful way to add period character to a bathroom. If they are installed in your house already, keeping them is highly recommended. Some modern toilet tanks have a Styrofoam insulation lining to solve the condensation problem, and retrofit Styrofoam insulation kits are available (though I have not heard good feedback on these). Another remedy is to add a mixing valve in the supply line to the toilet to fill the tank with warm water instead of cold. The tank water does not have to be hot, just tepid enough to prevent condensation. I have used this method myself and find it to be 100 percent effective. It requires a bit more plumbing to install the mixing valve, and it uses a bit of hot water every time the toilet is flushed, but that seems a small price to pay for a floor you will never have to replace.

PEX versus Copper

Black iron was the standard material for residential water supply lines from about 1900 to 1960, and rigid copper tubing has superseded it for most of the past sixty years. In the past several decades, however, cross-linked polyethylene (PEX) tubing has rapidly become the preferred choice for many professional plumbers and for do-it-yourselfers. This plastic tubing is flexible, relatively inexpensive, can freeze without bursting, and does not need to be soldered at joints. It is joined with either crimp or push-on compression fittings. If for no other reason than the elimination of open-flame torches for soldering, this tubing should be the first choice when installing or replacing plumbing in an old house.

I confess to being someone who, despite success with nearly all other aspects of home repair, has never been more than marginally competent at soldering copper tubing. I just do not have the touch. I get it too hot, and the solder flows right through the joint, or I do not get it hot enough, and the solder doesn't completely fill the joint. Invariably I struggle to get an acceptable result. The advent of cross-linked polyethylene tubing completely changed my attitude toward DIY plumbing. I now enjoy it. The last piece of copper tubing I soldered was the adapter fitting I installed inches above my water meter to go from copper to PEX. It has been all crimping from there for three new bathrooms and a new kitchen.

An additional advantage of plastic tubing is the unlikelihood of its being ripped out by thieves and sold to scrap dealers, as copper pipe too often is when houses are left vacant. Thousands of historic houses have been vandalized by thieves stealing copper pipes. They typically use cordless reciprocal saws or chainsaws to slice through plaster, wainscot, baseboards, flooring, and whatever else is in their way, and they can do an incredible amount of damage in a very short time. If you are installing new supply lines in a house, using plastic tubing will eliminate this temptation.

In the past, the high cost of buying and installing cast iron DWV pipes and copper tubing influenced the locations of bathrooms, the frequent goal being to supply all fixtures with a single pair of pipes and drain everything to a single soil pipe. The homes of the wealthy often stood out in part because of their multiple soil stacks to serve various bathrooms. The less expensive PVC and plastic tubing available today requires significantly less labor to install and allows more flexible plumbing arrangements for those of us who are not wealthy. This can be helpful for preserving elements of an original floorplan and other character-defining features of a house.

Gas Lines

Many houses were once plumbed for gas lighting, which was later replaced by electric lighting. Most gas lighting lines—typically threaded iron pipe—were long ago disconnected near the meter location. *Do not assume old gas lines are disconnected without verification!* More than a few people have discovered gas in lines they assumed to be disconnected. Trace the lines to where they enter the house and confirm that they are not connected to a meter. If they are, call the gas company or a licensed plumber/HVAC technician to determine if there is still gas being supplied

This 1850s gasolier has never been converted to electricity, as can be seen in the close-up view of the gas jet tube inside the frosted shade on one of the three arms. Note that each arm has a valve to control the flow of gas to the jet. Later gas lighting fixtures might have mantles on the flames to produce a more intense light.

to the pipes. *Do not cut into any gas pipe if you are not certain there is no gas in it.*

Gas lighting can be restored to a house that once had it if the lines remain in good condition. They will have to be pressure tested by a licensed plumber. Exterior gas lighting is more common than interior, particularly in historic cities like New Orleans. But gas restoration is sometimes done for the elaborate gasoliers with etched-glass globes in the formal rooms of a Victorian-era house. The open-flame gas jets that were once used in secondary spaces are rarely used today due to fire safety concerns, and most surviving gasoliers were long ago converted to electricity. With modern dimmers and LED bulbs, it is possible to mimic the color and brightness of gas lighting without restoring the fixtures to burn gas.

Gas lines are also used to supply fuel to cooking and heating stoves, water heaters, furnaces, boilers, and clothes dryers. These lines, too, should be inspected to ensure that they are in good condition and were properly installed. Unexpected and dangerous conditions can exist. I once lived on the third floor of a six-unit triple-decker apartment building. While taking up the carpet in one room, I noticed a faint odor of gas. I called in the landlord, who took up some flooring to expose the gas line leading to the kitchen range in the next room. There was a definite smell of gas, and more disturbingly, touching the gas pipe imparted the tingle of an electric charge to fingers. Further investigation revealed that the electrical system was grounded to the gas line rather than the water line in the basement, and a small amount of electricity was leaking into the ground wire somewhere in the building and charging the live gas line. Once discovered, the grounding issue was quickly remedied and the small gas leak under the floor was fixed as well, but the story could have ended far differently. Use caution around gas lines, and always be aware of the potential for explosive consequences from inappropriate work.

When gas lines need to be replaced or extended, threaded black iron pipe is still commonly used. Its installation is labor intensive, but once installed and pressure tested it can generally be trusted to remain safe for many decades. A newer material, plastic-coated stainless-steel corrugated tubing, is also in wide use now. It has the advantage of being flexible, so it can make long runs without the need for threaded pipe ends and fittings at every turn. Compression fittings are used with this tubing and with the flexible copper tubing that is sometimes used for short connections from a gas line to an appliance. Before you install new gas lines, confirm that your local building code and your insurance company both allow a homeowner to do this work. If not, hire a licensed plumber or HVAC technician to do it.

An antique cast iron clawfoot tub and a pedestal sink in an antique shop. Historic fixtures are not hard to find in many areas of the country and often are ready to serve another century or more.

Working with Historic Fixtures

Many historic houses retain period plumbing fixtures, and architectural salvage shops are full of them. These fixtures are often still functional or can be made so with a bit of reconditioning. As with many elements of an old house, plumbing fixtures were once designed to be repairable using interchangeable parts and simple tools. Some historic fixtures cannot meet modern safety or environmental requirements, but even these can sometimes be made to function (legally) by a clever plumber working with a flexible code officer. In other cases—early toilets without S-traps to keep sewer gases out of the house, for example—it is just as well you cannot use them.

In communities with strict water use ordinances, low-flow toilet tanks are often required by code. Period toilets that use three or more gallons per flush can generally be left in place but not installed in such communities. A flexible code officer may accept that a smaller plastic tank liner or stacked bricks in the old tank (without interfering with the float valve) will reduce the flush quantity enough to meet requirements. Keep in mind, however, that the historic toilet was engineered to use three gallons per flush, and a gallon-and-a-half flush may not be sufficient to clear the bowl fully. It is worth a try, but it might not work. High tank toilets have a better chance of success with this approach thanks to the additional velocity the water gains in its drop from the tank to the toilet.

The most common problems with an old sink are the drain hole size and the faucet hole spacing. If you are working with an old sink that is missing its metal

This is a typical late-Victorian bathroom (or bedroom in this case) sink. It has a marble top and backsplash with a marbleized ceramic bowl and separate faucets for hot and cold water with a centered fitting to which a rubber drain plug on a chain can be attached. Victorians were happy to mix hot and cold water in the bowl to achieve the desired temperature for washing, and we can still use this option today. Those of us who want water of the desired temperature to flow from a single spout, however, can get that with this sink and the right faucet, as described in the text.

drain stem and does not accommodate the standard sizes available today, you may be facing a long search for a replacement. Salvage shops and online auction sites are your best bets for finding what you need. In a slate, soapstone, or marble sink, a too-small hole can often be reamed out enough to fit a modern standard drain. I used an old heavy steel rasp to do this with a slate kitchen sink.

Old bathroom lavatory sinks often had separate faucets for hot and cold water, to be mixed in the sink bowl to the desired temperature. Today we are accustomed to a single spout giving us water premixed to the desired temperature. Accordingly, old lavatory sinks have two faucet holes, whereas modern sinks typically have three or one. Companies that reproduce historic fixtures make faucets that convert the two-hole setup to a single spout unit. These faucets bridge the space between the knobs with the spout at center and tend to have a somewhat Steampunk character that can work just fine in a historic bathroom. If you have a marble sink top with an undermounted porcelain bowl, you can drill a third hole in the marble at center to accommodate a modern three-part faucet.

Some old undermount lavatory sinks do not have overflow drains as modern code requires. If you have an exceptional bowl (perhaps hand-painted or transfer-printed ceramic) and are determined to make it work, consider mounting it with a small gap between the top of the bowl and the underside of the marble top (several washers on the screws holding the bowl to the stone will do this) with a second, modern sink bowl with drain (and overflow holes) mounted below the first, perhaps on the floor of the cabinet, to catch any overflow exiting the historic bowl through the gap. As with any nonstandard solution to a code issue, approval or denial may depend on the flexibility of the code officer. Any solution that clearly addresses the issue without creating new issues is worth a try if it will preserve something that matters to you.

A local plumber who is knowledgeable about traditional fixtures and willing to work with them is a treasure and should be treated like a close cousin. Consider indenturing your favorite child to him or her for training, so the knowledge will be preserved. If you are replacing fixtures that are missing or broken beyond repair or want to use historic fixtures in a new bathroom, there are local shops and mail-order companies across the country that specialize in the restoration of historic plumbing fixtures. Some do repairs only, and some sell restored items from their stock. Some of these companies will note on their websites which fixtures are not legal for use in states with very restrictive water-use laws, California chief among them. See the Sources list in this book.

Compatible New Fixtures

Thirty years ago, there were few options for anyone seeking reproductions of historic plumbing fixtures. Fortunately, this dearth was partially offset by the ready availability of historic fixtures being tossed from old houses and the relative ease of finding plumbers who were experienced with repairing traditional fixtures. I grew up near a plumbing company, Harry D. Bunker & Son, which had been in business since the 1880s. Harry Bunker (the second generation) was still working, though his son ran the business by that point. Harry never threw anything away, and he had rooms filled with old plumbing parts and fixtures and the tools to repair them. My best friend was Harry's grandson, and although we were not often allowed into the shop, I was fascinated whenever I could get in there for a peek at all the treasures. Like many businesses of this type, it closed down when Harry's son retired, and the contents of the shop were dispersed. Shops like this existed all across the country until recent decades, when built-in obsolescence largely destroyed the market for skilled repair people. Things are now designed to last a relatively short time and then to be replaced rather than repaired.

This is a new ceramic American Standard pedestal sink. The design is adapted from an American Standard sink of decades ago but accommodates a modern, wide-spread faucet arrangement with separate taps and a centered spout. It also has a code-compliant overflow drain. The faucet set looks traditional but contains ceramic disc valves and an integral drain stopper.

Fortunately for the restoration market, as these old resources were fading away, new companies appeared to produce new fixtures based on old designs, primarily Victorian. Eventually, even the big fixture manufacturers like Kohler and American Standard recognized an important new market segment and began to reintroduce some of their own classic designs with subtle changes to address modern code requirements. These now include designs up through the 1960s. Today it is possible to pick up a marble sink top with an undermounted porcelain bowl off the shelf at a big-box home improvement store.

Twenty years ago, the challenge was finding any reproduction plumbing fixtures. Today the challenge is finding ones that best fit the period and aesthetic you are going for. Victorian? Early twentieth century?

Moderne? Postwar? There are many choices for any period. It is important to pay attention to the quality of the reproductions, as less expensive and sometimes less well-made pieces are increasingly available. As is typical when versions of expensive or relatively rare items become more widely available, there can be a degradation of design as well as materials. Crisp-edged cast metal is replaced by muddy forms; carefully reproduced classical detail is crudely copied or overdone; brass internal parts are replaced with plastic.

Study original examples of what you are after in person, in photos, or in old advertisements. Become an educated buyer who will spot badly done knock-offs when you see them. Last and certainly not least, buy the best quality you can afford. Plumbing fixtures get a *lot* of use and have the potential to do significant damage to the house if they fail. Once again, water is the enemy of historic houses.

A final word of warning on plumbing: In recent decades, laundry rooms have increasingly moved from the basement to the upper story of a house. Not coincidentally, damage to houses from burst washing machine hoses has increased. I know several people who have experienced a washer hose bursting while they were away from home and ended up with tens of thousands of dollars in damage and months of repairs to flooring, walls, and ceilings. The chief culprit in this scenario seems to be the inexpensive rubber hoses that are frequently used for washer hookups. The packaging generally warns that the hoses need to be replaced every two or three years, but few people replace them as often as they should, if ever. Washing machines on upper stories make a lot of sense for convenience, but if you have one, use high-quality woven stainless steel washer hoses for the hookup. These are far less likely to fail. Installing a tile, sheet metal, or plastic overflow tray with a drain under the washer will provide even more protection.

Electrical

It is conceivable that the new battery technology being developed for automobiles will make its way into homes in time. Will people be willing to change out the rechargeable batteries in their refrigerator once a month if it eliminates the need to spend tens of thousands of dollars wiring a house? If the phone in your pocket is any indication, they will. Technological innovations appear to be nudging us toward a time when home appliances, media devices, and

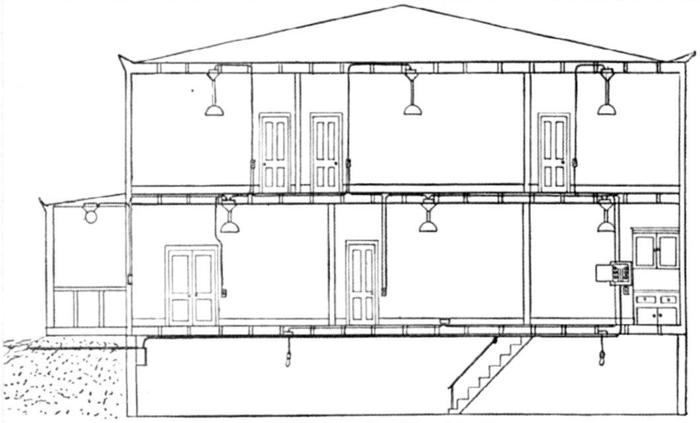

Sectional view of residence showing rigid conduit wiring.

An illustration from *House Wiring: A Treatise*, by Thomas W. Poppe, 1920. The book promoted itself as "describing and illustrating up-to-date methods of installing electric lights and power wiring, bell wiring and burglar alarm wiring. Intended for the electrician, helper, and apprentice. It aids in solving all wiring problems and contains nothing that conflicts with the rulings of the National Board of Fire Underwriters. It contains just the information needed for the successful wiring of a building."

lighting can be powered wirelessly, and that will be good for historic houses if it does away with the need to open up walls and ceilings to replace wiring every generation or two.

We are not there yet, however, and in the meantime we depend on a technology developed at the end of the nineteenth century to deliver electrical power via insulated wires from a central control panel (with fuses or circuit breakers) through walls and ceilings to the locations where it is needed. A mid-twentieth-century improvement on the original concept was the addition of a ground wire to the system, decreasing the risk of fire and deadly electrical shock. This technology remains in use because it is relatively simple to install (in new construction anyway) and works well.

While the technology of residential electrical wiring has not changed much in over a century, the use of electricity for modern life has. When residential electricity was first displacing gas and oil lighting as the American standard, the focus was on lighting—usually one fixture per room, installed in the center of the ceiling and operated with a pull chain. Sconces might be used in bedrooms, bathrooms, and hallways. There were few electrical appliances available, and a single outlet per room was often considered sufficient. Since those early days, the availability of electrical appliances and our appetite for them have both grown exponentially.

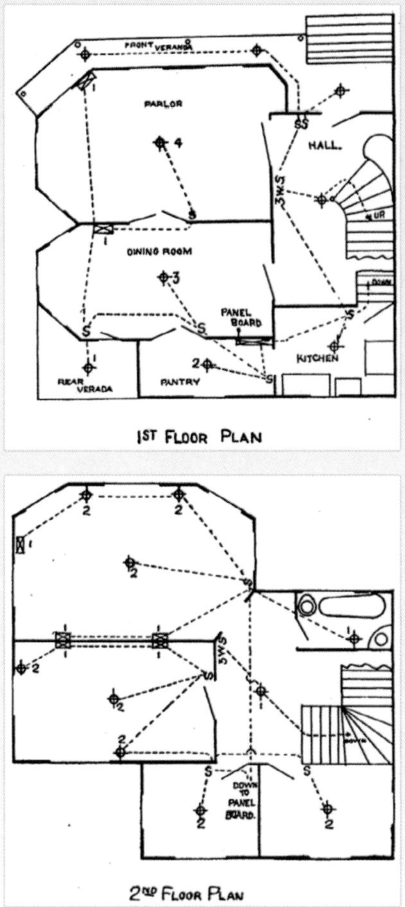

More illustrations from a 1920 manual on house wiring. The first story has two electrical outlets (indicated by a rectangular box with an X through it) to plug in lamps or appliances, neither of which is in the kitchen. The second story has five outlets. Houses with such limited wiring are not uncommon to this day, perhaps with an added circuit for a few kitchen outlets, installed when the first refrigerator arrived in the house.

Modern building code requires an electrical outlet every six feet, and even that is frequently not enough. Many of us plug eight-plug power strips into outlets and then additional power strips into the power strips. A small room in my house right now has a desktop computer, monitor, printer, speakers, desk lamp, television, video streaming device, DVR recorder, audio surround system, laptop computer, cell phone charger, floor lamp, table lamp, and an antique illuminated earth globe. In the holiday season, four window candles and a small lighted tree will be added, making nineteen electrical plugs in this 110 square-foot room. And then there are kitchens!

Contemporary life seems to require a lot more electrical outlets than our great grandparents needed. Since few historic houses have anywhere near enough outlets, adding them is a common part of rehabilitating an old house. Accomplishing this without doing any more damage to surfaces than necessary is the challenge.

When existing houses were first wired for electricity, outlets were often installed in baseboards, usually oriented horizontally. It was relatively easy to pull off a baseboard and drill a hole through the framing to the basement or joist bays between floors to fish wires through. Cut a rectangular hole in the baseboard, install a metal outlet box, pull the wires into the box, reinstall the baseboard, install an outlet and cover plate, and you're done. If a room was to get more than one outlet, as sometimes happened, the wires could be run behind the baseboard by drilling through the wall studs from bay to bay and cutting more outlet box holes in the baseboard as needed. This was a good way to add outlets to finished rooms with a minimum of disturbance to wall surfaces. Cutting vertically through horizontal plaster laths to install an outlet above the baseboard often causes the severed lath ends to break loose behind the plaster. Mounting outlets in the baseboard avoids this issue.

A twentieth-century electrical outlet installed in a baseboard from the 1850s. This method of installation required temporary removal of the baseboard but allowed outlets to be added to the room without disturbing the wall plaster.

Unfortunately, most current building codes require outlets to be installed 18 inches above the floor—well above the baseboard—and no licensed electrician will install them lower. Existing outlets in baseboards can be rewired, and most electricians are willing to do this work. Homeowners who add outlets to baseboards themselves may be in violation of code.

Be aware that an electrical box designed for "new work" is different from one for "old work." The former are intended to be attached directly to fully exposed wall studs, whereas the latter use screws or clips to attach to the plaster lath or gypsum board or wood of a finished wall. Metal boxes are generally attached with small flathead screws, though separate clips are available. Plastic boxes are attached with clips.

A careful do-it-yourselfer with sufficient knowhow can add a new circuit or even rewire a house. Safety comes first. The 110-volt current commonly used in American homes will kill you. The 220-volt current used for larger appliances will kill you quicker. *It is impossible to overemphasize the need to understand and respect the dangers of electricity.* That said, there is no danger in working with electrical lines while the power is disconnected. If you are working on a circuit, shut off *and remove* the circuit breaker or fuse. If your house still has fuses, they need to be replaced with a circuit breaker panel for the safety of your family (and probably to satisfy code). After shutting off the breaker and removing it, go back and test the circuit to be sure you turned off the correct one. Simple voltage testers are inexpensive and necessary for this work. If there is any doubt about whether you have shut off the power to the correct circuit, shut off the main breaker. That will cut off all electricity to the house.

Knob-and-Tube Wiring

The primary method for installing electricity in homes in the late nineteenth and early twentieth centuries was knob-and-tube wiring. This approach used porcelain insulators to carry two separated rubber- or cloth-covered wires, one of which was "hot" and the other neutral or common. The two wires usually ran several inches apart along joists and other framing members and were held an inch or so away from the wood by porcelain knobs. The two parts of a knob were nailed to the framing member, and the wire was clamped between them in a tight-fitting groove. When a wire needed to pass through a framing member, porcelain tubes were inserted in drilled holes, insulating the wire from the wood. Power flowed through the hot wire to the fixture and returned via the common wire. Because the two wires remained completely separated except at the fixture, the risk of a short and resulting fire was almost nonexistent. Insulated cable that carried the hot and neutral wires together in a single sheath came along later.

This knob-and-tube installation is in the basement of a late-nineteenth-century house. Because of its location, it was never covered by plaster, making it possible to see all the components of a knob-and-tube system, including the porcelain insulators (knobs) and the tubes to keep the wires insulated from each other and the framing. The light fixture is very basic in this utilitarian space, but it shows how any fixture would be wired. In a finished space, the wires coming through the plaster or wood ceiling would be covered by a metal canopy.

When knob-and-tube was the standard wiring approach, fixtures were usually equipped with pull-chain or turnkey on/off switches, thus avoiding the need to run wires through walls to reach wall switches. Electrical systems in new construction often did have wall switches, however, because they could be wired before the lath and plaster was installed.

There is a common belief today that knob-and-tube wiring went out of use because it was unsafe. In fact, as long as the wiring remains undisturbed, it is a remarkably safe system. The wires are physically separated, and porcelain is an excellent insulator. The only real threat came from rodents chewing through the cloth or rubber insulation on the wires and setting their nests on fire. I have found occasional mention of this in newspaper archives, but the tone of such notices suggests that this was uncommon. I believe knob-and-tube wiring went out of use because it was labor-intensive and therefore expensive to install. Once well-insulated two-wire sheathed cable became available, it was no longer necessary to install hundreds of porcelain knobs and tubes or drill two holes through every joist. The addition of a third (ground) wire to the cable in the mid-twentieth century decreased the possibility of deadly shock and is now required by code for most residential wiring.

Updating Wiring

If you need to replace or upgrade the electrical service from the street, have the new service run underground if at all possible. Run the cable TV line and landline phone wire through the same conduit and get them all off the face of the house. You may never plan to use a landline phone or cable TV, but if they are in place underground, a future owner will not have them strung up to the house and run down the siding. Cable TV and telephone company installers are only concerned with making your service work as quickly and easily as possible, and they will run wires all over the exterior of your house if given free rein. Insist that wires run under the house to reach first-floor rooms, and then up through a chase or other unobtrusive location to the upper story.

If you are replacing old with new wiring for some reason (as will be necessary if your system is not grounded), try to determine how and where the existing wiring was installed. Most houses more than fifty or so years old have had their wiring updated since it was installed. Much of the wiring should be visible in the basement and attic if those spaces are unfinished. In finished rooms, look for evidence of floorboards that have been taken up and put back down; baseboards that have been removed and put back; small patches in plaster near the junctions of ceilings and walls; chases added by boxing out a corner of a room or closet; and other evidence of surfaces disturbed and then patched or put back. If you can access the wiring from those same places, you will minimize new damage to the house. Adding new circuits, outlets, and fixtures will almost certainly require new holes in walls and ceilings and possibly removing more floorboards and trim elements temporarily.

This is a recent underground electrical service entry to a historic house museum. One buried tube carries electricity, one the telephone line, and one is a spare for any future need. In many households, this would be used for the cable TV line. This is a far less distracting attachment to the grid than the typical aerial wires connecting to a metal "mast" running down the side of the house. Many power companies require that the underground line extend up the exterior wall to a meter installed 4 feet off the ground, but you should insist on locating this meter off the principal elevation. Some power companies will mount meters on poles at the edge of the property or in a basement or outbuilding. In this case, the museum owns the property across the street, and the meters for both properties are mounted on a service building there, then run underground to the historic buildings.

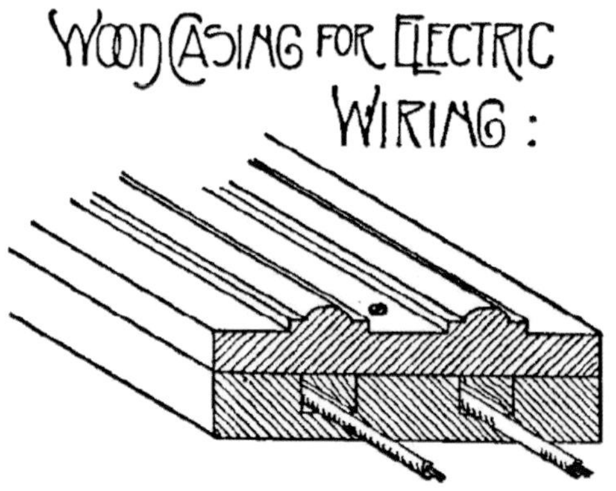

FIG. 151.

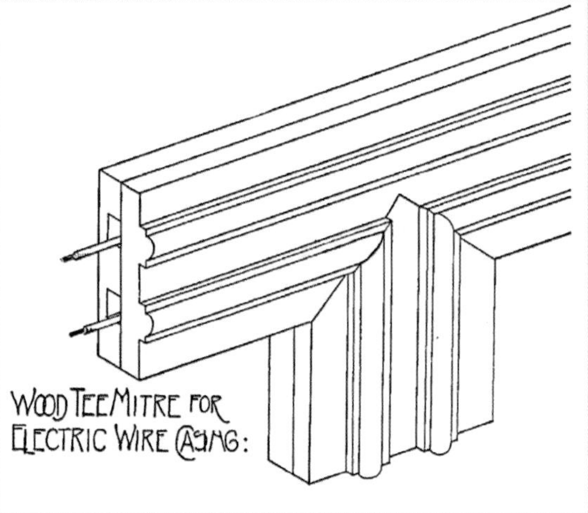

Historic illustrations of wood conduit used for surface-mounted electrical work. It could be stained and varnished to match wood or painted. A similar new wood molding, rabbeted to fit over a modern metal surface-mount conduit containing code-compliant insulated cable, would likely satisfy a local building inspector.

Sometimes the updated wiring is easy to identify because it was installed in surface-mounted conduit. Early work may be in carefully installed, unobtrusive molded wood conduit. More recent work tends to be in metal or plastic; this often detracts from the historic character of a space and should be replaced with a more sensitive approach.

Fishing wires through walls, ceilings, and floors is often slow and frustrating, though it does get easier with practice. You should have a fish tape, which is a stout, spooled cable—often a flattened steel wire 50 or 100 feet long—terminating in an eye to which a wire can be attached. The tape is stiff yet fine enough to poke through a tight space or conduit, and stout enough to pull an attached wire back through. Generally speaking, wires have to be pulled; they can't be pushed through a run of any length. Ideally, an assistant will feed the wire at the source end while you pull at the destination end.

Gravity is frequently your opponent when working on a house, but here it can be your assistant. If you're working alone, you can drop a strong string tied to a heavy washer or nut down through a vertical

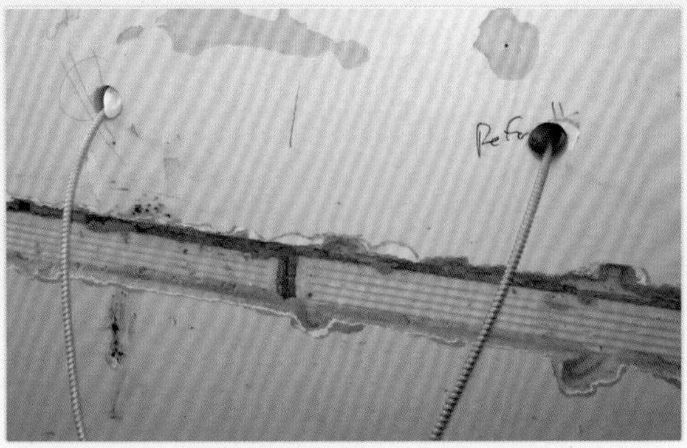

New wires have been fished into existing walls and pulled into the room through newly drilled holes. These walls are a type that is particularly challenging, being formed of hollow ceramic tile blocks that are mortared like brick and then plastered over. The electricians on this rehab job managed to fish the new wiring through the hollow cores of the tile blocks. This is gold-star work by an exceptional subcontractor!

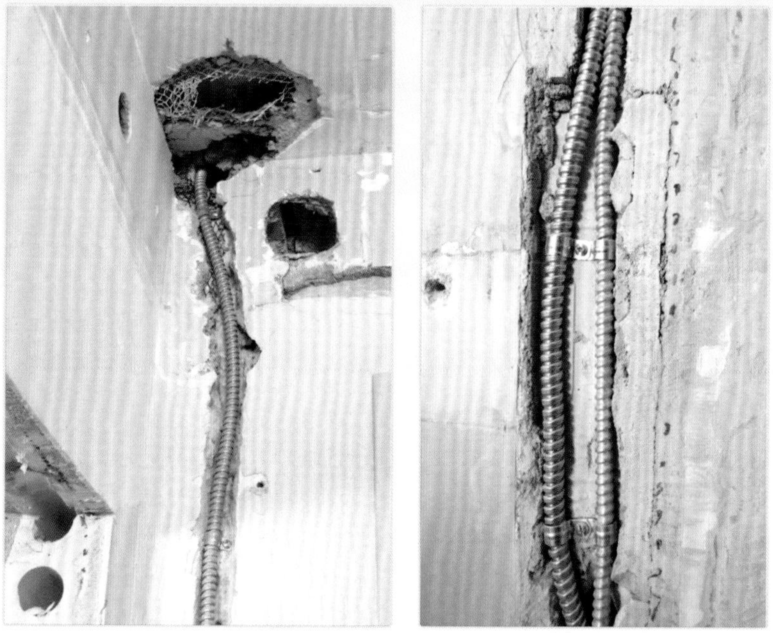

Here are examples of new conduit being run in plaster channels in a masonry building. When the choice is between channeled plaster or surface-mounted conduit, this is the better solution.

space (like a wall or chase) to reach an outlet or fixture hole or get under the floor below; tie the string to the wire end, then retrieve the string from above to fish the wire up.

You can often use existing wires to pull in their replacements if they were fished through walls and ceilings originally and are not attached to framing behind the finish surface. Do not rush ahead and pull out all the existing wiring before installing the new. This can be particularly helpful for getting new wires into wall bays filled with insulation. If your rewiring is being done by a professional electrician, the person who will be repairing the plaster (whether you or a professional plasterer) should cut the holes for new electrical (and plumbing); the holes are almost guaranteed to be neater that way. This will require coordination between tradespeople (or between a tradesperson and you), but it will result in less destruction and simpler repairs. Tradespeople may resist this level of coordination, but you should insist on it.

Masonry buildings pose particular challenges. Often these buildings have plaster applied directly to their brick, stone, or adobe exterior walls with no void to accommodate wiring (or plumbing or insulation). In adobe buildings, channels are typically excavated in the soft clay block walls for electrical wires and plumbing and then covered by patching the plaster finish on the wall. It is more complicated with brick or stone, neither of which is easy to channel into. A cooperative code enforcer might let you install metal conduit on the masonry surface and then hide it behind a thick baseboard or wainscoting in a channel cut into the wood. If you only need outlets—no switches or fixtures—recessed floor outlets accessed from below might solve the problem.

If the original finishes are gone or so deteriorated or damaged that they must be replaced, you might consider framing the wall further into the room to accommodate wiring, plumbing, and electrical work. On many of the historic tax-credit rehab projects I work on, the use of low-income housing tax credits (along with the historic credits) requires increased energy efficiency standards. In many masonry buildings, these standards can only be met by increasing the depth of the wall to accommodate insulation.

The usual approach involves a new stud wall set ½ inch off the masonry (or the old plaster if it will remain in place). Once wiring and any plumbing are in place (plumbing is not typical in exterior walls in this part of the country), foam insulation is sprayed into the stud bays, filling the 3½-inch stud depth and the ½-inch gap between the studs and the masonry to prevent thermal bridging. New gypsum board is installed as the finish wall, and door and window jambs are increased in depth to accommodate the new wall thickness. Historic trim—including window and door trim, baseboards, etc.—is then reinstalled on the new wall surface. Alternatively, the studs can be turned on their sides so that the increased wall depth is only 2 inches plus the gypsum board thickness (preferably with skim-coat plaster). The ½-inch gap behind the studs will allow wires to run past the studs without drilling through them.

Cornices and fireplaces can be a challenge when insulating walls in this way. A precast plaster, pressed tin, or wood cornice is put together from modular pieces and can be moved without extraordinary effort. The pieces will have to be carefully removed and stored while the wall is reconstructed, then reassembled on the new wall surface. A run-in-place plaster cornice is significantly more challenging, as it is bonded to the historic plaster. It is *possible* to remove and reinstall a run-in-place cornice, but this is a lot more involved than uncovering the screws and taking down the pieces of a precast cornice.

A fireplace that stands proud of the wall surface in a chimney breast is not a problem as long as its standout depth exceeds the wall thickness to be added. The relief between the chimney breast face and the wall on either side will be reduced, but their visual relation ship will be maintained. If the fireplace is flush with the old wall surface, you may need a mason to build the fireplace surround into the room or replicate it in front of the original. If the original is plastered brick, tile, or stone slabs, the mason can simply replicate the plaster or move the tile or stones forward with masonry infill behind. Increasing the depth of a brick surround is more involved. In all cases the mantel will have to be moved forward onto the new wall surface along with the other trim in the room. This is a lot of expensive work, and alternatives to building out the wall may prove a better choice.

If the exterior walls are a significant character-defining feature—if they carry historic murals, for instance—increasing the wall thickness is probably not practical. It is *possible* to move murals and other delicate finishes on plaster, but the cost could be astronomical. Given the relatively small percentage of heat lost through side walls, the payback period on moving wall murals to install insulation might be measured in centuries. On the other hand, a fully wood-paneled wall could be relatively easy to move forward.

If installing electrical outlets rather than insulation is the primary concern, it will be much simpler to install recessed floor outlets along the walls with wires run under the floor. Solid cast brass cover plates with lift-up covers to expose the outlets are available for floor outlets.

Low Voltage for Historic Lighting with LED Bulbs

The widespread availability of affordable LED bulbs to fit traditional sockets makes it possible to use lower voltages for lighting than was previously the norm. If you have a house with historic light fixtures and wiring in place—an intact knob-and-tube system, for instance—it may be possible to switch them over to a low-voltage system and use LED bulbs (which require far less voltage than traditional bulbs). This could greatly reduce the risk of fire from an accidental short in the wiring. You'd need a transformer to reduce the voltage in the wires from 110 to 12 or 28 volts, similar to the popular exposed-cable and under-cabinet lights now in use. Low-voltage wiring has been used for more than a century for doorbells, thermostats, and landline telephones, not to mention electric toy trains, but its use for lighting has been far less common.

Electrical outlets would have to be separated from the lighting circuits and rewired with modern 110-volt grounded circuits. RV and boat lights are often low voltage, and a range of LED bulbs are made for these applications that may work well for a retrofitted low-voltage home installation. An electrical engineer should be consulted to ensure the retrofit is done correctly and to reassure the code inspector, who may very well not have seen this done before.

Light Fixtures

There are many excellent reproduction light fixtures available for periods from the mid-nineteenth century, when gas lighting appeared, to the Mod styles of the mid-twentieth century. This includes Colonial Revival fixtures of the early twentieth century that often evoke the forms of Colonial and Early Republic candelabras, chandeliers, and whale oil lamps. The great variety of reproduction fixtures have made it possible to find appropriate fixtures for nearly any historic house.

Antique fixtures are also widely available, although restored pieces can be expensive and finding matching multiples can be a challenge. With patience and an ongoing search of flea markets, salvage shops, eBay, and classified ad sources it is possible to acquire antique fixtures at reasonable prices, often for less

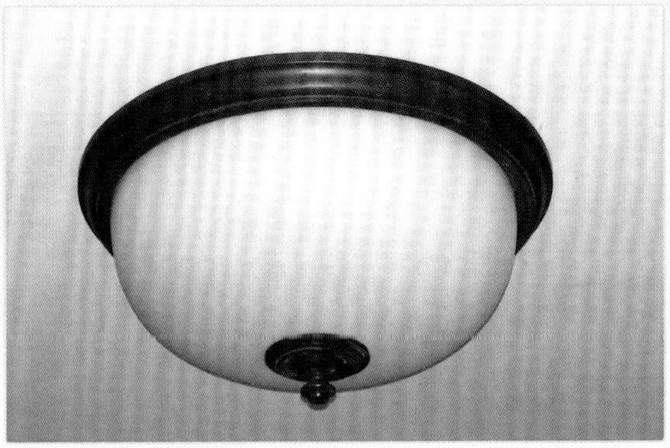

This 14-inch-diameter early-twentieth-century ceiling fixture was found in a dusty box in the back room of a low-end antique shop. The frosted white shade was filthy and covered with black marks, the porcelain sockets were cracked, and the brittle insulation on the wires was flaking off. On the other hand, it was all there, and the metal finish was excellent under the dust. And the price was $15. The shade cleaned up nicely with a long, hot soak in Oxy cleaner and some scrubbing, and the local big-box home-improvement store sold porcelain replacement sockets that were identical to the originals, screwed right in place, and cost less than $5. I had some appropriate replacement wire on hand.

than good-quality reproductions. The old fixtures generally need to be rewired, and you may have to find the fixture and shade separately. Two wall sconces and shades remained in Whitten House from when it was first wired in about 1898. Over a period of years, I have been able to acquire matching fixtures for additional locations. These include a new bathroom and two attic bedrooms that were never wired historically.

Many communities have a local specialist in rewiring historic light fixtures. The one I use is named The Lamp Shop and occupies a small downtown storefront. Half the shop is cluttered with restored light fixtures and assorted other antiques for sale, and the other half is filled with workbenches and shelves full of parts. It

My collection of antique and vintage light fixtures and shades resides in my tool room. Some of these have intended locations once restored; others were just too wonderful to leave at the flea market.

is a low-key operation; the cost of repair work is reasonable, and its quality is high. It might take five or six weeks for a fixture to be repaired, but it is worth the wait. Look for a place like this in your community.

Historically, many fixtures were installed by screwing a cast iron mounting bracket, or "frog," through the plaster into the wood lath over a hole drilled through the plaster and lath for the wire to pass through. A threaded rod or nipple screwed into the top of the frog to attach the fixture. After the electrical connections were made and taped to insulate them, a bell-shaped canopy was slid along the fixture's wire tube until it was tight against the plaster, then held in place with a set screw or pressure from a tube cap at the other end of the threaded rod. The bell canopy essentially functioned like a modern junction box but was outside the plaster rather than recessed into it. Modern code requires electrical connections to be made using plastic wire nuts in a junction box recessed into the wall or mounted to the surface. These often require a retrofit connection under the canopy for a historic fixture. Once the fixture is installed, it is impossible to tell (without removing it) if it is mounted with a frog or a recessed box. Reproduction fixtures use modern attachment methods and are fully code-compliant.

If you need to buy new fixtures on a limited budget, select simple, classic designs. The difference between a $300 schoolhouse-style fixture from a high-end reproduction lighting manufacturer and the $40 version at a big-box retailer is not going to be noticeable to most people. With the $40 fixture, you won't have six finishes to choose from and the metal will likely be thinner, but the effect will be virtually the same. Avoid elaborate "historic" fixtures from the big-box stores. The manufacturer may call these "Victorian," but chances are that "poorly designed and gaudy" is a more accurate description. Seek out historic fixtures at flea markets or online and get them rewired, or spring for accurate reproduction fixtures if the budget allows.

If you are planning to return your rooms fully to a period appearance, it is important to realize that in every pre–World War II period, lighting was considerably less bright than we are accustomed to. If you use accurately reproduced colors, patterns, and textures from a period but do not replicate appropriate lighting levels, you will not accurately recreate the period ambiance of the room. At the same time, you need to live a twenty-first-century life in these rooms. My preferred solution to this dilemma is lots of dimmers. In general, I keep light levels low in period rooms and brighter in the kitchen and bathrooms, with the option to turn light levels up or down to suit what is happening in a room at a given time.

I am also conscious of the color of lighting. Historic residential lighting was warm, whether the warm pale yellow of an oil wick flame, the orange-yellow of a gas light flame, or the warm yellow of an incandescent bulb. None of these resembled modern "cool" white light in the least. Historic rooms (and people) look better in warm light, preferably not too bright unless needed for a specific task. Fortunately, the manufacturers of modern LED bulbs have realized we don't all want to live in the ambiance of an operating room, and warm colors are now available. LED versions of classic early electric bulbs have come onto the market recently.

Switches and Outlets

The most common switches for early residential electrical systems were pushbutton, and they remain in use in some houses. The switch had two round buttons, one above the other. Push in the top one to turn the light on, and push in the bottom one to turn it off. These usually had brass cover plates, as did outlets. The ubiquitous flip switch commonly in use today came along in the 1940s. Early switches and outlets were typically very dark-brown or black. The pushbutton ends were often finished with mother-of-pearl.

The easiest way to be compatible with the historic character of a pre–World War II house when rewiring is to use pushbutton switches and dark-brown or black outlets with brass cover plates. Fortunately, UL-listed pushbutton switches are being produced today and include virtually every imaginable type of switch: two-way, three-way, four-way, dimmer, etc. These switches are more expensive than standard flip switches but contribute to the historic feel of a room.

A classic pushbutton light switch with a brass cover plate. These are now reproduced in UL-listed versions that include dimmers.

Often the best approach for flip switches is to camouflage them as much as possible. Painting or wallpapering cover plates to match the wall behind helps. Switches for lights that are not used regularly, such as lighting in glass cabinets, can be placed out of sight, perhaps inside the cabinet base. Avoid fancy ornamental cover plates if they are just going to draw attention to a contemporary functional element. In some cases, low-profile switches may be less apparent than standard switches. Of course, a more modern historic house should have fixtures and switches appropriate to its time period.

is your job to insist that the work be done as you want it.

Internet, TV, and Phone Cables

The many devices we now need to plug in are accompanied by a variety of cables and low-voltage wiring connecting them to service providers and to each other. We seem to be moving toward a wireless future that will eliminate much of this wiring. I look forward to the day. A decade ago, people ran hundreds of feet of low-voltage wires for built-in speakers through their houses, but most home speakers are wireless now. We recently installed a new wireless router and booster system at Whitten House. The boosters can be plugged into any outlet in the house, receive the data signal through the electrical wiring, and transmit it wirelessly to a nearby user. We now have the same internet speed in the kitchen as in the parlor, with no added wiring in our 1827 house. Given the pace of change, there is a fair chance that anything I write on this subject will soon be obsolete, so I'll keep it brief and general.

Unlike basic electrical wiring or plumbing, it makes little sense to undertake destructive methods merely to install low-voltage wiring in the walls and ceilings of a historic house. If surfaces are being opened up anyway for electrical or plumbing work, it makes sense at the same time to run data cables, TV cable, and the like from a central hub to desired locations. These cables may be obsolete in a decade; install the highest-capacity cables available in order to extend their usefulness as long as possible.

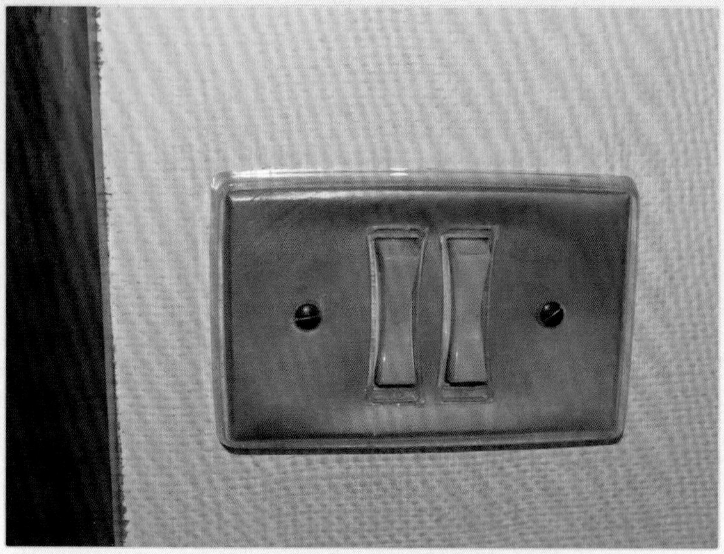

This switch and switch plate were installed in a 1957 Ranch-style house (see Weston House, page 243) as part of a low-voltage switching system that did not catch on fifty years ago and is now challenging to maintain as it wears out. Hopefully, as more postwar homes are recognized as historic, manufacturers of reproduction switches will recognize another market and add more contemporary designs to their offerings.

Whatever switches you use, do not allow your electrician to select the locations for switches or outlets without consulting you. Discuss each one and mark their agreed locations on the walls. Make clear that crooked outlets or switch plates are not acceptable. I have seen an electrical box cut into nineteenth-century decorative painting when moving it three inches would have spared the painting entirely. I have seen many outlets placed off-center between two doors when they could as easily have been centered. I have seen a box for a wall switch cut half into eighteenth-century wood wainscoting and half into the plaster wall above. (Try to find a cover plate for that switch!) Electricians are not usually intentionally destructive or sloppy; they are just focused on other concerns. It

Before installing any wiring that will require altering a character-defining feature of a house, even temporarily, investigate whether new devices or equipment are available to address your need without altering the house. When installing wires that you suspect will be obsolescent soon, look for unobtrusive surface-mount conduit to run them in. The conduit will look better than a half-dozen wires running around the edges of the floor, especially if it can be painted to blend into the surface it is attached to. A good general rule is "no long-term solutions for short-term problems if they will affect the fabric or character of the house in a negative way."

Solar Electric

No discussion of electricity in historic houses would be complete these days were it not to mention rooftop solar electric generation. This topic has been touched on previously in this book, and it probably doesn't need to be repeated that any solar collection installation should be located so as to minimize its visual impact on the house. The barn installation shown in

Although not on a residential building, this comparison of typical reflective blue-black solar collectors with the newer flat-black units shows how much less apparent the newer units are against a black roof.

the accompanying photo is a good example, invisible from either of the roads that intersect in front of the house. The solar electric industry has recognized that there is a market for less obtrusive collectors. Solar collectors that mimic roof shingles in appearance may soon become widely available, and flat-black units that draw less attention than the standard reflective blue-black units are already available.

Cassidy Hill Contemporary Unit in a Victorian House, 1890

This vernacular Victorian-era house was built around 1890 as a two-family home, but in the 1950s an apartment was added in the previously unfinished attic. It has long been common practice to fit an income-producing apartment into an urban historic house; unfortunately, the quality of materials and workmanship in many such renovations leaves much to be desired, and such was the case here. The attic space was cramped and dark, and the poorly built shed dormers added in the 1950s had caused structural issues for the entire roof.

In order to rehabilitate this unpromising space into a stunning contemporary apartment, the new homeowner, Ken, had to completely remove the 1950s work, exposing the framing and beginning with a blank canvas.

Ken was an experienced auto mechanic and welder but had little experience working on houses, and his partner at the time had no building experience of any kind. As the project got underway, it became evident that the house inspection conducted prior to purchase had missed a number of serious structural issues. With little choice in the matter, Ken began an on-the-job course in building rehabilitation, serving as general contractor, principal designer, laborer, carpenter, and cabinet installer. He received two years of nearly full-time volunteer help from his friend Ralph, a retired schoolteacher who had built his own home in the 1970s. He also got occasional help from his brother, who is a structural engineer, and other friends with building experience. Ken hired subcontractors for electrical, HVAC, plumbing, and drywall work and an architect to help with code issues.

Removing the poorly done 1950s renovations exposed numerous structural deficiencies. Because the attic had originally been unfinished, its floor had not been framed for the dynamic and static loads of an active living space. Compounding this issue, some floor joists had been hacked up for plumbing under the 1950s bathroom, further weakening them. These had to be replaced, and the rest of the floor joists were sistered with new joists to help carry the heavier loads of a finished living space.

Because snow melts more quickly from uninsulated roofs than from insulated ones, the roof rafters were undersized for the increased loads that would result from insulating to current code requirements. To remedy this, new rafters were sistered onto the existing ones, and the shed dormers were rebuilt with stronger framing members. A 4-inch by 6-inch tube-steel beam was used in one dormer to reduce the necessary header height from what a wood header would have required, allowing taller windows. To gain as much headroom and visual space as possible, the ceiling was carried up nearly to the peak of the gable. Ken designed and fabricated custom cable collar ties to reinforce the roof framing and establish a contemporary look for the space.

Maximizing space and light were primary design considerations throughout. The 6-foot-tall knee walls of the 1950s work were replaced with 2-foot-tall knee walls in the redesigned spaces, creating much more floor space. The perimeter spaces are only minimally useful at 2 feet tall, but they contribute to an overall sense of spaciousness. A safe and wine chiller were built into the eaves. The larger dormer windows, new skylights in the opposite plane of the roof (including a sun tube in the bathroom), and a glazed door to a new deck on the back of the house increased the natural light considerably. Ken calls this "the most striking of all the changes."

IKEA cabinets created a contemporary feeling in the kitchen, but installing them required serious problem solving, particularly where finish panels were to be attached to cabinet sides and backs. In order to maximize the use of space, a counter-depth refrigerator was chosen, and the upper cabinets had to be stepped down to accommodate the dormer pitch. The counter is 1¼ inches lower than standard against the outside wall to accommodate a taller window and to maximize the space between the upper and lower cabinets. Carrying this counter height into the peninsula created an issue with the gas range, which was designed for a standard 36-inch-high counter. By setting the range's adjustable leveling feet at their lowest setting and taking advantage of a slight slope of the floor toward the center of the building, it was possible to lower the range top to only ¼ inch above the marble countertop, which is hardly noticeable. The broiler drawer on the range has only 1/16-inch clearance at the bottom, but it's enough.

Ken says this about nonstandard installations of cabinets: "You will run into situations the instructions do not address. Even for a more typical instruction, using the directions would leave many places with the white cabinet bases and raw panel edges exposed." His solution was extruded aluminum trim glued over the problem areas, which turned a liability into a feature, both functionally and aesthetically. Suspending the range hood from the sloped ceiling required the fabrication of a five-sided plywood box, sloped to match the ceiling and firmly screwed to the framing. The range hood was then suspended from the box, with the stainless-steel cowl cut to fit the slope.

The kitchen, dining, and living areas occupy an open L-shaped space. Double pocket doors open from this space into a large bedroom, allowing light from the two double-hung windows in the gable end to reach the other spaces. A small office and laundry closet with stacking washer and dryer fill the space between the living room and the entry stairs at the rear. The bathroom is tucked between the kitchen and the rear gable wall.

Contemporary elements include the cathedral ceiling, IKEA cabinets, stainless-steel appliances, large ceramic and marble tiles in the kitchen and bath, and the cable collar ties. Traditional materials like white marble countertops and Victorian-inspired door and window trim relate the spaces to the historic house that contains them. The hardwood flooring is also traditional, but its installation at a 45-degree angle to the walls moves it into the realm of contemporary design. An antique arched stained-glass window—purchased from an architectural salvage shop and restored—is built into the wall above the double pocket doors. With sleek and shiny contemporary furnishings, the apartment is undeniably and stylishly of the twenty-first century, but its traditional elements harmonize with the historic house.

The vernacular late-Victorian house at Cassidy Hill is typical of thousands of simple historic houses across America.

The contemporary unit created in the attic made the most of the limited space.

The house in 1924. *(Courtesy City of Portland Planning Department, Historic Preservation Program)*

Looking from the living/dining area through the kitchen toward the door to the deck.

The new stair up to the unit uses balusters, newels, and railing of traditional design.

A round mirror over the vanity fits under the eaves.

In the new kitchen, the aluminum angle trim piece at the corner of the cabinetry hides the unfinished edge of the cabinet back panel and adds an aesthetic element.

Ken spent many hours creating the custom cable collar ties.

The view from the kitchen through the dining area and into the bedroom shows the river beyond. The sheet metal cowl for the range hood had to be custom cut to the angle of the ceiling.

The finished apartment as seen from the bedroom.

The Work

Installing the new hardwood flooring on the diagonal helps to create an illusion of spaciousness.

Added roof rafters and new plywood gussets joining them at the top.

A view toward the future dining area suggests how much structural reinforcement was required in the roof system. The steel box beam over the windows allowed taller windows than a wood header could have.

Molded trim and paneled doors relate the unit to its Victorian building.

Installation of the kitchen cabinets. The finish panel for the back of the breakfast bar is clamped in place while the adhesive cures.

Ceramic tile flooring being installed in the bathroom atop an electric radiant heat mat, visible at the far end.

The dining area dormer as demolition was getting underway.

Custom-fabricated brackets to be bolted to the rafters.

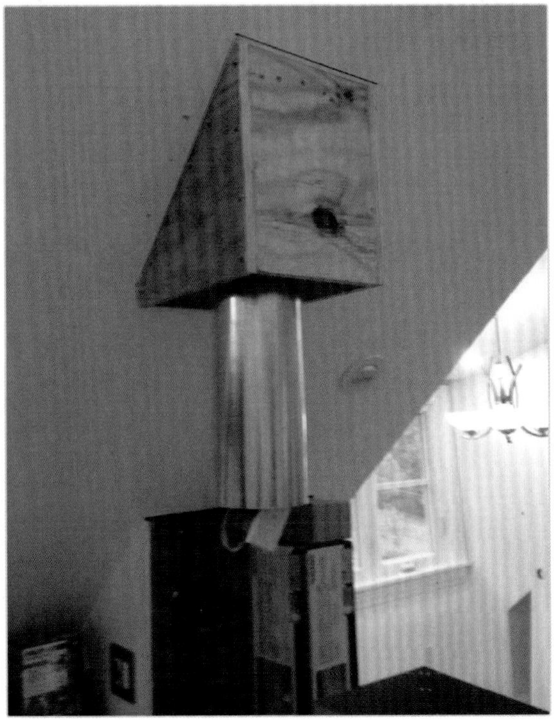

The new range hood hangs from a plywood box that attaches to the framing of the sloped ceiling.

PART FOUR

The Exterior Envelope

11. Roofing

12. Exterior Walls

13. Exterior Trim and Windows

The exterior of a historic house announces its style to the world and says something about its original owner's social status. The walls and roof surfaces are critically important for keeping the weather out and protecting the interior, even as they identify the house as historic.

The materials with which these things have been accomplished vary with climate, period, and style, but they are usually wood or masonry. The former materials include clapboard, shingle, lap board, log-cabin, and occasionally novelty wood siding materials. Masonry includes brick, stone, and adobe, which, unlike brick or stone, is unfired and soft and needs a covering to keep water from eroding it.

The more obscure exterior materials occasionally found on historic houses include wattle and daub and "tab" (oyster shell and lime). Concrete blocks found limited use for residential structures in the twentieth century; reinforced concrete found even more limited use. But wood and masonry were prevalent.

Whatever the exterior materials of your house, ensuring a weathertight envelope should be your first priority. Whether that means undertaking an immediate exterior rehab or doing a short-term emergency stabilization to buy time for a more permanent solution, keeping water and pests out of your house is critically important. Temporary solutions are generally better than rushed permanent solutions. The National Park Service publishes a *Preservation Brief* on stabilizing and mothballing buildings that contains excellent guidance: https://www.nps.gov/tps/how-to-preserve/briefs/31-mothballing.htm.

The Envelope as a System

The exterior materials of your house comprise a system designed to keep out the elements, the most destructive of which is water. Water will deteriorate wood, masonry, and most metals. Since it is nearly impossible to keep out *all* water, and since moisture within the building has to get out, the exterior envelope needs to be able to "breathe."

Nearly all traditional wood siding materials are applied in an overlapping fashion, with each piece partially covering the piece below it on a wall. This keeps out most of the rain and allows any water that does get blown between the siding pieces to drip back out. The natural expansion and contraction of wood as it is exposed to hot-cold and dry-damp cycles will cause microscopic cracks to form in the paint where pieces overlap. These cracks allow moisture from within the house to migrate out through the walls.

Masonry walls, whether brick or stone, also experience expansion and contraction due to weather conditions, and are more likely than painted wood to absorb water from their surface. The softer and more porous the material, the more water it will absorb. Brick and "soft" stones, including brownstone, will absorb more water than granite. The key to managing moisture in a masonry wall is the mortar used to hold the bricks or stones together. The mortar needs to be softer and less dense than the primary wall material to allow moisture to pass outward and evaporate from the surface.

The mortar needs to be softer than the primary wall material for another important reason as well. In a masonry wall, one material must be sacrificial to the other. As the two materials expand

and contract at different rates, one needs to "give" to the other. If the mortar is harder than the primary wall material, the bricks or stones will slowly be destroyed by expansion and contraction. Mortar needs to be the softer, sacrificial material so it can periodically be replaced, or repointed, without threatening the integrity of the structure. Crumbling brick or spalling stone is not so easily dealt with as deteriorated mortar.

Depending on your climate, more or less water will come into contact with your house as rain, snow, sleet, or fog. Nothing can be done to stop this, but you can affect what happens next. The goal is to identify the best ways to control the flow of water from roofs and walls and get it away from the house once it reaches the ground.

Roofs need to be covered with materials that water cannot penetrate, and the water needs a definite path off the roofing and toward the ground. This can be achieved with gutters and downspouts or by a drip edge. When the water gets to the ground, it needs to flow away from the house. This can be done with subsurface drains, with grading of the ground surface, or with a combination of the two. Go outside with an umbrella during a heavy rainstorm and watch how the water comes off your house and where it goes. If any of it ends up in the basement or cellar, or if you have soggy ground around the foundation shortly after a storm ends, you have a problem to deal with.

Wind-driven rain will also soak walls, windows, doors, and trim. This water, too, needs a pathway to the ground. Moisture-vulnerable materials such as wood and adobe need effective coatings to minimize the water they absorb. Appropriate flashing at the joints between architectural elements and materials is important for keeping water out of places it doesn't belong. In addition to allowing water to drain, the exterior envelope needs exposure to air and sun to allow moisture to evaporate. Plantings too close to the house impede airflow and block the sun, keeping surfaces wetter longer. This can contribute significantly to the deterioration of wall and trim materials.

Setting

An in-depth discussion of landscaping for historic houses is outside the scope of this book, but a few key points should be made. An appropriate setting will emphasize the historic significance of your restored house. An inappropriate setting will diminish it.

Fences can help establish an appropriate setting and are character-defining features if historic. Every effort should be made to preserve historic fencing where it exists. Wood fences are especially vulnerable if not kept painted. Once rot sets in, it is only a matter of time before the fence will have to be removed. A historic fence that is beyond repair should be replicated as closely as possible with the longest-lasting appropriate materials available.

Because wood fences are so exposed to the elements, they may be good candidates for replication with appropriate paintable substitute materials. This does not include shiny hollow vinyl fencing, however, because nothing screams "I am fake!" quite like shiny vinyl in a historic context. Composite material and PVC are possibilities if painted to achieve a traditional finish. Because these materials do not absorb or release water, paint holds up well on them if applied correctly and with an appropriate primer. Traditional carpentry techniques should be used. Essentially, you want a fence that looks just like a wood fence but lasts longer. Anything that snaps together or is obviously assembled without nails will look fake even with a painted finish. If you are in a local historic district, any change of materials for the fence may need approval.

Iron fencing is typically more durable than wood and can survive extended periods of neglect. Consequently, it is often found in seriously deteriorated condition. Even badly rusted iron, whether cast or wrought, can often be restored. Missing elements can be replicated, and attachments to masonry or earth can be repaired. This is generally not do-it-yourself work unless you have previous experience with metalwork. Thirty years ago, it was difficult in much of the U.S. to find craftspeople who could do custom iron casting. Now many areas where historic cast iron is prevalent have at least one ironworker who specializes in this work. If you see a nicely restored iron fence in your community, stop to ask who did the work or check with your local preservation organization or municipal historic preservation staff for sources.

The next three chapters will explore appropriate approaches for repairing and restoring roofs, walls, trim, and windows.

CHAPTER ELEVEN

Roofs

The original roofing on most buildings built before the early twentieth century was wood shingle in much of America. Old-growth wood shingles might last up to fifty years, especially if painted. The red roof of George Washington's home, Mount Vernon, is painted wood shingle. Metal roofing—soldered sheets of lead, lead-coated tin, or painted tin—was expensive and uncommon. The flat roofs of Victorian-era rowhouses were typically built up with layers of tar-impregnated paper, often covered with a layer of brushed-on tar and gravel.

Unlike most other exterior materials, the original roofing of a historic house is likely to have been replaced—usually many times, unless it is slate or clay tile. Many eighteenth- and nineteenth-century houses were on their third or fourth wood-shingle roofs when asphalt shingles and more affordable metal roofing came along. Still, roofing can be a character-defining feature. An excellent overview of historic roofing can be found in the National Park Service's *Preservation Brief 4*, "Roofing for Historic Buildings," available online at: https://www.nps.gov/tps/how-to-preserve/briefs/4-roofing.htm.

Roof Problem Areas

The most common leak locations on a roof include valleys, chimney edges, around dormers and skylights, and anywhere a tree branch or other object might damage the roof. Eaves can be a problem in cold climates due to ice dams. Clogged gutters or downspouts can also cause leaking around eaves, particularly if the gutters are integral. (Remember, not all leaks come from the roof. If you see a water stain on a ceiling below a bathroom, start by investigating the plumbing, not the roof.)

Failed valley flashing on the roof above this corner is the primary cause of the damage to the masonry wall and wood trim elements. A damaged gutter and missing downspout have also contributed to the problem.

Water often gets in around flashing. The purpose of metal flashing is to create a barrier between the roofing material and an adjacent surface such as a chimney, a dormer, or an intersecting plane of the roof in a valley. Step flashing, a common choice, is assembled from overlapping L-shaped pieces of lead, copper, tin, aluminum, or other metal that are interleaved with the shingle courses. A second layer of flashing, the counter flashing, is inserted into the joints of a brick wall or chimney or extended under the siding material of a wall or dormer and covers the upper portion of the step flashing.

This chimney is step-flashed and counter-flashed with lead. It is important that a compatible metal be used for both flashings (plus any nails driven through them) to avoid a bimetallic chemical reaction that could shorten the useful life of the flashing.

This combination is effective on chimneys and the sides of dormers, where pieces of flashing can be inserted horizontally between bricks or clapboards. Over time, however, expansion and contraction differences between materials can cause some of the joints in a step flashing to open, allowing water to penetrate to the roof sheathing. At first this might only happen in rainy and windy conditions, when wind pushes rainwater upward on the flashing. Eventually, however, all flashing fails and needs to be replaced, ideally when the roofing material is replaced. A roof with a few leaks can often be patched effectively for the short term, but once water starts finding a way in, the odds are good that a new roof will be needed before long.

Hot Roof? Cold Roof?

There has been a great deal of debate in recent decades about the pros and cons of a cold (vented) roof versus a hot (unvented) roof. Many people believe that a cold-roof system can help prevent ice dams and excessive moisture buildup in the roof framing in colder climates. Others believe that adequate insulation properly installed with a continuous interior vapor barrier makes a cold roof unnecessary. BuildingScience.com—a free online resource owned and operated by a consulting and full-service architecture firm in Massachusetts—has an excellent page discussing the pros and cons of vented and unvented roofs in various climate zones: https://buildingscience.com/documents/reports/rr-0404-roof-design/view.

For those of you fortunate enough to live where "ice dam" is a foreign term, a bit of explanation is in order. Ice dams form along eaves when the heat from within a house melts snow on the roof while the overhanging eaves remain below freezing temperature. Meltwater flows down the roof under the snow until it hits the cold eaves, where it refreezes and builds up. Once this buildup creates a dam, water pools behind it, insulated by the snow above so it does not immediately refreeze. If there is enough snow on the roof, and if temperatures remain below freezing so the eaves cannot thaw, the pool will eventually back up and begin to get under the shingles or slates and into interior living spaces.

A cold, or vented, roof incorporates a void between the insulation and the sheathing, and air vents at the bottom and top of this gap set up a flow of cold air. This keeps the roof surface cold enough that overlying snow will not melt until the outside temperature is high enough to melt it, at which point the eaves will melt as readily as the rest of the roof and no damming can occur. Advocates for unvented (hot) roofs point out that a properly insulated house should not allow heat through the roof to melt snow and cause ice damming, but insulation this effective is far easier to achieve in new construction than an existing building.

Another advantage of a cold roof is that its air circulation removes the water vapor that passes through the insulation. In theory an effective vapor barrier will prevent the escape of moisture from the interior, but this too is more feasible in new than

The ventilation space for a cold roof was created outside the original sheathing by installing 2 x 4s vertically (i.e., running from eaves to peak) on 16-inch centers to support a new layer of ¾-inch plywood sheathing. A ventilated metal drip edge at the eaves and a ridge vent at the top allow air to flow through the roof assembly, removing the heat and moisture that migrate into the space from the building's interior. Prior to the installation of the new roof layer, the existing roof was insulated with blown-in cellulose from the exterior to the full depth of the rafters above the historic plaster attic ceiling.

existing construction. Explore what is most effective for your climate zone. I live in an area that's prone to ice dams, but my historic house has had none since I installed a cold roof as described in Chapter 9, Insulation and Ventilation.

Roofing Materials

Wood Shingles

The replacement of wood-shingle roofs with asphalt shingles or metal roofing is often the most noticeable difference when comparing historic photos of houses with their modern appearance. Wood shingles are typically ½ to ¾ inch thick at their butt ends, creating strong shadow lines on a roof. They often have a noticeable texture, fuzzy when new and becoming striated with the grain of the wood after several decades of rain, sun, and snow erode the fuzz and some of the softer wood in the grain. Some twentieth-century houses were built with shake roofs—shakes being particularly thick shingles that are split rather than sawn, resulting in a highly textured surface. The split shingles used on seventeenth- and early-eighteenth-century roofs were probably replaced with sawn shingles on the second or third generation of roofing.

Original wood-shingle roofs are virtually nonexistent, though you may find portions of them enclosed by later additions. The house I grew up in was built in the mid-nineteenth century and expanded in the 1920s, enclosing one slope of the original roof in the enlarged attic. This retained its wood shingles and was still partially covered with pine needles in the 1960s. In Whitten House, shingles from the 1827 roof were used as shim shingles when the second story was added in about 1850; I have found a number of these as I have taken things apart to work on the house. If you plan to return to wood roofing, any evidence of the historic wood roof is useful for knowing what the thickness and exposure of the shingles was, and possibly their paint color.

If you install a new wood-shingle roof, use the highest-quality shingles available. Labor is a significant expense with wood-shingle roofing, either in dollars or in your own time and effort. Spending more on materials can make the difference between a 15-year and a 30-year roof. If you install the roof yourself, you are unlikely ever to want to do it again. Ever. Spring for the more expensive red or Alaskan yellow cedar shingles. As in the past, a painted wood-shingle roof will last significantly longer than an unpainted one. I recommend dipping the shingles in paint or solid-body stain before they are installed, which will coat all surfaces and prolong their useful life. Wood roof shingles on Second Empire and Queen Anne houses often

These workmen are installing a new wood-shingle roof on an outbuilding at Thomas Jefferson's home, Monticello.

This diamond-shaped wood-shingle roof is painted dark green, quite possibly its original color. The shingles on the left are recent replacements; those on the right are historic, possibly original if they have been kept well painted since the house was built in 1873.

had several courses (or more) of decorative, shaped shingles painted a different color from the field shingles. Green and brown were possibly the most popular colors for roofs, with dark red close behind.

Ten or twenty years ago, products intended to allow airflow under wood shingles were all the rage and were said to prolong the shingles' useful life. Contractors who install wood roofs regularly tell me they no longer use such products, having found that they don't prolong shingle life and may in fact shorten it. Most installers I know cover the sheathing with asphalt-impregnated building felt, adding a layer of polymer-modified bitumen ice-and-water shield along the eaves and in valleys. Building felt will prevent water from penetrating from the exterior while allowing interior water vapor to escape.

Some roofers install horizontal strapping to nail the shingles onto, spaced to allow the desired exposure. A narrow, screened gap is left between the shingles and the fascia board along the gable ends to promote airflow under the shingles, helping to keep them dry. I suspect other installation methods are prevalent in other areas of the country. Research local practices.

Shingle roofs must have a ridge cap to cover the point at which the shingles on each side of the roof meet. It is common practice today to install a row of horizontal cap shingles along each side of the ridge, one slightly overlapping the other. Based on many historic images I have studied, it was far more common historically for a ridge cap to be formed from two boards, typically 1 x 6s, cut to the angle of the roof and nailed together to form an upside-down V. This is still an effective method to cap a ridge and is less labor intensive than installing a shingled cap. A modern ridge vent, metal or fiber, is easily concealed by a board ridge cap without impeding its function. I screw the cedar boards together with deck screws and attach them to the roof with screws too, making removal easy if necessary to access the ridge vent.

On Tudor Revival and other romantic revival houses of the early twentieth century, wood shingles were sometimes built up in many layers and rounded

A typical wood-shingle ridge cap with overlapped horizontal shingles.

at the edges to mimic thatch roofing. If you have one of these unusual roofs, try to replicate it when the time comes to reroof. There never were many of them, and few survive.

Fire-safety concerns have prompted some jurisdictions to outlaw wood-shingle roofs in dense urban communities, where fire can easily spread from building to building, and in suburban communities where wildfire is a threat. Shingles that have been treated with a fire retardant may satisfy your local code official, but you should investigate whether the retardant used is likely to reduce the longevity of the shingles.

If you want the look of a wood-shingled roof in a jurisdiction that doesn't allow one, a clay, synthetic rubber, or plastic imitation might work. Clay tile roofs that mimic wood shingles have been used at Colonial Williamsburg since the first building restorations there in the 1920s. (See "Clay Tile/Terracotta Roofs" below.)

There are now plastic polymer and other imitation wood shingles that look passably convincing when the colors are right. These claim to be very long-lived with little maintenance. I have no personal experience with these products. Some are said to be recyclable at the end of their useful lives. Beware of exaggerated "wood grain" or unrealistic colors or sheens, and bear in mind that it is not possible to predict how the colors of a new product will hold up over decades in direct UV light. If possible, view an installed job on another house before committing to having a product installed on yours.

More information on wood-shingle roofs is available online in the National Park Service's *Preservation Brief 19*, "The Repair and Replacement of Historic Wooden Shingle Roofs," at https://www.nps.gov/tps/how-to-preserve/briefs/19-wooden-shingle-roofs.htm.

Slate Shingles

Used in Britain and Europe for centuries before the settlement of America, slate was valued for its fireproof qualities as well as its longevity. It remains one of the longest-lasting roofing materials in common use. Slate was imported from Europe to East Coast urban areas until the first half of the nineteenth century, when canals and railroads extended far enough inland to provide low-coast transportation for slate quarried in the U.S. There is some variation in longevity depending on the color of the slate and where it was quarried. Red and green slate tend not to last as long as black.

A properly installed slate roof can remain watertight and serviceable for a century or more without

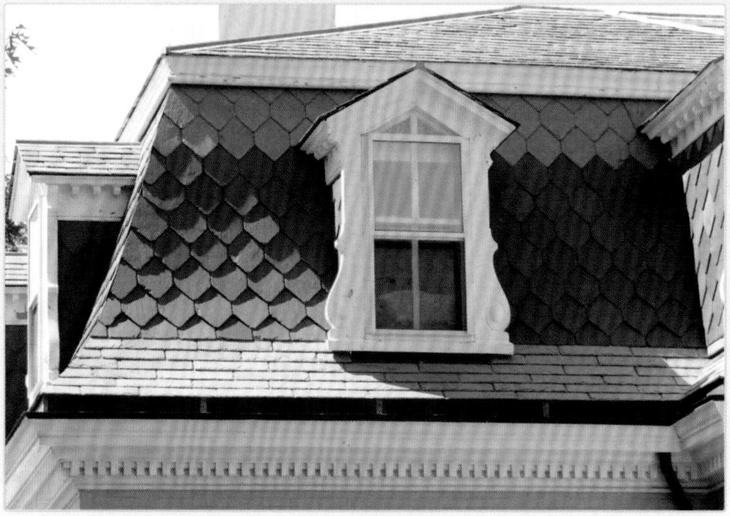

Slate shingles in two shapes and three colors are used on this Second Empire–style mansard roof.

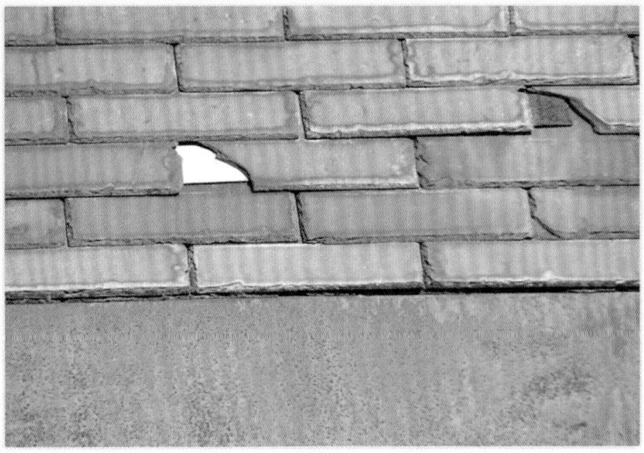

This roof shows where sheet metal was inserted under damaged slates to fill gaps. This can be effective for temporarily stopping a leak, ideally with the metal painted to blend in with the slate. The unpainted and rusty condition of the sheet metal eave flashing suggests that this roof has been neglected a long time.

major work. This is a great thing for the generations who do not have to replace the roof every twenty to thirty years, but it is not so great for the generation that owns the house when the roof finally comes due for re-slating. Slate roofing is expensive—worth every penny, but expensive. The job is rarely undertaken by do-it-yourselfers because of the skill required and the danger involved.

When a slate roof fails, it is often because the nails holding the slates in place rust through, leaving individual slates unfastened. If more than a few slates are slipping from their positions and falling off the roof, it is probably time to reroof. There are methods for reinstalling or replacing a slate without replacing the entire roof (see below), but they are not going to save the roof if there is widespread nail failure.

The other major symptom of aging in a slate roof is the failure of flashing, which can occur before the slate nails start to fail. Usually it is possible to remove only enough slate to replace the flashing, then reinstall the removed slate. However, it is worthwhile to inspect the nails while the staging is erected for the flashing. Staging can be a considerable expense; if the nails are within ten years of failure, it may make sense to reroof while the staging is in place.

It is important to find an experienced slate roofer who has previously worked on roofs like yours. The patterned, colored slate often found on a mansard roof is more complicated to install than a simple gable roof in black slate. Also, the multiple dormer windows usually found on mansard roofs require experience and skill to flash properly. When you are spending a lot of money to install a roof material that should last a century—possibly longer with modern stainless-steel nails—you want it done right.

Also important is finding slate to match what you have. Rarely does all the slate on a roof need to be replaced when the nails fail. Often, most of it can be

This slate roof is failing, with nails rusting through and slates falling to the ground. At this point, there is little alternative but to replace the roof.

removed and reinstalled. Some slates will break, however, and matching replacements will be needed. The

These red slate shingles have been carefully removed and stacked, awaiting reinstallation.

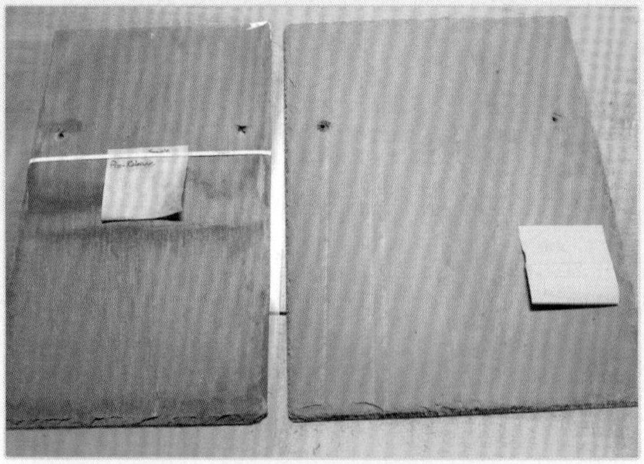

An acceptably close match (right) has been found for the color of this original slate shingle (left). They will be ordered in the same size as the original.

longer the roof has been neglected, or the more damage it has suffered from a falling tree or other calamity, the more replacements you will need. The first place to look is in the basement, attic, or barn, where you may find replacements remaining from a previous reroofing. If you strike out there, you need to look for slate salvaged from other roofs or new slate. You want to match both the color and the size of the historic slate. If that proves impossible, try to consolidate the original slate on the most prominent planes of the roof, and use the new slate on less visible areas. If you have dormers, bay windows, or other small sections of roof, using the new slate there may help disguise the mismatch.

A sad reality for many homeowners faced with a failing slate roof is that reroofing in slate is simply unaffordable. In this circumstance, consider re-slating the most prominent parts of the roof and using another material elsewhere. The owners of the Hench House (featured on page 103 and on the cover of this book) were faced with this dilemma. Parts of the roof had failed long before, and the resultant water damage inside the house was extensive. The house was a wreck inside and out, and the budget was

This proprietary stainless-steel hanging system uses real slate that is half the length of traditional slate shingles. The installation is faster, and less slate has to be purchased and transported to the site, reducing costs. Systems such as these do not have the centuries-long track record of traditional slate-hanging methods, and this roof appears to be holding up less well than a slate installation should after just a decade in use.

Two examples of asphalt shingles designed to mimic slate roofing. The second appears to have been installed incorrectly, leaving wider bands of black between the shingle strips than between individual shingles in each strip. The slate effect would be better if properly installed. Also, the overlapping shingles on the ridge are clearly asphalt. Copper ridge flashing would be more effective.

not unlimited. Few budgets are. The Hench House owners chose to retain slate on the corner tower roof and use asphalt shingles on the remainder.

There are now proprietary stainless-steel hanging systems that do not overlap the slates to the depth of traditional slate installations. This reduces the amount of slate needed for a roof by approximately half, and the labor for installation is also greatly reduced. The weight of the roof is substantially reduced as well, which can be helpful if the framing was undersized originally. I oversaw a roof installed with this approach during my time as a municipal historic preservation staffer and have watched that roof over the ensuing ten years. The photo here, taken recently, seems to show some lost ridge flashing on the dormer, several displaced slates, and one that has fallen. A properly installed traditional slate roof would not have these problems ten years after installation.

When replacing slate with asphalt, select a shingle that mimics the shapes and lay-flat profile of slate. A number of asphalt roofing manufacturers produce shingles with something of the character of slate if not the smooth texture and sheen. Avoid architectural asphalt shingles, which mimic the texture of wood shingles. If the slate being replaced is patterned, try to find asphalt shingles that mimic the pattern or at least the colors. Bands of color are easier to mimic than diamond or other elaborate patterns, but shapes, too, can be passably imitated.

There are clay tiles that mimic slate. They are installed in much the same way as slate, so there is not much reduction in labor costs, but they may be more affordable in your region, particularly if local contractors are more familiar with tile than slate. (See the "Clay Tile/Terracotta Roofs" section below.) There are also plastic polymer and composite imitation slate shingles that work passably well when the colors look right and the texture is not overly exaggerated. As with wood-shingle imitations, these claim to be very long-lived with little maintenance, though I have no personal experience with them. As mentioned, some such products claim to be recyclable—though what the plastics recycling market will look like in the future is anyone's guess. Beware of exaggerated slate texture or unrealistic colors or sheens, and bear in mind that it is not yet known how the colors will wear over the long term in direct UV light. If possible, view an installed job before committing to having a given material installed on your home.

The National Park Service's *Preservation Brief 29*, "The Repair, Replacement and Maintenance of Historic Slate Roofs," is an excellent online resource: https://www.nps.gov/tps/how-to-preserve/briefs/29-slate-roofs.htm.

This modern faux wood grain asphalt shingle is on a Tudor Revival gatehouse. Shingles like this were popular in the early twentieth century but are rarely seen today, though they are still produced. These are particularly appropriate for houses built from about 1900 to 1940, including Tudor Revival, Colonial Revival, and Craftsman houses.

Asphalt Shingles

Asphalt shingles have superseded wood shingles as the most common residential roofing in the United States. Three-tab shingles were formerly the standard, but thicker and more textural architectural shingles have become popular in the past several decades. Architectural shingles in a color that mimics weathered wood shingles are a compatible choice for a house built before 1900 or so, and even more so for a house built before 1860. Through the Victorian era, painted roofs became more popular, often with decoratively cut shingles in varied colors. If you are doing a period paint scheme on the exterior of your house, consider a roof shingle color that contributes to the scheme.

A cedar ridge cap has been used with asphalt architectural shingles in a wood-like color. A typical modern ridge cap of overlapping asphalt shingles would have made the roof look less like a wood-shingled roof. The ridge cap also conceals a ridge vent.

After the turn of the twentieth century, black, red and green three-tab shingles were popular for most styles of houses. Other shingle patterns, including diamonds and scallops, and shingles with a faux wood grain were also available in the early twentieth century. Most of these are unavailable today, while others may be discoverable with some effort.

On houses in some of the romantic revival styles of the early twentieth century, like Tudor Revival, asphalt shingles were sometimes built up in many layers and rounded at the edges to mimic thatch roofing. If you have one of these unusual roofs, try to replicate it when the time comes to reroof. There never were many of them, and few survive. Wood-shingle versions of these roofs were installed as well (as mentioned above) and are even more rare. The use of asphalt to replace a slate roof is covered above under "Slate Shingles."

If you are using architectural asphalt shingles to mimic the color and texture of a wood-shingled roof, consider using cedar boards for the ridge cap as described above under "Wood Shingles." This will help considerably in creating the character of a wood-shingled roof; a typical asphalt shingle cap is a giveaway.

Metal Roofs

The history of metal roofing goes back to the colonial era for some homes of the wealthy. Because of its cost, however, it did not become common until well into the nineteenth century. The earliest metal roofs were soldered sheets of lead or copper. In the nineteenth century, terne metal (sheet steel plated with a tin alloy) and painted tin became more available and affordable in the rapidly industrializing United States.

Roofs covered with long sheets of tin with standing crimped seams came into use in the nineteenth century and are still available. On high-end buildings, the metal might be copper. Metal roofs seem to be enjoying something of a revival in northern New England in the early twenty-first century. Many colors are now available with baked enamel coatings that do not need to be painted. It is best to stick with colors that were common on roofs in the period of the house, or alternatively to something recessive and neutral like dark bronze. Standing-seam metal may not be the best choice for an elaborate roof, like many Queen Anne–style roofs. Standing seams intersecting at odd angles or awkwardly applied to curved or conical roofs are unlikely to feel compatible.

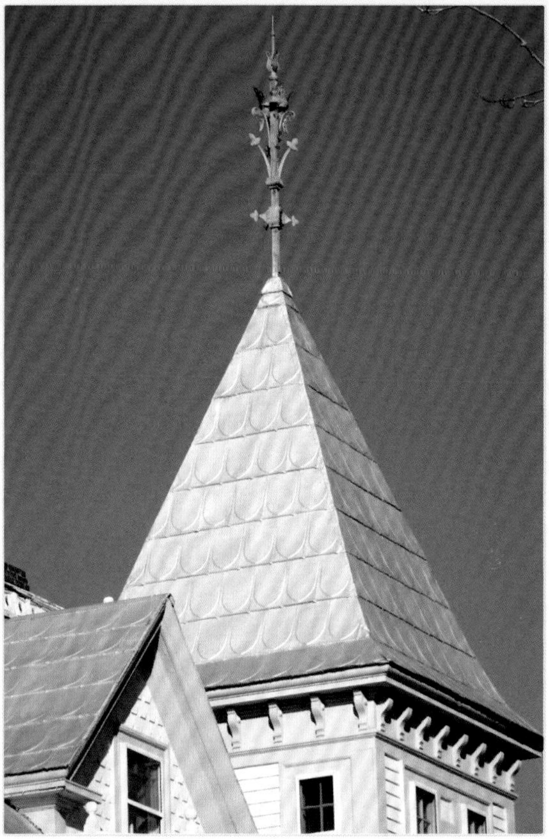

Interlocking stamped metal roofing panels like these were produced in many designs in the early twentieth century. These make a very long-lasting roof if kept painted. Some legacy manufacturers are still producing this roofing, and new versions have been introduced in recent years, often with baked enamel coloring.

After the Civil War, the use of steam presses and drop presses to emboss patterns into sheets of tin became widespread, and a golden age of metal-shingle roofing resulted (along with a lot of tin ceilings). Many patterns of roofing were available, including imitations

This old standing-seam metal roof may date to the early twentieth century and will continue to keep the house dry into the future if it is kept painted to prevent rust.

of wood shingle, clay tile, and slate roofing, and fortunately some are still produced or are being produced again. New patterns are also being produced.

Most historic standing-seam and metal-shingle roofs were intended to be kept painted; unfortunately, many have not been, leading to rust and eventual failure. If you have one of these roofs in salvageable condition, keep it painted with alkyd-based paints. Water-based paints do not belong on materials that rust unless a good coat of alkyd primer is applied first. Even then, an alkyd finish coat is preferred. Beyond traditional paint, some companies have developed modern elastomeric coatings for historic architectural metalwork, including roofs. These coatings look like paint but remain flexible to accommodate the expansion and contraction of roof components in varying temperatures. They can be combined with reinforcing fabric to repair damaged or weakened sections of roofing.

Corrugated tin was a common roofing material on modest homes and outbuildings for decades and is still appropriate there. Like modern standing-seam roofing, it is now available with baked enamel coatings and does not need to be painted.

Clay Tile/Terracotta Roofs

Clay tile roofing is an ancient building material, dating to twelve thousand years ago in Asia and appearing in the Middle East not long after. It was used by the ancient Babylonians and Egyptians as well as the Greeks and Romans, who introduced the material to Europe. Even in regions with abundant supplies of wood for shingles, clay offered the advantage of being fireproof. It also offered longevity, lasting a century or more with periodic maintenance.

Clay tiles came to America with European settlers. The first such settlers were probably from Spain,

This modern interlocking-panel metal roof is made with a long-lasting baked enamel finish. All red pigments are prone to fading in UV light, and a red roof will likely turn pink eventually, at least on the sides that get direct sunlight.

settling in Florida and subsequently in the Southwest and California as Spain expanded its colonial rule northward from Mexico. There is evidence of clay tile use in the early English settlements at Roanoke Island, North Carolina, and Jamestown, Virginia. The Dutch brought clay tiles with them to the Hudson River valley in the early seventeenth century and were manufacturing them there by 1650. Although Dutch influence waned after the British took control of the colony in 1664, several tile manufactories remained in operation in the New York area in the mid-eighteenth century. Both French and Spanish builders used clay tile roofing in New Orleans during their periods of control there.

Terracotta pan tiles, also called barrel or mission tiles, are the most common form of clay tile roofing in the U.S. today. This type of roofing was brought to America by Spanish settlers and remains most popular in the areas they first colonized, from Florida to California. It is widely available in a variety of colors and sizes.

The widespread availability of wood in the eastern and southern parts of the American colonies likely limited the use of clay tile outside cities such as New York and Boston, where concerns about fire led to early building and fire codes that encouraged tile or slate roofs. Spanish missionaries established a clay tile works in California using Native American labor in 1780. Unlike in the East, clay tile became a widely used roofing material in the West and Southwest.

This romantic Tudor Revival building from 1932 was roofed with clay tiles resembling those common in England during the Tudor period. Now covered in lichen, they add greatly to the character of the building. Replacement with another material would likely diminish the charm of the building considerably.

In the East and South, early tiles were typically flat and rectangular, much like slate roofing. They were hung by nailing into sheathing or battens through two holes at the top of the tile, with each row of tiles overlapping the one below by approximately half and with vertical joints offset from row to row. As with slate, nails must not be driven tight but remain slightly raised so the tile can expand and contract with temperature changes and move slightly if walked on. In the early twentieth century, similar clay tile roofing was used for Tudor Revival buildings, often resembling historic British tile roofing. Later flat tiles were of interlocking design, eliminating the overlap and reducing the weight of the roof considerably.

Clay roof tiles mimicking the appearance of wood shingles have been used at Colonial Williamsburg since restoration of the eighteenth-century city began in the 1920s. Tiles can be an effective and long-lasting alternative to wood shingles, but their greater weight must be taken into account. An engineer should confirm that your roof structure can take the load.

Overlapping flat tiles can successfully replicate the appearance of wood-shingle roofing, as Colonial Williamsburg in Virginia has done since the 1920s. It is critical to locate tiles with an appropriate striated surface to resemble wood grain and with the right variations in color if the goal is to mimic a natural (not stained or painted) wood roof. Clay tiles that imitate the appearance and character of wood-shingled roofs are made today. If using these, beware of tiles with an exaggerated wood grain appearance or too much uniformity of color or size. The most common giveaway for imitation materials, the thing that makes them look fake, is trying too hard to look natural. Whether it is vinyl siding, composite decking, PVC trim boards, or clay roof tiles, the finished surface will not appear more real if its "grain" is more obvious than the real grain. Less is more.

Unlike most historic American building traditions, Spanish-style clay tile roofing did not spread from east to west, but north from Mexico into the West and Southwest. The tiles commonly used in the Southwest and California are pantiles, or pan tiles—rounded, red, unglazed terracotta tiles commonly called mission or barrel tiles. These spread to all parts of the United States with the rise of the Spanish Colonial Revival and Mission styles during the early twentieth century.

Clay tiles are most commonly terracotta red, the natural color of clay with a high iron-oxide content, but they have been and still are produced in other colors. Some clays are more naturally tan or buff, and color can be applied with or without a glaze. Browns, greens, black, blues, and various shades of red have been produced. Older tiles may show more color variation because they were fired in less controlled circumstances than modern kilns provide. Uneven heat resulted in some tiles being darker than others.

A clay pan tile roof from the early twentieth century in Maine.

This clay tile roof installed in the 1930s remains in good condition. When the time comes for repairs, it will not be difficult to find matching roof tiles, as a number of cartons of the original tiles remain stored in the basement. If even more are needed, the manufacturer identified on the tiles is still producing them as of 2019.

Roof tiles are generally long-lived (particularly in arid climates) and impart a distinctive character. When a tile roof fails, it is often because the nails holding the tiles in place rust through. If more than a few tiles slip from their positions, it is probably time to reroof. As with slate roofs, the other common symptom of an aging tile roof is the failure of flashing, which can occur before the nails start to fail. This problem can usually be addressed by removing enough tile to replace the flashing, then reinstalling the removed tile. Investigate the nails while the staging is in place; staging can be expensive to erect, and it may make sense to reroof now if widespread nail failure looks likely in the next ten years.

I have had no personal experience with installing or maintaining tile roofing, but if you have a terracotta roof that needs attention, the National Park Service's *Preservation Brief 30*, "The Preservation and Repair of Historic Clay Tile Roofs," is an excellent place to start your research. It can be found online at: https://www.nps.gov/tps/how-to-preserve/briefs/30-clay-tile-roofs.htm.

As with any specialized work, if you're hiring a contractor, seek out one who has experience with historic tile roofs. Contemporary installation techniques are often different from traditional methods, and a contractor who has done a hundred new tile roofs may not have the knowledge to do yours correctly.

Concrete Tile

Concrete tile is similar in many ways to terracotta tile but often lighter in weight. Historically it was less expensive, and it became a popular roofing material in many areas in the early twentieth century. It is very durable if properly made. (Many surviving Roman buildings, like the Coliseum, are built of concrete.) There are concrete tile roofs installed a century ago that are

This concrete tile roof remains in good condition more than eighty years after its installation in the harsh climate of Maine.

doing fine, including the roof of the Kowalski-Policht House featured on page 453.

Concrete tiles are still produced, though matching the dimensions and color of historic tiles may require diligence or a custom order. (The owners of the Kowalski-Policht House have had to replace a few pieces, but a supply of extra tiles left in the basement by the original builder eliminated the search for matching replacements.) I have no personal experience with installing or maintaining this type of roofing.

Membrane Roofing

Sheet Membrane

Modern EDPM thermoset membrane and thermoplastic membranes like PVC and TPO have largely displaced more traditional roofing materials for flat or slightly pitched roofs. These sheet materials are installed with overlapped and bonded seams and form a continuous roofing surface, typically installed with flashing to prevent water or wind penetration at the edges.

Although a modern material, this type of roofing is acceptable for nearly any application on a historic house where the roofing cannot readily be seen from the ground. For locations with some visibility, such as the roof of a bay window, it is recommended to use a dark membrane. For flat roofs that are not visible at all, white may be preferred for its heat-reflective qualities, particularly in a hot climate.

If a flat or slightly sloped roof is being insulated with rigid insulation above the sheathing, membrane roofing can be applied with no additional layer of sheathing over the insulation. This is best done on roofs with a parapet to hide the additional thickness of the roof, but it can sometimes be done successfully by tapering the outer 18 to 24 inches of insulation down to the level of the historic roof at the edge. Covered in a dark membrane, this taper generally will not be noticeable from the ground if the eaves are high enough.

Spray-On Membrane

Spray-on membrane is popular in some regions, and it too can be applied over rigid or sprayed-on insulation. It is often used on adobe houses in the Southwest, where the parapet hides the thickness of the insulation. This is one case where the increased thickness of an insulated roof mimics the thickness of a historic roof construction technique that has gone out of use: clay roofs built up to a foot or more thick over exposed log rafters (vigas) and split- or sawn-board sheathing (latilas). These thick clay roofs moderated the intense heat of the sun in an arid climate where the scant annual rainfall was not enough to turn them into liquid mud.

Tesla Solar Roofs

The Tesla Solar Roof shingle went into production in 2017. Installations began on the West Coast and, as of early 2019, are projected to move eastward as installers are trained. Judging from photographs and current information on cost—which reportedly is competitive with high-quality asphalt shingles—it seems possible that this will be a game-changing technology. If an affordable solar roofing product that looks as good as or better than asphalt roofing becomes widely available, it would not be surprising to see it dominate the market for residential roofing within a few decades. The slate lookalike is particularly good in the photos I have seen.

The question may be whether Tesla can produce enough of it and train enough installers to make that happen. As with any new and unproven technology, time alone will tell if it lives up to its potential. I very much hope it does. The lingering tension between historic preservationists and solar advocates could be largely erased by a product like this, bringing energy-efficient roofing to buildings that are already green by virtue of their sustainable materials and embodied energy. Though I don't recommend removing a slate roof in good condition to install a new solar roof, this is a promising technology.

Gutters and Downspouts

Gutters and downspouts (also called leaders) have been used for thousands of years to channel water off roofs in a controlled way, and they can be a character-defining feature of the exterior, particularly if they form a part of the eave trim. Many historic houses originally had wooden gutters built into the eaves, often appearing as a crown molding when viewed from the ground. Many of these have subsequently been roofed over with new metal gutters outboard of the eave or with a drip edge and no gutter. If there is evidence of patching in the underside of the wood crown molding at one or both ends where a downspout might have attached, there is probably a gutter under the roofing material that deteriorated and was roofed over. Integral wood gutters are also sometimes

This house had integral wood gutters that were later roofed over, as evidenced by the remaining outlet for the downspout extending from the crown molding profile at the eave. Later roofing was simply extended over the gutter and allowed to drip directly to the ground.

A new copper half-round gutter with a round downspout installed on the front porch of a Greek Revival house. This traditional type of gutter is more appropriate to historic houses than modern aluminum box gutters and square downspouts.

This house retains a functioning integral wood gutter that serves as a crown molding at the eaves. A wonderful and unusual feature of this house is the carved granite end block for the gutter, which retains the cross-sectional profile of the original gutter and tells us that the existing gutter is a replacement. Hopefully the next time this gutter is replaced, the owner will have the replacement milled in the original profile. Note how the downspout is painted to blend into the brick wall behind it.

removed. The result can be boxy, awkward fascia boards at the eaves where a graceful crown profile originally transitioned the wall to the roof.

If your house had integral gutters that have been removed or roofed over, give serious consideration to restoring them. They can be replicated with stock gutter of the appropriate shape and size or custom-milled if necessary. Wood gutters should be treated periodically with shingle oil to prolong their useful life. This can be done as part of the annual clea-nout but probably only needs to be done every three to five years. If the wood appears to be drying out and checking, it is time to treat it with the oil. It is also possible to have custom copper gutter made to mimic the profile of a wood gutter. This can be a good approach if the capacity of the wood gutter is inadequate in heavy rain. Once installed and painted, the copper will be nearly indistinguishable from wood.

Gutters are an excellent way to manage water; the work they do rewards their maintenance requirements. Restoring integral gutters or installing new gutters will often dry out a damp basement. The water still needs to be directed away from the foundation when it gets to the bottom of the downspouts, but that is a more manageable problem than directing water away from the foundation along the entire eave drip line. Connecting the downspouts to underground PVC drain pipe running to daylight or to a subsurface dry well prevents any roof water from seeping under the house. If the grade slopes away from the house where the downspout reaches the ground, the natural flow of the water away from the house may suffice. Rain-collection barrels are another option, especially if you need water for gardening. Historically, water was commonly collected from roofs for kitchen and household use, with downspouts connected to a cistern in the cellar. A hand pump in the kitchen would draw water up from the cistern.

Modern aluminum box gutters and rectangular downspouts are not typically appropriate on houses more than 75 years old. Half-round metal gutters are

The original wood gutter of Whitten House, roofed over since at least the 1930s, was replicated in Spanish cedar and is being installed here on the new bay window added to the back of the house. When the house was reroofed, the roof edge was returned to its original position to allow the replicated gutters to be installed and used. Note that the exterior of the gutter has been painted, but the interior was treated with shingle oil, which will protect it without creating a film that could trap water under it as paint would eventually do.

Two approaches to staging for reroofing. Staging from ladder jacks, as in the top image, can be appropriate for wood-shingle, metal, and asphalt roofing, as these materials are comparatively light. More substantial staging is generally required for slate or clay tile, as in the bottom photo. Fall protection is always necessary for roof work, whether harnesses tied off with rope or full railings and netting as in the bottom photo.

more compatible, whether copper or painted galvanized steel. Downspouts should be either smooth or corrugated round. Traditionally, galvanized gutters and downspouts on a wood house were painted to match the trim or wall behind. On a masonry house, they might be painted black or green. Copper could be painted or left to weather.

Drains and Grading

With or without gutters, establishing and maintaining a positive grade slope away from the foundation all around the house is critically important. Even if you are collecting and managing every drop that hits the roof, the rain and snow that falls around the house needs to be directed away from it. If it is not possible to create sufficient grade slope to accomplish this—as is often the case due to site constraints, paving, or other issues—subsurface drains are the solution. These are typically a perforated PVC pipe laid in a trench and covered by gravel and filter cloth, with soil or more gravel on top. The pipe runs to daylight or to a dry well where water can collect quickly in a storm and slowly be absorbed into the surrounding soil. These are often called French drains in New England and go by other names in other parts of the country. Information on designing and installing a subsurface drainage system is readily available. Remember, water destroys far more old houses than fire does.

Satellite Dishes

Satellite dishes do not belong in any visible location on the walls, roof, or chimney of a historic house. They are not compatible with historic character and are likely to cause some degree of damage to roof materials when mounted or later removed. If you cannot get television reception via cable, DSL, or an antenna hidden in the attic eaves (our solution at Whitten House), find a location for the dish that does as little visual and physical harm as possible—ideally on a rear ell or outbuilding.

The installer may be unhappy about having to run a longer cable in the least conspicuous manner possible. If challenged, you could tell him or her that your local historic preservation commission requires it. The prospect of a lengthy bureaucratic review is often enough to convince a balky installer to cooperate. His priority is to do the installation as quickly as possible and get on to the next one; your priority is to preserve the house you have devoted enormous resources to restoring. Stand firm.

John H. and Jeanette Davis House

Shingle Style, 1883

The Davis house was the second Shingle Style design by noted architect John Calvin Stevens, one of the style's originators. The Shingle Style is associated mainly with summer resort architecture of the late nineteenth century, but Stevens used it as well for year-round suburban homes like this one. His own Shingle Style house was located on the same Portland, Maine, block.

John Hobart Davis served in the Union forces during the Civil War and returned to Portland at age 21 to work in the banking business. He remained at the same bank from the end of the war until his death at age 75. His wife, Jeanette, was a school teacher early in their marriage. When they built their new house, Davis turned to his friend Stevens to design it.

Stevens had trained in the office of architect Francis Fassett and opened a Boston office for the firm shortly after becoming a junior partner in 1880. The Boston office shared a building with architect Ralph William Emerson, who had designed the first Shingle Style cottage, Redwood, built in Bar Harbor, Maine, in 1879. Stevens and Emerson became friends, and Emerson's work strongly influenced Stevens. The Davis House was among Stevens's first designs after returning to Portland.

The Davis house was featured in the *Scientific American Architects and Builders Edition* issue of August 1893, a publication devoted to promoting good architectural design from the U.S., Great Britain, and Europe. The magazine description stated, "The underpinnings and first story are built of various kinds of rough stone, laid up at random, with brick trimmings. The second and third stories are covered with shingles and painted a dull red. Roof shingled and painted similar.... [The] interior throughout is trimmed with whitewood, the first floor being finished in cherry. Lobby and hall have paneled wainscotings and ceiling beams. Hall also contains a paneled divan and staircase of unique design, which is lighted by delicate stained glass windows." Also mentioned were a parlor fireplace trimmed with tiles, the dining rooms, paneled wainscot, and "a pleasant nook with seats and fireplace built of brick, with hearth of same." On the second floor were three bedrooms and a bath, with two additional bedrooms on the third floor. The house cost $3,400 to build.

By 1917 the house had passed to other owners, and John Calvin Stevens was engaged to design a two-story addition on the west side and a large sunroom on the rear. The two-story flat-roofed addition contained an extension of the dining room on the first floor and a sleeping porch on the second. Indoor-outdoor living spaces like sunrooms and sleeping porches were all the rage in the early twentieth century. At some later point the sunroom was removed.

In 1951, the original octagonal bay window at the northwest corner was removed and a boxy one-story addition was added in front of the boxy two-story addition. Also, a portion of the raised front yard was removed to create a garage in the basement. The raw concrete retaining walls of the new driveway were completely at odds with the original character of the house, with its mix of stone and brick and highly textured shingled wall surface. The flat-roofed form of the additions was quite at odds with Stevens's original saltbox roof form. The resulting appearance suggested that two very different houses had been sliced in sections and shoved together.

When its current owner Ruth and her late husband Gene bought the house in the early twenty-first century, the house had lost much of its original charm and character. The dull gray paint was peeling, and even the bulbous turned post at the entry porch had been replaced with an unpainted pressure treated 6 x 6 post. The interior retained many historic features, as did the National Register Historic District neighborhood the house was located in. In fact, the Davis house had lost more of its historic character than any of its neighbors. Recognizing the home's potential, the new owners decided to take on the challenge of healing its wounds as a retirement project.

As described in Chapter 2, local architects Barba & Wheelock did a masterful job of designing a few simple changes to the exterior to make the awkward additions feel like a part of the house. Historic photos of the turned porch post were not clear enough to establish its profile with certainty, but a matching post was located across town on a house designed by Fassett & Stevens several years before Stevens designed the Davis house, and this was replicated by a custom millwork shop to replace the pressure treated 6 x 6. The house was repainted in its original colors using the published 1893 description. The concrete retaining walls were veneered in half-brick with a granite cap, relating them to the original brick and stone elements of the house, and a compatible carriage house door replaced the modern paneled overhead garage door. The wood steps from the sidewalk to the entry porch were replaced with matching granite, again relating to the original first-story materials of the house.

Inside, finishes were repaired and renewed, and the electrical and plumbing systems were updated. The cherry trim in the first story had long since been painted, probably after the additions darkened the original rooms, so this trim was freshly painted in white, setting off the cherry doors that remained unpainted. A new deck off the back of the house provided convenient outdoor living space just off the kitchen.

The thoughtful work done to integrate the additions with the original house, and the attention to detail on both the exterior and interior, returned the Davis House to its position as one of the most attractive houses on the street, all but one of which were designed by John Calvin Stevens between 1883 and 1930.

Floor plans were published in the *Scientific American Architects and Builders Edition* of 1893. Such documentary information is valuable for understanding how a house has changed over time. *(Courtesy Greater Portland Landmarks)*

This 1924 tax photo shows the two-story addition on the west side of the house and the large greenhouse addition at the rear.

The granite for the new steps matches one of the stones in the original masonry at right.

Two views of the house as built. *(Courtesy Greater Portland Landmarks)*

The dining room with inglenook fireplace.
The parlor is in the distance at left.

New cedar shingles match the patterning of the originals and were painted in the original colors.

Sunlight on a restored gable highlights the staggered and plain shingle patterns of a Shingle Style home.

Architect John Calvin Stevens likely prepared this drawing in 1883 to "sell" the design to the clients at a time when the Shingle Style was just emerging.
(Courtesy Maine Historical Society)

The new brick patio behind the house is a sheltered outdoor living area in the heart of Maine's largest city.

The rear garden and patio.

The front hall from the parlor after restoration.

Before restoration, the boxy additions and incompatible concrete retaining walls of the driveway cut were prominent.

The front façade as the project neared completion.

The new deck off the kitchen, at the rear of the house. A traditional railing design helps relate the deck to the historic house.

A new custom-turned porch post based on historic photos. The original had been replaced by a 6 x 6 pressure-treated post.

The nearly completed new entry porch has a simple but compatible railing treatment and a "brick" pattern lattice skirt.

The concrete retaining walls—shown here during the renovation—were not compatible with the house.

Samples of the flat and corner veneer brick used with granite capstones to cover the concrete retaining walls.

This nineteenth-century post designed by the Davis House architect was a model for the replicated post in the photo above.

CHAPTER TWELVE

Exterior Walls

Walls keep the outside where it belongs, protecting us from cold, heat, and precipitation. They are less susceptible to water infiltration than roofs, but they are not impervious. When wind-blown rain soaks a masonry wall, the masonry may absorb quite a bit of water. Properly painted wood siding is less absorbent, but water can find its way into the sheathing or wall cavities through any uncaulked or unflashed cracks around trim. The framing at the base of a wall can be soaked by water splashing off steps (a common cause of sill rot under doors) or the ground if the house sits low. It pays to be aware of these vulnerabilities, because simple measures can often be taken to cure them. Simply caulking cracks around trim or installing a gutter or rain diverter over a doorway can prevent a lot of damage in the long run.

Maintaining Masonry Walls

Identifying the materials

Masonry buildings are typically built of stacked natural stone or manmade masonry units held together with mortar. In rare cases a stone building might be "dry laid" without mortar, but this is atypical in

American houses built after European settlement. Commonly used stones for historic homes in the United States include limestone, brownstone, granite, and less commonly, marble. Manmade masonry units include fired clay brick, unfired clay brick (adobe), cast stone, concrete block, and terracotta. Mortars include clay mud, lime mortar, natural cement, and Portland cement. The particular combination of materials used in a house often depends on where and when it was built, particularly if it was built before railroads reduced the cost of transporting heavy materials such as stone or brick from quarries or manufactories to distant locations.

A key to understanding historic masonry buildings is that masonry construction is a sacrificial system composed of two materials that expand and contract differently in varying conditions of temperature and humidity. To accommodate this difference, one of the materials (the mortar) is designed to yield to the other (the brick or stone). The mortar is therefore softer, more porous, less dense, far cheaper, and can easily be renewed through periodic repointing. It is the sacrificial component in this two-component construction, and over a period of 80 to 100 years, it will break down and fall away until it needs to be replaced, or repointed, to maintain the wall's structural stability and water resistance.

A complicating factor when maintaining or repairing historic masonry is that common modern materials are not the same as they were historically.

The mortar in this historic wall is softer than the brick and will yield and deform as the two materials expand and contract. The mortar is designed to be sacrificed so that the integrity of the brick is maintained. Over many decades, the mortar will erode to the point of requiring renewal, or repointing. The mortar on the left needs repointing, while that on the right has been repointed recently. New mortar should match the old in color, texture, and hardness.

This wall was spot repointed with modern Portland cement mortar (at right) that does not match the historic mortar in color, aggregate, or hardness. This was on a historic tax-credit project, and I had to tell the mason to cut it out and redo it with appropriate mortar, after first preparing several samples of mortar mixes on the wall so the best match could be selected.

Standard modern Portland cement mortars are much harder than historic Lime or natural cement mortars, as are the modern bricks they are intended to be used with. If modern mortars are used to repoint or repair historic brick, the brick will become the sacrificial element and will be destroyed over time. *Do not* repoint historic brickwork with standard off-the-shelf bags of mortar mix from a home-improvement store. Bagged mixes of high-lime, Type O mortars are produced but may have to be special ordered. Types M, S, and N have high Portland cement contents and should be avoided for work on most historic buildings.

The walls of most brick buildings are constructed with several layers, or wythes, of brick bonded together internally. In buildings of several stories, the lower stories typically have thicker walls. The pattern of brick visible on the exterior generally shows how the wythes are tied together, as you can see the exposed ends of the *bond bricks* that cross the wythes at right angles. Nearly always, the bricks of the inner wythes are softer than the outer, which were fired at a higher temperature to withstand exterior exposure.

Sometimes the inner and outer wythes become partially separated due to broken bond bricks within the wall. This can be caused by water infiltration, impact damage, or by too few bond bricks in the original construction, placing too much stress on the bond bricks that were used. Traditionally, repairing this condition required careful removal of areas of detached face brick and rebuilding with new bond bricks. Today, stainless-steel helical anchors are often used to tie together the inner and outer wythes without having to remove the exterior wythe. An engineer should be consulted on the needed number and placement of anchors.

A labeled mortar sample. Colored mortars were popular during the late nineteenth century and can be particularly difficult to match. The color card matches the historic mortar, which has been cut away for the sample. This match is close but not quite there yet. *(Courtesy Sutherland Conservation & Consulting)*

This wall was repointed with a modern mortar that was too hard for the historic bricks. Unable to expand against the mortar with changes in temperature and humidity, the bricks have begun to crumble from the internal pressure. The hard mortar also prevents moisture in the wall from passing outward and escaping, which creates additional problems. The modern hard mortar must be cut out and replaced with soft mortar, or the wall will be destroyed.

Any masonry wall that appears unstable or has been significantly altered with new openings or other structural changes (including foundation problems) should be inspected by an architect or structural engineer. Masonry is heavy, and a compromised wall can quickly become dangerous while under repair. It is critical to follow the scope and order of work laid out by your engineer. On one project I worked on several years ago, the mason failed to follow the engineer's instructions while repairing a second-story brick wall, and the entire wall ended up on the sidewalk, taking out the staging and a streetlight. Fortunately, no one was hurt.

Most historic masonry is loadbearing, with wood joist ends embedded in the walls for support and roof timbers resting on top of the walls. Water infiltration can cause these connection points to be weakened by rot in the wood or deterioration of the masonry. A structural engineer experienced with historic masonry buildings should determine the best way to effect repairs in this situation. Often, the historic framing does not meet modern code, and your local building inspection office may require drawings stamped by a licensed engineer for more than the most basic repairs.

Another possible complication is a chimney supporting structural timbers, as was quite commonly done historically. If such a chimney is active, it may be worthwhile to have an engineer determine if the wood should be separated from the masonry by a few inches (as modern code requires), and if so, how to accomplish that. A timber cut back two inches and supported by a steel bracket embedded in the masonry might be enough to satisfy the building inspector if an engineer says it will work. Remember, a building inspector is unlikely to require complicated and expensive structural alterations if the historic plaster remains in place and the framing cannot be seen.

Repointing Masonry

Repointing is the most commonly needed masonry repair. It is important that this be done by someone who understands the difference between historic and modern masonry. It is possible for a homeowner to become proficient at repointing with a moderate amount of practice; the several steps involved are not difficult. Do not start with your front façade. The work must be done carefully, especially cutting out the old mortar to a depth sufficient for the new mortar to bind to the brick or stone. This was traditionally done with a chisel and hammer but is now most often done with a pneumatic chisel or a grinding wheel, either of which can do significant damage to bricks if not carefully controlled.

The photo at left shows three custom natural-cement mortar mix samples in front of the historic mortar to be matched. The original aggregate mix included black sand taken from a nearby beach, a source no longer available. Sample #1 was closest but needed a bit more adjustment. Because the amount of visible aggregate varied around the building, and only selected areas were being repointed, the aggregate was shipped separately by the mortar company so it could be mixed in on-site, adjusting to match the surrounding original material. The second photo shows a section of the final new mortar mix in place.

The photo at left shows an area that was repointed with inappropriate hard Portland cement mortar and is now cracking and failing. The second photo shows the same area repointed with the custom natural-cement mortar shown in the previous pair of photos.

Latex or silicone caulking is never an appropriate substitute for mortar between masonry units. Not only will it look bad, it will make repointing in the future more difficult.

The National Park Service offers an excellent Preservation Brief on *Repointing Mortar Joints in Historic Masonry Buildings* at: https://www.nps.gov/tps/how-to-preserve/briefs/2-repoint-mortar-joints.htm. If you own a historic masonry building, you should read this free publication.

Caulking is not—and never can be—a substitute for repointing a wall. Caulking will trap water in the wall and will stick tenaciously to the stone or brick surface, making it difficult and expensive to remove when someone ultimately repairs the wall correctly. It also looks terrible.

Matching Brick or Stone

It is often necessary to replace some damaged brick or stone when repointing. Other times, matching material is needed to close up an added window or door opening. It is important to find brick or stone that matches the surrounding wall in size, color, and texture. The texture will be affected by the "flash" of the brick, a result of the manufacturing method used (water-struck or sand-struck). Locating matching brick can take some hunting through brick or stone yards and may require locating a supply of old brick from another building. Several brickyards specialize in producing new brick that looks like old brick. Resist the temptation to use misshapen, mis-colored "olde" brick unless you have a twentieth-century house that was built with them. Take one of your bricks with you when searching or bring samples back to hold against the historic bricks for comparison. Matching stone can be more difficult, depending on the size of the stones in your house. If the house predates railroad lines, chances are the stone came from nearby.

When filling an added door or window opening, remove the half bricks around the edges so the new work can "tooth" into the old. Make sure your mortar joints match the widths of the historic joints and are tooled in the same manner. The density and color of

The original window in this location was removed and a large doorway created in the 1960s. During a recent rehabilitation project, the door was removed and a new window matching the original was installed after infilling the enlarged opening with matching brick. Because the new brick and mortar are well matched and the edges toothed in, it is difficult to tell that the wall was ever altered.

The sample card at right has ¼-inch-thick slices from new bricks mounted so they can easily be transported to a project and held up for comparison with the historic brick. This is a pretty good match. Sample cards are not always available, and it is sometimes necessary to carry full bricks to the site for matching. Replacement bricks need to be a good match to the originals in size as well as color.

The wood shutters in this window opening disguise the fact that it has been infilled and there is no window there. This is a good alternative to filling a window opening flush with the wall and removing all evidence that the opening ever existed.

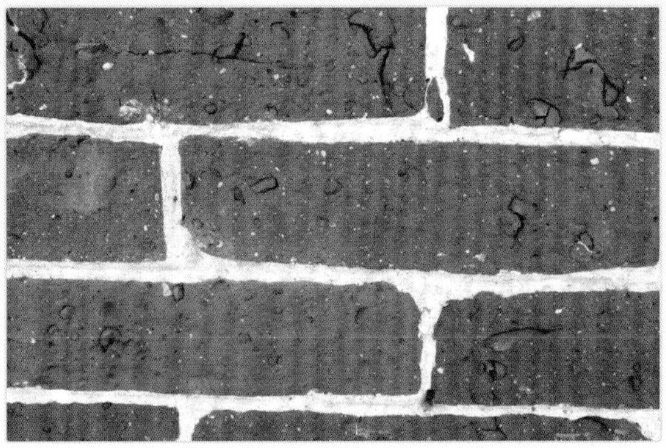

This exterior wall from about 1810 was "cleaned" with aggressive sandblasting, which removed the hard, outer surface of the brick, exposing the soft core. A wall in this condition looks bad and will absorb water like a sponge. Do not do this.

the mortar must match the historic mortar. If you are filling in a historic opening (not recommended, but sometimes necessary), set the infill brick back in the opening an inch or so, thus retaining the "memory" of the historic opening. If a window opening is to be sealed on a house with shutters, consider sealing the opening with wood infill painted black and mounting a pair of shutters in the closed position over the infill.

Cleaning a Masonry Wall

Masonry surfaces commonly need to be cleaned, and this project is probably more often done by a homeowner than repointing is. A brick or stone building may look like a rugged object that need not be cleaned with care, but historic masonry is often more fragile than it seems and can be damaged by pressurized water or cleaning chemicals. The misuse of water repellents or sealers is another danger to a masonry wall; as with repointing with the wrong mortar, sealing moisture into the wall can lead to damaged bricks or stones. The National Park Service addresses these topics in a Preservation Brief, *Assessing Cleaning and Water-Repellent Treatments for Historic Masonry Buildings*, that can be downloaded for free at https://www.nps.gov/tps/how-to-preserve/briefs/1-cleaning-water-repellent.htm.

Sandblasting was once a common way of cleaning masonry walls, but this practice was particularly destructive to brick buildings. When a clay brick is fired, the outside surfaces form a skin that is harder than the interior. Sandblasting often removes this skin, leaving the interior exposed to the elements. Over time, the soft brick erodes and crumbles in the wind, rain, and snow. Yet another National Park

Service Preservation Brief, *Dangers of Abrasive Cleaning to Historic Buildings*, offers information on this topic at: https://www.nps.gov/tps/how-to-preserve/briefs/6-dangers-abrasive-cleaning.htm.

Brownstone Restoration

Brownstone was once such a popular urban building material that the middle decades of the nineteenth century are sometimes referred to as "the brown decades." This sandstone—warm brown in color and relatively soft and easy to cut and carve—made an appealing and effective facing for the richly detailed exteriors of Italianate and Second Empire buildings. The underlying structures are nearly always of brick, with a brownstone facing on the public elevations. Sometimes the bricks in the secondary elevations are covered with brown stucco that is scored to look like blocks of brownstone. The resultant rowhouses are so ubiquitous that they have come to be called "brownstones," and people who restore and live in them call themselves "brownstoners." In eastern cities, entire neighborhoods are made up of streets lined end-to-end with brownstone rowhouses.

A row of Brooklyn brownstones.
(Courtesy Marc Sirois)

Unfortunately, this soft stone, like marble, proved highly susceptible to damage from acid rain, which came to be understood as a serious environmental threat in the 1970s. Caused by emissions of sulfur dioxide and nitrogen oxide from industrial, commercial, and domestic sources, this acidic brew was deposited on buildings by rain, fog, dew, snow, and ice. The burning of coal was a major contributor, and coal was the fuel that powered the industrial revolution while heating America's shops, offices, and homes and cooking its meals. Over time, taller smokestacks had been built to disperse coalsmoke away from its communities of origin to distant downwind locations, including rural and wilderness areas. By the time the United States addressed the problem seriously, starting in the 1980s, the damage from acid rain was widespread. The "cap and trade" program established to reduce emissions was very successful, and there is little ongoing damage from acid rain today. The skies are clearer, and rivers, lakes, and forests are recovering—but for brownstone buildings, the damage had been done.

Brownstone can also suffer from delamination. As a sedimentary stone, it formed from layers of sand that were deposited on ancient lake and ocean bottoms. Over millions of years, deeply buried layers were compressed into solid stone and bound together by clay particles, which are smaller and "stickier" than sand, but the "grain" between the layers remains, much as wood grain does. This layering may be barely detectable or clearly evident, depending on the source quarry. When water is absorbed by the porous stone, it can

On this side elevation of a brownstone house of the 1860s, scored stucco imitates the stone blocks in the public-facing elevation. The use of a less expensive material on this secondary elevation has been revealed a century and a half later by a loss of stucco, likely due to water infiltration.

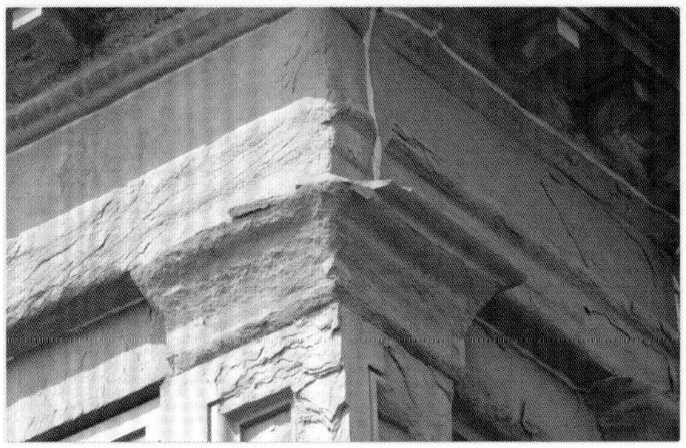

Brownstone trim eroded by acid rain. Although the skies are cleaner today and the threat of damage from acid rain much reduced, many buildings suffered serious damage and loss of detail.

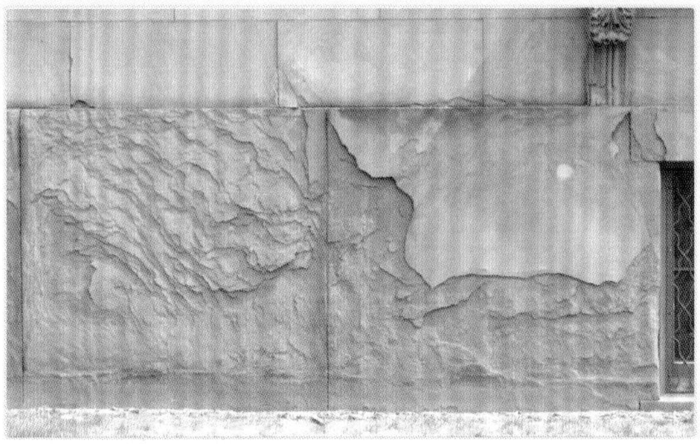

These large blocks of brownstone show significant damage from delamination.

cause the clay binding to swell—especially in cold climates where the water freezes—and this can cause layers of the stone to separate. The surface layer—which, of course, includes any carved ornamentation on the building—may be sloughed right off.

Many brownstone buildings suffer from both acid rain damage and delamination, each condition accelerating the other.

Acid rain damage to brownstone (and other soft stone) exteriors has been addressed in several ways since the 1970s. Various coatings, colored to match the stone, have been tried with greater or lesser degrees of success. Some historic quarries have reopened, allowing repairs with matching stone to be made on some houses. Other restorers have replicated damaged elements in wood and finished them with sanded paint to mimic the appearance of brownstone, an approach with historic precedent, as it was sometimes used on the secondary elevations of brownstone houses when they were built.

New York City has America's greatest concentration of historic brownstone buildings, many of which are in local historic districts. Work on the exteriors of these buildings must meet the city's historic preservation standards, creating a need for guidance that the city has addressed by specifying a recommended approach. In short, the procedure is to remove delaminating or loose stone by chiseling back to solid stone, then rebuild the surface with a cementitious material—duplicating all missing carved detail by sculpting the cement mixture—and finish with a color-matched coating that replicates a brownstone surface.

The experience of New York City cannot be rivaled elsewhere, and their Landmarks Preservation Commission is a recognized authority on the subject. It is worth quoting their required approach to restoring, or refacing, sandstone (including brownstone) buildings in more detail:

Recommended Sandstone Repair and Resurfacing Specifications

These specifications can be found in Section 2-14 of the Landmarks Preservation Commission's Rules, available at: http://www.nyc.gov/landmarks. Please note that these specifications also apply to the most common stone surfaces found in New York City, such as limestone. While there is no specific rule for resurfacing or repairing other masonry surfaces, LPC's staff may approve proposed restoration methods that are consistent with best preservation practices.

The most recent research on sandstone repair recommends taking the following steps when patching or resurfacing sandstone.

Patching Recipes Slurry Mixing Tips

1. Prepare the surface: Cut back all deteriorated surfaces to be repaired to a sound base with a toothed chisel to remove all loose stone and provide a rough surface.

2. Key the surface: To create a mechanical key or holding mechanism for the patch, undercut the edges of the patch to form a slight dovetail and drill 1/2-inch-diameter holes ½ inch deep, spaced two to three inches apart in staggered rows. The angle of the holes should be varied.

3. Apply the patching material using the following five steps, recipes, and mixing tips:

- Wash the surface: Wash the prepared surface with water and a soft brush.
- Apply the slurry coat: Apply with a brush and rub vigorously into the surface.
- Apply the scratch coat: Press the first scratch coat into the slurry coat while the slurry coat is still moist. Each scratch coat should be scored before initial drying to provide a key for subsequent coats. No coat should exceed 3/8 inch in thickness. Scratch coats consist of the materials in the mix below, mixed by volume.
- Apply the finish coat: Apply finish coat after the patch has been built up to the required thickness. Only this last coat is formulated to match the color and texture of the stone being repaired (see below).
- Apply the surface finishing: Finish the surface to match the original stone tooling or existing condition. Possible surface treatments include damp sponging (stippling), dry toweling with a wooden float, and acid etching with diluted hydrofluoric acid, all executed while the patch is partially cured to leather hardness.

Patching Recipes

SLURRY COAT

1 part white Portland cement

2 parts type S lime

6 parts sand

Mix with water

SCRATCH COAT

1 part white Portland cement

1 part type S lime

6 parts sand

Mix with water

FINISH COAT

1 part white Portland cement

1 part type S lime

2-3 parts sand

3-4 parts crushed stone

Dry pigments

Mix with water

Slurry Mixing Tips

All measurements are parts by volume:

- All ingredients should be combined dry and then mixed with potable water.
- Use dry pigments (natural or synthetic stable oxide pigments) when crushed stone is not sufficient to give a color match, and be careful not to exceed recommended maximum amounts, as too much pigment reduces strength and will give unstable color.
- The best brownstone patching contains actual crushed stone, and you may want to consider using stone removed from the area being repaired or old stone with the same qualities. The crushed stone should be ground and passed through a 16-mesh screen, and washed thoroughly.[3]

Custom patching mortars can also be sourced from Cathedral Stone's Jahn Restoration Mortars or Edison Coatings' Custom System 45.

The Morse-Libby Mansion, known as the Victoria Mansion, in Portland, Maine.

One of the most historically significant brownstone houses in America is the Morse-Libby House, or Victoria Mansion, in Portland, Maine. The Italianate

[3]https://www1.nyc.gov/html/lpc/downloads/pdf/pubs/App_Guide_Restoring_Arch_Features.pdf

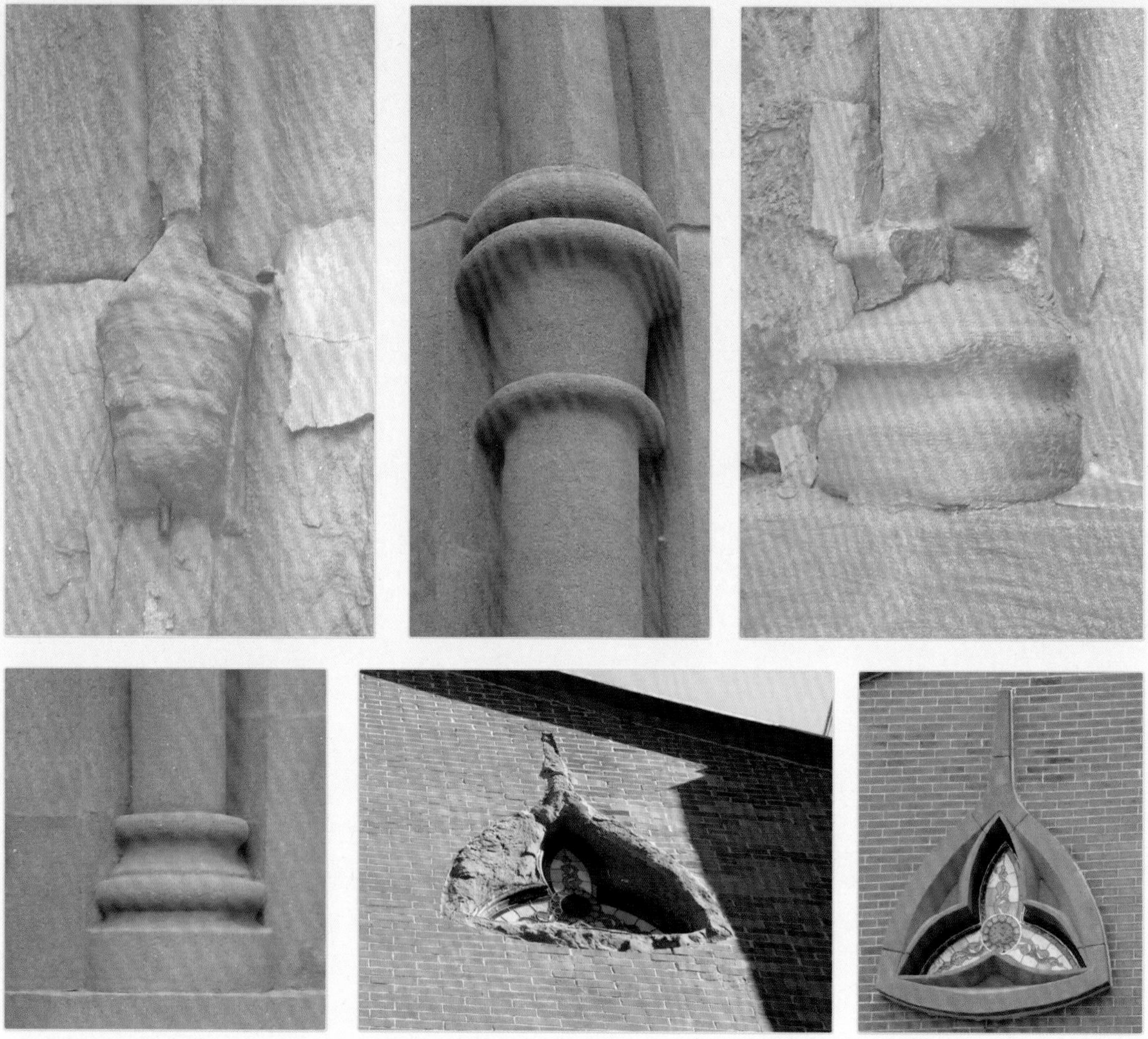

These before-and-after photos show significantly deteriorated carved brownstone trim elements restored with Cathedral Stone's Jahn Restoration Mortars. In the hands of a skilled craftsperson, products of this type can achieve remarkable results. *(Courtesy Sutherland Conservation & Consulting)*

Villa–style house was designed by New Haven, Connecticut, architect Henry Austin and constructed between 1858 and 1861. It contains the only surviving intact interiors by noted German-American designer Gustav Herter. It was Herter's first commission in the United States, before he was joined by his brother to form the Herter Brothers design firm in New York City. The house has been operated as a museum since the 1940s, and its caretakers have extensive experience with various approaches to restoring damaged brownstone. The Connecticut quarry where the mansion's stone originated was closed for many years, and

A section of the Victoria Mansion porch replicated in mahogany and painted with sand paint to mimic the original brownstone. The inset shows a close-up of the sand-painted wood (left) against the brownstone (right).

An unrestored area of corner quoins with carved vermicelli ornament on Victoria Mansion.

no matching replacement material was available. In the 1980s, the front porch and other eroded architectural elements were replicated in wood and finished with sanded paint, an approach that had been used originally to imitate brownstone on the rear porch of the house.

By 2004, when a restoration of the eroded and delaminating stone on the tower of the Victoria Mansion was undertaken, the original quarry had reopened. With matching stone available, EYP (Einhorn Yaffee Prescott, an architecture/engineering firm with historic conservation expertise) of Boston employed a hybrid approach. Some badly eroded or delaminated stone was removed and replaced with matching pieces of new stone, while several experimental methods were used to reattach loose carved elements on the quoins with ceramic pins and stainless-steel bone screws. The Architectural Conservation Laboratory (ACL) of the University of Pennsylvania researched the comparative strengths and weaknesses of these methods. The project demonstrated that replacement in kind of deteriorated masonry elements and reattachment of loose detail can both be done successfully.

It should be clear by now that brownstone restoration is not typically a do-it-yourself project. As a homeowner, however, it is important that you understand the causes of the deterioration and be knowledgeable about appropriate repair methods and materials. Many cities do not have New York City's protections, and in the hands of a sloppy contractor you could end up with a miscolored, smeary coat of stucco that obscures more detail than it restores and makes your house look like it is melting. The only

Restored area of corner quoins and newly carved replacement brownstone window trim.

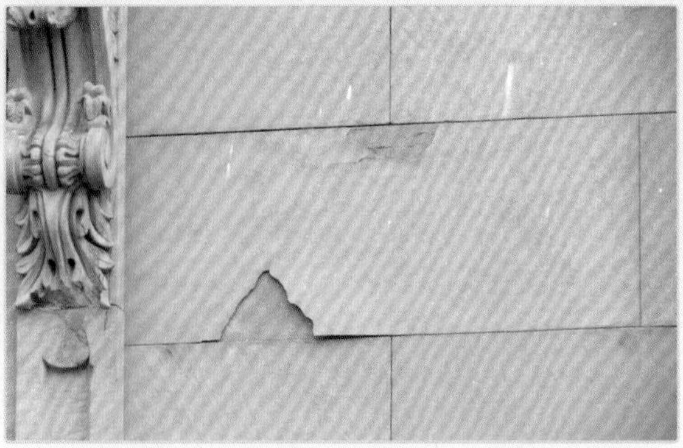

Unrestored area with delaminating blocks of brownstone.

Detail of restored quoin with reattached and replaced vermicelli ornament.

Restored area with replaced block of brownstone and newly carved replacement trim elements.

A view of the west side of the tower after the brownstone restoration project.

saving grace of a bad stucco job is that it won't bond well to the deteriorated surface of the stone and will eventually fall off. Find a reputable and well-qualified contractor who can do the work appropriately.

Terracotta and Cast Stone Restoration

Terracotta and cast stone are manmade materials that are used like natural stone, either as a wall covering or as trim on a brick building. As with brownstone, limestone, marble, granite, and other stones used for buildings, terracotta or cast stone is usually a facing on a brick wall rather than a stand-alone structural wall material. Terracotta is a fired clay material, often with a fired glaze finish. Cast stone is made from cement and stone aggregate and often closely resembles natural stone. Both are formed in molds that include any desired ornamental features. Thus, three-dimensional detailing is cast in the mold rather than formed by carving as in natural stone.

Terracotta is usually anchored to underlying brick or a steel frame with metal clips or rods. When the material is used as a decorative trim accent on a brick building, the joints are filled with mortar. When a building is clad with glazed terracotta, the joints need to be pointed with mortar or caulked to keep moisture

Terracotta trim ornaments a window on this Queen Anne-style house.

from getting between the castings and attacking the metal attachments. Water can also cause deterioration of the castings themselves if it penetrates the surface glaze, particularly if it then freezes and expands in the underlying pores.

Cast stone trim is used on this door surround and the window to the left.

This cast stone column base can be restored with appropriate materials like Cathedral Stone Products Jahn restoration mortars.

This cast stone entablature has been damaged by water from a failed gutter above it. While it might be possible to restore this with appropriate restoration mortar, given its exposed location and the probability of water hitting the top surface in the future (even with the gutter repaired), this might be a better candidate for replacement with a new cast piece carefully matched in color and aggregate size and exposure.

Cast stone is attached to a supporting brick or steel-framed structure like terracotta, and the joints are filled with mortar. For all practical purposes, cast stone is used like natural stone, whether to add ornamentation to a brick building or as cladding for the entire building. It is not unusual for a building to be clad with natural stone at the base and cast stone above the first floor.

Terracotta ornamentation is expensive to reproduce, so damaged terracotta is often replaced with custom-made fiberglass pieces today, particularly if the damaged pieces are glazed. Matching the color and sheen of the historic material is as critical as matching the dimensions and detailing. This is work for skilled and experienced experts.

Cast stone is usually replaced with new cast stone, carefully matching the aggregate size and color, dimensions, and detailing of the original concrete pieces. Pieces that are not structurally failing can sometimes be repaired with modern patching mortars like Jahn restoration mortars made by Cathedral Stone Products, Inc. Jahn installers have to be certified by the manufacturer. In the hands of a skilled mason, these materials can produce nearly invisible repairs.

Brick Veneer

In the twentieth century, applying brick veneers of a single-wythe thickness to wood-framed walls became a common way to create the appearance of a brick building. The brick in this instance supports nothing but itself, with the help of metal ties attached to the underlying wood. In essence, a brick veneer wall is siding. This construction looks good and combines the reduced maintenance needs of a brick exterior with a wood-framed wall's ease of insulation. A brick veneer is also considerably less expensive than a load-bearing masonry wall.

Many postwar houses were built with brick veneers on all or part of the exterior. If your house has brick veneer, repointing may be needed. If the metal ties between the brick and the wood wall are rusted, areas of brick may be coming loose. In some unfortunate cases, no air gap was left between the brick and wood, or no weep holes were installed in the lower part of the wall. The frequent result is rotted wood sills and studs from moisture trapped between the brick and wood. Sometimes this condition can be addressed by rebuilding the wall in sections from the interior, this time leaving an air space. In other cases the brick will have to be taken down, the wall structure repaired or replaced, and the brick relaid with an air space and weep holes. Because veneer brick is nonstructural, a do-it-yourselfer may safely tackle repairs or repointing.

When only one wall (or a portion of one wall) of a postwar house is brick or stone, there is a good chance it is a masonry veneer attached to the wood frame of the house.

Adobe

Adobe construction uses large unfired clay bricks protected by a coating of clay or cement on the building's exterior and by clay or plaster on the building's interior. Adapted from Native American construction techniques by seventeenth-century Hispanic settlers in the American Southwest, adobe is only suitable to areas with twenty or fewer inches of rainfall annually. Adobe walls were historically built one to three feet thick for structural reasons and to moderate the daytime heat and nighttime cold in a desert climate. Wood timbers are used for lintels above door and window openings, which tend to be small. Round logs, called vigas, traditionally supported a flat roof of clay shoveled onto smaller branches or boards laid across the logs. Openings in the parapet surrounding the roof allowed water to drain through wood troughs called canales. These were designed to carry water away from the water-soluble, stucco-covered wall.

Craftspeople still work in adobe in the Southwest, with some adaptations for modern building codes.

New adobe bricks made of clay and straw, covered to protect them from rain.

The unfired clay bricks are readily available. Steel reinforcement is often incorporated into walls today, and flat roofs are covered with modern membrane material rather than clay. Do-it-yourselfers can tackle basic repairs to adobe construction, though unstable walls or roofs will probably require input from an architect or structural engineer. Like all masonry, adobe is heavy, and falling adobe has the potential to cause serious injury.

The National Park Service has an excellent Preservation Brief on *Preservation of Historic Adobe Buildings* found at https://www.nps.gov/tps/how-to-preserve/briefs/5-adobe-buildings.htm.

Stucco

Stucco coatings are applied to brick, stone, and wood-framed houses as well as adobe. Early stucco was lime-based in much of America outside the Southwest, but since the early twentieth century it has usually been a cementitious compound (using either natural or Portland cement) applied by trowel, often over a metal lath that is nailed to the underlying structure, particularly on wood-framed buildings. If moisture has penetrated the wall to any significant degree, this lath or the nails holding it in place may corrode and fail. Sooner or later, this will cause the stucco to separate from the building and start falling to the ground. Even if the lath is in good condition, portions of stucco may loosen or be damaged from other causes.

Identifying the cause of the damage is the first step toward halting and repairing it. Is the corner of the house next to the driveway vulnerable to being hit

These craftsmen are applying stucco, the traditional finish surface used to protect adobe bricks.

Stucco was a popular exterior finish for Mediterranean Revival, Craftsman, Tudor Revival, and Southwest-inspired styles.

by cars? Is a tree branch hitting the side of the house in high winds? Is failed flashing allowing water to penetrate around a dormer window? Identifying the makeup of the stucco and understanding how it was applied is the next step.

If stucco is a commonly used material in your region, you may be able to find a plasterer or mason with enough experience to know the typical mixes used historically. If no such expert is available, a sample of the stucco can be analyzed by a lab to identify its ingredients and proportions. You want the composition of any new stucco to match the old as closely as possible so it will expand and contract with the historic mortar. Color and texture must also match well enough that the repair does not stand out, though this is less important if the historic stucco is painted.

If you're painting stucco, mineral paints may work better than modern film coatings because they bind chemically to the surface and impart a matte finish that resembles uncoated stucco. Many colors are available. Historic stucco is often unpainted and merely needs to be cleaned. Cleaning solutions for historic masonry are available but should be applied cautiously so as to avoid unintended chemical reactions with the lime, cement, or other components of the stucco.

If the damage is extensive, it is wise to call in a professional. Most masons learn their trade under experienced tutelage, starting with simple tasks and slowly gaining experience in the work over a period of years. The learning curve for a do-it-yourselfer is steep. Some homeowners learn to be decent masons. I know one who became a noted restorer of historic masonry after discovering he loved the work, but it took years.

The National Park Service's Preservation Brief on *The Preservation and Repair of Historic Stucco* can be found at https://www.nps.gov/tps/how-to-preserve/briefs/22-stucco.htm.

Caulking

No matter what masonry material is used, it will need waterproof joints where it meets doors, windows, the roof, and applied trim elements. Most commonly, these are made of wood or metal. Good-quality caulking with appropriate backer rod should be used in all joints. Not all caulking types are appropriate for all materials, and new products come along with regularity. Many contractors prefer polyurethane caulks for masonry applications, while others prefer silicone. Neither of these can be painted. Latex caulking can be painted but is generally considered inferior for longevity. Research what is best for your circumstances. The caulk you choose should match the mortar color.

Sometimes lead or copper flashing is incorporated into the masonry to seal joints, particularly on top of windows and doors if there is wood trim extending beyond the masonry surface. Recessed windows and doors commonly have a "brick mold" molding that joins the frame to a relatively even masonry surface such as brick or stucco. Do not use caulking in place of missing mortar.

Attachments

When it is necessary to attach a heavy structure such as a portico or porch roof to a masonry wall, through-wall fasteners with interior bearing plates should ideally be used. If this is impossible, the attachment should at least be made into the inner wythes or backing brick of the wall. Do not rely on a face brick or veneer stone surface to support a heavy load that has the potential to pull away from the building. Lighter loads—such as shutters, gutters, downspouts, and house number signs—always attach into the mortar joints, not into the brick or stone.

Wood

The methods and techniques used to repair, replace, and maintain historic wood siding are readily available in other books, including Michael Litchfield's *Renovation: A Complete Guide*. Here we'll focus on why the details matter so much when undertaking a sensitive rehabilitation/restoration as opposed to a "fix-it-up" renovation.

Identifying the Materials

Wooden buildings are typically covered with a siding that is fastened either directly to vertical framing elements or to an intervening layer of board sheathing.

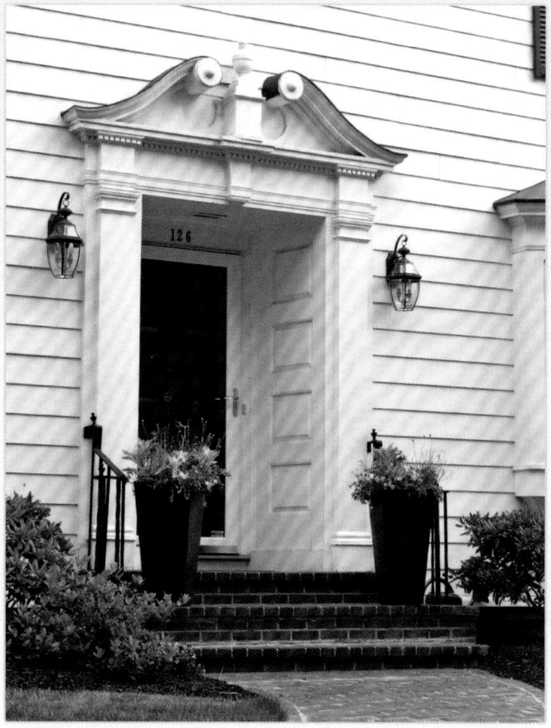

Clapboards, the most common wood siding material, are cut to taper across the grain from bottom to top and are installed horizontally, with each course overlapping the course below. From the seventeenth through the early twentieth centuries, clapboards were typically 6 to 8 inches wide; wider clapboards found some use in the later twentieth century. The amount of clapboard exposed in each course varies with the locale and period of construction. Early houses tended to have tightly spaced courses, sometimes with only an inch or two of exposure at the bottom of the wall, increasing to three or so inches from the height of the first-story windowsills to the top of the wall. Later buildings may have a consistent 3- to 4-inch exposure from sill to eaves. Clapboard exposure is a character-defining feature and should be replicated when historic siding is replaced.

Narrow clapboards were common in New England in the seventeenth and eighteenth centuries. The clapboards on this eighteenth-century house are nailed with hand-forged nails.

Before the mid-nineteenth century, clapboard ends were often joined with scarfs rather than butts; the builder used a draw plane to taper the ends to

These recent replacement-in-kind clapboards are scarf jointed, as can be seen in this photo. Ideally, the joints would be more staggered on adjoining courses, but sometimes it doesn't work out that way. These clapboards were primed on all sides with alkyd primer before installation, which is always recommended.

The façade of this 1820s house has flush board siding to resemble stone. The sides and back are clad in less formal clapboard. This was not uncommon.

overlap, thus creating a wider joint that was more water resistant and could be attached with a single nail through both boards. Later, butt joints became the norm in much of the country.

Flush board siding was used on more formal houses, at least for the front facades. Often the flat-faced boards have beveled edges so that each overlaps the one beneath, permitting water to drain by gravity if the joint opens up. Shiplap siding is another form of overlap, as is tongue and groove. Because this siding was intended to appear monolithic, like a smooth ashlar stone building, minimizing the appearance of the joints was often the goal.

In one variation on flush board siding, the boards are grooved to look like stone blocks, often with projecting quoins at the corners. Hiding joints is not the goal with this type of siding, and vertical grooves are made at intervals to mimic joints between blocks. This treatment most often appears on Georgian, Federal, Gothic Revival, Greek Revival, Italianate, and Colonial Revival houses.

Shingles are another common wood siding material on historic houses. Shingles are boards that taper from bottom to top, parallel with the grain, and are installed vertically in horizontal rows. As a rule, they appear on less formal houses or less visible elevations than clapboard or flush board siding. It is not uncommon to find shingles on ells or sheds attached

This Williamsburg, Virginia, building has board siding grooved to resemble ashlar stone block.

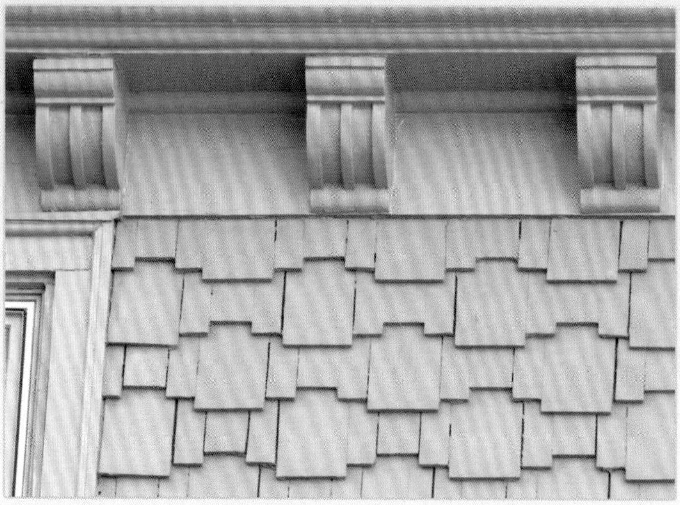

Decoratively shaped and patterned shingles were frequently used for portions of Queen Anne and Shingle Style houses. Simpler, square-cut shingles were used on rural houses in the seventeenth and eighteenth centuries and on outbuildings into the twentieth century.

These shingles were installed in the 1960s to replace worn original shingles on this 1896 Shingle Style cottage. With no paint to protect them in an exposed coastal location, they are ready for replacement again.

to houses with clapboard siding. In fact, it is possible to find houses with formal flush boarding on the front façade, clapboard on the sides and rear of the main house, and shingles on the ell. In their playful use of historic materials, some Colonial Revival designers used shingles on the facades of rather formal houses. After the mid-nineteenth century, the bottom edges might be cut into shapes to create patterns on the wall. Shaped shingles are particularly common on Shingle Style and Queen Anne houses.

Novelty siding is a generic name for a variety of milled siding profiles, typically with tongue-and-groove edges. These range from smooth, flat boards with tapered edges to form a V-groove at the tongue-and-groove joint to false "log" sidings. These materials appeared in the late nineteenth and early twentieth centuries and may be a character-defining feature of houses from those eras. Twentieth-century

These new red cedar shingles are replacing the worn 1960s shingles on the cottage shown in the previous photo. Shingles generally have a shorter expected useful life than clapboard, in part because they are more often left to weather or treated only with stain, and not paint, which would protect them better.

This mid-twentieth-century asphalt shingle siding with a stylized faux woodgrain is as character-defining for a house of that period as narrow wood clapboard is to an eighteenth-century house. Finding new materials of this type can be challenging. Your nearby big-box home-improvement center is unlikely to carry it, but diligent searching on the internet may turn up a source.

houses may have more unusual materials, such as asphalt shingle siding. These may prove more challenging to replace in kind than some much earlier materials simply because a restoration market for them has not yet become established.

Whatever the historic siding material on your house, it is very likely a character-defining feature and should be preserved or replicated closely if it cannot be saved. The proper replacement of siding requires matching the dimensional elements of the historic material and the details of its installation. Replacing historic wood clapboard having a 3-inch exposure with new wooden clapboards having a 4-inch exposure will change the character of the building. Getting the dimensions and details right gets more complicated if you are using a substitute material, as detailed below.

Assessing the Siding Condition

Assessing the condition of a wood siding is straightforward. Be aware that the condition can vary widely between walls and even between the top and bottom of a given wall. Start with any areas of peeling paint, often a sign of wet and possibly deteriorated wood. Use an awl to probe the wood; if it penetrates easily to a depth over 1/16 inch, the wood is probably soft and rotten and ready for replacement.

Rot often shows up first in inside corners where wings or ells join the main house. The junction of the roofs will channel a lot of water into a corner like this unless prevented by well-functioning gutters. Rot is also more likely in siding that's screened by shrubbery. Shrubs that are too close to the house prevent the siding from drying between rains; they should be moved at least 3 feet from the foundation.

Work your way methodically around the house, using a ladder to reach higher areas if necessary. Siding deterioration is less common in the higher portions of walls except where trees or vines are preventing moisture from evaporating; or bay window roofs, balconies, or other projections are allowing rain to splash onto the siding; or snow accumulates on a horizontal surface adjacent to the siding.

In-Kind Siding Replacement

Rarely is a house siding so deteriorated that it must all be replaced. Contractors sometimes recommend

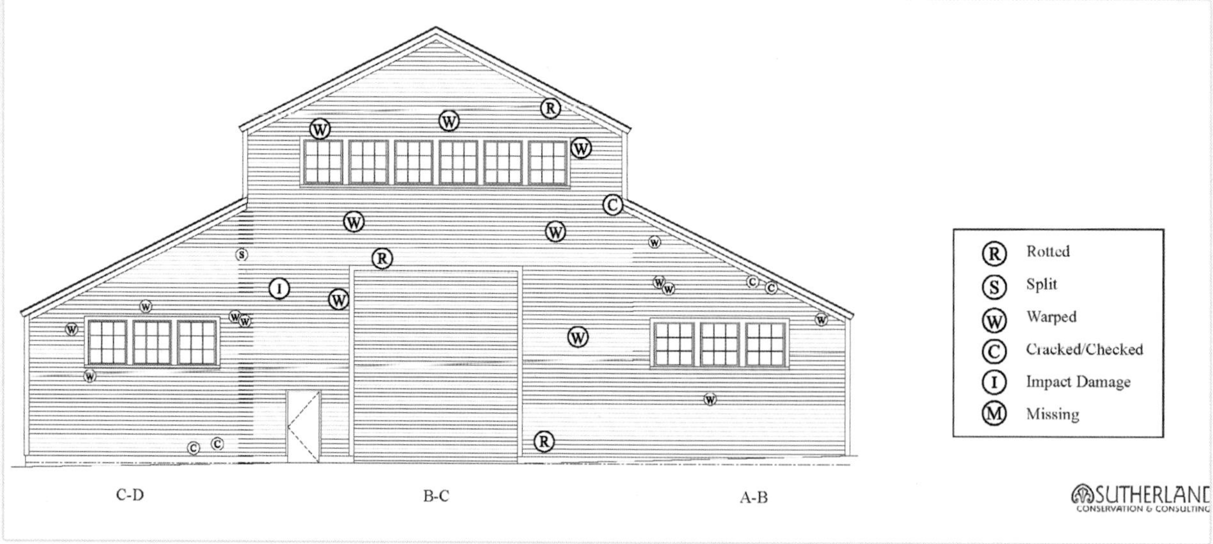

This drawing is taken from a report our firm did to document the condition of the siding on a wood building constructed in the 1940s. The on-site assessment was conducted from the ground and from a lift, marking the condition of individual siding boards onto elevation drawings with a pen. These presentation drawings were made from those on-site worksheets. For most purposes, the worksheets suffice to give a good understanding of existing conditions. Lacking elevation drawings, you can make your notes on a photo of the wall printed on paper in grayscale.

this, but selective replacement of deteriorated pieces is preferable whenever possible. Shingles may be the exception, as they are prone to splitting over time and were often stained instead of painted, providing less protection from the elements (particularly ultraviolet light from the sun).

Most readily available modern replacement siding is inferior to the original material. If you have to replace clapboards, locate radially sawn new clapboards. This method of cutting triangular clapboards from a round log creates surfaces with the edge grain of the wood facing out on the wide sides of the board. This is a denser surface than flat-sawn clapboards have. If possible, buy red cedar clapboards. Wood shingles should be replaced with red or Alaskan yellow cedar. Modern white cedar lacks the density and longevity of old-growth cedar or even pine.

When replacing wood siding (or trim), *always* prime all surfaces before installation, including the cut ends. This keeps moisture from migrating into the wood from the back side or ends and goes far toward making up for the difference in quality between old-growth and modern wood. Shingles are usually dipped in stain before installation. Lengths of aluminum gutter wrapped in chicken wire, mounted with drywall screws to a wall covered with plastic sheeting, and pitched toward a bucket make a good drying rack for stained shingles. Dip the shingles in stain, place upright on the chicken wire rack over the gutter, leaning against the wall, and let the excess stain drain back to the bucket.

With any replacement siding material, it is important to match the dimensional qualities of the original as closely as possible.

Removing Inappropriate Siding

It is not uncommon to find a great old house that has had its exterior covered with a manufactured siding material at some point in the past fifty years. A plethora of cover-up materials have been used since the mid-twentieth century in the effort to achieve a "maintenance-free" exterior. These include asphalt, asbestos, stamped metal, fake brick, fake stone, and vinyl. Rarely do these sidings successfully imitate their natural counterparts. In general, it is preferable to remove them and return a house to its intended appearance. Fortunately, most were installed with as little work as possible, meaning that historic trim and siding frequently remain behind them.

These materials generally require a relatively flat surface for application. Consequently, portions of trim that stand proud of the wall—including the "ears" of window sills—are often hacked away. Removing the vinyl, aluminum, asphalt, or other manufactured siding will reveal what has been lost. If molding has been removed from a corner board or eave brackets from a frieze, areas of unpainted wood will provide evidence of the width of the missing element. The challenge will be to determine the profiles and details of the missing elements so they can be replicated and restored. If the trim element returned to meet the siding, there will be a profile of the element preserved in the paint or in the historic siding itself, as shown in the Compatible Bay Window Addition example project on page 162. Exterior trim is covered in more detail in the next chapter.

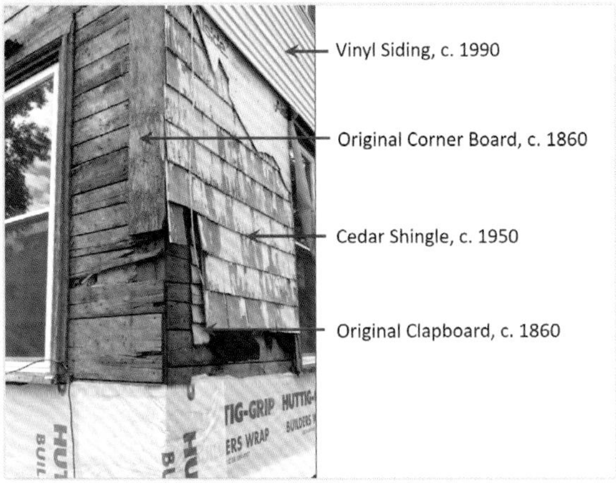

This Italianate-style house of about 1860 is being stripped of its three layers of siding. The original clapboards were covered with cedar shingles in the mid-twentieth century. Shingles covered the corner boards as well. In the 1990s, vinyl siding was installed over the shingles and the window trim, requiring removal of moldings and notching of sills to accommodate it.

New Vinyl?

It may seem odd to follow a section on removing vinyl siding with a section on installing the same material, but sometimes this is a reasonable choice. *How* it is installed can greatly reduce the negative impacts on the historic character of the building.

Vinyl siding can be a reasonable choice on walls that have limited visibility from the public way and are likely to present an unusual maintenance challenge due to regular exposure to harsh weather and difficulty of access. Consider, for example, the ocean-facing side of a cupola on a coastal cottage; cheek walls on dormer windows; a house side that can only be accessed from an uncooperative neighbor's property, or one that stands above an embankment that is too steep to support standard ladders or scaffolding. In such instances, thoughtful installation of vinyl siding may be appropriate, particularly if the historic materials are deteriorated to the point of demanding replacement.

How can vinyl siding be installed thoughtfully? These guidelines will help:

- It must be smooth and not have a fake wood grain embossed on the surface. This stippled surface will always be a giveaway, even from a distance, because light strikes it differently from smooth-painted wood clapboards.

- Vinyl siding comes in various thicknesses. The thinner (and usually cheaper) the material is, the less likely it will be to remain flat once installed. Wavy surfaces are not a traditional characteristic of wood clapboards. Use thicker, better-quality vinyl siding.

- Vinyl has a sheen that is different from a painted wood surface. Paint your vinyl to match the rest of the house siding. With an appropriate primer, exterior paint will bond very well to vinyl siding, and because the material does not have moisture passing through it, like wood siding, the paint is likely to last longer than on wood.

- Use the longest pieces of siding available. Each length is several "clapboards" in height and must overlap the adjoining lengths, and each overlap creates a vertical line that is more visible than a single-clapboard-width butt or scarf joint on wood siding (in which the joints are intentionally staggered from one course to the next). Longer pieces of siding will result in fewer overlaps.

- Vinyl siding is installed with a vinyl J-channel at the edges. This channel keeps water from penetrating behind the siding and allows space for horizontal and vertical movement as the material expands and contracts with changes in temperature, but it is also a dead giveaway, adding a nonhistorical trim element wherever the siding meets trim. This can be minimized by rabbeting a space in the back of the trim for the J-channel to sit in. The siding will then appear to intersect the trim as it does with wood clapboards.

- Preserve historic wood trim in place or replicate it exactly. Vinyl trim elements do not accurately replicate historic trim elements and should not be used. Neither is cladding historic wood trim in aluminum recommended. Molding profiles cannot be accurately copied, and cladding can trap moisture in the wood, rotting it where it cannot be seen and addressed.

- Do not install lengths of vinyl siding tightly against the inner edge of the J-channel, especially in cool weather. The material must have space to expand when warmed by changing weather conditions, and if there is no allowance for this expansion, it will buckle along its length, creating another dead giveaway for vinyl siding.

- Darker paint colors will cause the siding to expand more in sunlight than lighter colors. This becomes more of a problem when the siding is tightly installed end-to-end.

Kowalski-Policht House, 1924

Chicago Bungalow

The Kowalski-Policht House is a classic Chicago bungalow. These single-family houses were built in the early twentieth century as the city of Chicago expanded rapidly across the prairie, swallowing farmland to house its burgeoning population on newly developed lots served by streetcar lines extended from the urban core. More than 80,000 bungalows were built in Chicago between 1910 and 1930. Describing this phenomenon for a National Register historic context statement, architectural historian Daniel Bluestone of the University of Virginia wrote:

> These new bungalow neighborhoods represented a major innovation in Chicago urbanism. Here a new style of house, unprecedented in the previous century, provided Chicago homebuyers of moderate means with extraordinary levels of domestic comfort made possible through innovative systems of heating, plumbing, and electricity. … The bungalow played a crucial role in fostering home ownership among the expanding ranks of Chicago's middle and lower middle classes. Chicago's bungalows emerged as a local appropriation and variant of a house style that was national in scope.

The word bungalow came to America from India by way of Great Britain, where it was first used to describe one- or one-and-a-half-story houses reflecting the aesthetic of the Arts & Crafts movement. In the US, the Arts & Crafts or Craftsman style first took root in California, where wood-framed bungalows were built on large lots in rural and suburban settings. The Chicago bungalow is distinguished by several characteristics that were adopted in response to the long, narrow lots of Chicago's burgeoning suburbs and the rigid fire safety code enacted by the city after the Great Fire.

Usually 25 to 37 feet wide and 125 feet long, with alley access to the rear, these Chicago lots called for long, narrow houses placed close together. In the dense urban neighborhoods of the northeastern U.S., such lots were filled with adjoining multistory rowhouses, but in the streetcar suburbs of Chicago, one-and-a-half-story single-family homes were shoehorned onto the lots. These were nearly always built of brick with limestone trim for fire safety, and the typical bungalow's full-width front porch was replaced with a smaller entry porch to one side, tucked under the main roof. The front wall is often pushed out into a broad bay window to light the living room. The windows are usually tall, with transoms above to maximize the amount of light let in. Colored glass accents are common. A single dormer is often located in the broad, low-pitched hipped roof facing the street. The roof's deep side overhangs may cover bay windows that extend from the side of the house to increase light in the dining room or bedrooms. In plan, the living room occupies the front of the house, beside the entry porch, with the dining room and kitchen extending toward the rear on one side of the house and bedrooms and a bath lined up on the other side. A steep enclosed stair accesses the basement and attic, where bedrooms could be added.

The Portage Park neighborhood where the Kowalski-Policht House is located was developed in the 1920s, largely with similar story-and-a-half bungalows. Most of these houses cost between $5,000 and $6,500 when new. In 1920 the neighborhood was predominantly working class, with nearly 50 percent of the men working in manufacturing, 16 percent in the trades, and 15 percent in clerical positions. The neighborhood's population remained stable through the middle of the twentieth century before declining somewhat for several decades. More recently it has increased again as the popularity of Chicago's bungalow neighborhoods has rebounded and new residents have moved in to restore and preserve the historic houses.

The current owners, Howard and Laurie, owned a Victorian-era frame house for 22 years and knew they wanted another old house when they moved in 2014. They also knew they wanted the advantages of big-city life, including access to public transportation, but did not want to live in a downtown apartment. Chicago's bungalow neighborhoods were the right fit for them, and the Kowalski-Policht House held particular appeal.

The previous residents had done important maintenance tasks in their 43 years of ownership, including having the exterior repointed and the concrete tile roof rehung. The electrical system and some of the plumbing had been updated in the past decade as well. Their remodeling of the kitchen and bathroom in the 1960s, shortly after moving in, had not aged well, but most of the original tile work in those two rooms had survived. Much of the door hardware in the house had been replaced, but original light fixtures survived in the living room, dining room, and front hall.

While structurally sound and largely intact, the house needed attention to its finish details, including twenty-nine art glass windows. The hardwood floors needed to be refinished, including the original maple floor in the kitchen, which was covered by three layers of vinyl flooring. Once uncovered, repaired, and sanded, the floors were given a low-sheen satin finish. Most of the interior woodwork retained its original finish, and although it is somewhat darker than Howard and Laurie would prefer, they touched it up and left it alone. The plaster was largely intact but suffered from extensive water damage in some areas; it was repaired and not replaced. The historic art glass windows were stripped and repainted on the exterior, and four sashes with cracked art glass panes were professionally restored. New storm windows were installed to protect the historic glass and to improve the energy efficiency and comfort of the house. Howard and Laurie received an award from the Historic Chicago Bungalow Association for the restoration of the windows.

Their goal in the remodeled kitchen and bathroom was to restore some of the lost historic character of the spaces without, as Howard stated, "trying too hard with the Arts & Crafts character, as some people do." The prior owners had expanded the kitchen to incorporate the original back porch, and this space was retained for the new kitchen. Some areas of the 5-foot-5-inch-high white subway tile wainscot had been damaged or removed in the previous renovations, and restoring these areas was a high priority. Through a restoration tile expert in St. Paul, Minnesota, Howard and Laurie found a Chicago-area craftsman who was skilled with historic set-in-mortar tile. By salvaging tiles from locations where new cabinets were to be installed, they were able to restore all the visible wainscot.

New reproduction subway tile from Subway Ceramics was selected in green for the backsplash above the new soapstone countertop. Howard and Laurie chose frameless Shaker-style cabinets, considering them to have a feel that is at once vintage yet contemporary. The red birch of the cabinets harmonizes with the finish on the historic woodwork without trying too hard to imitate it. Since Howard is a serious cook, maximizing counter space was a priority, and they also wanted to minimize wall cabinets. They achieved both aims by installing more base cabinets with pull-out drawers and overlying counter space. Fewer wall cabinets allowed for more windows to flood the space with natural light. A new base cabinet with soapstone countertop was installed below the original shelf unit in the pantry.

A small room off the kitchen, originally a bedroom, is now an office and breakfast room. Paint colors tie it to the kitchen, helping them to feel like functionally related spaces.

The bathroom retained its original basket-weave tile floor and period Kohler cast iron tub and porcelain towel bars, soap dish, and toothbrush holder, but the sink had been replaced in the 1960s with an oversized (for the room) cabinet/sink unit. Laurie and Howard updated the bathroom plumbing, reusing the historic porcelain tub/shower control knobs, and replaced the sink cabinet with a compatible new Kohler pedestal sink, creating more space in the room. The classic 1960s Elkay Lustertone stainless-steel lavatory sink from the cabinet was moved to a new half bath in the basement.

Laurie and Howard decided to continue using the original radiators and the boiler installed in the 1950s for heat while installing a separate ducted air-chiller system in the unfinished attic for cooling through the ceiling. This approach avoided vertical duct chases through the house and took advantage of the natural rising of hot air and falling of cold air to achieve more efficient heating and cooling. The heating system was zoned, and new wi-fi–controlled thermostats were installed, though the historic metal thermostat was left on the wall for character. New red birch covers were made for the radiators to replace the existing painted 1950s covers, and traditionally styled cast iron grilles cover the outlets and intakes for the chiller. The attic was already insulated, so they focused on air sealing around windows, doors, and eaves to minimize air infiltration.

Outside the house, the wood windows and trim needed to be repainted, and the face brick and limestone needed to be cleaned of accumulated dirt and a century of soot from coal-burning furnaces. Seeking a safe and responsible way to clean the surfaces, Howard finally settled on using Prosoco Enviro Klean Restoration Cleaner, a hose, a scrub brush, a pressure washer, and a lot of elbow grease. This project was all DIY, with no professional help other than consultation from a local masonry restoration firm.

Behind the house, the original wood-framed garage that opened onto the service alley was in poor condition and needed to be replaced. The simple hipped-roof form of the original structure was replicated, and a double-hung window design used on the back of the house was duplicated for the new garage. Between the house and garage, a 14-foot-square flagstone patio was created to form an outdoor living space off the kitchen. Landscaping around the patio used dwarf species of plants to prevent overgrowth. An urban garden was created in the small front yard using plants that are native to the area.

Favoring a spare aesthetic in furnishing and decorating, Howard and Laurie limited the palette for the house to four colors and the furniture to what is necessary for a functional home. The furniture is compatible but not necessarily antique. For the dining room, they refinished a large 1920s oak table that had come from a parochial school. The surviving original light fixtures were rewired and rehung, and they selected new period light fixtures where needed, some with an Arts & Crafts sensibility. The replaced door hardware was replaced again, this time with salvaged historic pieces. Picking up on the octagonal bay windows, they sought out hardware with an octagonal motif.

With restoration work largely completed, Howard and Laurie are now planning to add a master bedroom suite in the unfinished portion of the attic. They are determined to avoid the insensitive "pop top" upward expansion that is altering the character of many bungalows. Utilizing the unfinished space while meeting modern code and preserving the character of the house is a challenge, and they are taking time to find an architect who can make it work.

The side entry porch is a distinguishing characteristic of the Chicago bungalow.

The bathroom before restoration.

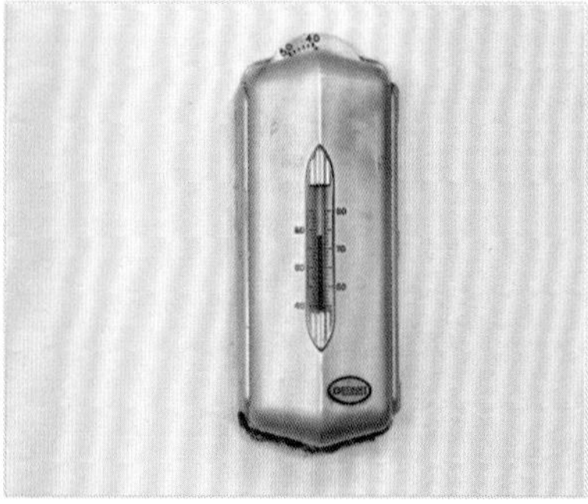

Although no longer functional, the thermostat that controlled the boiler for 90 years was left in place.

The house before restoration.

A pre-restoration interior view.

The obscured glass in the dining room's restored art glass windows disguises the close proximity of the neighboring house. The original light fixture was rewired and returned to its location.

View from the kitchen toward the dining and living rooms. The 5-foot-high subway tile wainscot is original.

The restoration retained the kitchen's original tile wainscoting, hardwood floor, plaster, doors, and windows while adding compatible new birch cabinetry, stone countertops, and stainless steel appliances.

Much of the original door hardware had been removed over time. Howard and Laurie located vintage replacements with an octagonal motif related to the form of the bay on the front of the house.

The kitchen before restoration.

The dining room before restoration.

The new garage replicates the pyramidal hipped roof of the original building.

This small bedroom off the kitchen is used for a home office and breakfast room.

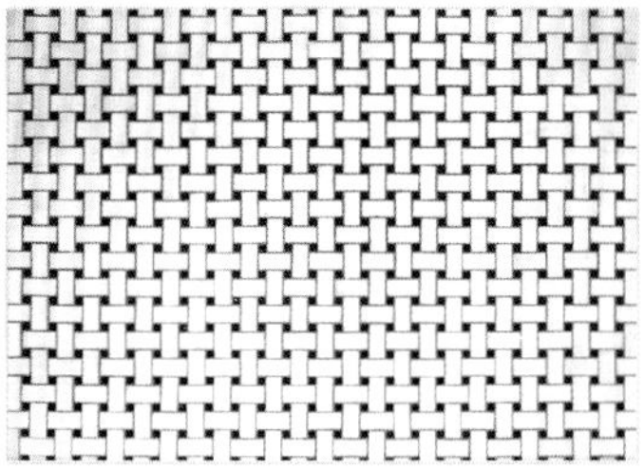

The front façade after restoration. The multifaceted bay window on the front of the house is one of several forms of projecting living spaces found on Chicago bungalows. The concrete tile roof and hipped dormer in the attic are also common features.

The original basketweave tile floor in the bathroom was retained.

In this pre-restoration view of the kitchen, multiple layers of modern flooring are hiding the original hardwood.

The new kitchen features a reproduction green subway tile backsplash.

The restoration replaced the plumbing behind the walls but retained the bathroom's original tilework, tub, and water control knobs.

The front bay. Laurie and Howard won a Chicago Bungalow Association Driehaus Award for the restoration of the home's art glass windows.

This 1920s illustration of a Chicago bungalow shows a rounded front bay. More than 80,000 bungalows were built in Chicago in the early twentieth century. *(Courtesy Chicago Historical Society)*

The front bay before restoration.

A new chiller unit in the attic cools the house, while the existing boiler in the basement provides heat.

The floral motif in an art glass window.

The fireplace alcove in the living room is one of many Craftsman-style details in the house.

Daylight floods the house through the restored bay at the front of the living room.

The Work

Before (below left), during (above), and after views of the exterior face brick, which the owners cleaned using Prosoco masonry restoration products.

CHAPTER THIRTEEN

Exterior Trim and Windows

While usually occupying far less square footage than siding or roofing, the trim, doors, and windows of a historic house are essential to its identification with a time and place. Narrow corner boards speak to the character of a Federal-style house, and wide-paneled corner boards speak to an Italianate style. Small windows with small panes of glass suggest a time when glass was often imported and very expensive. Large walls of plate glass speak to the postwar era when American manufacturing was booming and houses opened to the world outside. Whether one is looking at doors, porches, bay windows, eave brackets, or gingerbread spindlework, these elements are always character defining.

Trim

Wood trim is like wood siding when it comes to maintenance, repair, and replacement, with the additional challenge of matching any molding profiles and carved details on elements to be replaced. Historic house trim is often made of thicker boards than today's nominal 1-inch-thick stock, which is typically planed to ¾ inch. Most lumberyards carry 5/4-inch-thick (five-quarter) boards, which are 1-1/16 inch thick when planed and often match historic 1-inch trim boards. Red cedar and Alaskan yellow cedar are good choices for replacing old-growth exterior trim material, as is Honduran mahogany. Priming all surfaces before installation is important. Properly flashed joints contribute greatly to the longevity of the wood, especially if it is not a rot-resistant species. Do not use pre-primed finger-jointed pine boards for exterior work. They may look fine in the big-box store, but they will not last. I have seen them rot in less than a decade.

Missing Trim Elements

Historic photos can be very helpful for determining the locations, profiles, and details of missing elements. Often the photos will not be as clear or close-in as you could hope, but their shadow lines and highlights can nevertheless provide clues to shape and depth. Incised or carved decoration is often apparent even from a distance. Portrait photos of people outside the house will often show a door or window in the background with clearly visible details.

If there is little evidence of trim profiles left in paint marks or other physical evidence and no historic photos can be located, the next-best approach for determining what would be appropriate is to look for similar houses in your community and copy their details. True "one of a kind" houses are rare, even if

This graphic was developed as part of a campaign to educate the public in Vermont about the long-term effects of incremental change to historic buildings. From left to right, the images show a classic wood-framed Italianate house with a front porch, double doors, window hoods, a bracketed cornice, and an ornamental chimney slowly devolving into a characterless building of no determinate style. This illustrates perfectly why trim elements on historic houses matter. *(Courtesy the University of Vermont Historic Preservation Program and the Vermont Division for Historic Preservation)*

The window trim on this 1860s house was stripped off when the exterior was covered with asphalt siding in the 1950s. Using the 1924 photo shown here and evidence left by paint marks, it was possible to partially replicate the missing trim. Unfortunately, the many brackets on the bay windows proved too costly to reproduce. *(Historic photo courtesy City of Portland, Department of Planning and Development)*

you have an architect-designed house. Architects and builders of old tended to use the same moldings and trim details on multiple buildings, as they do today. If the house dates to before the middle of the nineteenth century, its details are even more likely to match other houses by the same builder, since moldings in that time were created on-site with hand planes, and carpenters had a limited number of plane profiles to work with.

Some local historic district regulation discourage the copying of elements from other buildings out of concern that what might result is a "false sense of history," as the Secretary of the Interior's Standards describes this. It makes sense to discourage, say, a Federal-style door surround from being copied on a Queen Anne house, but can a case be made to copy it on another Federal home? If you're presenting to a local review board, you might characterize what you are doing as "compatible new work based on local traditions" and promise to date all the new materials on their back sides before installation so future restorers can identify what was added.

It is always worthwhile to search the attic, basement, shed, or barn for leftover moldings from the original construction that were saved for future repairs. It is surprising how often they turn up.

Custom milling of molding profiles is covered in the interior finishes chapter. Such details make a difference. If you have purchased a historic house because its character appealed to you, try to retain that character as thoroughly as you can. Replacing rotted 8-inch turned porch posts with 8-inch turned porch posts of a similar but different profile will preserve a home's character much more effectively than 6-inch turned posts or 8-inch square posts. If the originals have a highly distinctive profile, however, custom-milled replacements with an exactly matching profile are the ideal solution. If you can find no evidence of the historic profile and details, opt for simple and compatible new work.

This Federal-style door surround looks okay at first glance, but a closer look shows that its moldings have been altered in a way that degrades its authenticity. When the vertical side boards were replaced (with clearly inferior wood), the two distinct moldings at the transom line were discarded and replaced with a poorly installed single molding that matches neither of the originals. A similar molding was added below the surviving original molding at the top to hide the joint between the new boards and the old work. This change eliminates a degree of refinement in the detail of the primary entrance, where maximum effort would have been expended to make a positive impression in the early nineteenth century. Fortunately, the profiles and locations of the original moldings are documented in the coped clapboards that once met them. The moldings could be reproduced from this information, restoring the door surround to its original appearance.

New brickmold has been copied from a piece of the original by a millwork shop to replace rotted and missing molding on the windows of this building.

Given that custom knives can cost $75 to $200—depending on the millwork shop and the size and complexity of the molding—it may be cost-prohibitive to have a knife cut for reproducing a short piece of molding. As shown in the photos starting on page 467, I have had success casting short sections of molding to solve this dilemma. In the example shown, I needed about 10 inches of a small Federal-style molding to replace short pieces thrown out when a door surround was fitted with a screen door in the mid-twentieth century. Fortunately, the long length of molding that ran across the top of the opening had been stored in an upstairs closet along with other bits and pieces that were removed over the years—one more occasion to be grateful for that frugal "save everything" Yankee mentality.

Epoxy Fillers

Often the best solution for rotted wood trim elements, whether typical or unique, is to repair them with epoxy fillers. You need a filler that's formulated to expand and contract with the wood through the seasons; otherwise, it will come loose eventually. Begin with a consolidant (a low-viscosity resin) that is absorbed like a sponge by the remaining wood fibers around the deteriorated area. This strengthens the surrounding wood and makes a surface to which

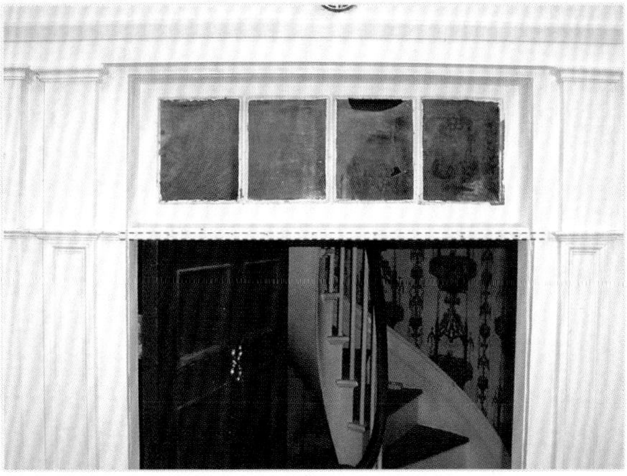

The dashed red lines show where the original molding was removed from this door surround to install a screen door in the mid-twentieth century. The molding was a continuation of the molding that appears on the pilasters at both sides of the door surround.

Fortunately, the original length of molding (shown here) that ran between the door and the transom was stored in a closet and could be returned to its original location. All that was missing were two short sections of the remaining door-surround molding and a small, curved piece of molding that had broken off a half-round pilaster.

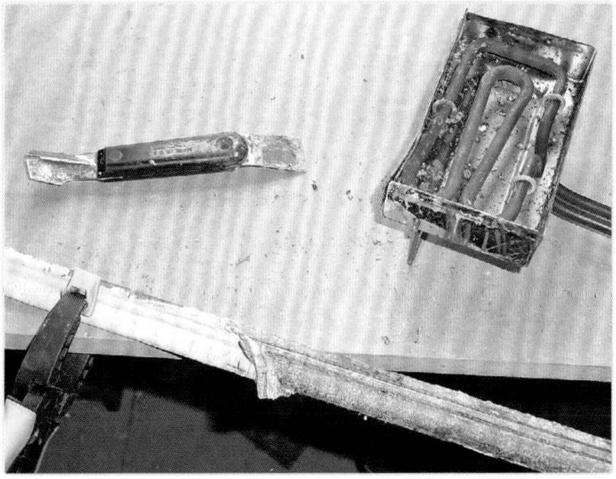

I first stripped the paint off the surviving molding to reveal its detail and dimensions. The door surround had already been stripped of a heavy paint buildup.

Once the original molding was stripped, sanded, and primed, I placed it on waxed paper, surrounded a section of it with a dam of plasticized modeling clay, and sprayed the section with a mold-release agent.

filled epoxy will adhere. Then mix and apply a compatible epoxy filler, which should be viscous enough to retain the approximate shape you need. In this way it is possible to create repairs that retain (or restore) structural integrity to supporting elements as well as to merely decorative trim.

Unfortunately, the filler you need for this cannot be purchased at a big-box home improvement store or at most hardware stores. The leading manufacturer, Abatron, sells direct via mail order and also through major online retailers.

In a disposable double boiler made from a tin can (with pliers for a handle), I melted more modeling clay and poured it over the molding. (Pressing the clay into the molding unheated would have required more pressure and increased the odds of it sticking to the wood.)

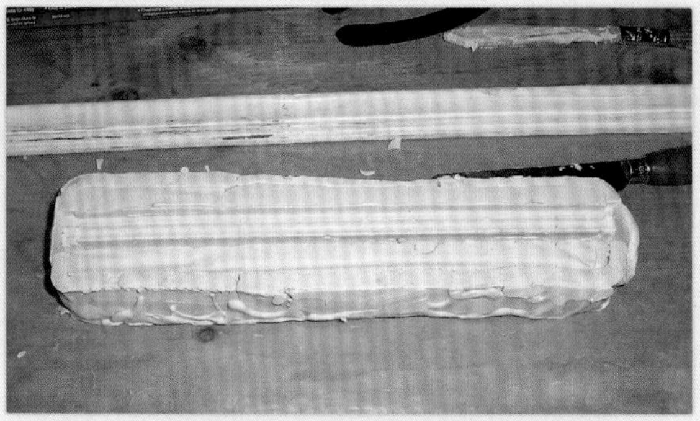

Once the modeling clay cooled and hardened, I lifted it off the molding, turned it over to expose the mold, and sealed the mold's ends with small patches of clay.

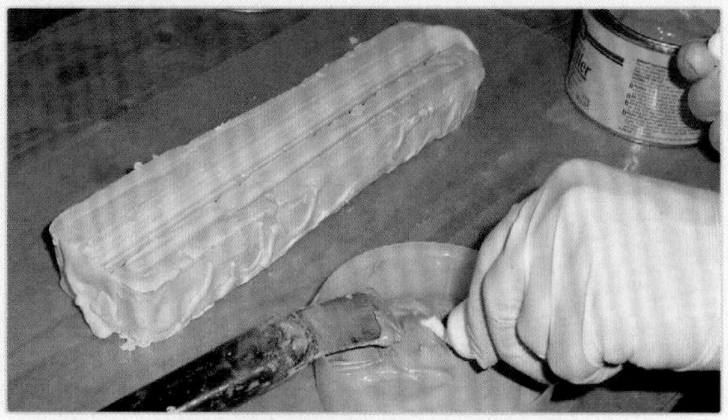

I then cast the replacement molding pieces in this mold using two-part epoxy wood filler, which is being mixed here. Standard epoxy filler is not ideal for most exterior restoration work, as it does not expand and contract with changes in temperature and humidity as specialty restoration epoxy wood fillers do. This product worked for this application. *(Andrew Jones photo)*

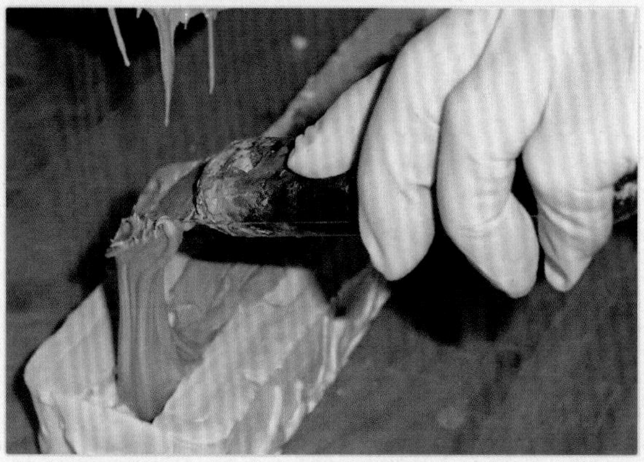

I slightly overfilled the mold with the epoxy. *(Andrew Jones photo)*

With an orbital sander, I vibrated the work board to "settle" the casting material into the mold and help release air bubbles from it. *(Andrew Jones photo)*

Once the casting material was well vibrated, and before it set up and began to cure, I scraped the excess off the top. *(Andrew Jones photo)*

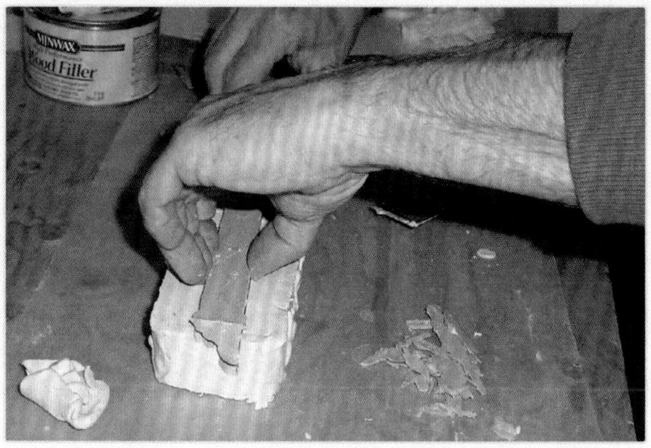

When the casting material cured, I lifted it from the mold. *(Andrew Jones photo)*

The casting wasn't perfect, but it accurately replicated the profile and dimensions of the original molding.

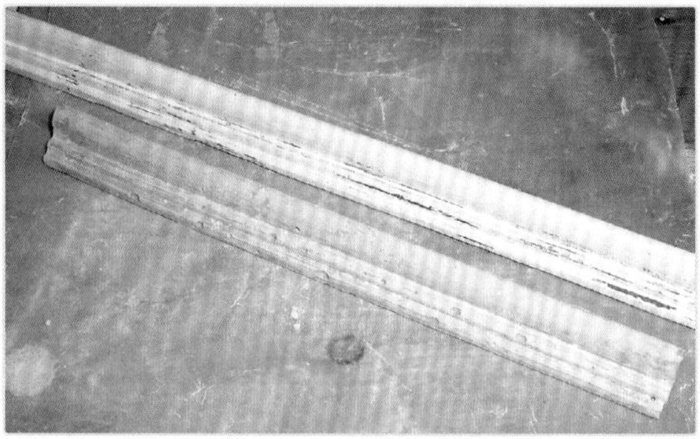

With some trimming of excess material, sanding, and filling of small air pockets with more epoxy filler, it began to look more like it should. Another round of filling and sanding completed the job.

Here the first section of missing molding has been installed between the original molding still in place at left and the original molding returned to its place at right. I cut the miter with a fine-toothed blade on a miter saw and secured the piece with a thin application of the same epoxy wood filler it is made from. The casting material is more brittle than wood and would have had to be predrilled for nails, which I would have done for larger pieces, but epoxy adhesive seemed adequate for these small pieces. After thirteen years in commission, it has caused no problems.

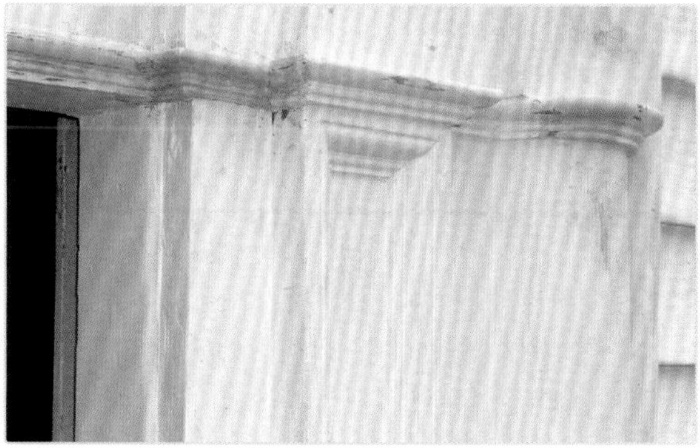

Here is the other side of the door with the replacement pieces in place. To cast the missing curved piece for the half-round pilaster, I formed a piece of the same clay mold around a can of the correct diameter.

The completed project shows off its delicate Federal molding and its shadow line crossing the door surround uninterrupted.

Dutchman Repairs

A more traditional way to repair a hole in a piece of wood is with a "Dutchman" patch—a rectangular piece of wood just a bit larger than the area to be repaired. A same-sized rectangular hole is cut or chiseled around the damaged area, then the patch is glued in place and planed. The patch block needs to be slightly thicker than the board into which it is inserted so it can be planed exactly flush with the surrounding surface. To prevent the patch from sliding through the hole, you can cut the hole with stepped or angled ends

These photos show a repair of rotted wood with a modern epoxy restoration product formulated to expand and contract with wood. The first step is to remove any remaining wood that is loose. Then a consolidant is applied to restore structural integrity to deteriorated wood fibers around the area of loss. Next, the two-part epoxy is mixed and applied, filling the voids and achieving the approximate shape needed. Once cured, it can be sanded to the final shape, then painted to blend with the surrounding trim. *(Courtesy Abatron, Inc.)*

(see the illustration on page 519, in the next chapter). Traditionally, insert holes were formed with a hand chisel, but nowadays a plunge router and a template made from ½-inch-thick MDF can be used, squaring up the corners with a chisel. Angled hole edges still need to be cut with a chisel, however.

This beautifully restored porch was in poor condition, with substantial rot in the floor and in the bases of the four massive columns, as can be seen in the inset photo.

Porches and Steps

Porches can be highly significant character-defining features, particularly for Queen Anne–style houses and other styles that prominently feature porches. Unfortunately, they are also among the most exposed elements and are consequently susceptible to deterioration. Since they are not primary living spaces, they are sometimes considered expendable when resources are limited. Wooden steps are even more exposed and susceptible to the ravages of time, being close to the ground and subject to the wear and tear of shoes and boots.

As with exterior trim, focus on getting the details right when you're working on porches and steps, and be even more aware of where water goes after it falls on a porch. Keeping the porch environs dry—be it with gutters and downspouts, positive grading, or subsurface draining solutions—is the most important thing you can do to give a porch a long life. Pitching the deck to drain away from the house is equally important.

This stair and landing provide a good example of simple and compatible new work where no evidence remains of the original elements.

Adequate airflow under the structure is also important. You don't want to trap moisture rising from the soil. Various types of wood lattice and skirting have historically filled the space between the ground and a porch deck. These materials were often thicker than the 4-foot by 8-foot panels of diagonal lattice available

The first photo shows this porch as it appeared when I showed up to inform the contractor that he was doing unapproved work in a historic district, having ripped out and discarded the original columns and bases on this fine Italianate-style house. Fortunately, the hand-carved capitals of the columns remained. Such visits were never an enjoyable part of my job as a municipal employee. Although the columns were gone, a 1924 tax photo documented their appearance, and the pilasters against the building remained, with base moldings that matched the now-missing column bases. Working with the contractor and owner I was able to locate a source for round fiberglass columns that provided a fair match for the originals, which were unusual in having no entasis (taper) along the length of the column. With these replacement columns set up on bases copied from the surviving pilaster bases, the porch was restored to its correct appearance. Note the simple metal handrail in the center of the wide granite steps, an excellent way to add a handrail to wide steps that did not originally have one.

off-the-shelf at today's home-improvement centers. Making the effort to construct a more appropriate-to-the-period solution can do a lot to add curb appeal and make the age of your house apparent.

Because porches and steps are so exposed and susceptible to deterioration, it may be reasonable to consider substituting synthetic materials for wood elements here. See the discussion below of substitute materials for finished surfaces. If the questions asked there can be answered in the affirmative, substitute materials may be acceptable.

As for the underlying framing, this may be in fine shape in a porch or steps built of quality materials a century ago if the underside has remained dry. If so, there is no need to touch the framing. If the framing is rotted, however, it is often prudent to rebuild it with pressure-treated wood. As mentioned elsewhere, modern woods lack the density and natural resins of old-growth woods and, generally speaking, are less resistant to decay. Pressure-treated wood is significantly more resistant and is suitable for work that will not be visible.

Porch decking is particularly vulnerable to wear and precipitation. The standard off-the-shelf solutions from the big-box home centers are pressure-treated wood and modern composite decking, but neither is great for replacing a historic wood deck. Appropriate woods include cedar, redwood, mahogany, and ipe (Brazilian walnut), but some of these come with concerns about the sustainability of their harvesting in

This fabulous Colonial Revival porch is skirted with traditional square-lattice panels for ventilation. Unfortunately, the plantings are too close to the porch and are trapping moisture that is causing paint to peel.

Unlike dense, resinous, old-growth wood, modern softwood decking is not rot-resistant or long-lasting even when painted, as can be seen in this front step landing that is less than 20 years old.

This newly replaced deck features a modern composite decking material that has been carefully detailed to resemble a wood deck. It has a subtle imitation wood-grain finish that provides slip protection when wet. It would not be my first choice, but it's a reasonable compromise.

tropical rainforests. Pressure-treated wood is the least expensive but also the worst of the choices. In theory, it can be painted or stained after being exposed to the elements for a year, but within a year the surface is generally full of checking cracks, and getting a good-looking and long-lasting finish on it is almost impossible. I prefer one of the better-quality woods that can be finished to a level appropriate to the era of the house. For early houses, the porch decking might be unfinished and left to weather naturally. For Victorian-era houses, it will be painted.

I have come around somewhat to the use of composite decking materials that do not have a pronounced fake wood grain and come in colors appropriate to the house in question. These materials are examined next.

Substitute Materials

Substitute materials, many claiming to be superior to the historic materials they hope to replace, are widely available. Modern materials may imitate traditional

The original cornice had long since rotted away when this rehabilitation project started. Replicating the missing element in wood was beyond the project budget, but the formal Colonial Revival building looked sadly incomplete without it. A solution that fit the budget and returned the building's dignity was to replicate the original cornice form (documented in historic photos and the original architect's drawings) in sheet metal. A brushed paint finish helps to disguise the material. Nearly 30 feet off the ground, it is unlikely to betray itself to a casual viewer. Even in a zoomed-in photo, you have to look very closely to see the rivets at the outside corner joint.

materials or exude a contemporary character that does not attempt to look like another, older material. The latter are generally inappropriate in a historic house. Here we will concentrate on new materials that imitate historic materials. Whenever one of these is considered, several questions must be asked:

1. Is the material dimensionally correct for the period of the house? Many of these materials are thinner than the historic materials they claim to replicate, creating awkward transitions.
2. Does the material have comparable durability to the material it will replace?
3. Is the material's texture or pattern reflective of the period?
4. Is its finish appropriate to or compatible with the period?
5. Will it accept an appropriate finish?

If the answer to any of these questions is no, the material is probably not the best choice.

One modern material that has become widely available and popular for new construction is cellular PVC (polyvinyl chloride) trim boards and moldings. Some manufacturers of this material have made a strong push to sell it in the home restoration market. Note that this is not high-density rigid polyurethane or a PVC-coated product, which have soft open-cell interiors and cannot be cut or molded like wood. This material is a cousin to the PVC piping used in plumbing, and like that material (and wood) it is possible to glue joints. This material is expensive. Honduran mahogany may be cheaper.

The principal claim made for cellular PVC is that is can be cut and shaped like wood yet will not rot. These claims are true, but it is also true that cellular PVC expands and contracts more than wood in response to temperature variations; long lengths and dark colors (which absorb more heat) can be particularly problematic, requiring sizable gaps at the ends to allow for expansion. Many carpenters disguise these gaps with scarf joints and other such approaches, but the gaps cannot be eliminated without causing the boards to buckle when they expand. The gaps must be carefully caulked with paintable acrylic or polyurethane caulk to keep water from getting behind the boards. Avoid silicone-based caulks—they don't adhere well to vinyl or take paint well. Keeping these gaps caulked can be a maintenance issue.

My conclusion after observing this material for a decade or so is that it is a good choice for areas that will regularly be wet, such as trim around porch bases or outdoor steps, and areas where regular maintenance will be a challenge, such as trim on a cupola. If the historic materials are deteriorated beyond reasonable repair, if the cellular PVC can be shaped to match the historic materials, and if it will get a painted finish like wood elements, its use in these circumstances seems appropriate. Because moisture cannot migrate through the material and push paint off from behind (as happens with natural wood), properly applied primer and paint will adhere well and yield a finish that is indistinguishable from wood when viewed from any distance. Some paint manufacturers, including Sherwin Williams, produce paints specifically for cellular PVC. Follow all preparation instructions to get a well-bonded finish.

Like many substitute materials, cellular PVC trim is available with an embossed fake wood grain. Avoid this. No carpenter or homeowner in the past wanted the painted finish on a house to look rough and grainy. The goal of spending hours planing and sanding finished surfaces was to avoid such an appearance. Nothing gives away a substitute material faster than a fake wood grain.

Do Not "Fix" Historic Work

Restorers sometimes have the urge to fix a flaw in the original construction that may not be a flaw at all. For example, I once checked in on a project where an experienced carpenter was replacing severely rotted wood columns and pilasters on a Federal-style portico with custom reproductions. The new columns and pilasters had just been installed when I arrived, and I noticed that the placement of the two pilasters against the building did not align with paint residue from the old columns on the brick wall. When I asked about that, he proudly told me he had fixed a mistake made by the original carpenter in 1805 by centering the columns under the entablature. I invited him down to the sidewalk to look up at the columns. Due to an optical illusion that was understood and adjusted for by the nineteenth-century carpenter, the centered columns appeared off-center, throwing the entire portico out of visual balance. Carpenter's guidebooks of the nineteenth century (many of which have been

In this restoration of a Federal-style portico, the deteriorated original portico has been removed to the contractor's shop to be replicated, and the installation of the framework for the new portico roof is underway.

This close-up shot shows the paint marks where the 1805 roof and pilaster intersected the wall.

This is how the new columns and pilasters appeared when first installed. If you look closely to the left of the pilaster against the brick, you can see the faint remains of the paint mark showing where the original pilaster was located. The new pilaster (and corresponding column at the front) were offset about an inch from where they belonged, having had their placement "corrected" by the carpenter.

Here the new roof trim has been installed and is being supported by temporary props until the reproduced pilasters and columns arrive.

The finished portico with the pilasters and columns relocated to their correct placements. The difference is subtle but necessary to correct for an optical illusion. Ancient Greek builders had figured out how to make this almost imperceptible adjustment, and that knowledge was common among builders into the nineteenth century. Lessons like this are still contained in the builder's guides that were popular then, some of which are available in reprint. Note also the custom restored cast iron railings, a continuation of the fence surrounding the property.

reproduced and merit study) carefully instructed carpenters in matters like this, but most modern carpenters have no idea how to work with the elements of Classical design. The twenty-first-century carpenter moved the columns and pilasters, but he can't have been happy about it.

Paint

The purpose of paint is to protect wood from the elements. Stain serves the same purpose less effectively. Water is enemy number one, but ultraviolet light can

Period advertising can be a good source of inspiration for appropriate color schemes for nineteenth- and twentieth-century houses.

also do significant damage to a wood surface. Old-growth wood is often remarkably resilient when exposed to the sun for long periods, modern wood much less so. In either case, keeping wood painted is the best way to preserve it.

Some masonry buildings were painted as well. Indeed, this practice was common in some regions to protect relatively soft brick exteriors from moisture. Whitewash coatings on brick and stone buildings were also used in some areas, apparently more for aesthetic than practical reasons.

This fabulous Queen Anne–style house is Wilderstein, now a museum house in New York's Hudson River Valley. The house went many decades without paint, as can be seen on the left of this image from 2007. The old-growth wood survived exposure to the sun and weather remarkably well and has been restored with limited replacement. The new paint duplicates the historic colors.

Historic Colors

I will go out on a limb and state that a historic house always looks best in a color scheme from the time it was created. By "looks best," I mean that it looks as it

The Hench House and its neighboring white-painted twin provide an excellent illustration of the difference a period-appropriate color scheme can make on a historic house.

was intended to look by its builder and original owner—which is the point of restoration. That said, I am not among those who want to regulate paint color in historic districts. Exterior paint is temporary, and even an excellent paint job will have to be redone within twenty years. If someone wants to express his or her taste in the color(s) of a house and the result is not particularly sympathetic with the house's style or period, don't worry: It is only a matter of time before someone else paints it differently.

There are excellent books on the topic of appropriate paint schemes for historic house exteriors. Several are listed in Chapter 20 and are well worth the time and expense. An understanding of how paint was used historically can guide your choices and placement of color.

Unlike architectural features, paint is relatively inexpensive and easy to change. For this reason, many local historic district commissions do not regulate paint colors, though some do. Check with your local city hall to find out. In most cases where color is regulated, there is an approved palette to choose from. The identified colors may be made by a particular manufacturer but can usually be mixed to match with any good paint brand.

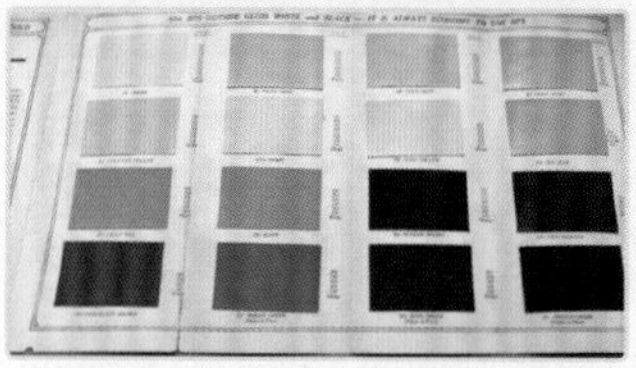

This paint manufacturer's brochure includes actual paint samples that have been protected from UV light. Such samples in historic brochures provide excellent documentation of period colors. Alkyd-based whites behave in the opposite fashion, tending to yellow without UV exposure; leaving such samples exposed to daylight for a week or so will allow them to return to their intended appearance. Note that many of the colors have suggested locations for use on the house, such as "trim," "body," and "porch ceiling."

As a rule, period-appropriate colors will complement a historic building better than this year's "in" colors or the ever-popular white. "Safe" colors are rarely what historic character calls for unless you're dealing with a postwar house. Eighteenth- and nineteenth-century homes often used rich colors on their exteriors. Victorian-era homes sometimes had elaborate paint schemes with multiple colors on various architectural elements. It is easy to overdo this, however. Study period examples to understand how and where different colors of paint were applied.

If you want a period-appropriate color scheme, you will need to research what colors were commonly used for house exteriors at the time in question. Excellent books on the subject include *Paint in America: The Colors of Historic Buildings,* by Roger W. Moss, and *Victorian Exterior Decoration: How to Paint Your Nineteenth-Century House Historically,* by Roger W. Moss and Gail Caskey Winkler. Period advertising can be an excellent source of information about popular colors and their placement on houses after the middle of the nineteenth century, when color printing became widespread. Do not depend on the "Historic Color

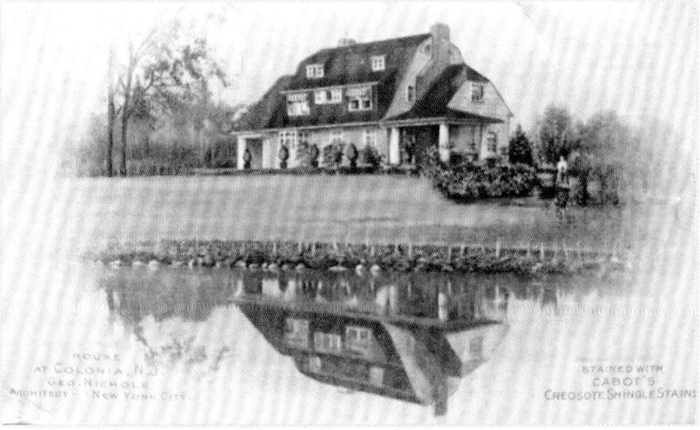

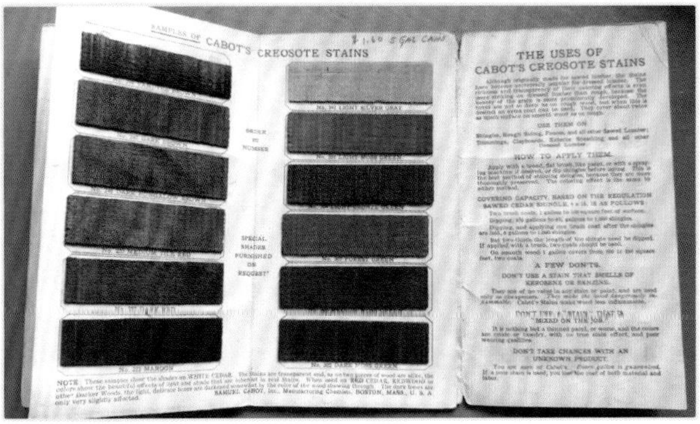

This Cabot Creosote Stain advertising piece contains actual pieces of wood shingle with stain colors applied. Applied exterior paints and stains are often affected by the elements, dirt, and overlying coats of paint and may give false colors when exposed by scraping or sanding. Short of microscopic paint analysis by a lab, an advertising piece like this one can be your best source for understanding the original colors.

This historic photo shows how much more interesting this Italianate-style house was in period colors than its current, mostly white paint scheme. Although the historic photo is black-and-white and cannot tell us what the colors were, we can see that at least three colors were used, and we can see where. With this information alone, it would be possible to select period-appropriate colors of similar tones (degrees of dark and light) and return the house to an approximation of its historic appearance. Microscopic paint analysis could allow a return to its exact historic colors.

Collections" of paint manufacturers, which usually make no distinctions between interior and exterior colors. Do your research elsewhere, then turn to these collections to match the colors you know to be appropriate for where you want to use them. Even black-and-white photos can be helpful for identifying how colors were used on different parts of a building.

In addition to the print and online guidance for selecting period-appropriate colors, museum houses can be another good source, provided their interpretation is based on research and not on the tastes of historical society members. Historic paint research has advanced significantly in the past several decades, and recently restored exteriors of well-run museum houses provide the best opportunities for seeing accurate historic finishes. Many older museums, like Colonial Williamsburg, have worked hard to reinterpret colors based on recent research. Research has shown that the colors of the eighteenth and early nineteenth centuries were bolder and more vibrant than previously supposed.

Paint Analysis

Early historic paint documentation was generally of the "scratch and match" variety. Whatever could be seen of the earliest layer after scraping through the overburden determined what color to restore it to. What this did not account for were the changes that can take place in paint pigments and binders over time. This method was responsible for the inaccurate 1920s restoration colors at Colonial Williamsburg, which became hugely influential in the interpretation of thousands of eighteenth-century homes.

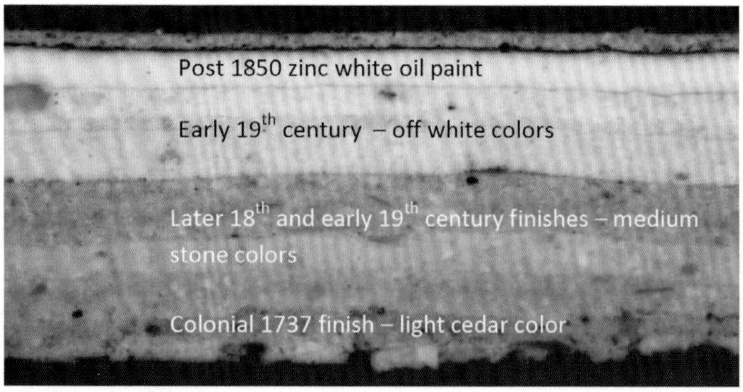

This paint sample under magnification comes from the front door of the John Hancock House, which stood on Beacon Hill in Boston from 1737 until 1863. Its demolition was one of the signal events that led to the development of a historic preservation movement in America. The front door was saved and today belongs to the Bostonian Society. In 2018, the society partnered with the restoration carpentry program at Boston's North Bennett Street School to replicate the door's elaborate surround (demolished with the house), allowing the door to be displayed in context. Our firm was engaged to conduct cross-section microscopy on the door to document its history of finishes. Using the physical evidence of the paint layers and documentary research into the ownership of the house and lives of its residents, it was possible to identify probable dates for the various layers. A change of ownership or marital status is frequently followed by a redecoration or a finish update. Using this information and knowledge of common colors and treatments for relevant eras, it was possible to assign likely dates to the layers. *(Courtesy Sutherland Conservation & Consulting)*

If your goal is to exactly replicate an earlier paint scheme, historic paint analysis using cross-section microscopy can look into the layers of paint left on the surface and identify the colors and makeup of the paint, including the binder and pigments used. This work must be done by a conservator or paint-analysis specialist with the right laboratory equipment and knowledge of historic finishes to interpret what is seen in samples at 100x to 400x magnification. Our firm does this work, and it can be fascinating to see how paint schemes changed over time and to uncover the original design intent.

Differences in thickness of the layers of dirt between paint layers can suggest the length of time between paint jobs. Differences in the number of paint layers on various architectural elements can document changes. If one window frame has fewer layers of paint than another, it may have been added later. If most of the paint layers on a door do not match anything else on the house, the door may have come from another house or it may have intentionally been painted a contrasting color. A good conservator will take samples where the door hinges meet the casing or other trim to look for evidence of colors overlapping from one element to the other. It is generally possible for a skilled conservator to make a clear determination on questions like these.

The most common use of microscopic paint analysis in historic homes is for documenting early colors for accurate restoration to a particular period of significance. Fully documenting the paint history of a house through microscopic analysis is expensive, but using it selectively to understand key elements is less so.

Painting Prep

It is often said that prep work is 50 percent of a good paint job, but I say it is more like 75 percent. If you do not have a good surface for paint to bind to, a paint job will not last.

Working with paints, old and new, is covered in detail in Chapter 18. Paint will last a lot longer on exterior wood siding if moisture is prevented from migrating through the siding from the building's interior. If you see new paint lifting off the siding in long strips, or large areas in which a thick buildup of old paint is flaking off, moisture escaping from the interior might be to blame. Historically, house interiors were much drier than is typical today. Daily showers, cooking, laundry, and humidifiers cause the average house today to produce a lot of moisture that will be drawn toward the exterior whenever the outside temperature is lower than the interior. Modern houses are built with a layer of plastic on the inside of the exterior studs to prevent moisture from getting into the walls, but this is not generally possible in a historic house, particularly if saving plaster is a priority (and it should be). Good ventilation is the best answer for removing excess moisture from a house. Installing and using bathroom vents and kitchen range vents and avoiding excessive use of humidifiers will improve the health of a house and help exterior paints last longer.

Paint Removal

Unlike interior painting, for which washing and sanding are the standard preparatory steps, exterior surfaces are generally scraped to remove loose paint, and the remaining paint edges are then feathered into the surrounding surface by sanding. Sharp scrapers and lots of sandpaper and elbow grease are essential for this work. File your scraper blades frequently to keep them sharp.

Many historic house exteriors have received numerous coats of paint over a century or more of life. Most of the layers will be lead-based alkyd (oil) paint, which continues to harden indefinitely and

This worker is using a heat gun with a diffuser to remove paint from a Queen Anne-style door hood. Since this paint likely contains lead, which may be vaporized by the heat if it gets hot enough, he is wearing a respirator to avoid breathing lead fumes or particles. Lead paint safety is covered in Chapter 18.

A thick buildup of paint layers on the exterior of a house will eventually reach this stage. The chemical properties of linseed oil guarantee it. Paint like this is very easy to remove with a sharp paint scraper. A tarp should be spread under the area being scraped to catch the falling paint chips, and lead-containing paint needs to be properly disposed of. A light sanding with 150-grit sandpaper will make this wood ready for primer. If the wood has been exposed to the sun long enough to become very dry where the paint has already flaked, it may be advisable to apply a first coat of primer thinned 50-50 with paint thinner, and then give it a quick second sanding after it dries before applying a coat of full-strength primer.

eventually becomes brittle and starts to break apart or crackle. Additional coats of paint may bind the crackled paint back together, but ultimately the first layer will begin to separate from the wood and flake off the building. A coat or two of modern acrylic latex paint often accelerates this process, as a latex topcoat expands and contracts differently from the alkyd paints beneath, creating a tension that pulls on the older paint. Once the lowest layers of paint start to break free from the wood, applying new paint layers over the old becomes a losing battle. At that point it makes sense to strip the buildup of old paint back to bare wood and start the process over.

For large expanses of wood siding, the two most common methods of stripping paint are mechanical scraping and softening the paint with heat. Liquid and gel strippers are less often used on expansive wood surfaces due to cost, though they are the only really effective option on masonry walls. High-pressure water sprayers are not appropriate for paint removal on wood, as they will thoroughly soak the wood and make it nearly impossible to get the wood back to an acceptable moisture level for painting before the next rain wets it again. Even in the desert it is unwise to soak wood before applying paint. Sand blasting is completely inappropriate for paint removal on historic wood trim or siding. Just don't do it.

The best approach often depends on the condition of the paint you need to remove. If it is severely crackled and in the process of removing itself chip by

My friend and neighbor Mark has been systematically stripping the exterior of his 1852 Greek Revival house, starting in the gable and working down and around the house. Three years into the project, here he is on the home stretch and able to work under cover on the wraparound porch.

chip, a good, sharp paint scraper and elbow grease may do an effective job quickly. Powered versions of mechanical scraping are available, most using a rotary

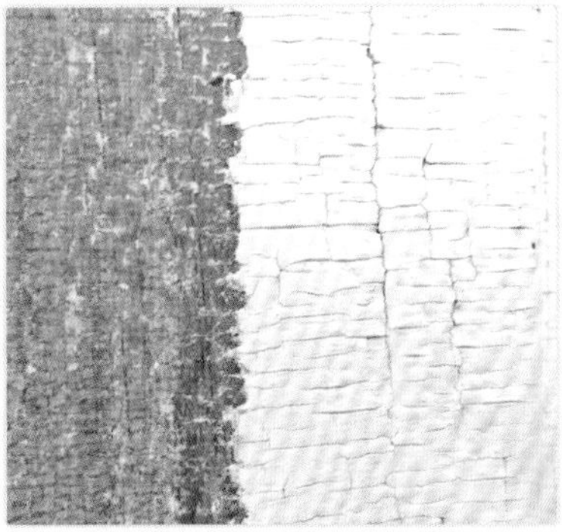

This close-up shows a buildup of many layers of paint that has begun to check or alligator, indicating that the bond to the wood is failing. At this stage, stripping back to bare wood and starting over is the only way to achieve a long-lasting paint job. The good news is that it will be good for another 150 years before it has to be done again.

Mark fabricated the rolling pole rig for working here on the porch, where there is a suitable surface to roll the platform on. It holds the infrared paint stripper unit in place and allows it to be swung back out of the way to scrape the wood once the paint is softened. At the end of a run, the heater unit is moved down the pole and the process repeated.

This wall-mounted sliding support rig was used on most of the house, where the pole rig could not be used. In the peak of the gable and other areas where neither rig could be used, the heater unit was handheld.

The infrared paint stripper unit can be rotated to a horizontal or vertical orientation. Once an area is heated, the unit is moved to the left or right to start softening the paint of the next area while the first is scraped to remove the softened paint. When the heating time is properly calibrated, the paint will come off easily without scorching of the wood.

This transition area from wall to window trim and porch ceiling has been stripped with the heating unit handheld. A sheet-metal shield protects the new paint on the previously restored ceiling.

This section of clapboard and corner board has been stripped but not yet sanded.

The stripped paint falls to the floor and almost immediately hardens as it cools. A drop cloth catches the mess and makes it easy to clean up. Lead paint must be disposed of properly, as discussed in Chapter 18.

This area against window trim is sanded and ready for primer. The nail heads need to be dabbed with a primer made for ferrous metal before the wood is primed. Even with an alkyd wood primer, nail heads can rust through water-based finish coats unless this preliminary step is followed.

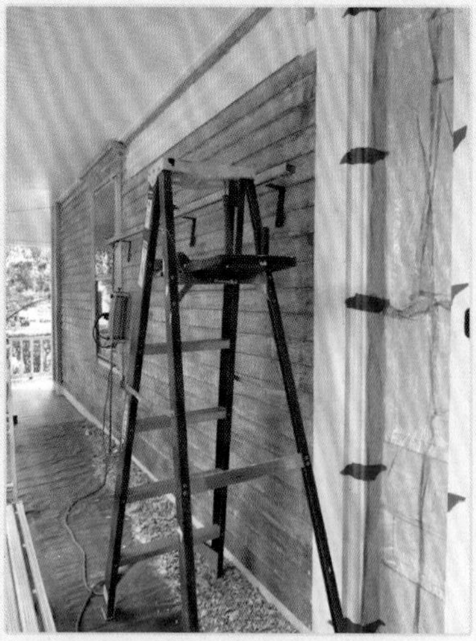

Mark has found that professional-strength soy gel paint stripper is more effective for areas with molded profiles, such as this paneled area beside the door. The plastic prevents the gel from drying before it has a chance to soften the paint layers.

An area of clapboard and corner board after sanding. Note the tight grain of the absolutely clear old-growth pine clapboards and trim. Wood of this quality is almost impossible to find today at any price.

This is Mark's preferred soy gel stripper. Professional-strength stripper is worth the extra effort to locate and have shipped to you if necessary. Unlike toxic traditional paint strippers, soy (and citrus) strippers are nontoxic, biodegradable, and, in short, better for you and the environment. As always, stripped lead-based paint must be disposed of properly.

A stripped, sanded, primed, and repainted section of wall. Mark's next project is to strip, repair, and repaint all the window shutters for the house. He is using a professional steam stripper unit to strip the windows for rehabilitation (as described below), but the shutters will not fit in it. He is considering making a dip tank of Styrofoam insulation and a plastic liner just large enough to submerge one shutter at a time in soy gel and cover it to prevent the gel from evaporating.

Paint can minimize the incongruous appearance of modern elements like the electric meter, and telephone and cable boxes on this house. They can't be made to disappear, but they are less apparent when painted to match the background.

blade. If the built-up layers of paint are still bound to the wood in large areas, however, heat may be the best approach.

Open-flame torches should be avoided if you don't want to accidentally burn down the house you are restoring. Electric heat guns and pads will soften paint so it can easily be scraped off, but these heat only a small area at a time, making it a daunting task to strip a house. Modern infrared paint strippers are designed to heat a large area at once, speeding up the task. The marketing for these units claims that the infrared heat will not reach a temperature that can vaporize lead and release its molecules into the air, but I still recommend using lead-safe practices, including a respirator. This subject is explored in Chapter 18.

An infrared paint stripping head mounted to an aluminum pole rack can strip large areas of wood siding efficiently. The softening paint often falls off the wood at the touch of a scraper, and will quickly become dry and brittle as it cools on a drop cloth on the ground, making it easy to clean up and dispose of properly. The wood should then be sanded lightly and primed for the new paint. Note that infrared strippers may be less effective on moldings and carved surfaces. It may be necessary to switch to a gel-type stripper for these areas.

Remember, while its aesthetic values are important, the most critical purpose of paint is to protect wood from the elements. Good-quality, well-applied primer and paint will answer both purposes best. I have yet to be convinced that a thin coat of paint sprayed on the exterior of a building will weather as well as a thicker coat applied by brush; on the contrary, thin sprayed coats seem to fail much sooner. Labor—whether your own or hired—will be a big part of the investment, so you might as well justify that labor by using a high-quality paint that will last longer. An annual cleaning and touch-up will further extend the life of the job.

Doors and Windows

Doors and windows are important character-defining features of a house and can constitute 25 percent or more of the exterior surface. Historic exterior doors were intended to make a statement about the house and its owner. The front door, in particular, was traditionally the focal point of the façade. Historic wood

windows are designed to be repairable, unlike modern replacement windows. Replacing historic windows should be the last option when considering improvements for energy efficiency. Even good-quality wood replacement windows have an expected life of just 25 to 30 years. Given their high cost, they rarely recoup their cost before needing to be replaced again. Less expensive vinyl windows, with an expected life of only 10 years, are even less likely to pay for themselves with energy savings.

Doors

Preserving doors—especially front doors—should be a high priority in a historic house. Fortunately, like historic windows, stile-and-rail paneled doors were designed to be repairable. The vertical elements (stiles) and horizontal elements (rails) are usually held together with pegged mortise-and-tenon joints. Panels sit in grooves cut into the stiles and rails, with the ability to move slightly in response to seasonal changes in humidity and temperature. Moldings are either formed into the edges of the stiles and rails (typically earlier) or applied and held in place with small nails into the stiles and rails. Simply drilling out the pegs allows all or part of the door to come apart so that damaged elements can be repaired or replaced as needed.

This Italianate entryway of about 1860 retains most of its original elements. Unfortunately, far too many doors like this are being torn out and replaced by stamped steel or fiberglass doors in the name of energy efficiency. Good-quality weather stripping, properly installed, would provide nearly the same benefit without loss of a key character-defining feature of the house.

It is not uncommon for a door panel to become stuck to the stiles and rails by layers of paint, creating tension that cracks the panel during seasonal changes of temperature and humidity. Such panels can often be repaired in place by cutting through the paint seal around the panel with a utility knife (if you aren't

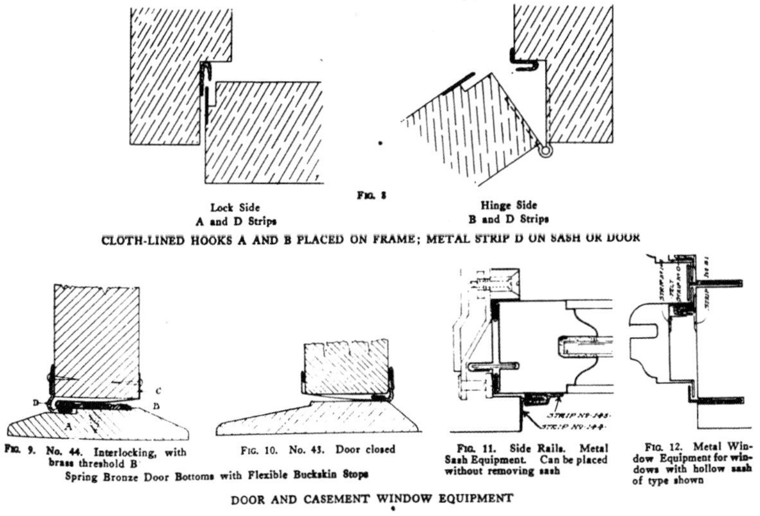

Fig. 8
Lock Side
A and D Strips
Hinge Side
B and D Strips
CLOTH-LINED HOOKS A AND B PLACED ON FRAME; METAL STRIP D ON SASH OR DOOR
Fig. 9. No. 44. Interlocking, with brass threshold B
Fig. 10. No. 43. Door closed
Spring Bronze Door Bottoms with Flexible Buckskin Stops
Fig. 11. Side Rails. Metal Sash Equipment. Can be placed without removing sash
Fig. 12. Metal Window Equipment for windows with hollow sash of type shown
DOOR AND CASEMENT WINDOW EQUIPMENT

This 1913 illustration shows the sort of high-quality metal interlocking weather stripping that can make a historic door as tight as a new one, with minimal impact to historic character. Primarily made in bronze today, this weather stripping lasts a long time.

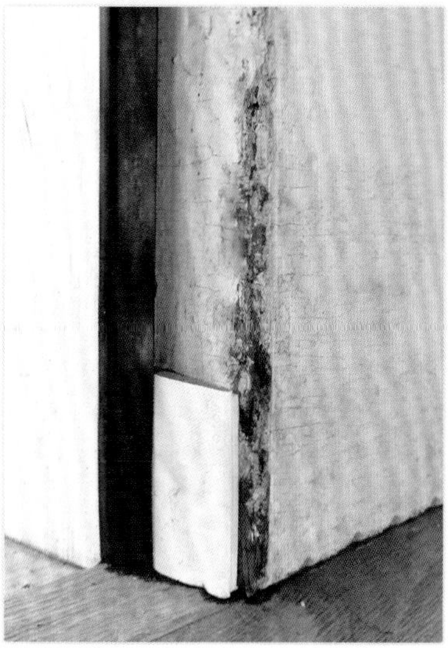

This modern neoprene weather stripping is friction-fitted into a groove cut in the doorstop and compresses slightly to form a seal with the door edge, hidden from view when the door is closed. A companion weather strip is available for the bottom of the door, fitted into a groove cut there. The small foam pad at the bottom of the door return blocks any gap between the side and bottom weather stripping. It will not last as long as bronze weather stripping but is easily replaced when worn out and is a more affordable option.

stripping the paint off the door) and gluing the panel back together as shown on the next page.

Storm doors are a traditional means of increasing the energy efficiency of doors during the cold-weather months. Historic storm doors were frequently rotated seasonally with screen doors, often using matching hinges so the hinge pins could be pulled and the doors swapped in minutes. Aluminum- and wood-framed storm doors with glass panels are available, which keep the primary door visible through the glass. Usually the glass panels can be replaced with screens in the warmer months without removing the door.

Some houses have truly exceptional doors, but most have simpler designs. Even if your house has a relatively ordinary six-panel door, no modern off-the-shelf door is going to look the same. Painted steel does not look like painted wood up close—and a door, by definition, is seen up close. A manufacturer may claim that you can stain a wood-grained fiberglass door to achieve the appearance of a stained and varnished wood door, but it isn't true. You will only get a door that looks like stained fiberglass, and it will probably cost you a lot of money. If your historic door is beyond repair, have it replicated accurately by a millwork shop or experienced carpenter.

This aluminum channel with vinyl bulb weather stripping is the least expensive option and does a good job if installed carefully so the bulb is in contact with the door all the way around. It is also the most visible, however. Be sure to use a white bulb with light-colored doors and trim and a brown bulb with dark colors. The aluminum channel can be painted to match the trim, reducing its visibility. A similar channel with a vinyl sweep seals drafts at the bottom of the door; it screws to the face of the door.

After cutting through the built-up paint around this cracked panel to free it, wood blocks were screwed in place to provide a clamping surface, and glue was spread in the crack, which was then closed with clamp pressure. The clamps were left in place until the glue cured. The screw holes and the vestiges of the crack were then filled and sanded before the door was repainted.

These Italianate doors have simple wood-framed storm doors with glass panels to allow light in and visibility out.

This exceptional Gothic Revival door could not be replaced with an off-the-shelf door without a significant loss to the historic character of the house. The same is true, if less obvious, of more ordinary historic door types. It should be noted that varnished exterior finishes need to be renewed more frequently than paint. Marine-grade varnishes are recommended.

Windows

The key difference between historic and modern wood windows is in the quality of the wood. Windows before World War II were generally made with old-growth wood from trees that had grown slowly in a dense forest environment. Their slow growth resulted in a dense cellular structure with a high resin content that resists decay. Modern wood is grown quickly in managed forests that are kept intentionally thin, allowing light to penetrate and speed growth. The cell structure of this wood is less dense and the concentration of natural resins is much lower, making the wood more susceptible to decay.

This wonderful Colonial Revival window is a character-defining feature, and its preservation is important to the character of the house.

That is why modern wood intended for outdoor use is frequently pressure-treated with toxic chemicals to keep damaging organisms from destroying it quickly. While well suited to structural uses, pressure-treated wood does not hold paint well and is not well suited to finish work. A handful of manufacturers are producing new mahogany-sashed windows with characteristics similar to historic wood windows, but these are very expensive.

The replacement window industry has successfully convinced a large portion of the American public that replacing old windows will make a significant difference in heating and cooling costs. One selling point is the use of "insulated glass" in place of a traditional single glazing (one pane of glass). A double-glazed window *is* more efficient than a traditional single glazing. The "insulation" is a thin layer of air trapped between the two panes, and heat will transfer more slowly through this sandwich of glass and air than through a single pane of glass alone. However, the same advantage can be achieved by installing a storm window over a historic window.

Another marketing pitch is that replacement windows will be much tighter than "drafty" old windows. This, of course, depends entirely on the condition of the historic windows and the quality of the installation of the replacements, which can be just as drafty if not properly fitted and sealed. Broken sash cords

This illustration shows the extreme simplicity of the Unit Sash Balance.

Please note that all the moving parts are encased. There is nothing to rub on the woodwork and it is equally impossible to interfere with the action of the spring by squeezing the balance in a tight mortise.

The Pullman Unit is the only balance having this feature.

The Pullman Unit is the perfect balance. You can consistently specify Pullman Units, knowing that they will give your clients longer service and quieter and better working windows than the old fashioned cords and weights.

PULLMAN MFG. COMPANY, ROCHESTER, N. Y., U. S. A.

Page Six

The Pullman Company, established in Rochester, New York, in 1889, still makes these spring tape window balances. They continue to use their 1938 catalog image since the product is unchanged. Various sizes are made to accommodate different window sizes and weights.

Here, new spring tape balances have been installed to replace cords, pulleys, and weights. The weight pockets were then insulated.

on historic windows are another major selling point for replacements. What is not mentioned is that sash cord is replaceable, whereas the balance systems in the replacement windows are not. Often when historic windows are rehabbed, the sash cord is replaced by brass chain or spring tape balances to eliminate the risk of the cord breaking in the future. Spring tape sash balances have been around for nearly a century and often can be fit into the existing cord pulley opening. An advantage of these balances is that the weight pockets can be fully insulated.

What the advertising departments of the window manufacturers don't tell you is that your historic windows can be tightened up and weather stripped as well, making them just as tight as new windows. When combined with a storm window, a repaired, tight historic window can match the efficiency of a modern replacement window, often at a lower cost and with a much longer life expectancy.

An excellent and widely distributed article on the topic is "What Replacement Windows Can't Replace: The Real Cost of Removing Historic Windows," by Walter Sedovic and Jill H. Gotthelf. It is available on the Washington State Department of Archaeology and Historic Preservation website at www.dahp.wa.gov/sites/default/files/WhatReplacementWindows-CantReplace.pdf and at a number of other online locations found by searching the title. It includes a useful worksheet for comparing the real costs of restored versus replacement windows.

Window Rehab

A window rehabilitation industry has grown up in the United States over the past several decades, consisting of craftspeople who specialize in the repair and rehabilitation of historic wood windows. Combining traditional woodworking techniques with modern methods for removing old lead paint and glazing compounds, they can make historic windows as good as or better than they were when new. For many people, the removal of old lead paint is a highly desirable aspect of window rehabilitation. Because traditional windows were built of replaceable parts that are joined with pegged mortise-and-tenon joints, damaged or deteriorated elements are replaced with parts salvaged from other windows or newly milled from old-growth wood to match.

This homeowner has set up a window rehabilitation shop in his attic and is systematically working through the nearly fifty windows in his house. A sash ready to be stripped of paint to start the process is in the left foreground, and a pair of completed sashes are on the other worktable. The finished window at the left rear is on a handy rotating stand made with a Lazy Susan mechanism. The ability to rotate a sash while painting or glazing is a time-saver.

This photo was taken on a cool autumn morning when the window on the right had been tightened up and the window on the left had not been. The difference in condensation on the exterior storm window clearly illustrates how much more airtight and efficient a traditional wood window can be made with a little time and low-cost materials. This solution is not equivalent to full window rehabilitation and will not last as long, but if money and time are tight, it can increase comfort and save a lot in heating costs.

Most of the tools and materials needed for low-cost window tightening are shown here. A hammer, utility knife, scraper, and small pry bar are essential. Sandpaper, a staple gun, and a screwdriver may also be useful. Neoprene bulb and vinyl V-strip weather stripping are the materials.

Generally, new weather stripping is incorporated into the rehabilitated windows, either traditional interlocking bronze or modern vinyl or neoprene. The rehabilitation of historic windows by these craftsmen can be as expensive as replacement windows, but the resulting window will last much longer. The "Window Rehabilitation in Photos" feature at the end of this chapter shows the process in detail.

While full rehabilitation of historic windows (by yourself or by expert specialists) is ideal, there are less expensive actions that can dramatically improve the operation and efficiency of historic windows. The windows in the accompanying photographs were tightened up and made fully functional with about an hour of time per window and less than $5 in weather stripping materials. These materials will not last nearly as long as bronze interlocking weather stripping, but those in the photos have been in use for eight years now and will be significantly less expensive even after one or two replacements.

Using the pry bar (if nailed) or a screwdriver, remove the window stops from both sides and lift out the bottom sash. If it is weighted, tie the cords around something like a pen to prevent them from being pulled all the way into the weight pocket. Next remove the stops between the two sashes, which are usually either friction-fitted or held by small nails. With the stops out of the way, remove the upper sash, again keeping the cords from disappearing into the weight pockets. This is the time to do any necessary basic repairs to the window, such as replacing cracked or broken panes of glass or deteriorated or broken sash cords. (Information on these basic repairs is available from many sources.) Using the scraper and sandpaper, clean up the areas of the window that intersect other elements, including the sides, bottom, top, and meeting rails. Any paint buildup should be smoothed out. When these steps are done, wipe down the meeting surfaces of the window and install the weather stripping as shown in these photos. Number 1 is the bulb and #2 is the V-strip, both of which come with an adhesive strip for attachment. I usually add staples to the bulb for longevity. When reinstalling the inner stops, make sure they are snug to the bottom sash but not so tight as to prevent it from sliding.

Full paint removal by a do-it-yourselfer is possible using a professional's steam box (these are available for sale) or other paint-removal method. However, if the paint buildup does not interfere with operation and the paint is still well attached to the wood, encapsulating it with new paint will meet lead safety guidelines. The basics of wood window construction and repairs can be found in Preservation Brief #9, *The Repair of Historic Wooden Windows*, by the National Park Service, available online at www.nps.gov/tps/how-to-preserve/briefs/9-wooden-windows.htm.

Storm Windows

Storm windows are a highly effective traditional means of making single-glazed windows more efficient. Wood-framed storm windows came into use in the nineteenth century. In the homes of the wealthy, they were sometimes built in as a second full window sash set within the depth of the wall. More commonly, however, they were added on the exterior over primary windows that had become loose and drafty with age. Because these storm sashes were made after changes in the glass production process made larger panes possible, they often have fewer, larger panes than the primary windows they cover. Special hardware for hanging storm windows allowed them to be installed and removed annually. Some types of hardware allowed the lower end of the sash to swing open several inches to admit air to the house in warmer weather without removing the storm window.

Historic houses often had louvered shutters to reduce solar gain in summer, prevent the fading of natural-dyed fabrics, keep out insects prior to the availability of wire screening, and for privacy. Because storm windows prevented the shutters from being closed, the storm windows were removed for the warm-weather months. Disused wood storm windows can sometimes be found in the attics or sheds of

This 1870s mansion has both interior storm windows and louvered shutters that slide into a recess in the wall.

historic houses in colder climates. New wood storm windows are being produced by a number of manufacturers today and remain an option for owners of historic (or new) homes. It is also possible to make your own wood-framed storm windows. A friend outfitted his historic house with wood storm windows using glass salvaged from aluminum-framed storm sashes that he found at the local dump.

After World War II, aluminum-framed double- or triple-track storm windows largely dominated the market. These have a fixed frame permanently attached to the exterior window trim and sliding top and bottom sashes that allow airflow in the warm months without having to be removed and stored. A screen rides in the third track of the triple-track version. Until recent decades, these storm windows were produced with a natural aluminum "mill finish" that was not particularly compatible with historic houses. However, the metal can be painted to make it less apparent. A great many historic homes still have these storm windows in place and operable. They may need some attention to function smoothly and easily, but often there is no reason they cannot continue to do the job they were designed for.

This traditional wood storm window has its light (pane) size and configuration carefully matched to the primary window behind it. Wood storm windows should have weather stripping where they meet the window trim to ensure a good seal. They should also have notches on the bottom to allow any condensate water to escape.

Most traditional storm windows of this type swing out at the bottom to allow ventilation without removing the storm window entirely.

These low-profile storm windows from Allied Window are designed to minimize their visual impact on historic buildings.

As the historic home restoration/rehabilitation movement has grown in the U.S., some manufacturers have started to produce less obtrusive storm windows for historic buildings. These "low profile" storms stand out less prominently from the window frame than traditional aluminum storm window frames. If you are installing new exterior storm windows, this design is recommended. Manufacturers will produce custom sizes and shapes for unusual windows. Always match the placement of meeting rails on storm windows to the primary windows they cover.

Special glazing is available to protect the interior against UV light, and low-E (low-emittance) coatings can improve efficiency. However, these coatings can impart a tint that is unlike traditional clear glass, and you should ask to see an installed example in natural daylight before agreeing to one. If it is noticeably different in appearance than nearby traditional glass windows, you may want to keep looking. In the wide range of these coatings, some maintain a far more traditional appearance than others.

Interior storm windows have also grown in popularity for the restoration/rehabilitation market. These, of course, have the advantage of being invisible from the exterior and minimally apparent from the interior on windows with drapes or curtains. An interior storm window is typically a sheet of Lexan or a similar acrylic in a thin frame of aluminum or plastic; a sprung frame or a perimeter foam gasket air-seals the storm panel. If the frame is painted to match the window trim, the unit blends in even without a window treatment. Lightweight panels can be removed and stored in the warm months, or the panels can remain in place year-round if air-conditioning is used.

Some manufacturers of single-glazed wood-sash windows (yes, they are still made!) attach a sheet of glazing in a narrow aluminum frame to the exterior of each sash, over the muntins that separate the panes in the primary window. This provides a second layer of glazing to create an insulated window but does nothing to prevent air infiltration around the sash, which is often a significant cause of heat loss through windows. For this reason a full storm window, exterior or interior, is far more effective.

Vacuum Insulated Glass (VIG)

As of this writing, a new product known as vacuum insulated glass or glazing (VIG) has potential for rehabbing historic windows. Whereas standard insulated glass is between ½ and ¾ inch thick, VIG achieves an equivalent insulating benefit with a mere ¼ inch or so of overall thickness.

Standard insulating glass slows the transmission of heat with one or more layers of air that are usually at least ¼ inch thick. Air transfers heat much less efficiently than glass; the air gap must be wide enough to insulate but not so wide as to allow air to circulate across the gap. VIG goes one step further, because a vacuum transfers no heat at all. Consequently, the vacuum gap in a VIG pane need only be 1/64 inch.

VIG was expensive as of 2018, but a number of companies were scaling up production, and prices will likely come down. Judging by the thinness and lightness of the pane I examined, it could replace traditional glass in most historic wood window sashes. Currently there are two drawbacks other than cost, one minor and one more important. The minor issue is that tiny beads of glass are distributed in a grid pattern between the panes of glass to prevent the panes from touching one another when the vacuum is created, which would, of course, eliminate the insulating effect. These are only noticeable up close. Of greater concern is the fact that a hole has to be made in one of the panes for the air to be sucked out and the vacuum created. This hole then has to be plugged, and every pane in a multi-light sash will need such a plug. The sample I examined used a ½-inch-diameter black rubber plug. In a one-over-one or two-over-two sash, the two or four plugs would probably not be terribly noticeable. In a six-over-six sash, however, the plugs could be distracting. Hopefully less conspicuous plugs will be developed; sealing the holes with molten glass or another clear material would be ideal.

Using VIG in storm windows while leaving historic glass in the primary sashes is another possibility. This would limit the number of plugs and keep the character of old, wavy glass in the historic sash.

Appropriate Replacement Windows

Few elements match windows as character-defining features. Windows often constitute a quarter to a third of a wall's area and convey telling details about the age and style of the house. If historic windows are beyond reasonable repair, or if new windows need to be added, matching windows must be found. Largely because the National Park Service has

The window on the right is a replacement window carefully matched to the original on the left. With care, it is sometimes possible to get a very good match from the major wood window replacement companies. Camouflaging a mix of old and new windows behind matching storm windows disguises minor discrepancies in dimensional details and molding profiles.

quite rigidly applied their rehabilitation standards for commercial historic tax credit projects, a number of window manufacturers offer good replacement windows for historic buildings. There is no one-size-fits-all replacement window, however. Replacement windows that are advertised as being "approved by the National Park Service" were only approved for replacing windows they closely matched in dimensions and details.

Vinyl windows are almost never a good choice for a historic house. The material requires a bulky vinyl frame for the sash, which has to be installed inside the original wood-framed opening. This reduces the height and width of the opening by 4 inches, thus reducing the amount of light reaching the interior and altering the proportions of wall to glass. Further, because vinyl windows are almost always bright white, they are guaranteed to look out of place on anything but a bright white house. Painting the vinyl a

This house demonstrates the effect of vinyl replacement windows on a historic house. The original wood twelve-over-twelve windows survive in the attic to compare with the two-over-two vinyl windows installed on the bottom two floors. It is easy to imagine how much more character the house displayed when it had twelve-over-twelve windows on all floors.

dark color will generally void the warranty, because dark colors cause the vinyl to absorb heat and expand excessively in sunlight, possibly separating the sash corner joints. Vinyl windows typically have a ten- to

This attempt to make do with standard vinyl window sizes—using a two-over-two double-hung window and filling the leftover space at the top with another window—adds an awkward, bulky horizontal element in an opening intended to be tall and elegant. The thick frame of the vinyl window reduces the opening width by 4 inches, and the horizontal frame elements take 8 inches from the height. The result is a window that would be more comfortable on a manufactured home; the fine Italianate surround might be wondering what happened to it. The vinyl siding doesn't help.

twenty-year expected lifespan and rarely outlive the payback period of their purchase and installation cost. Once they fail, they are not recyclable. Vinyl windows might be appropriate in a narrow alley between two buildings where air circulation is poor, access for maintenance difficult, and visibility from a public way low or nonexistent—but apart from such specialized circumstances, vinyl windows are not recommended.

The keys to selecting replacement windows are to match the widths of the stiles, rails, and muntins as closely as possible; match the historic pane pattern (two-over-two, six-over-six, etc.); match the muntin profiles; and preserve the glass area by not placing a new frame within the historic opening. Often the best solution is a sash-only replacement. These typically have thin, modern, sprung-balance tracks on either side that only slightly reduce the width of the opening. If the historic window frames are rotted or damaged beyond repair, it will be appropriate to remove

When this former school was converted to a municipal office building, its large, multi-light, double-hung windows were replaced by much smaller vinyl one-over-one double-hung windows that altered the character of the building and reduced the natural light.

This historic view of the same building shows the original large windows that flooded the interior with light and contributed to the character of the building.

This recent photo was taken while the building was undergoing rehabilitation for conversion to senior housing. It shows the reductions in window size and light transmittance experienced with the previous, inappropriate window replacement.

The restoration of appropriately sized and detailed windows in the recent rehab significantly improves the appearance of the building, restoring the proportions and character of its fenestration.

the old frames and install new ones of equivalent dimensions. This, of course, requires removing the trim from the interior or exterior to access the frames.

Do not install a too-small standard window in a larger opening with fill around it. It will never look right from the interior or the exterior. Most window

These newly custom-made double-hung windows of traditional design await installation. Most of the double-hung windows in this cottage were replaced by standard wood casement windows in the 1960s. The casements were failing, and the owner opted to return the windows to their original style and sizes, copying the details from one surviving original window. Historic photos and evidence in the framing confirmed the sizes and light configurations for the windows to be replicated.

manufacturers serving the restoration market produce custom sizes and shapes. If your house needs these, spend the money to get them made. It may be worthwhile to look beyond the large window companies; local millwork shops or craftspeople might be able to produce custom windows at a lower price. If only a few of your windows need to be replaced, a local window restoration company may have matching historic sashes in stock, allowing you to replace old with old. Again, matching their details closely is important. There are companies producing new windows with rope- or chain-weighted balance systems, just like historic windows. These are not cheap, but they can be an excellent option.

Stained, Colored, and Art Glass

Needless to say, any stained, colored, or art glass in a house is an important character-defining feature, and its preservation should be a high priority. Repair or replication of missing pieces is not generally undertaken by do-it-yourselfers, though there are exceptions. Fortunately, there are experienced craftspeople and artists who specialize in this type of work. Your local preservation organization would be a good place to start looking for a specialist.

This etched ruby glass (it is dark-red, though it appears nearly black in the photo) is located in the sidelights of the front door of the house where Harriet Beecher Stowe wrote Uncle Tom's Cabin. Finding artisans to repair or replicate glass like this can be challenging but is possible.

Screens

Wire window screens became common in the late nineteenth century. Previously, insects were often provided free access to houses through open windows. Louvered window shutters or blinds provided limited protection, and sometimes a loosely woven cotton mesh fabric was stretched on a wood frame to provide screening. Once efficient production methods were developed for wire screening, its use spread quickly to all but the poorest homes, for screen doors as well as for windows. Iron and copper were at first the most commonly used materials. Iron screening had to be painted to prevent rust, usually black or dark green.

Aluminum became widely available in the postwar years and largely replaced iron and copper for window screens by the 1960s, though copper continued to be used occasionally in screens for expensive homes. The aluminum was usually left in its natural silver-gray color. Nylon screening arrived on the market shortly thereafter and is the standard material for residential window screening today, typically either black or dark gray.

On double-hung windows, screening was traditionally installed in simple wooden frames fitted under the upper sash, filling the space between the jambs from side to side and between the sill and upper sash from top to bottom. Full window screens covering both sashes were uncommon, though they are the industry standard with new and replacement windows today. A full-sized screen obscures the upper sash as well as the lower and leaves the window looking like a dark hole in the wall, especially if black nylon screening is used. Half screens are strongly recommended and need to be specified when ordering.

The most common way of providing screening to a rehabilitated historic window is with a modern exterior storm window as described above. If you are mounting interior storm windows that are removed in the warm months, you can insert simple, traditional, wood-framed screens under lower sashes.

With casement windows, screening is usually installed on the interior. You need a convenient way to remove the screen to access the window hardware. Hinged screens are one solution.

Glass Block

Glass block achieved postwar popularity both for exterior walls and interior partitions. Laid up in mortar as with brick or concrete block, they create a solid wall that transmits light while remaining opaque for privacy. Glass block can be a key character-defining feature in a modern house.

A broken glass block can be removed by cutting away the surrounding mortar. Finding a matching block can be a challenge. Different manufacturers used different surface patterns in the glass, which create subtly different effects as light hits or passes through the blocks. The differences can be quite noticeable when one or several non-matching blocks are placed into a pattern.

Architectural salvage shops might be a local source of matching blocks. Failing that, glass blocks are now being produced for the restoration market. You will need to request samples, and hopefully one of these will prove a good match. If you have several damaged areas of glass block, it may make sense to consolidate the good historic blocks in one or more openings and use new block in the opening or openings that remain. This is especially effective if the new and old blocks can be located on different elevations so it isn't possible to see them at the same time. When you replace a block, be sure to match the historic mortar as well.

Most new blocks are available with an interior thermal break between their faces, preventing the heat losses that period glass blocks suffer from. The energy efficiency of old glass blocks can be improved greatly by installing insulated glass panels over the expanse of blocks, either on the exterior or the interior. A more extreme option, which will preserve the appearance of the glass block on the exterior but cover them permanently on the interior, is to spray the interior surface with white paint and build an insulated wall in front of it. This might be a good solution where changes to the interior require a solid wall, such as for new kitchen cabinets.

Window Rehabilitation in Photos

The following photos show how historic wood windows can be rehabilitated by professionals and many homeowners. Windows rehabbed in this way will be as good as the day they were built, ready to serve another century or more. Most of these photos are courtesy of Bagala Window Works.

This 150-year-old wood window sash is in pretty good shape, but it could be disassembled by drilling out the wood pegs that hold it together if it were necessary to replace a deteriorated or broken element. Often deteriorated pieces can be repaired with an epoxy wood filler system like Abatron (see page 470 in this chapter).

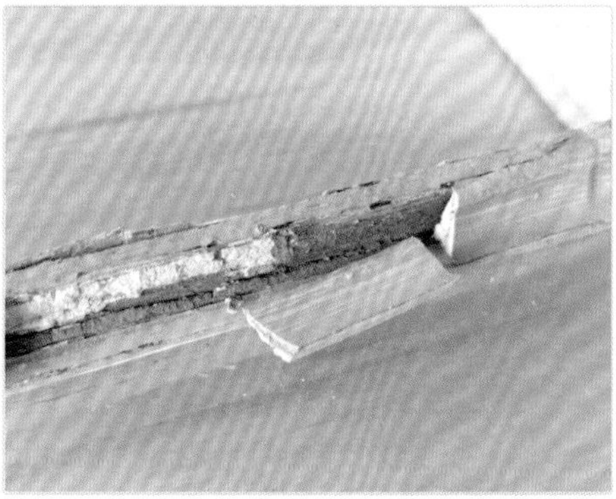

Dried window putty that is chipping off is par for the course on historic windows that have not been kept painted. It will all have to be removed and replaced in the rehab process.

The Steam Stripper was developed by Marc Bagala for window restoration and is available to professionals or homeowners.

This sash is coming out of the Steam Stripper after being exposed to steam long enough to soften paint and putty for easy removal, but not long enough to saturate the wood.

Once a sash is removed from the steam, paint and putty scrape off easily and can be disposed of in a lead-safe manner. The rehabbed window will have no lead contamination remaining.

After the metal points holding the glass in place beneath the putty are removed, each pane of glass is carefully removed to be cleaned and stored for later reinstallation.

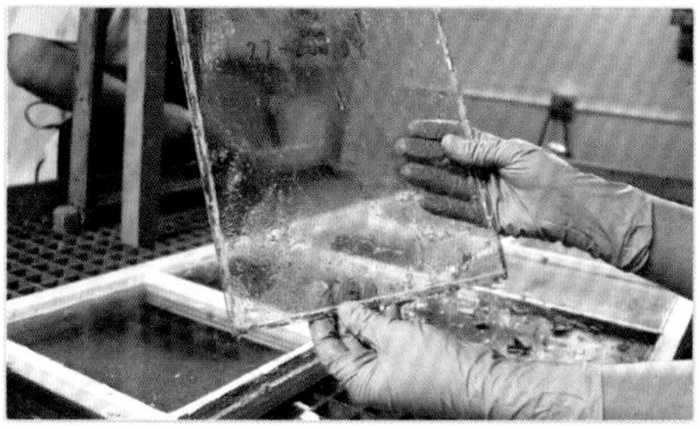

Historic glass is often wavy, possessing a character unlike modern, perfectly flat glass. If panes are broken or missing, it is important that their replacements be salvaged or reproduction wavy glass to match the remaining existing panes.

The glass has been removed from this sash, and paint removal has begun.

Here, a hand sanding block is being used to remove paint residue and to smooth any roughness in the wood after stripping.

A detail of the same sash with the paint removed and the wood sanded.

Triangular metal points (seen on the pane at right) are pressed into the wood around the edges of each pane to secure them in place.

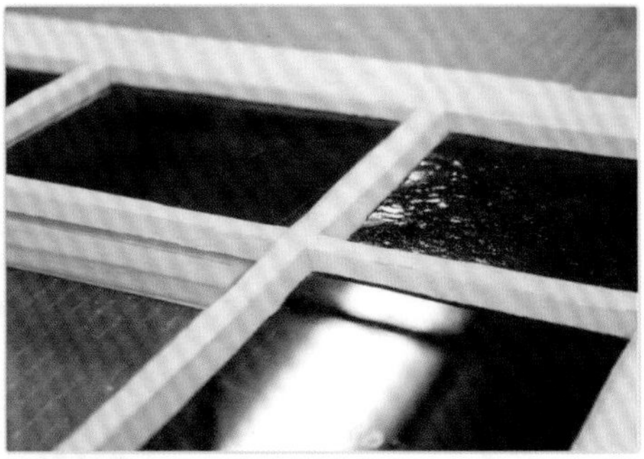

Here is the exterior face of the sash after putty has been applied to cover the points and smoothed to a narrow taper. The putty must set up and form a skin before it can be painted.

This turn-of-the-twentieth-century sash has been primed. Now putty is being applied in the recesses to bed the glass into.

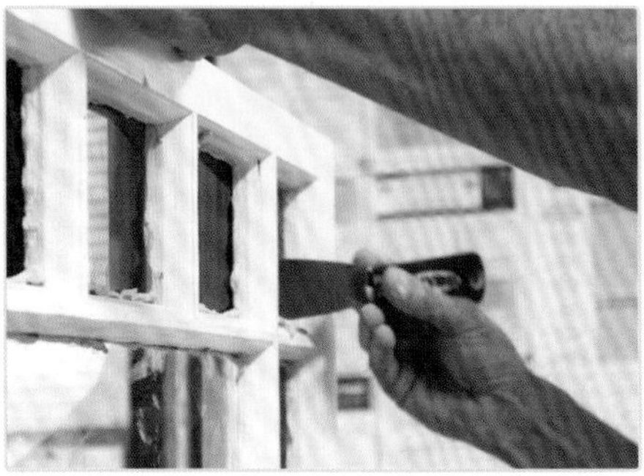

With all the panes set in the sash, a putty knife is used to remove any excess putty from the interior face of the sash.

Here is the exterior side of a rehabbed window after painting.

A colored glass pane is bedded in the putty.

Once the bulk of the excess putty is removed, a brush and whiting are used to remove the oily residue from the interior glass surface.

A completely rehabbed sash is often reinstalled with high-quality bronze weather stripping. Grooves have been cut in the sides and bottoms of these sashes to allow the weather stripping to engage with the sash and seal out air infiltration.

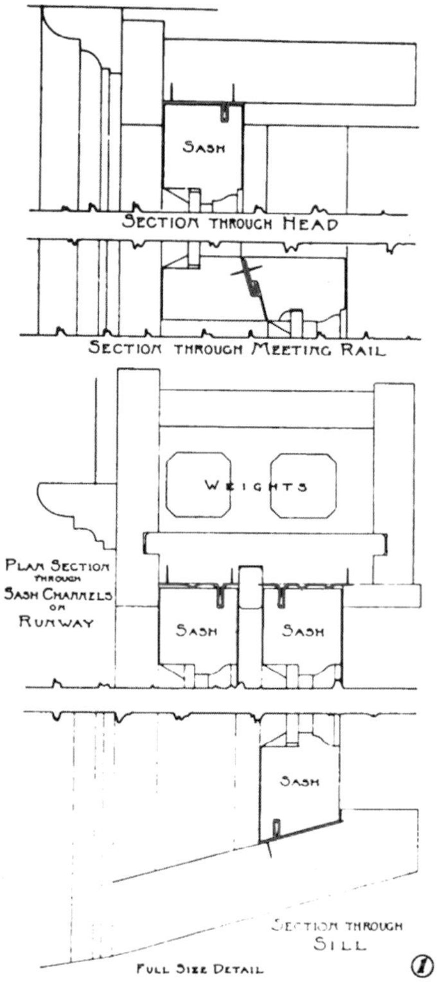

DETAILS OF STANDARD "CHAMBERLIN" CORRUGATED WEATHER STRIPS AS APPLIED TO DOUBLE-HUNG WINDOWS

This illustration from 1913 shows where bronze weather stripping is installed to ensure a good air seal on both sashes. The weather stripping has been highlighted in red for clarity. The same approach is used today.

Here is new bronze weather stripping on a recently rehabbed window. This window has a neoprene rubber bulb for air sealing on the bottom.

The bronze weather stripping on the bottom of this rehabbed window will last longer than the neoprene bulb shown in the image on the previous page.

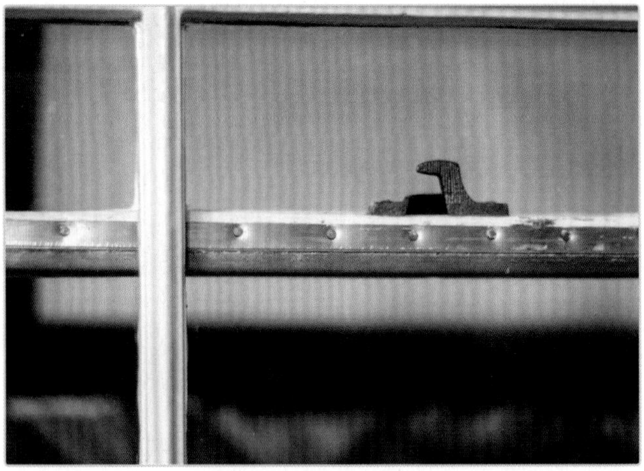

Weather stripping at the meeting rails slides together when the sash is closed. A traditional sash lock ensures a good seal.

A completed sash ready for reinstallation.

PART FIVE

Interior Finishes

Imagine walking through the front door of a Queen Anne–style house with an exquisitely restored exterior—the Hench House on page 103, for example—only to be confronted by rooms that are plain sheetrock boxes with gray commercial carpet instead of the Hench House's wonderful period interior with richly detailed Victorian rooms. Imagine finding no hardwood floors, no molded trim, no rich Victorian colors. Your sense of the house *as a historic house* would be greatly diminished. This admittedly extreme example highlights the effect of getting the interior finishes wrong. No element of your house will be looked at as closely or as often as its interior finishes. The spaces we sit in, eat in, cook in, prepare for the day in, sleep in, and sometimes work in are experienced closely and repeatedly. They are also the spaces our guests visit and observe.

As illustrated in the chapters on identifying historic styles and designing appropriate exterior alterations and additions, the details are often what distinguish one style from another and make each house unique. One of the most important details to get right is the materials used. The following pages will discuss various building materials and their use in historic homes of different periods and styles. Repair and replacement strategies and substitute materials will be explored. Here, as elsewhere, I will skip information on construction methods and techniques that is readily available elsewhere, focusing instead on the application of those methods and techniques to historic homes. I will not cover how to hang and tape gypsum board, for example, but I *will* describe the repair of historic plaster.

The following chapters take us through floors, walls, ceilings, and trim before concluding with information on painting, wallpaper, and other decorative and protective finishes.

CHAPTER FOURTEEN

Floors

Flooring is nearly always a character-defining feature in a house. The subtle ripples of hand-planed eighteenth-century wide pine floorboards, the tiger-stripe effect of quarter-sawn oak, the plush effect of modern wall-to-wall carpeting—floors are important to the character of the space. Flooring also gets a lot of wear and tear, particularly in high-traffic locations, and for this reason at least some of the flooring has often been replaced in a house more than fifty years old. As discussed in Chapter 2, it is important to start with an understanding of what the current material is and whether it is original or a later addition. Then a decision can be made about whether it should be retained or perhaps restored to an earlier treatment.

Wood Flooring

The most common flooring material in historic homes is wood. American wood floors are typically laid in two layers: a subfloor and a finish floor. Rough-sawn softwood boards are the most common subflooring, laid perpendicular or at a 45-degree angle to the underlying joists. The finish floor can be softwood or hardwood, laid perpendicular to the joists and often nailed through the subfloor into the joists. Parquet floors laid in patterns of variously colored hardwoods may be present in high-end homes. Narrow hardwood flooring came into vogue in the late nineteenth century and remained popular for many decades. Sometimes the original softwood finish floor was removed and the new floor laid directly on the subfloor, but it is not unusual to find a hardwood floor on top of two layers of original flooring in an older house. It is also not uncommon to find additional layers of more modern flooring on top of that. See Chapter 6, Demolition, for information on removing modern layers of flooring. Wood flooring is generally laid with the "crown" up so that any cupping will raise the center of the board, not the edges. Quarter-sawn boards are an exception to this, as their method of sawing results in a cross-grain board without a crown.

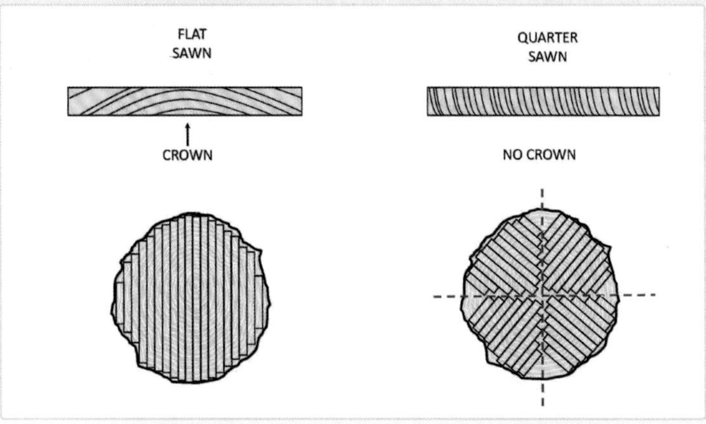

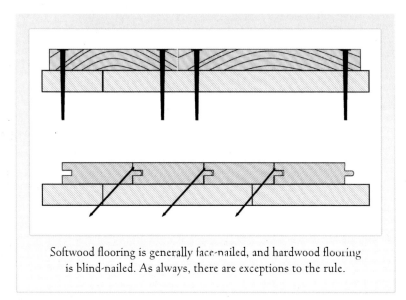

Flat-sawn boards have a crown, which should always be facing up for flooring. Quarter-sawn boards are more dimensionally stable and can be laid either side up.

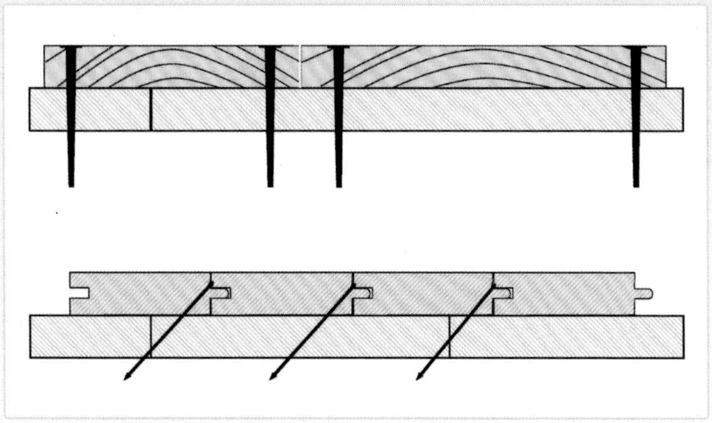

Softwood flooring is generally face-nailed, and hardwood flooring is blind-nailed. As always, there are exceptions to the rule.

Damaged wood floors can often be repaired if they are not rotted or worn paper thin. If a historic floor is beyond repair, replacing it in-kind is the preferred option. This will best retain the historic character of the room. I know of several homeowners who were able to salvage matching historic floorboards from their own attics to replace floors that were beyond repair. In Maine, where I live, floorboards salvaged from old houses that are being torn down or renovated can be found through classified ads or in architectural salvage shops.

Resawn historic lumber is another option. Small mills in many parts of the country are sawing new boards from old beams or trees salvaged from rivers or lakes, where they sank during log drives a century or more ago. This old wood is far denser and more durable than modern wood due to the slow-growth characteristic of old-growth trees. Many of these mills use vertical band saws, which leave marks resembling those of historic boards that were hand-sawn or cut by a water-powered mill. These marks will have to be planed off the finish side of the boards, as they were on the original boards, but you will have the satisfaction of knowing that the side that will never be seen again looks right.

A softwood floor is more challenging than hardwood to replace with new wood, as there is less difference in quality between new and old-growth hardwood than between new and old-growth softwood. Quarter-sawn hardwood is particularly tough. Softwood will need a good multicoat protective finish, either paint or poly/varnish, and even then it will be vulnerable to damage from dragged furniture, dropped objects, and stiletto heels. Newly sawn old-growth wood can occasionally be found. Logging of these trees, which have stood for hundreds of years, is restricted in most areas, but severe storms, forest fires, or pest damage can result in old-growth trees being salvaged and milled into lumber. This wood is preferable to new wood when replacing damaged wood in an old house.

Of the many flooring options available today that were not available historically, some may be appropriate to an old house in some circumstances. We'll return to this later in the chapter.

Freshly cut wood is still "green" and needs to be dried thoroughly before it is used. As it seasons, losing moisture to the surrounding air, it shrinks. Traditionally, freshly sawn lumber was stacked with spacers to allow air movement around all sides of every board, then left for months to season. Most wood today is kiln-dried at the mill, lowering its moisture content to 6 or 7 percent by weight, but once it leaves the mill it can gain moisture back from the air—a lot of moisture if exposed to rain or snow.

It's a good idea to test new wood with a moisture meter. Hand-held moisture meters cost from $20 to several hundred dollars, but a top-end meter is unnecessary for home-rehab purposes. As a rule of thumb, a moisture percentage between 7 and 8 is ideal for flooring in most areas of the U.S., but up to 11 percent may be fine in coastal locations, where the relative humidity is higher. The wood should be kept in the room where it will be used for up to a week before installation to acclimate to the microclimate of the room.

Softwood Floor Repairs

In general, softwood floors are earlier than hardwood, which didn't come into widespread use until after the development of steam mills in the mid-nineteenth century. Some exceptional houses had hardwood floors at an earlier date, either because the owner was wealthy or the house was built in an area where hardwood was prevalent. Continuing to generalize, floorboards also became narrower over time. When lumber was sawed from trees by hand or by a slow, water-powered mill, minimizing waste was a priority. For this reason, early softwood floorboards were sometimes cut with the taper of the tree trunk, creating long, trapezoidal boards. When every other board was reversed end to end as the floor was laid, the centers of the boards remained more or less parallel with each other and with the side walls. This characteristic is not always immediately apparent if one is not

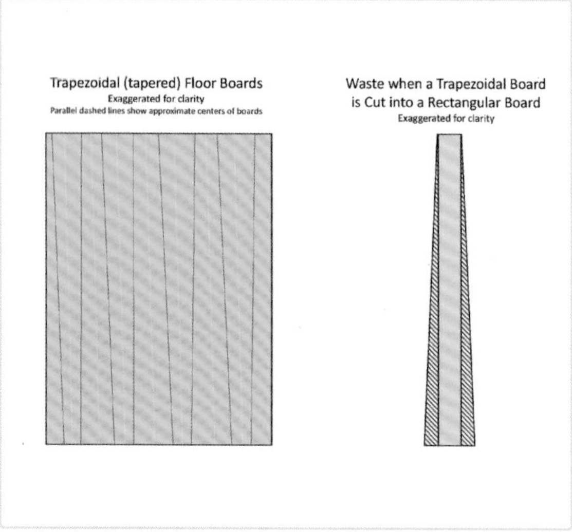

Early softwood flooring is often trapezoidal in shape. If you are replacing a damaged board in a floor like this, you will need to cut the replacement board to match the shape of the opening.

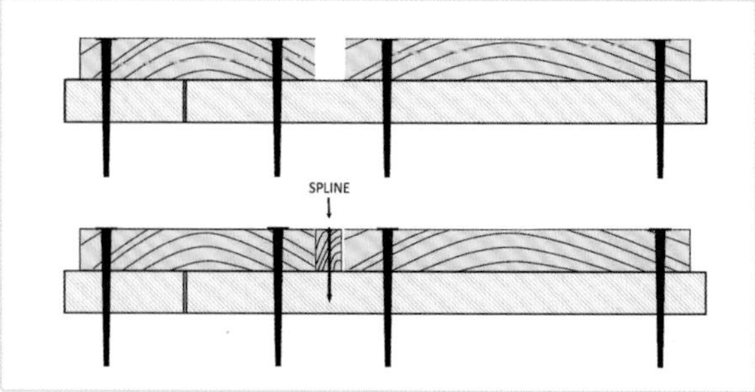

Splining can be used to fill gaps between boards, typically on softwood floors.

This 1850s pine floor was splined in the 1940s.

looking for it, but it will become obvious when you attempt to replace a damaged trapezoidal board with a rectangular one.

Historic softwood floors may show damage from neglect, misuse, or alterations to the house or simply from accumulated wear and tear. A high-traffic floor may be worn through to the subfloor.

Splining

The seasonal expansion and contraction of boards due to changes in temperature and humidity may have resulted in gaps between boards that no longer disappear when the boards expand in summer. In some cases, these gaps have been filled with wood filler, plaster, caulking, or other materials which fail over time, leaving irregular gaps. If the house is now air conditioned, the boards may never expand in width as they once did, leaving permanent wide gaps. Splining is the installation of narrow strips of new wood between boards to fill these gaps.[4]

Because the gaps between boards are rarely of consistent width even from one end to the other of a single

[4]Confusingly, splining is also the term used for inserting a thin horizontal piece of wood into dados cut in two adjacent floorboards as a bridge between the two. This sort of spline is essentially a loose "tongue" inserted into two "grooves" in an early form of tongue-and-groove flooring.

gap, each spline needs to be cut to fit the gap it will fill. If the splines are shorter than the gaps, their butt joints should be staggered from one gap to the next. In a house where the floors still expand and contract seasonally, it is best not to cut and install splines when the gaps are widest; otherwise, the next time the floorboards swell, they could crown excessively and even split if they have no remaining gap to expand into. The splines need to be finished to match the floorboards, whether painted or stained and clear-coated.

Patching Holes

Holes are cut in floors for a variety of reasons, plumbing and the installation of heating system equipment being two of the more common. When pipes, ducts, registers, and other such insets are removed, the holes they leave behind need to be filled.

The traditional way to repair such a hole is with a "Dutchman" patch. Essentially, this is a rectangular piece of wood just a bit larger than the area to be repaired. Cut a rectangular hole of the same dimensions around the damaged wood, then glue your patch in place. The patch block should be slightly thicker than the damaged board so it can be planed flush with the surrounding surface after the glue sets. To prevent the Dutchman from just sliding through the hole, the ends of the hole can be cut with stepped or angled edges. Traditionally, a hand chisel was used to form a rectangular hole within a board. Today a plunge router and a template made from ½-inch-thick MDF can be used for this, though the corners still need to be squared up with a chisel. Angled edges still have to be cut with a chisel.

Smaller holes can be filled with a good epoxy wood filler that is formulated to expand and contract with the wood through the seasons. Any filler that cannot do this is going to come loose eventually. As of this writing, the filler you need for this cannot be purchased at big-box home improvement stores or at most hardware stores. The leading manufacturer, Abatron (see Chapter 13, page 470), sells its products via mail order and through major online retailers.

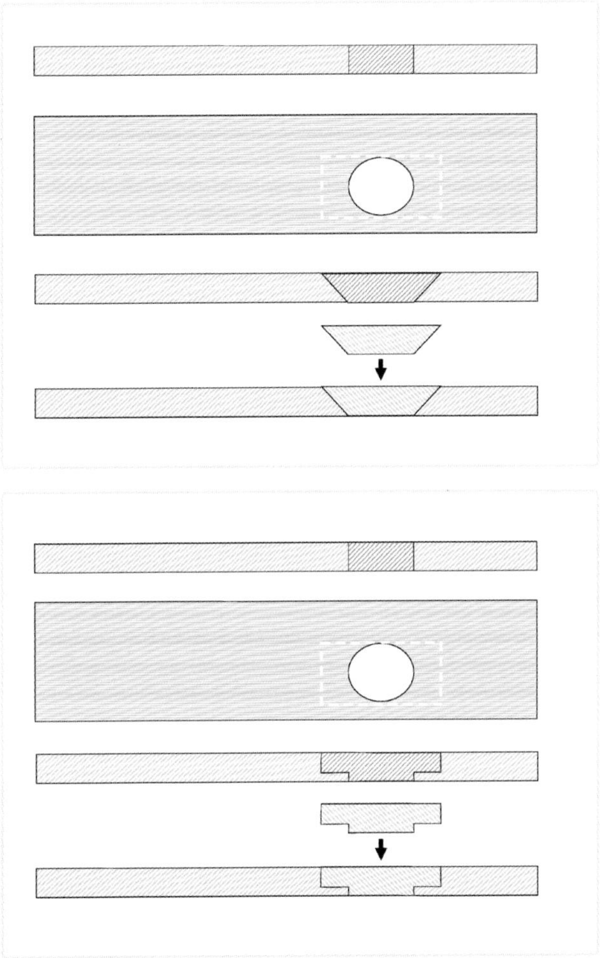

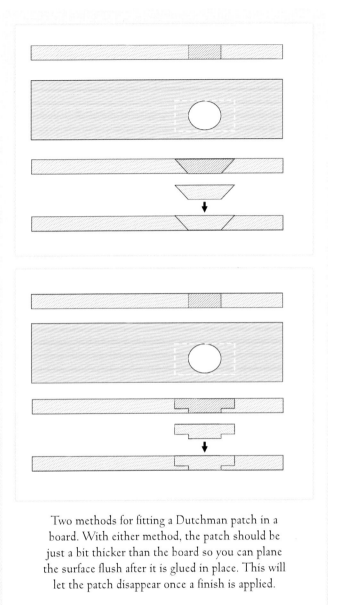

Two methods for fitting a Dutchman patch in a board. With either method, the patch should be just a bit thicker than the board so you can plane the surface flush after it is glued in place. This will let the patch disappear once a finish is applied.

Replacing Boards

Sometimes a board is so badly worn or damaged that replacing it is the only reasonable option. However, a dead-flat machine-planed new board will stick out like the proverbial sore thumb in a floor of gently rippled hand-planed boards with a century of wear. To avoid this, you can either find an old board or buy a rough-sawn new board and plane it by hand to match the character of the others. Architectural salvage shops may have old floorboards, and they can often be found in online classified sites. You might even find what you need in your own house. Start by looking in the attic. More than one old house now has an attic floor of new boards because the originals were reused to repair floors in the occupied part of the house. At Whitten House, I salvaged floorboards for replacements elsewhere in the house from the small rooms that became bathrooms with tiled floors.

The subtle ripples of a hand-planed floorboard cannot be matched by a modern machine-planed board. Replacing such boards in-kind means finding old boards to reuse or hand-planing new material with an antique plane like the one shown here.

Keep in mind that softwood expands and contracts more than hardwood. Depending on ambient conditions at the time you install the floorboards, you may need to leave space for side-to-side expansion. Winters are dry and summers humid in coastal New England. I will fit a softwood floorboard tightly in summer, knowing it will shrink during the winter, but will leave gaps for swelling if I am laying it in winter.

Entire Floor Replacement

Sometimes a floor is so damaged or worn that it makes sense to replace the entire floor. Resist the temptation to install a floor that is "nicer" than what was there historically. Replacing wide pine boards with 3-inch hardwood boards may give you a beautiful floor, but it will also change the character of the space. Be especially cautious about replacing a historic softwood floor with modern flooring options such as 12-inch-square or 12-inch by 24-inch marble tiles. While appropriate to some stylish urban houses of certain periods, marble floors are incongruous in most early houses built with softwood floors. If you must have the look of marble, a floor cloth (see below) painted with a faux marbleized finish over the board floor would be appropriate in a pre–Civil War house.

When reusing old floorboards, you will have to deal with the nail holes remaining from their previous use. The correct way to remove nails from salvaged boards is to pull them through from the back. This prevents splintering around the nailhead on the face of the board, though pulling the nailhead through the board also enlarges the hole. Old softwood floors are most often face-nailed with square nails, which are still produced. If a reused board happens to have nail holes that align with the framing, it is often possible to re-nail with larger nails through the existing holes. More often, however, it will be necessary to nail in new locations. The old holes should be filled with a good epoxy filler that will expand and contract with the wood, like that made by Abatron. I have also used large (16-penny) finish nails to attach reused boards to the framing, with the original nails returned to the old holes for appearance. The finish nails disappear when they are countersunk and filled. The pattern of visible square nailheads is part of the character of an old floor, and it is important to use similar nails, either new or salvaged, for replacement floorboards.

Softwood Floor Finishes

Wood floors during the early settlement period often remained unfinished for decades. Oil paints were not widely available and were expensive due to the skilled labor involved in grinding the pigments and mixing the oil and white lead on site. Unfinished wood floors were often cleaned by spreading fine white sand over the floor and allowing the abrasion of being walked on to clean the floor, literally sanding the wood to get a clean surface. Another method was to wash floors with a scrub brush using lye soap in hot water. Lye was usually homemade, leached from the ashes left from cooking and heating fires. It is caustic and has a bleaching effect on wood, especially softwoods like pine.

This Southern yellow pine floor at Colonial Williamsburg is unfinished, as it would have been in the eighteenth century.

Late-eighteenth- and early-nineteenth-century wall-to-wall carpeting was woven in strips that were sewn together to fit the room, like this reproduction of a mid-nineteenth-century design. Such carpet was often installed over unfinished softwood flooring.

Even after painted floors became the norm, some rooms retained unpainted floors under wall-to-wall carpeting or floor cloths. Carpet came in narrow (26- to 33-inch-wide) strips that were sewn together to match the width of the room, sometimes with a border pattern around the edges. The carpet was generally tacked down along the edges. A row of small tack holes in the floor around the perimeter of a room is often evidence of early carpeting. Like painted floors, carpeted floors became more common as the nineteenth century progressed. Floor cloths were made by sewing together lengths of canvas, sizing it, and painting it with oil paints, usually with several coats of varnish to make a durable finish. These could be solid colored or have decorative painted patterns. Trompe l'oeil marble blocks were a popular painted

This painted floor cloth has a marbleized design imitating a stone floor. Note the unpainted pine floor visible around the edges of the room.

It became common in the twentieth century to strip the paint from softwood floors and varnish them. Varnished Eastern white pine floors in northern New England came to be called "pumpkin pine" after the amber color of resinous old-growth pine, though pumpkin pine is not a species. While lovely, shiny varnished pine floors are not historically accurate for the eighteenth and early nineteenth centuries. Another downside of stripping painted softwood floors is that the most common stripping method is with a large drum sanding machine, which removes all evidence of original hand-planing and the wear pattern a softwood floor develops over time. These markings, though subtle, contribute significantly to the historic character of such a floor. Drum sanding of later hardwood floors is fine, as these were mill planed perfectly flat when new.

If you have painted floors, consider keeping them that way. For an authentic historic feel, consider

motif, as were imitation carpets. Reproduction period carpeting is available today in a variety of patterns, some still produced on antique looms in England. Floor cloths are also being made again and are not particularly difficult to make at home.

Painted floors became more popular in the second quarter of the nineteenth century and continued to be used widely into the twentieth century. Probably the most popular early floor color was ochre, a warm brown-yellow. My belief is that this color was popular because it resembled a freshly cleaned pine floor, the gold standard of good housekeeping for previous generations. Other traditional colors for painted floors were Spanish brown and dark green. Black floors appeared in historic Colonial and Federal homes during the Colonial Revival period, when many of these houses were restored. Antique oriental rugs and braided or hooked American rugs were set off nicely by these floors. Black-painted floors can be striking but show off dust remarkably well.

The original ochre paint color is clearly visible where later layers of paint have chipped away on this early-nineteenth-century staircase. This color was widely popular for floors and stairs from the late eighteenth century well into the nineteenth century.

These historic pine floorboards were likely originally covered with carpet in this front parlor of a large Federal house (see Alden House, page 565). Later painted, the floor was a dull gray when the house was restored in the 1960s. At that time the floors were stripped and varnished, giving them the "pumpkin pine" look that has been popular since the Colonial Revival movement began at the end of the nineteenth century.

going back to the historic paint color. If you are less concerned about historic authenticity, however, your color choice will be wider. Prep well and use two coats of a high-quality floor and deck enamel, preferably one that's alkyd (oil) based. Alkyd paints are much harder and more wear-resistant than even the best acrylic/latex paints, and nowhere is this characteristic more important than on floors. See Chapter 18, Paint and Paper, for more detail.

If you decide to strip and varnish a painted softwood floor, try to do so in a way that preserves the planing ripples and wear marks. This means skipping a huge, powerful drum sander in favor of one of the several preferred methods of removing paint covered in Chapter 18. Once the paint is off, sand the floor with a palm sander using 150-grit sandpaper, moving in the direction of the grain to preserve ripples. The goal is not to flatten the board but to finish removing the paint residue and provide a good surface for the new finish to bond to. Keep a nail punch and hammer nearby to tap down any nailheads that stand proud of the floor surface. You want them flush with the finish surface or very slightly recessed. The floor should then be vacuumed, and any remaining residue removed by a quick pass over the entire floor with 220-grit sandpaper. Follow with another round with the vacuum and a wipe-down with a tack cloth to get the last of the dust. Remember, the clear finish you are putting on the floor will place any remaining sanding dust on permanent display, and light falling across the surface from a window or lamp will highlight it further.

A satin-finish oil-based polyurethane is strongly recommended. Apply it with a brush, because the brush strokes will push the coating into the pores of the wood for a better bond. The slower drying time of oil poly will allow it to fully penetrate the pores of the wood, bringing out the color and richness of the grain much more than water-based poly. Once the first coat is fully dry, another sanding with 220-grit sandpaper followed by another vacuuming and tack cloth wipe-down will prepare the floor for the second coat. High-traffic areas should absolutely get a third coat following the same procedure, and all areas will benefit from a third coat.

Some refinishers use an oil-based polyurethane for the first coat, to bring out the color and grain, and a water-based poly for subsequent coats to save time, but this approach sacrifices as much durability as it saves in time. Traditional marine varnish can be used in place of the polyurethane. Whatever material you use, follow the manufacturer's recommendations for time between coats and remember that the finish will not be fully cured for several weeks to several months, depending on material, weather, and other variables. Don't drag furniture across the floor, place rugs, or dance around in stiletto heels until it is fully cured.

Keep in mind that any patches or replaced planks you installed prior to refinishing may not match the colors of the flooring you stripped. Apply your clear coat finish to a small area of the existing floor to see what the color will be when finished. Using small pieces of the same wood used for patching, use stains or dyes followed by poly to find a match to the existing floor colors. It will typically take several samples to find the right stain or dye, as it is hard to predict how they will look once applied to the wood and covered with poly.

If you have pine floors that have already been refinished with varnish or poly, perhaps long ago, and the finish is once again badly worn, it is best to remove most or all of it if you can. Follow the methods outlined above for removing paint. If an existing varnish or poly finish is only slightly worn, it can be refreshed by "screening" the surface with a rotary sander (the very type I've just recommend against above). Using the nylon screen material in place of sandpaper allows the machine to remove dirt and wax buildup and abrade the finish sufficiently for a new coat of varnish or poly to bond well. Vacuuming and wiping with a tack cloth are necessary before brushing on the new clear finish.

Professional floor refinishers will generally want to use their big sanding machines for removing paint and old clear finishes, and they will want to use a water-based poly for all coats. This is natural; time is money, and the faster they can do a floor, the more money they will make. Or they may simply not realize or appreciate the distinctive characteristics of a hand-planed softwood floor. The historic character and durability of a new finish on your floors is not their primary concern, but it should be yours. Insist that the job be done correctly.

Hardwood Floors

Hardwood flooring was much less common during the colonial period than in the later nineteenth century, when it became ubiquitous. It might be found in early homes of the well-to-do or in areas where hardwood trees were especially abundant. Early hardwood floors tend to have wider boards than later hardwood floors, although even the earlier boards were narrower than typical wide pine boards. After the middle of the nineteenth century, steam-powered mills began to produce hardwood flooring in narrow widths (3 inches or so) with tongue-and-groove edges for interlocking. Oak and chestnut were the most popular though by no means the only hardwoods used across

This freshly refinished hardwood floor has been coated with a satin-finish polyurethane for a more subtle shine than a full gloss finish would have imparted. The floor likely did have a high-gloss finish originally, along with the extensive woodwork in the room. The surviving original finish on the woodwork has become less glossy with time, however, and matching the floor to the current woodwork finish is appropriate.

the nation. Some woods are associated with particular regions. By the early twentieth century, lighter-colored maple came into popularity, generally milled 2½ inches wide.

Hardwood Repairs

Finding replacement wood that matches the color and grain of an existing hardwood floor often requires some searching. Start close to home—in your home, in fact. Can material be salvaged from the floor of a closet or room that will be getting new flooring anyway? Perhaps you are planning a new bathroom with a tile floor in a room that currently has hardwood flooring. Will new cabinets be going into a kitchen with hardwood flooring? If so, you can salvage the hardwood and replace it with plywood underlayment where the cabinets will cover it. A future owner may curse you if they want to remove the cabinets, so leave a note explaining that the flooring was used to restore more visible floors in the house. They'll get over it.

If there is no salvageable flooring in the house, look for matching material at nearby architectural salvage shops. Historically, materials like hardwood flooring would often come from local suppliers and be used in many homes within a community. You may find matching material from another house of the same period in your town.

If you strike out on locating old material to use for repairs, you'll have to find new wood that will blend in when stained and clear-coated or oiled. Take a sample of the old flooring to a hardwood supplier and look to match the grain and coloring. Old stains and finishes complicate this matching, so sand the back of your sample to expose fresh wood and try to match that. If your old wood is quarter-sawn, the new wood should be as well. Once you find a suitable material for repairs, buy more of it than you need and store the extra in a dry spot in the attic or an outbuilding. You or a future owner will someday be grateful for the foresight.

Repair methods and techniques for hardwood are similar to those for softwood, as described above. One difference is that tongue-and-groove hardwood boards are interlocked and blind nailed. The drawings

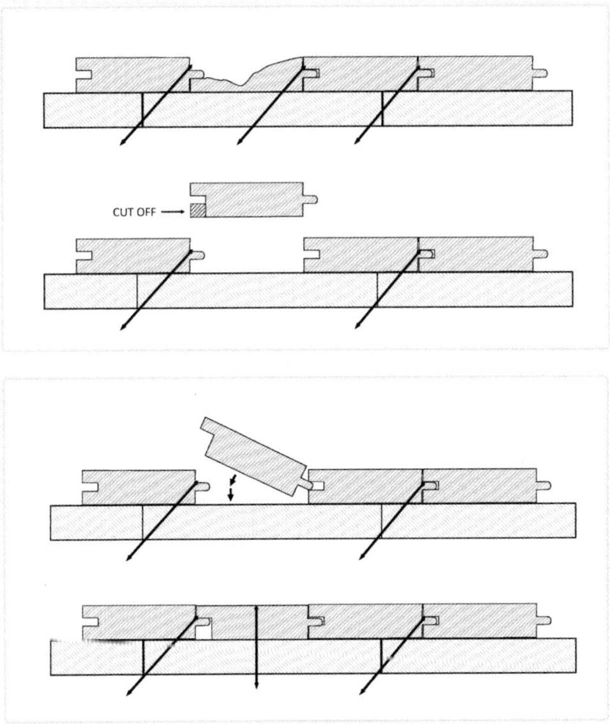

The steps for replacing a damaged or missing board in a hardwood floor are shown here. After removing the broken board and its nails, a new matching board is cut to length and the wood below the groove is cut off. The new board is then tipped into place, forcing the tongue into the adjacent groove, and the opposite edge is pressed into place. A bit of tapping and patience may be required. Once the new board is in place, it is nailed to the joists through the face with 16-penny finish nails, sunk below the surface and filled with color-matched wood filler.

illustrate the method to replace these. The replacement board will have to be face nailed. The nails should be countersunk and covered with wood filler.

Hardwood Finishes

Hardwood floors were much less commonly painted than softwood. If your house has hardwood floors that appear to have always been painted (no clear finish under the paint layers), refer to the information above on painted softwood floors.

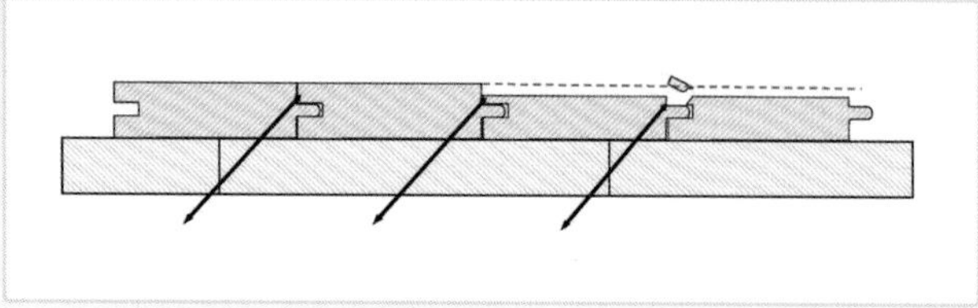

This drawing shows the danger of oversanding a tongue-and-groove hardwood floor. Most floors of this type can withstand one or two sandings without an issue, but after that there is a good chance the thin strip of wood left over every tongue will be vulnerable to breaking from being walked on or even from seasonal expansion and contraction of the wood.

If your hardwood floor has a clear-coat finish (varnish, shellac, or polyurethane) that is dull and worn but the wood itself is in good condition, removing a buildup of floor wax (and embedded dust and dirt) may be all that is necessary to restore the luster to the wood. Wax can be stripped by hand, but this is no fun. Using a wax stripper with a nylon pad on a drum sander is much faster. The object is not to remove the varnish, only the wax buildup. With the old wax removed, a fresh coat of wax will restore the shine without the grime.

If the existing clear coat is worn through or yellowed, the next level of intervention is to screen the floor using a sanding screen on a drum sander. The goal this time is to remove the finish but not to sand away any wood. If there is minimal wear into the wood, some spot sanding by hand can remove any surface grime from these areas. If you can successfully remove nearly all the clear coat and very little wood, you may be able to apply a fresh clear coat without needing to stain the floor. Staining and clear coating are covered below.

If the wood is damaged and the finish is in bad condition with scratches, stains from water or other liquids, and so forth, the recommended approach is to sand a thin layer of wood off the entire floor. A floor requiring this level of intervention may well need repairs as well as refinishing, and those are discussed below. Floorboards can only lose so much thickness before the top of the "groove" side becomes vulnerable to breaking. If the floor has been aggressively sanded in the past or less aggressively sanded several times, you may not have sufficient remaining thickness for another refinishing.

Some hardwood floors were never finished with varnish or shellac, but were oiled instead. Oil brings out the colors and grain of the wood and protects it from water without creating a hard, shiny surface. Oiled floors were sometimes waxed, but more often not. Oil would have to be applied periodically to prevent the floors from drying out and becoming dull. Oiling remains a finishing option. Tung oil, walnut oil, and other oils that do not form a film like linseed oil are used, and commercially produced floor oil preparations are available.

Whether or not to stain the wood before clear-coating or oiling depends on the wood and the house's period of significance. Nineteenth-century oak, chestnut, or similar floors are likely to have been darker than the turn-of-the-twentieth-century maple or birch that was installed over softwood floors in many older homes around that time. Research your style and period to determine what is most appropriate. In general, medium to dark tones will look "older" than light tones. Many nineteenth-century houses had floors laid in several contrasting woods, to create patterns with darker and lighter colors. These should not be stained, as the intent was to create an effect with the natural tones of the woods.

Floor Cloths

In the nineteenth century, painted canvas floor cloths were used in hallways, kitchens, pantries, and other areas where a waterproof and easily cleaned floor covering was needed. These forerunners of linoleum and vinyl sheet flooring could be homemade

This front entry hall floor in the Farnsworth House museum in Maine is covered by a restored nineteenth-century floor cloth that imitates a carpet design, in place since the 1850s. The runner on the stairs is a compatible replacement for the original stair carpet.

This reproduction canvas floor cloth under a dining room table has a design that imitates a period carpet with border, as floor cloths frequently did historically.

or factory-produced. Typically, they were made by covering thick canvas with several coats of oil paint followed by several coats of varnish. Several lengths of canvas might be sewn together to get the width needed, and the edges were turned under and sewn prior to painting. When fitted wall-to wall, several layers of newspaper were sometimes laid over the wood floor to provide a cushion under the canvas. Wall-to-wall floor cloths were often tacked around the edges with carpet tacks to prevent them from moving. Some floor cloths were painted in solid colors, but the majority were painted with patterns or trompe l'oeil effects, such as faux marble blocks or imitations of carpet patterns.

Surviving antique floor cloths are rare but are sometimes preserved under later carpet, in an attic, or in a barn. If the canvas backing remains in reasonably good condition, the cloth may be restorable even if dirty and worn. Even if one turns up, a reproduction might make more sense for any area that is going to be walked on regularly. An antique floor cloth will be better preserved in a guest room or other occasionally used space.

This is a partially cleaned piece of the restored hallway floor cloth shown on the previous page. Originally used in an adjacent cross hall, this piece is stored in the attic of the house and was used to test cleaning methods for the larger cloth that remained in place. It shows what is possible if the underlying canvas remains in reasonably good condition.

Carpet

It might seem that carpet should be considered "decoration or furnishing" rather than flooring, but it often determined what flooring was to be used, particularly if wall-to-wall carpeting was intended to be used. If a rather formal room of the nineteenth century has a rough softwood floor that was not painted historically, the room was probably carpeted when new. The same would be true for a 1950s living room with only plywood underlayment for flooring.

The history of carpet production and use is far too complex to delve into here; someday I'll write *Decorating Your Historic House* to tackle that. However, wall-to-wall carpeting came into use far earlier than many people realize. By the mid-eighteenth century, factories in England were weaving 27-inch-wide patterned carpet that could be sewn together in strips to create a repeating pattern from wall to wall. Border designs were also produced, to be used as a "frame" around the repeating pattern. This carpet was expensive and found in the homes of the wealthy and upper-middle classes. American mills began to copy the English carpets in the early nineteenth century, producing flat ingrain and similar carpets. The quality of the domestic carpets was initially lower but caught up quickly.

French carpets, such as those from Aubusson, were part of the shift in demand from English to French production for household furnishings toward the middle of the nineteenth century. Wealthy women who ordered their dresses from Paris increasingly wanted furnishings from France as well. With time, less expensive domestic carpets became available to the middle-middle and lower-middle classes. As wall-to-wall carpet became more common and affordable, the wealthy began to prefer large oriental rugs and carpets surrounded by gleaming hardwood floor. Floors from this period are

Ingrain carpet like this reproduction was woven in 27-inch-wide strips on mechanical looms and stitched together for wall-to-wall carpeting. Flat-woven, this style of carpet is reversible.

French-style carpets became widely popular in the middle of the nineteenth century, and the naturalistic floral and architectural patterns were quickly adopted by English and American carpet manufacturers. *(Courtesy the Carol M. Highsmith Archive, Library of Congress, Prints and Photographs Division)*

By the late 1920s, the Colonial Revival trend of using oriental rugs had caught up with this Late Georgian to Federal-style house built in 1794. The rugs were laid over earlier rush matting, which was a popular summer floor covering in the nineteenth century. *(Federal Hill, John Keim House, Fredericksburg, Virginia; photo by F. B. Johnson, between 1927 and 1929; courtesy Library of Congress, Prints and Photographs Division)*

sometimes found with hardwood around the edges of the room and unfinished pine in the center, where it was intended to be covered by a carpet.

With the Aesthetic Movement came a widespread craze for oriental rugs, sometimes layered two or three deep on the floor. Colonial Revival interiors exhibited a more restrained use of antique oriental rugs, along with hooked and braided rugs that had an "early American" feel. Wall-to-wall carpeting went out of favor in the early twentieth century, not to return in a substantial way until the postwar period, when solid color and mottled cut pile carpets competed with Berber carpet. In the 1960s long cut pile, or shag, became the must-have carpeting.

Many historic types of carpet are now reproduced, some on the same antique English looms that produced them historically. If there is evidence that a room had carpet historically, and your goal is to fully recreate the historic character of the space, then reproduction carpet is the way to go. See Chapter 20 for manufacturers and outlets of reproduction carpet.

Shag carpeting is used in this Frank Lloyd Wright house built in 1950, which has furnishings chosen by the architect. *(Castle Rock, Lowell and Agnes Walter House, courtesy the Carol M. Highsmith Archive, Library of Congress, Prints and Photographs Division)*

I picked up an 1840s ingrain carpet for the parlor of Whitten House at an estate auction. It is in somewhat delicate condition, and the resident cats love to hang out on it, so we keep the doors to the room closed except during holidays. Whitten and Maggie are enjoying the parlor here. The carpet gets a careful cleaning annually with a vacuum nozzle held at a sharp angle to reduce the suction on its delicate fibers.

Historic carpets do turn up at auction and at shops. If you find one in good shape that fits your room, consider using it if it is not going to get excessive wear in that location.

Stone Flooring

Tile and stone flooring are covered along with wall tile in Chapter 15. Here I'll mention terrazzo, which is a faux stone flooring composed of concrete and aggregate of real stone chips. Depending on the coloring of the concrete and stone chips used, a variety of effects can be achieved. The material is most often found in commercial and institutional buildings starting in the early twentieth century, but also was occasionally used in large houses of that period. It can be poured in place for a seamless floor or installed in premade slabs.

Damage to terrazzo comes in the form of cracks from movement in the structure below and impact damage from above. This often occurs during

A classic late-nineteenth- or early-twentieth-century combination of a white hex tile field with square colored tile border and marble baseboard.

Terrazzo flooring is composed of colored concrete and real stone chips, often marble.

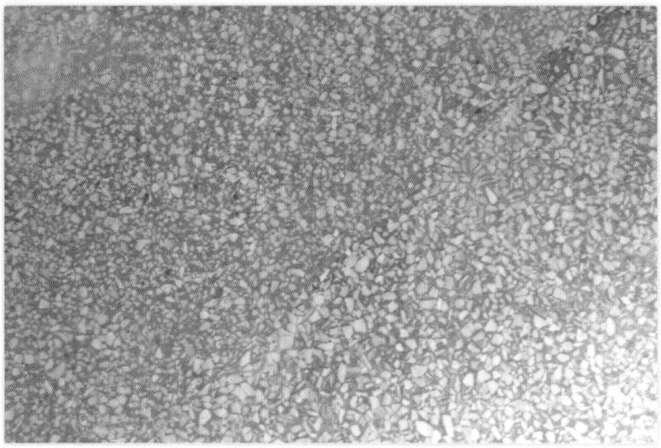

A terrazzo repair (upper left portion of photo). While not an exact match, this repair will disappear from only a few feet away, particularly once the room is furnished.

renovations, when floor plates for new walls are nailed into the terrazzo or new plumbing chases are cut through the floor. A skilled floor person will be able to match the consistency and color of the concrete as well as the size, color, and distribution of the aggregate to repair damaged terrazzo. It is recommended that several sample panels be prepared and set on the floor to be compared side by side. Because the distribution of the aggregate can vary across a room or from room to room, it is a good idea to move

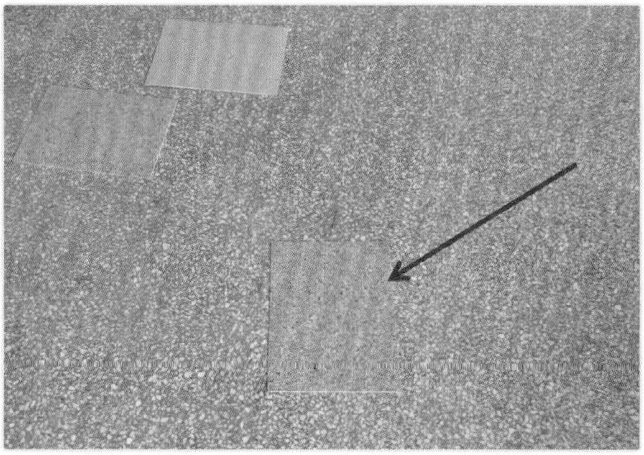

On this project, the terrazzo flooring was too damaged in some rooms to be reasonably repaired. A vinyl sheet flooring was found that was a close enough match to be acceptable in those spaces. Here, three samples of the vinyl material are shown against the original terrazzo. The sample indicated by the arrow was selected.

the samples around. Aim for a happy medium if there is a wide range in the distribution.

Linoleum

A product of the nineteenth century, linoleum was invented in 1864. Made of canvas, linseed oil, and powdered cork, it was the first sheet flooring material that arrived in rolls and was cut to fit a space from wall-to-wall and glued down. It represented an advance on the painted canvas floor cloths that had been in common use for decades. Unlike the thin floor cloths, whose canvas backing was protected only by several layers of paint and varnish, the cork and linseed oil coating of linoleum was an eighth of an inch thick with the same color all the way through. Linoleum was much longer lasting than a floor cloth. Originally available only in natural brown, it was soon manufactured in colors. Decorative designs could be created by cutting pieces of various colors and assembling them like mosaic tile, either on site

It's the **Gold Seal Inlaid Linoleum** that works magic all through the house!

This magazine ad from the 1950s shows linoleum on the floor, countertop, and cabinet fronts. Although the use of linoleum waned in the 1950s with the rise of vinyl flooring products, it is now regaining popularity as an all-natural product.

A sample card for linoleum available from one manufacturer. With increasing interest in the houses of the mid-twentieth century, more products from that period are likely to become available to serve the restoration market. Some of these patterns will be appropriate for kitchens, pantries, and bathrooms dating back into the nineteenth century.

Detail from a printed "linoleum rug."

or in a factory. This inlaid linoleum was often used in place of ceramic tile, particularly in entryways, pantries, and kitchens. By mixing colors in the manufacture of the product, it was possible to create "marble" patterns, which were popular for bathrooms. Linoleum floors were usually waxed to give them a shine and protect the surface from abrasion.

In the early twentieth century, a thinner linoleum-like product made of bitumen-impregnated felt paper became available. Often mistakenly called "linoleum rugs" today, rectangular pieces of this thinner material were printed to look like oriental carpets or hooked rugs and placed in rooms with a border of painted or varnished wood floor or of the same material with a printed wood grain pattern showing around the "rug." Unlike inlaid linoleum, the colors were only printed in thin paint and varnished (much like floor cloths) and thus were prone to wear off. In northern New England it is not uncommon to find these "linoleum rugs" in historic homes that have remained in the same family for several generations.

Linoleum was largely displaced by sheet vinyl flooring in the second half of the twentieth century, but it has experienced a resurgence in recent years as an ecofriendly flooring made from renewable natural materials. It is available as a glue-down sheet material or as snap-together tiles for a floating (adhesive-free) floor installation in a wide range of colors and patterns. The options for finding appropriate linoleum flooring for your house are broader today than they've been for decades.

Rubber Flooring

Rubber tile was developed in the 1890s and became a popular, resilient flooring for public spaces as well as residential kitchens, bathrooms, and hallways. Rubber tiles were sound-deadening, water-resistant, and easily kept clean. By 1929, Goodyear was producing thirty standard colors and three thicknesses of floor tile. Tiles were available as solid colors, marbleized, or paisley. Many of the styles were interlocking, allowing for a floating installation (i.e., one without adhesives). Decorative patterns were frequently created with varied shapes and colors of tile. A checkerboard center of marbleized tiles surrounded by a border in other colors was a common design. Later, rubber tile floors were glued down.

This ad for Goodyear Rubber Flooring tiles appeared in the September 1953 issue of *Better Homes and Gardens* magazine. The copy reads, in part, "Now your own home can enjoy the luxurious cushioned comfort, the lasting beauty of *flooring that almost never wears out!*"

Surviving rubber tile floors can be cleaned with mild soaps and cold water. A small amount of ammonia can be added to the water. A sponge mop is recommended. Do not soak the floor in water; the water might penetrate seams and loosen the adhesive. Be very cautious of any cleaning products beyond mild soap and ammonia. Many solvents will soften rubber and leave it gooey. Test any product in an inconspicuous area and rinse it well with clean water. Wait at least 24 hours after testing to be sure no damage occurs before continuing. Many rubber tile floors were waxed historically, and a coat of wax remains an excellent way to protect the flooring. It will also restore shine to a floor that has dulled. A water-based emulsion floor polish should be used. A hard paste wax can be applied over the water-based emulsion floor polish for a more durable finish. Do not apply paste wax directly to the rubber tiles, as solvents in the wax may damage the tiles.

Damaged tiles can be replaced with new rubber tiles if a good color match can be achieved. Custom colored tiles can be produced but tend to be expensive. Interlocking tiles with complex shapes can be cut from larger square tiles. If the modern tile is thinner than the historic, it can be shimmed with another thin flat material underneath.

Vinyl and Vinyl Asbestos Flooring

Vinyl flooring was developed in the 1930s and became very popular after World War II. Many postwar houses were built with vinyl tile flooring in kitchens, bathrooms, family rooms, and entryways.

A wide range of colors and patterns was produced. Marbleized patterns were probably the most common, frequently laid in a geometric pattern made up of two or more colors. These early vinyl tiles were made in many sizes. The 9-inch-square size appears to have been the most common for residential use, but 12-inch-square tiles (the standard modern size) were also used.

A 1967 ad for Azrock vinyl asbestos tile. Vinyl (polyvinyl chloride, or PVC) was discovered in the 1920s by a Goodyear Rubber chemist searching for an adhesive that would bond rubber to metal. It failed as an adhesive but went on the become the most versatile plastic of the century, eventually replacing rubber for many uses. Its use in flooring products began after World War II and became dominant by the 1980s.

This is a detail from a 1955 ad for John-Mansville Terraflex vinyl asbestos tile, "the original vinyl asbestos tile." The ad claims that the product "has practically eliminated one of the most unpleasant of chores – scrubbing the floor!" and emphasizes the simplicity of "do it yourself" installation.

Unfortunately, many of the early vinyl tiles contained asbestos, which poses some hazard if shoes abrade the tiles, producing dust. Like lead paint, asbestos that remains undisturbed poses little danger. Encapsulation is an accepted approach to keeping potentially hazardous materials undisturbed. For lead paint, this means a new coat of paint that does not contain lead. For vinyl tile that is still well adhered to the subfloor and not broken at the edges, encapsulation might be accomplished with floor wax or several thin coats of polyurethane. Loose tiles that are intact can usually be glued back in place. Alternatively, the asbestos-containing tiles can be replaced by new

12-inch-square tiles cut down to 9-inch squares. As when removing any potentially hazardous material, check with local officials for removal and disposal requirements. In most areas, there are firms that specialize in this work.

Cork Tile

Cork tile was also developed in the late nineteenth century and became popular for its resiliency and sound-deadening qualities. It is made of ground cork heated under pressure in a cast iron mold to a temperature at which the natural resins in the cork liquefy and bind together the granules. It was, and is, made in three colors—light-, medium-, and dark-brown—with the degree of darkness determined by the temperature to which it is heated. After World War II, other resins were added to improve the durability of the product. It was laid in a waterproof adhesive, often mastic, and patterns could be made from the different shades of brown. Originally it was installed as it came from the presses and waxed for protection after being sanded smooth. Later, resins were applied to the surface at the factory to give the tiles a protective coating. They were usually waxed after installation even with the lacquer or urethane finish.

Cork can be cleaned with gentle methods like those outlined for rubber tile above. Paste wax will protect the surface once it is cleaned. Cork tile is still produced, and replacement pieces can often be found to match the color of a historic cork. If not, stain or a tinted polyurethane might get you to a good color match. Wax will help new tile blend with old.

Modern Materials

Some modern flooring materials imitate traditional materials; others have a contemporary character and make no attempt to look like another, older material. In general, new materials that look like new materials can be considered inappropriate in a historic house. Here we will concentrate on new materials that imitate historic materials. Before one of these is used to replace missing historic materials, several questions must be asked:

1. Is the material dimensionally correct for the period of the house? Many of these materials are thinner than the materials they claim to replicate, creating awkward transitions.
2. Is the texture or pattern reflective of the period?
3. Is the finish appropriate to or compatible with the period?
4. Does the material have comparable durability to the historic material it will replace?
5. Will the material accept an appropriate finish?

If one or more of these questions cannot be answered in the affirmative, the material is probably not the best choice.

Laminate

Modern laminate flooring is generally made of fiberboard or cork and covered with a printed paper pattern that is sealed with a clear finish. Some have a wood grain texture in the surface that mimics the texture of real wood with varying degrees of realism, notwithstanding the fact that real wood floors are generally sanded as smoothly as possible, without

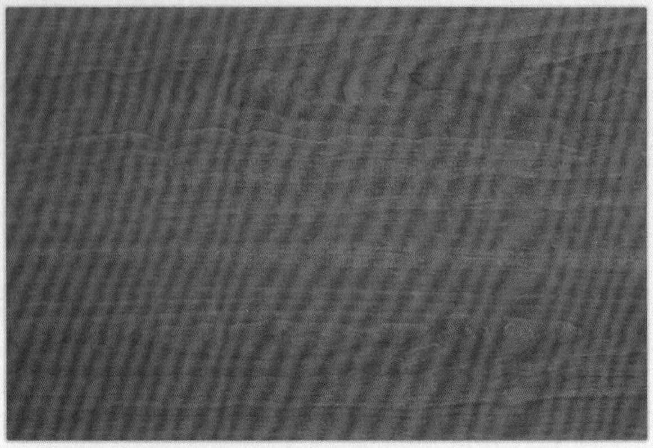

Perfectly imitating the appearance of wood (with a printed photograph of actual wood), laminate flooring may be appropriate for limited uses in historic houses but cannot be considered equivalent to real wood in durability or longevity.

texture. Laminate flooring comes in "planks" that snap together and float on the subfloor.

Shortly after this flooring hit the market in a big way, in the early twenty-first century, I read an editorial about that and other new substitute materials in *Old House Interiors* by founding editor Patricia Poore, who was also a founding editor of *Old House Journal* in the 1970s and is one of my preservation heroes. My recollection of the editorial is that Ms. Poore had visited a friend who had installed a new laminate floor in her kitchen that looked virtually identical to a real wood floor. She was pondering whether a flooring that looked just like wood, and in fact was printed from a photograph of real wood, could be considered appropriate for an old house. It challenged her beliefs as a preservationist.

I do not recall her conclusion, if she stated one, but have come to my own conclusion about laminate flooring now that it has been around for several decades. It will never replace real wood flooring for durability and longevity. Water destroys it relatively quickly, and abrasion that removes the printed paper pattern cannot be repaired. In my experience, five years is about the average time one can expect a typical laminate floor to look good—provided it never gets really wet. This is not to say it has no uses in preservation-restoration. I installed it over the concrete slab in the phase 1 kitchen of Whitten House (see the Compatible Kitchen example project on page 138). When the 1960s addition is removed and the new kitchen built on the footprint of the historic woodshed, real wood flooring will go into it.

Other Modern Materials

Other modern flooring materials that mimic the appearance of traditional solid wood include engineered wood (a real wood veneer attached to interlocking planks for a floating floor); vinyl planks (glue-down planks with a printed wood grain pattern); wood grain ceramic tile planks (6 or 8 inches wide by 24, 36, or 48 inches long) that are installed like any other ceramic tile but with thin, color-matched grout lines to minimize their visibility; and sheet vinyl with wood grain patterns (which has been around since the 1960s, when it replaced linoleum with wood grain patterns). Sheet vinyl with faux wood grain will never fool anyone but can be fun in a postwar house. Of these flooring types, engineered wood has the best chance of passing for a solid wood floor. However, it will be more susceptible to damage due to its relatively thin surface veneer, and it cannot be repaired or refinished like a solid wood floor.

Ridlon Camp, circa 1908

Rustic Style

In the mid-nineteenth century, well-to-do East Coast urbanites began to "rusticate" during the summer, escaping the stench and diseases of increasingly overcrowded cities during the hottest months. Some went to the seashore, rather quickly evolving from rusticating in modest cottages to summering in marble mansions that they continued to call cottages. Others, however, took to the woods. In upstate New York, the camps they built in the Adirondack mountain region were often architect-designed, with influences from Swiss chalets and the rustic follies of European estates. The Great Camps of the Adirondacks could have a dozen bedrooms (sometimes supported by steel beams hidden in logs) and a separate building to house staff.

Most of the camps built for rusticating (and hunting and fishing) in Maine took their inspiration from a more local source—the crude log cabins built by logging companies to house their crews during the winter logging season. The Rangeley Lakes region of Maine became famous for its fishing in the 1880s, and hundreds of relatively modest camps were built in the area, particularly after the railroads made it more accessible. Earlier in the nineteenth century, many Maine loggers had gone west to establish the logging industry in the upper Midwest, and similar recreational camps emerged from the logging camps there later in the century.

In Maine there were commercial camp operations with a dozen or more cabins available for rent and a central dining hall and other facilities for group activities. Many other camps were single-family cabins for private use, but a few private compounds likewise contained multiple cabins. The Ridlon Camp is an example of one of these, and the cabin featured here appears to have been the Ridlon family's personal cabin, originally called Wellesmere. It occupies the most advantageous spot on the site, with the best view of the lake, and is the only cabin in the camp to have a gambrel roof (providing more space for bedrooms upstairs) and a wraparound porch.

Frank Ridlon was a successful pioneer in the manufacture of electrical dynamos and motors; he owned an electrical supply company in Boston specializing in maintenance of central power station and railway equipment and was involved with street railway (trolley) and power supply companies. He was a regular visitor to the area well before he owned his own camp and had a family connection to the owners of the nearby Mooselookmeguntic House hotel.

National sporting magazines and several local newspapers tracked the comings and goings of the rich and famous "sportsmen" in the little rural towns of the Rangeley Lakes region during the period. The June 6, 1903, edition of *Forest and Stream* reported that "the Mooselookmeguntic House and cottages have been taxed to their utmost capacity to accommodate the host of sportsmen from different parts of the country." Frank Ridlon was listed among "a few of the Massachusetts fishermen guests at the hotel." In May 1906, the *Maine Woods* reported him "among the first parties that arrived at the lake" for the season.

The first mention of Ridlon buying or building his own camp appeared in the *Philips Phonograph* on September 28, 1900, where a report on activity at the Mooselookmeguntic House included a note that "Mr. Frank Ridlon of Boston ... returned to his home Monday morning after a stay of nearly two weeks. Mr. Ridlon expects to buy near Bald Mt. Camps and build a handsome cottage for his summer home, the plan of which has almost been completed." Four years elapsed before Ridlon bought land, however; in September 1904, according to the local registry of deeds, he purchased ten lots from his brother-in-law whose sister-in-law owned the nearby hotel. A number of camps already stood on the lots, but there is documentation of Ridlon having two camps built over the winter of 1908-09, and Wellesmere was likely one of them; the second may have been the wood-framed building to house servants that appeared on the property around the same time. Supporting the idea that Ridlon used the multiple cabins on his property as a private compound, *Maine Woods* reported on August 25, 1910, that "Mr. and Mrs. Frank Ridlon of Boston at their attractive camp entertain many guests and add to the pleasure of friends by taking them for a sail in their launch."

Like many of the camps of the era, Wellesmere featured a large living space open to the rafters at the front of the cabin and overlooked by a balcony that provided access to several upstairs bedrooms. A larger bedroom was located at the rear of the first floor, along with a bathroom and kitchen. A corner fireplace in the living area vented to an angled chimney rising from the roof. A second chimney on the side of the camp served the kitchen stove. The wraparound porch provided a large covered outdoor living space overlooking Mooselookmeguntic Lake. Like many camps of the early twentieth century, a hint of the Arts & Crafts style appeared in the large shingled piers supporting the porch roof.

Frank Ridlon did not get to enjoy his new camp for long; he died in December 1911 at the age of 73. His widow and daughter continued to use the camp on a less expansive scale, apparently staying in Wellesmere and renting out the other cabins beginning in 1912. A number of the cabins were leased to the Mooselookmeguntic House and rented to guests along with the hotel's own cabins. In 1938, the hotel owner purchased all the Ridlon camps, and they remained a part of the hotel property until the hotel was destroyed by fire in 1958. A subdivision plan filed with the registry of deeds in 1960 suggests that the property was broken up and the camps sold off around that time. By the 1970s, Wellsmere was known as Redlawn Camp (perhaps a corruption of Ridlon); most of its wraparound porch had been replaced by a contemporary open deck with bench railings, and its original windows (including dormer windows), interior fireplace, and other elements had also disappeared.

When the current owners, Dan and Debbie, bought the camp in 2012, it had lost much of its historic character and was in need of sensitive restoration. Dan says, "Our mission or vision was pretty simple—bring back or preserve the original character and charm. And of course, try and slip in some modern-day creature features without compromising the old-school look." They worked with Mark Gordon of Rangeley Builders, a local contractor with extensive experience working on Maine's historic (and new) log camps. Speaking of the work done to restore the historic character of the camp while updating "creature comforts," Dan said, "Mark's a rock star—he did it!"

The most apparent change is the restoration of the porch on the front of the camp. When the project started, the existing open deck projected farther from the building than the historic porch had. This presented a quandary that became an opportunity. The quandary was that the large deck overlooking the lake was a nice feature, but restoring the original appearance of the camp would require reducing its size. The solution was to restore the porch roof and piers in their original locations while retaining the larger deck, extending the historic design of the shingled railing around the larger deck to relate it to the historic building. At a glance, the porch appears much as it does in historic photos while accommodating larger groups. Other changes to the exterior include new windows matching the historic two-over-two double-hung window design, with a third window added in the center of the gable to increase natural light inside the camp, and building a storage shed into a portion of the porch on the side of the camp. The 1960s aluminum sliding glass door to the deck was replaced with a more compatible wood sliding door. A subtle but functionally important change was the addition of insulation on the exterior of the historic roof, made necessary by the introduction of a central heating system for the camp. Exposed log rafters and decorative shingling on the interior side of the roof would have been entirely hidden by interior insulation.

Inside the camp, a new kitchen occupies the corner where the original fireplace had been before its removal by previous owners. Simple painted beadboard cabinets are set off with a custom frieze and crown molding at the top. Peeled and varnished lengths of branch were applied to the frieze, providing a rustic touch to tie the new cabinetry to the log and beadboard interior. A stone hearth and traditionally styled propane gas heating stove with open front was installed in the opposite front corner, providing heat and the ambiance of a fireplace in the living room. The flooring in the kitchen and living room area was beyond saving and was replaced in-kind with new wood flooring. Where the flooring was salvageable, it was refinished. The first-floor bathroom was redone, and closets were added to the bedroom. A new forced hot air central heating system was installed (the first for the camp) with cast iron floor registers to maintain the historic character of the space.

On the second story, the beadboard wall between the two bedrooms was moved approximately three feet and a new wall built six feet from it, creating space for a new bathroom and closets between the bedrooms. A loft for kids to play and sleep in with log railings like the historic railings on the original stairs and balcony overlooking one bedroom was built above the bath and closets; it is reached by a ladder. Future plans call for placing the camp on a concrete foundation, replacing the log posts on stone slabs that now support it.

After restoring the camp's historic character and preparing it for another century of life, Dan and Debbie renamed it Ridlon Camp to honor the original owners.

The original log-and-branch railings on the stairs.

The camp in the 1970s after losing its front porch, chimney, dormers, and original doors and windows. *(Courtesy Gary Priest)*

The living room area with original beadboard and log walls and the decorative shingle pattern between the rafters on the cathedral ceiling.

The porch roof was recreated in its original form and location while the deck expansion of the 1970s was retained.

This view from the kitchen toward the living room area shows the enhanced natural light from the added window in the gable.

The new bathroom between the upstairs bedrooms.

New light fixtures in the first-floor master bedroom impart a rustic character.

Taxidermy mounts are a common feature of historic log camps built by city dwellers who came to the woods to hunt and fish.

This deck and bench railing from the 1970s was at odds with the historic character of the camp. *(Courtesy Rangeley Builders)*

This frieze with applied branch "bead" relates the modern cabinets to the rustic camp.

Ridlon Camp in about 1910, showing the original porch. Several of the other cabins and outbuildings owned by Frank Ridlon are visible. *(Courtesy the owners)*

A view of the added bathroom and closets on the second floor, with children's loft above.

An antique padlock excavated on the property is used as a decorative detail.

The dining area and kitchen as seen from the living room. The painted modern cabinetry adds subtle color to the space.

The camp not long after it was built, likely around 1910. *(Courtesy Rangeley Builders)*

One of the two bedrooms under the eaves in the second story.

The Work

The new stone hearth for the gas heating stove being constructed in the living room. *(Courtesy Rangeley Builders)*

Rebuilding the porch deck prior to reconstructing the missing portion of the roof. *(Courtesy Rangeley Builders)*

Adding a third window in the gable. *(Courtesy Rangeley Builders)*

A diamond-shaped shingle detail in the wall of the portion of the porch that was enclosed to make a shed. *(Courtesy Rangeley Builders)*

Lifting a beam into place for the reconstructed porch roof. *(Courtesy Rangeley Builders)*

Another view of the porch reconstruction. *(Courtesy Rangeley Builders)*

CHAPTER FIFTEEN

Interior Walls

Historically, interior walls were often divided into three horizontal parts. From floor to ceiling, these were the baseboard and dado or wainscot; the field; and the frieze or cornice. This tripartite wall division has its roots in classical architecture and is related to the base, shaft, and capital of a classical column. "Correct" classical architecture prescribed the proportions of the elements relative to one another and to the height of the space, and builders' guides of the eighteenth and early nineteenth centuries often advised carpenters on those ideal proportions.

Increasingly wide latitude was exercised in practice, however, and by the end of the nineteenth century fidelity to classical proportions in American interiors was largely restricted to architect-designed homes in the Colonial Revival and Neo-Classical styles. In the second half of the nineteenth century, the tripartite wall division was often expressed with

This re-created eighteenth-century room at Colonial Williamsburg shows the tripartite wall division common in primary rooms of the period, with a paneled wainscot (or dado), a field of smooth plaster, and a denticulated cornice where the wall meets the ceiling.

paint or wallpapers on plaster walls bordered by a wood baseboard at the floor and sometimes a wood or plaster crown molding at the ceiling.

Plaster

Since the majority of interior wall surfaces in American historic houses are plaster, we will begin with that amazing material. Historic wall plaster is most often composed of a natural binding agent (typically lime) into which sand is mixed with a fiber of some kind to help prevent cracking. It is applied wet to some form of lath or directly to a solid (but rough) masonry or wood surface.

Historic lime or gypsum plaster is a high-quality material that makes an excellent wall surface. It has sound-deadening qualities that are absent from modern substitutes, and it provides a good base for paint or wallpaper. Because historic plaster was applied by hand, it often has subtle modulations in the surface that give character to a room.

The back side of a plaster wall shows clearly how the wet plaster was forced through the gaps between the wood lath when it was applied, forming "keys" that grip the wood and hold the dried plaster tightly against the lath.

Wall Plaster Repair and Replacement

Plaster is applied wet and cures through a chemical reaction into a dense material that can withstand mild abuse. Because it is rigid, however, it can be affected if a house settles or suffers from structural issues. Plaster can move and adjust with a house if the movement happens very slowly over a period of years, but faster movements can cause cracks and sometimes areas of loose plaster.

If at all possible, save your historic plaster! If not, you will spend significant money or effort to remove and replace it with an inferior material. I recently read a newspaper article about a newly rehabbed house in which the contractor was quoted as saying, "All old plaster eventually dries out and crumbles, so you have to replace it." This is not true. The chemical reaction that occurred when the plaster cured produced heat that drove all moisture out of the material. Old plaster has been bone-dry since shortly after it was installed. Prolonged exposure to water can cause cured plaster to soften and crumble, but dryness will not. Excessive water exposure is generally localized, related to a plumbing or roof leak. An entire house with plaster damaged by water exposure is a rare occurrence. Plaster crumbles when it is abused in some way, not because it is old and dry.

Gypsum board (frequently called sheetrock or drywall) has largely replaced plaster in new construction because it is cheaper, faster to install, and does not require highly skilled craftspeople. As with most things, however, faster and cheaper requires a tradeoff, and what is traded off in this instance is the character of a hand-applied surface. Subtle surface modulations that register in the brain as organic and human are replaced with dead-flat surfaces that feel factory-produced because they are. Unless your installer is better than average, you also get visible seams at regular intervals and a noticeable pattern of screw holes. And because the material comes in large sheets—4 by 8 or 4 by 12 feet—it acts as a drum on the wall, transferring sound between rooms. These are the reasons why new high-end houses are still plastered.

Wall plaster is generally simpler to repair than ceiling plaster, because slathering a wet mixture overhead is a demonstration of gravity in action. Methods for repairing cracks and reattaching loose plaster in ceilings are covered in Chapter 16 but can be applied to wall plaster as well.

Walls are more likely than ceilings to have holes in the plaster and lath (from electrical work or plumbing) that need patching. A house can be rewired without removing plaster, but a number of small holes will have to be cut to fish wires through wall and ceiling cavities. Holes also result when switch boxes are moved and when new plumbing is installed. The accompanying photos show the repair of a hole created by moving a light switch.

Water damage has exposed how the plaster in this house was applied to both brick and wood lath. This is typical of houses built in the eighteenth and nineteenth centuries, and sometimes even later.

Because many plaster walls were intended to be covered with wallpaper, they were not finished with a hard, smooth, finish coat of plaster for painting. If you find unpainted plaster behind wallpaper, chances are it has been wallpapered since it was new. If you choose to paint these walls, you will need to

A switch controlled the light in an adjacent room and was behind a large piece of furniture, making it inconvenient to use. I relocated it to the room it served, leaving a hole to be patched. The plaster around the switch box had been repaired in the past, likely when the switch was installed, but was not in great condition.

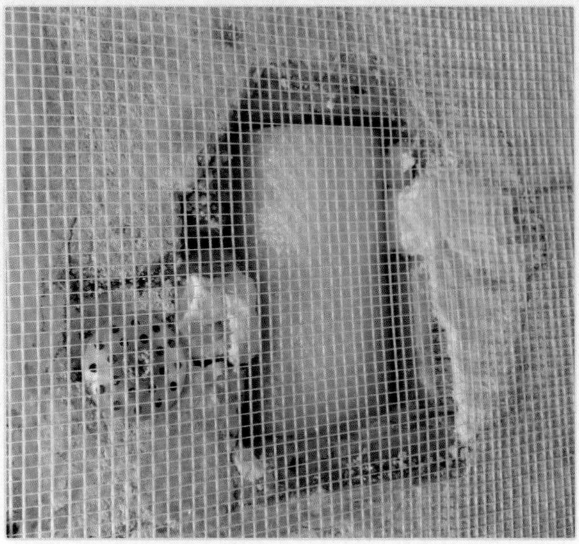

I installed a plaster washer to the left of the hole to secure a loose piece of plaster, then placed a pad of folded foam packing material in the hole as backing for the new plaster. (Wadded paper was often used for this historically.) I then cut a piece of fiberglass mesh large enough to cover the entire area.

I buried the mesh with Durabond setting plaster, tapering the edges of the Durabond gently back to the wall plane. Because time was tight in this instance, I applied a single thick coat instead of the normal two thinner coats. After the plaster set, I sanded it and then skim-coated the patch (and the rest of the wall) with softer joint compound.

With nearly all the painted-over wallpaper removed from this 1827 wall, the plaster is found to be in basically sound condition, with evidence of past patching and some newer holes and gaps along the trim to be patched. There are many small holes and irregularities across the wall surface as well.

Here the ceiling has been replastered (as described in the next chapter), and the holes and gaps in the wall plaster have been filled. The joint between the top of the wall and the ceiling was reinforced with fiberglass mesh joint tape, as were the vertical gaps along the window trim.

As a final step in preparing the wall for tinted sizing primer for new wallpaper, the entire surface is skim-coated with joint compound applied with a 6-inch putty knife, filling the small holes and irregularities. The goal is not a new coat of plaster (as was done on the ceiling) but a solid surface that retains the character of hand-applied plaster. Once the joint compound is dry, the wall will be lightly sanded with a 150-grit screen on a pole sander. Since little joint compound is applied, little sanding will be needed.

skim-coat the plaster to fill the many small voids and other irregularities in the surface. Unlike the thick skim coat used on badly damaged plaster (see the section on ceiling plaster in Chapter 16), this skim coat should be very thin. You are filling voids to create a smooth surface rather than creating an entirely new surface. Joint compound works well for skim-coating; I generally apply it with a 6-inch putty knife held with my middle and pointer fingers behind the blade to apply pressure as I scrape the knife across the plaster. Small nail holes and other damage will be filled as part of this application.

I generally apply 2-inch-wide paper or fiberglass mesh tape in corners when skim-coating wall plaster. There are usually cracks at the corners that have been covered by wallpaper but will be exposed as a painted surface, and joint tape prevents these from reopening. (The application of joint tape is covered in Chapter 16.)

A carefully applied skim coat needs little sanding. I usually go over the surfaces quickly with a sanding screen on a pole sander.

Once the walls are primed, some touch-up spots may become apparent. Simply skim over the primer, hit with the sanding pole when dry, and spot-prime over the new compound with a brush. A National Park Service *Preservation Brief on Flat Plaster* can be found online at www.nps.gov/tps/how-to-preserve/briefs/21-flat-plaster.htm.

Ornamental Plaster and Composition Ornament

Starting in American urban centers in the late eighteenth century, plaster and composition, or compo, were used to ornament the interior walls of houses. Plaster of Paris (usually without sand) was used instead of lime plaster for this work. Most ornamental plaster was cast in molds and installed with screws that were then hidden with plaster along with the seams between pieces. Plaster moldings could also be run in place, usually crown moldings at the junction of the wall and ceiling. A National Park Service *Preservation Brief on Ornamental Plaster* can be found online at www.nps.gov/tps/how-to-preserve/briefs/23-ornamental-plaster.htm.

Compo ornament used a mixture of glue, resin, linseed oil, and chalk or whiting (and sometimes other ingredients) to form one of the first widely used thermoplastics. This material, when heated, could be pressed into molds to create castings with fine detail. Once cooled, the material became very hard and dense, but could be made pliable again with heat for use on curved surfaces. A National Park Service *Preservation Brief on Composition Ornament* can be found online at www.nps.gov/tps/how-to-preserve/briefs/34-composition-ornament.htm.

All the ornament above the paneled wood wainscoting is cast plaster in this room built in 1899. Here, a sample area has been done to show the original color treatment, based on microscopic paint analysis. The wood wainscoting was originally stained and varnished, but the rehab budget did not allow for stripping the later paint and refinishing the wood. This is not the final shade of brown selected for the wainscoting.

Both of these products are still commercially available; some companies have been using the same molds since the nineteenth century It is also possible for a local plaster restorer or a homeowner willing to

The newly cast pieces on the left were made to match the historic crown molding casting on the right. These are cut-offs left after the new castings were shortened to fit a space where the historic casting was missing or too damaged for repair and reinstallation.

experiment and practice to create custom castings to match surviving pieces.

Wood Walls and Wainscots

Wall paneling and wainscoting could have been included in Chapter 17, Interior Trim, but since they often cover much or all of a wall, I am discussing them here.

The two types of wood paneling typically found in houses are historic solid wood stile-and-rail paneling and modern plywood sheet paneling. Either may be used for the full height of a wall or as wainscoting. (A wainscot may also be composed of horizontal or vertical boarding, as addressed below.)

Solid wood paneling was common in the Colonial era and reemerged in the Shingle and Colonial Revival styles. Paneled wainscoting can also be found in houses of the Victorian era, but full wall paneling was less common during that period.

Solid wood stile-and-rail paneling with raised panels in the 1755 Tate House Museum in Maine. Note the width of the large panel over the fireplace, cut from a single board of Eastern white pine.

Modern sheet paneling was developed after World War II and was used in new construction and as a cheap and easy cover-up for old plaster in existing houses. Ironically, style-and-rail paneling in early houses was often covered by plaster later in the nineteenth century in much the same way. No doubt there are houses with sheet paneling covering plaster that is covering solid wood paneling.

Modern rosewood sheet paneling was used in this 1957 bathroom along with other characteristic materials of the period, such as mirror tile and metallic wallpaper.

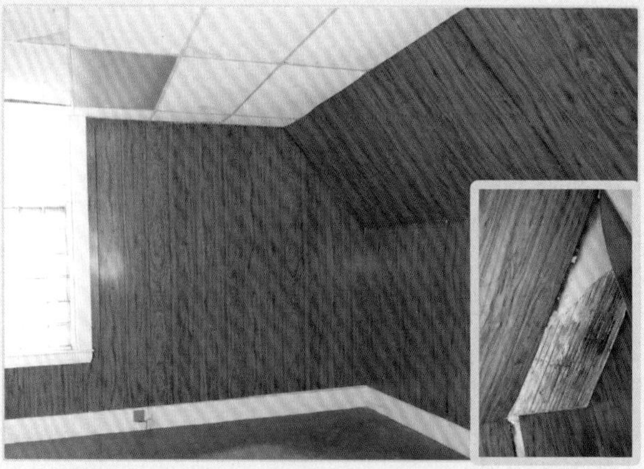

Here is a typical example of modern sheet paneling over damaged historic plaster. The inset image shows evidence of plaster damaged by a roof leak before it was covered. The paneling butts the historic baseboard and door and window trim, which were left in place—a common practice. In rooms with sheet paneling, it is not unusual to find suspended ceilings like this one hiding the historic ceiling plaster—not an appropriate treatment for the space.

A Victorian-era paneled wainscot at Historic New England's Eustis Estate in Massachusetts.

Wainscoting

Wainscoting comes in panels, wide horizontal boards, and vertical boards. Paneled wainscoting was installed in finer homes throughout the historic period, with variations in panel form, size, and detailing depending on when the house was built. Horizontal board wainscoting typically spans the gap between the top of the baseboard and the windowsills with one very wide board on each wall of the room. This wainscoting is typically found in houses built before the middle of the nineteenth century.

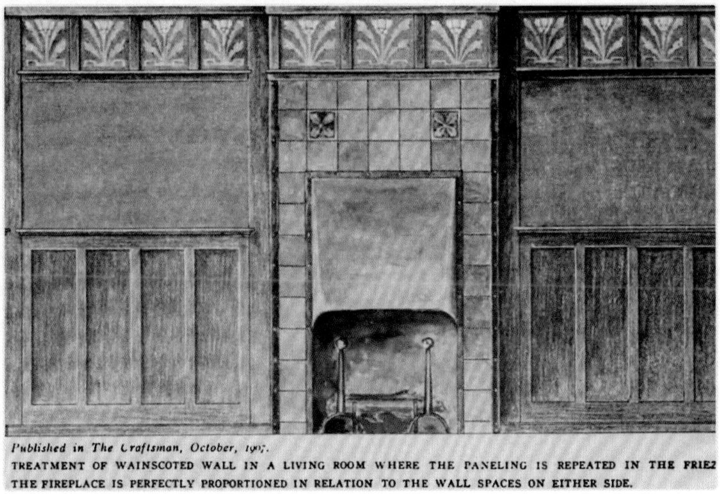
Published in The Craftsman, October,
TREATMENT OF WAINSCOTED WALL IN A LIVING ROOM WHERE THE PANELING IS REPEATED IN THE FRIEZ
THE FIREPLACE IS PERFECTLY PROPORTIONED IN RELATION TO THE WALL SPACES ON EITHER SIDE.

Board-and-batten vertical wainscoting became popular with the Craftsman style in the early twentieth century, as shown in this illustration from Gustave Stickley's 1909 book *Craftsman Homes*, originally published in *The Craftsman* magazine.

Wide horizontal board wainscoting in Whitten House. The wainscot board is 22 inches wide, and the baseboard is 10 inches. The wainscot cap continues as a windowsill, a typical detail of the era in the Northeast.

A Victorian-era paneled wainscot in a room of the 1870s.

Vertical board wainscoting, most often beadboard, came into popularity around the mid-nineteenth century. Vertical board wainscoting with battens covering the joints often rose to five feet or higher in Craftsman-style houses.

Wainscoting went out of common use in the later twentieth century, though sometimes a vestigial chair rail was used, particularly in Colonial Revival houses.

Repairing Stile-and-Rail Paneling

Solid wood stile-and-rail paneling is nearly always a character-defining feature of a room. During the Colonial period, housewrights showed off their skills on elaborate paneled interiors for grand houses, and less elaborate paneled walls were common in middle-class homes, typically on the chimney wall. Other walls might have a wood wainscot below the windows with plaster walls above. Paneling was typically painted.

When wallpaper became affordable and popular in the early nineteenth century, many "old-fashioned" paneled walls were plastered over or had the sloped edges of the panels filled in with strips of wood so they could be papered. Sometimes these infilled paneled walls emerge when a room is stripped of multiple layers of wallpaper. Exposing them may be an appropriate treatment provided the rest of the room was not completely remodeled in a later style. (A Colonial paneled wall is going to be out of place in a room with heavy Italianate trim on the windows and doors, hardwood floors, and a large Victorian gasolier lighting fixture. It is one thing to represent different periods of a house's history in different rooms but quite another to intermix them in a single room. That is not to say it *can't* be done successfully, but it isn't easy to pull off.)

A common repair need for paneled walls is the result of the mid-nineteenth-century introduction of cast iron woodstoves into houses that had previously been heated with fireplaces. All too often, an 8-inch-diameter hole was cut through the large panel over the fireplace to install a thimble in the chimney for the stove pipe. Sometimes these holes are still exposed, often with a metal thimble cap over them; other times they have been poorly patched. A bad patch just above a fireplace—the focal point in the room—is going to be noticed. Your options include installing a good Dutchman's patch in the panel, replacing the panel with a single wide board shaped to reproduce the original panel, or hanging a painting over the problem to hide it.

If you opt for panel replacement, it will probably require removing part or all of the stile-and-rail framework from its location (it is usually nailed in place), drilling out the small dowels that hold the stiles and rails together, and partially disassembling the framework. It is only necessary to open up one side of the rectangle the panel needs to slide into. The old panel is likely bound in tightly by paint, and it will be necessary to score along the joints with the stiles and rails to break the seal. The same is true of the junctions of stiles and rails that need to come apart. This will likely be a slow and tedious process; it needs to be done carefully to avoid damaging the framework the new panel needs to slide into.

Another common issue is cracks in historic panels. The stile-and-rail design is intended to allow the panels to expand and contract with changes in temperature and humidity; that is why the panels sit in grooves cut in the sides of the frame pieces without any mechanical attachment. Over time, however, a buildup of paint can "glue" the panels to the frame, creating stresses in the panels that cause them to crack along the grain as they try to expand or contract.

Cracked panels can often be repaired without removing them from the framework. The technique—which is the same as that used to repair cracked door panels (see the photo on page 488)—requires breaking the paint bond around the edges of the panel with a sharp utility knife, then clamping and gluing the panels back together. Clamping blocks will have to be temporarily screwed to the face of the panel, being careful not to screw into anything behind the panel. (There is usually a gap there, but not always.) Place the clamps on the blocks and exert just enough clamping pressure to begin to move the panel pieces in their grooves; then apply glue to the broken edges before tightening the clamps further to pull the pieces completely together. I have had good luck with urethane glues (e.g., Gorilla Glue) for panels that are to be repainted, but traditional wood glue will also work. Avoid getting glue in the grooves around the panel or you'll be creating the next pressure point for future cracks. Once the glue has fully set up, remove the clamps and blocks and fill the screw holes and any remaining evidence of the crack with a good-quality wood filler before sanding and repainting.

If the panel has been cracked for any length of time, there may be caulking, putty, spackle, or paint in the crack from earlier efforts to fill it. This will have to be removed before repairing the panel, or the pieces will not mate properly.

Repairs to vertical board wainscoting are addressed below in the discussion of full-height vertical boarding on walls.

Replacing or Covering Modern Sheet Paneling

Thin plywood sheets with a veneer of finish wood and grooves to imitate board seams became widely popular as an interior finish in the postwar years. This material was used in new construction and to cover old plaster in existing houses. For historic houses built in the 1950s and 1960s, it is a suitable and appropriate material. It is usually at odds with the character of older houses, however.

In prewar houses, removing or covering sheet paneling is usually recommended to return rooms to their historic character. Removal is generally the better approach and is not difficult. (See Chapter 6, The Demolition, for suggestions.) You will probably find layers of wallpaper behind it, and somewhat damaged plaster behind that, but plaster covered by paneling is often in fair to good condition and easily repaired. Occasionally the plaster was removed before the paneling was installed; in such instances the paneling will have been nailed to the wood lath or, if the lath was also removed, to the studs. (Because Victorian-era window and door trim was typically installed on top of the lath, remodelers seldom pulled the lath with

The "make-do" kitchen at Whitten House before phase 1 rehab (see "Example Project: A Compatible Kitchen" in Chapter 3). The sheet paneling was the original wall covering in the room added in the 1960s.

Here's a 1969 advertisement for Georgia Pacific plywood sheet paneling. The pitch reads: "Finally, paneling that captures the romance of the Old World!"

The first step in covering the paneling with lining paper was to fill the grooves in with Durabond setting plaster, applied with a 6-inch putty knife and lightly sanded with a 150-grit screen on a pole sander after setting.

Plywood paneling is straightforward to cover when it is well attached to the walls and reasonably free of holes, and this approach is still feasible even when the paneling is loose or has a lot of holes. Just filling the grooves with spackle or joint compound and painting the smoothed surface will greatly improve its appearance in a historic house, but this treatment will not last; the expansion and contraction of the wood will not be matched by the filler material, and over time the filler will start falling out of the grooves and the seams between sheets. A much better approach is to patch any holes, fill the grooves level with the surface,

Here the lining paper has been pasted up and painted. The stiff paper comes in 36-inch-wide rolls and is pasted like wallpaper. Joints can be butted or overlapped and cut through for a flush, tight seam.

A close-up view at the edge of the lining paper where a slate backsplash will cover the paneling.

the plaster.) If you have sheet paneling with no plaster behind it, you can either remove it and install new gypsum board or skim-coat plaster (as discussed elsewhere in this chapter), or you can cover it up.

and cover the entire wall with lining paper—a wide, heavy, white paper that is used as a base for high-quality wallpaper jobs—before painting.

I recommend filling the grooves with a setting plaster like Durabond, the application of which is covered in the plaster repair section of the next chapter. A setting plaster (which has bonding agents) cures quickly and exhibits minimal shrinkage compared with joint compound. Small holes can usually be patched with fiberglass mesh drywall joint tape and setting plaster. I often squeeze a plastic airbag (used in shipping) into the space behind the hole as a backer for the setting plaster. Once applied, the lining paper will hide the mesh.

A large hole should be cut even larger (counterintuitive though that sounds) into a square or rectangle that overlaps a solid nailing surface on each side. Cut a piece of plywood or paneling of the same thickness to the same length and width and nail it in place with small paneling nails. Apply fiberglass mesh drywall joint tape over the joint between patch and wall, then apply a very thin coat of setting plaster to bed the tape. Further filling or coating is unnecessary unless you have large gaps.

Lining paper is pasted to the wall like wallpaper. If carefully installed, the seams will disappear when it is painted. This surface will not have the character, sound-deadening, or other qualities of historic (or even new) plaster, but it will look far more appropriate than sheet paneling once it is painted or wallpapered.

Repairing and Preserving Modern Sheet Paneling

In postwar houses where sheet paneling is an original material, the issue is not incompatibility but preservation. The paneling was often attached to a gypsum board backer with adhesives that are losing their grip after fifty or sixty years. Two methods for dealing with this issue are to reattach the paneling with new adhesive or nail it to studs or joists (if it is on a ceiling) with small, dark paneling nails in the paneling

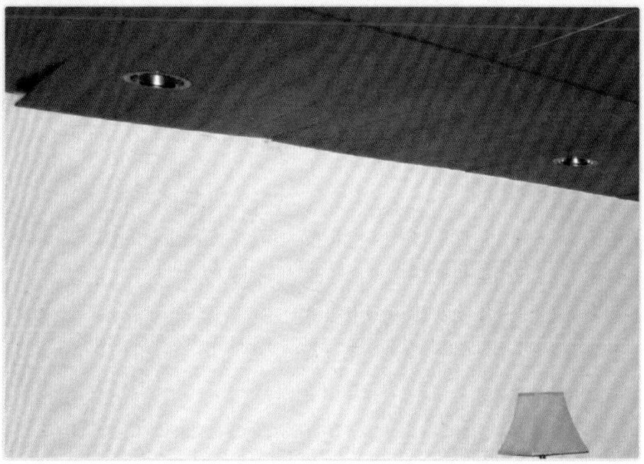

This sheet-paneled ceiling was installed with adhesive in the 1950s. The butt ends of the sheets were covered with matching battens, but the long edges were butted without battens. With the adhesive failing, the paneling is starting to come loose and sag in spots. Careful nailing through the dark grooves into ceiling joists may effectively reattach the ceiling without the mess of taking it down and regluing it.

grooves. Since modern framing is regularly spaced (typically on 16-inch centers for the efficient use of 4-foot-wide materials), it is easy to know where to nail once you find one stud or joist.

Water staining can be another issue with sheet paneling, many types of which are highly absorbent and retain stains or color bleaching after they dry out. This can be a problem around sinks and tubs or along the bottom of a wall where water can wick into the paneling (in a basement, for instance).

Many of these products were originally stained, with little or no varnish or polyurethane coating to protect them. Such paneling often dries out and becomes dull after fifty or more years. Products such as Scott's Liquid Gold can add oil back into the wood veneer and restore richness to the surface. Depending on the dryness of the paneling, several applications may be needed. This product became widely available

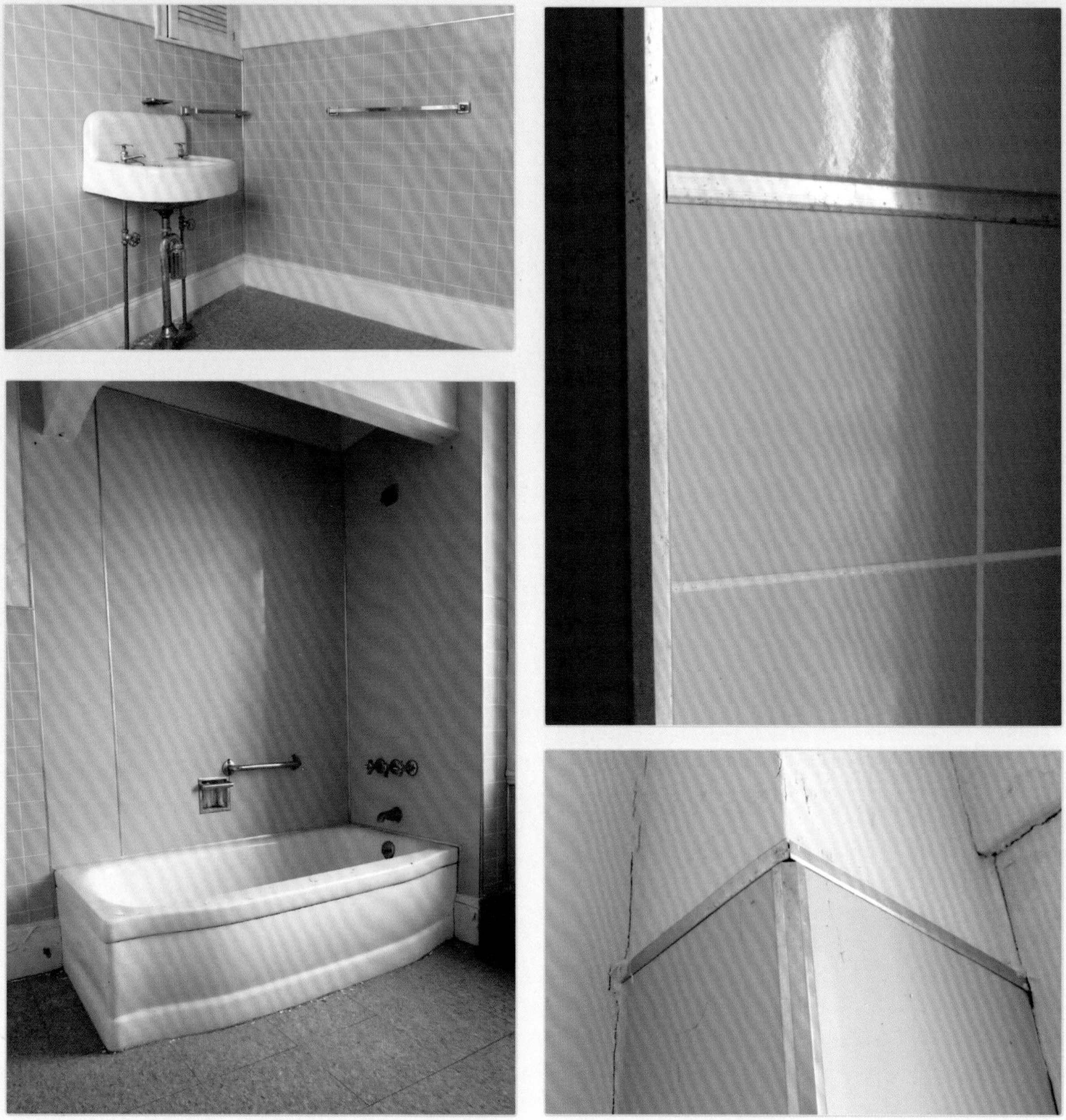

Masonite hardboard "tile" and plain sheet material used in a bathroom of the 1960s.
A variety of extruded aluminum trim shapes were used to finish the edges.

during the postwar period; I remember using it annually to wipe down the stylish paneling in my grandmother's new house during the 1970s.

Exposed sheet paneling (versus paneling that will be covered as described above) is harder to patch than plaster or gypsum board. When otherwise good paneling is marred by one or several holes from doorknobs or other causes, a full-height strip of the paneling can be replaced with a matching section from an area that can be sacrificed. Do the closets have the same paneling? Is the paneling in one room so damaged it needs to be replaced anyway? Having located a matching section to patch with, hold it up next to the damaged area to be sure it will blend in. If it will, cut inside the grooves of the damaged panel—using a long straightedge and a sharp-bladed utility knife—to remove a vertical strip that includes the damage. If the paneling is backed by gypsum board, you can cut a strip just wide enough to remove the damage. If the paneling is nailed directly to the studs, however, you'll have to cut on the studs to the left and right sides of the damage, hoping that you have a groove over each stud to make your cuts in.

The damaged piece may come out when a few nails are pulled, or you may have to peel it away bit by bit from the construction adhesive that holds it in place. Once the damaged piece is removed, cut a replacement strip to the same width, again making your cuts in the grooves. (Because all these panels have the same dimensions, you can cut essentially the same strip from multiple pieces.) Attach the replacement strip with nails or construction adhesive. Note that you may have to remove a piece of baseboard to access the full height of the paneling. If your measurements were accurate and your cuts straight, the replacement piece should be undetectable to the casual viewer. If a doorknob caused the original damage, install a door stop immediately.

Sheet paneling is still manufactured, so it is possible to replace all the paneling in a room with new paneling that comes close to matching the original if the original is beyond saving.

Another postwar sheet material is Masonite hardboard, which was available with a smooth finish or with a faux tile pattern (a colored surface and white "grout" lines) embossed in the face. Extruded aluminum moldings were produced to join sheets of this material and to finish the edges. At least one manufacturer is making this material as of this writing. It appears not to be made in the popular colors of the mid-twentieth century (turquoise, pink, etc.), but it is paintable. Given the increasing interest in midcentury housing, period colors may become available again at some point. Extruded aluminum moldings are also still made.

Beaded and V-Groove Board Walls and Wainscots

Beaded boards before the era of steam-powered sawmills were flat boards with a bead hand-planed along one edge. These were generally installed vertically in a wall of similar (though often of variable width) boards. The bead created a subtle ornamentation and

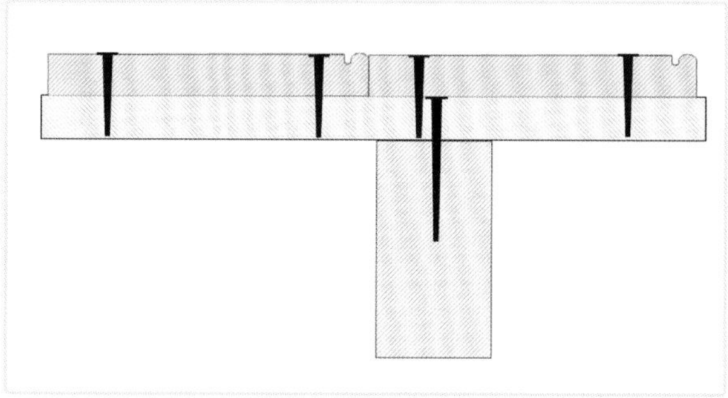

This type of beadboard, often pine, is simpler and typically earlier than other beadboard types. Board widths can be up to 20 inches and often vary on the same wall. The bead is usually hand-planed, and the boards are usually face-nailed into horizontal strapping for vertical installation or directly to studs for horizontal installation.

helped disguise the board seams. Sometimes the boards were shiplapped.

The advent of steam-powered sawmills led to the common Victorian-era beadboard configuration: ¾-inch tongue-and-groove boards in typically narrow widths (around 2½ to 3 inches), with a bead along one edge and an angle on the opposite edge that butted the bead on the adjacent board. Double- and triple-width boards were also produced, creating the appearance of multiple single boards on each wider board. By the early twentieth

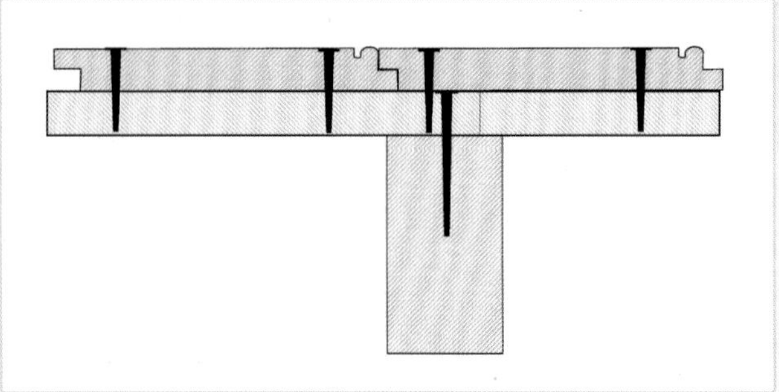

This variation on the previous example has shiplap joints or, sometimes, overlapping beveled edges.

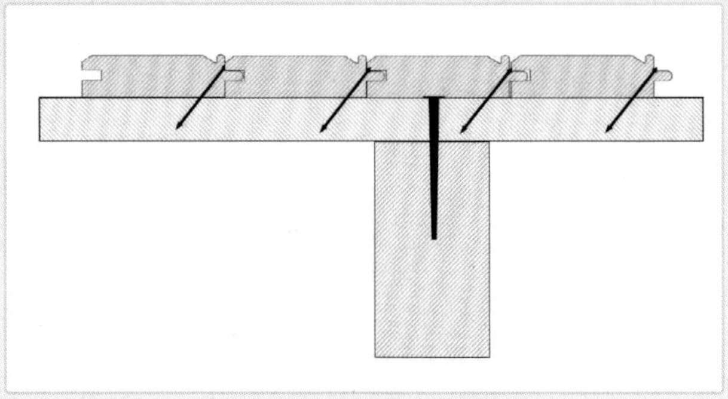

Typical Victorian-era beadboard, usually hardwood and produced in a steam mill. The boards were blind-nailed to horizontal strapping for vertical installation or directly to studs for horizontal installation.

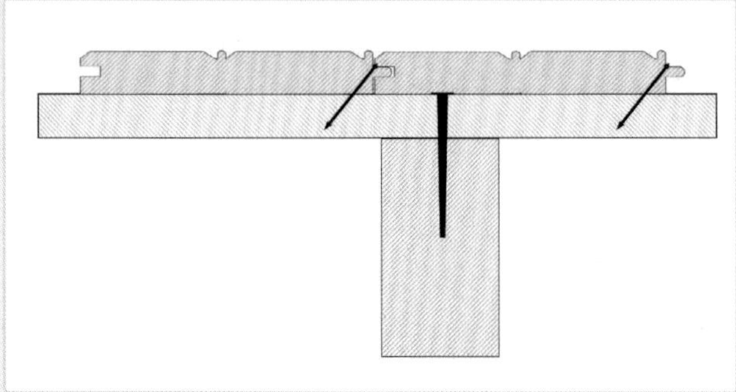

This wider variation on a typical Victorian hardwood beadboard was installed the same way. It was sometimes milled identically on both sides, making it reversible when salvaged for reuse. This allows an originally exposed surface with lead paint to be encapsulated within a wall.

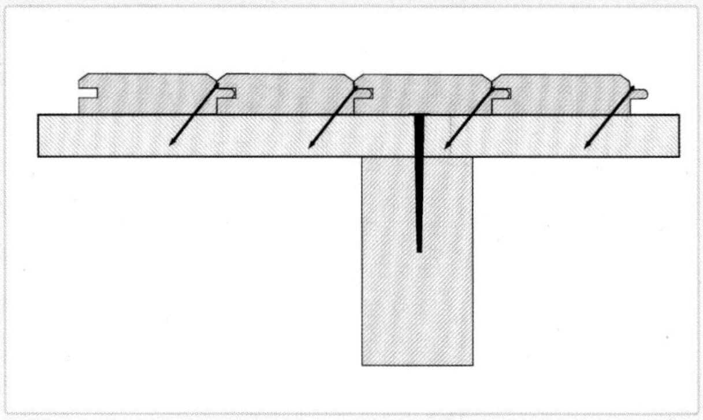

V-groove board, a variation on beadboard, is installed the same way.

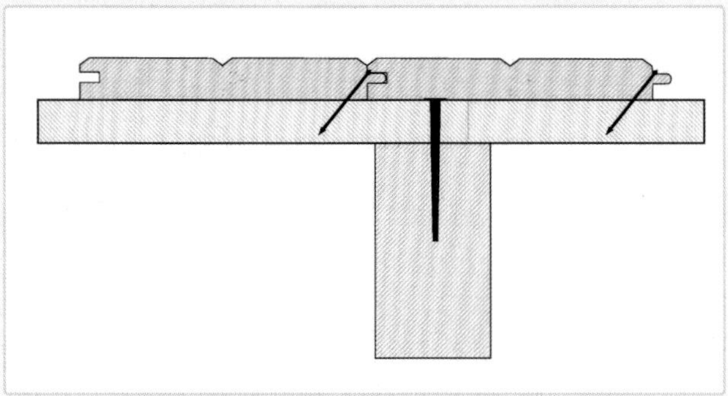

Like beadboard, V-groove board can be found in wider variations.

century, 3/8- and ½-inch-thick versions of the boards became available.

From the late nineteenth well into the twentieth centuries, beaded boards were used widely for secondary spaces in formal homes and could also be found in primary spaces in less formal homes. Beadboard was particularly popular for kitchens, pantries, bathrooms, back entries, attics, and sheds. It was installed as wainscot and was applied to ceilings as well as walls; it was also used for cabinet doors. Vertical beadboard, whether for wainscot or full wall coverage, was attached to 1 x 3 horizontal strapping. Horizontal beadboard was usually nailed directly to the wall studs. Victorian beadboard was most commonly finished with stain and varnish but was sometimes painted. A lot of period beadboard that was originally varnished has subsequently been painted.

Beadboard is available today in a variety of widths and woods. A regional building supply chain in my area has its own millwork production facility and offers a half-dozen beadboard widths in pine, fir, oak, and poplar. This can be important if you need to replace missing beadboard with a matching material. As with milled wood moldings, a millwork shop can custom produce beadboard to match your historic material if necessary. With the right equipment, you can mill it yourself.

Beadboard is also available as a modern MDF material in 4-foot by 8-foot and 4-foot by 12-foot sheets. I chose this material for the downstairs bathroom at Whitten House because it did not require the removal of the historic plaster and installation of horizontal strapping as traditional vertical beadboard would have. I shimmed out the historic baseboard ½ inch, set the sheet beadboard on top of it, and nailed it through the plaster and lath into the studs. I milled a nosed wainscot cap for it with my router table as described in Chapter 3, "Example Project: A Compatible Bathroom." This substitute material looks very much like traditional wainscot once painted but will be easily identified and dated by future restorers. MDF beadboard is only suitable for a painted surface.

Traditional beadboard should be finished (stained and varnished or primed and painted) before installation. The boards will expand and contract across their widths, depending on temperature and humidity, and if not finished before installation, they can expose a narrow strip of unfinished tongue between every board when they shrink. I recommend leaving the second coat of varnish or paint until after installation, however.

Sheet beadboard has joints every 4 feet rather than every 3 to 6 inches, and the material shrinks and expands less than natural wood. When painting, apply a good latex caulk in the joints after priming, ideally during dry weather when the sheets are contracted.

"Pickwick" Pine Walls

Pickwick pine is a molded tongue-and-groove board that became popular in the mid-twentieth century as a "Colonial" wall treatment, though there is no colonial precedent for it. It is typically installed vertically, blind-nailed into horizontal strapping. It was sometimes painted but more often finished naturally with stain and varnish or wax. Often, it is made with knotty pine for a more rustic appearance. It is still made, but finding an exact match for the profile of older boards may be a challenge. If blind-nailed, it can be removed and reinstalled without visible evidence. If you need to replace damaged boards and cannot find matching replacements, consider consolidating the old boards

Pickwick pine installed in the 1960s.

on some walls with new boards on others. If the stain color and finish are carefully matched, slight differences in molding profiles from wall to another should not be noticeable.

Barn Board Walls and Wainscots

Always suitable in a barn, barn board was also sometimes used for interior walls or wainscot in the mid-twentieth century. Finding matching barn board to replace damaged pieces may be challenging, as the coloring and weathering of salvaged barn boards vary with the species of wood and the history of exposure to sun and elements. If a match cannot be found, consider consolidating the existing boards on some walls with new salvaged boards on other walls. Try to match the color as closely as possible.

Fiberglass and Glass Block Walls

Fiberglass found limited use as a finish surface during the postwar period. With decorative designs incorporated into the material, it was sometimes used for light filtering in skylights and partitions, much as stained glass was used in earlier periods. It was also sometimes used for translucent panels in cabinet doors.

It might be challenging to find a matching replacement for a damaged fiberglass panel. Materials produced in the 1950s and 1960s do turn up in warehouses as new old stock, however, and the availability of reproduction materials from this period will probably improve as postwar houses are increasingly recognized for their historic character. Only thirty years ago, reproduction Victorian-era materials and

Detail of a fiberglass partition in the rec room of the Weston House, 1957 (see page 243).

hardware had very limited availability, but today there are hundreds of manufacturers of these products for the restoration market.

Glass block was occasionally used for interior partitions and in bathrooms. There is a section on "Glass Block" in Chapter 13, Exterior Trim and Windows.

Exposed Brick Interior Walls

Exposed brick on interior walls was historically rare. An exposed brick wall would have appeared unfinished and unsuitable for a residential space. Not until the second half of the twentieth century did exposed brick come to be considered a desirable interior finish by designers and design publications. The public embrace of exposed brick in old buildings continues into the twenty-first century, particularly in converted industrial spaces. Unfortunately, many people fail to make the distinction between former mill spaces, where exposed brick is historically appropriate, and formal rooms in brick houses, where exposed brick is utterly at odds with the historic character of the rooms.

That said, the "gut rehabs" common in the 1980s and 1990s left thousands of historic houses with interior walls stripped of plaster to expose rough brickwork surfaces, and some people continue to do this. If you buy a house with exposed brick walls, you will face a choice between leaving it that way or putting gypsum board or plaster back in place. Just bear in mind that an exposed brick interior wall is a contemporary treatment.

If you want exposed brick in living spaces, you will need to do something to control the ongoing slow shedding of mortar particles and lime dust. Frequently, exposed brick walls have been sandblasted to clean them, which exposes the soft interior of the brick and can result in brick dust being added to the mortar dust exfoliating from the wall.

In the late twentieth century, exposed brick was often covered with a thick coat of polyurethane to address this issue. Such walls now tend to be yellowed with areas of cloudy whiteness due to moisture trapped behind the polyurethane. Trapped moisture is never good for a masonry wall, especially brick. Unfortunately, if you have a house with this condition, it will likely worsen over time and negatively affect the integrity of the structure. Removing the polyurethane is the only solution; a low-pressure grit-blasting with an appropriate material may be the best approach. See Chapter 14 for more information on historic masonry walls and National Park Service publications on appropriate methods for cleaning and sealing these surfaces.

Ceramic Tile Walls

Ceramic tile has been used in American houses since the eighteenth century, when blue-and-white Dutch tiles were sometimes installed in fireplace surrounds.

Blue-and-white Dutch tiles were used for fireplace surrounds during the colonial era.

As the nineteenth century progressed, the production of ceramics for the American market greatly

This geometric tile—solid-color tiles cut and laid in a pattern—is on the entry porch of Historic New England's Eustis Estate house in Massachusetts. In America, such tiles are known as encaustic tile along with the multicolor patterned tiles known by that name in the UK.

A 6-inch by 6-inch encaustic tile made by the Minton pottery in Staffordshire, England. Minton developed the method for producing multicolored tiles with different clays, providing depth to the color so it would not wear off over time as it would on transfer printed or glazed tiles.

expanded in the Staffordshire region of England. This included tiles of different sorts: hand-painted, transfer printed, encaustic, relief, etc. Some tiles were glazed and some were not. Later in the nineteenth century American tile manufactures, including Low Art Tile Works (Chelsea, Massachusetts), Providential Tile Works (Trenton, New Jersey), and the American Encaustic Tiling Company (Zanesville, Ohio), established successful potteries that competed with the British firms. By the end of the nineteenth century, American tile manufacturers were established from New England to California.

Because wall and floor tiles are essentially the same material and are installed and repaired in similar ways, both will be addressed here to avoid redundancy.

Ceramic and stone floor tile became popular during the Victorian era, particularly for vestibules, entry halls, and fireplace hearths. Stone tile in checkerboard patterns had been used primarily in high-end homes during the colonial and Federal periods. The earliest and simplest ceramic floor tiles were solid-colored and usually unglazed. Two colors might be laid in a checkerboard pattern, or tiles in several colors could be cut and arranged into geometrical patterns. Inlaid, or encaustic, tiles were a medieval style of tile "rediscovered" by proponents of the Gothic Revival style in the nineteenth century. They were formed of multiple clay colors in a single tile, so that the colors were not just a surface finish and would not wear off underfoot as painted or printed designs could do. Once modern dust-pressed production methods were developed in England, reducing the cost, the popularity of these tiles spread to America, and potteries here began to produce them as well.

There was also an increase in the use of wall tile during the mid-nineteenth century, with English firms once again supplying much of the tile at first. Following the 1876 Centennial Celebration world exposition in Philadelphia, where millions of Americans saw exhibits of British tile, a number of American manufacturers entered the "art tile" market.

"Sanitary tile" became popular just before the beginning of the twentieth century following the introduction of indoor plumbing and a better understanding of the causes of disease. The colorful tiles of the Victorian era gave way to the white tile commonly called subway tile today. With radiused corners and thin grout lines, their surfaces were easy to keep clean. Subway tiles made their way into kitchens and bathrooms across America as well as subway

Four of the Minton tiles arranged into a 12-inch-square pattern.

These encaustic tiles were produced with embossed patterns to resemble small pieces of mosaic tile. The grout applied between tiles also fills the recesses left by the embossing, thus creating the illusion that many more pieces of tile were used than is actually the case. A true mosaic tile floor would not have wide grout lines separating the floor into rectangular sections. Note that in the dark red and floral element to the left of bottom center, the tile has worn sufficiently to remove the "mosaic" embossing, leaving areas of solid color.

stations, hospitals, and other commercial and institutional settings. The Associated Tile Manufacturers, a trade organization, established and published standard dimensions that were followed by most American manufacturers, imposing a uniformity that made it possible to mix and match different brands. Small, unglazed tiles were common on floors, often in patterns of black and white.

In the early twentieth century, the Arts and Crafts movement introduced muted, earthy colors to bathroom, kitchen, and fireplace tiles. Matte finishes were popular by the 1920s. Stylized floral and other simple decorative motifs were shared with textiles, wall stenciling, and stained glass of the period.

Nine Minton 6-inch by 6-inch tiles in four designs with a Fairies motif, arranged into an 18-inch square. Such tiles could be used linearly for a fireplace surround or arranged into larger patterns for walls or wainscoting. Minton and other companies produced tiles with many motifs, including themes from literature and nature.

The Aesthetic Movement drew much inspiration from Japanese design and motifs, as can been seen in these "Naturals" pattern tiles from the Low Art Tile Company, produced through a process patented by the company in 1878, just before these tiles were produced and installed in the Eustis Estate.

Square tiles began to appear more frequently in the ensuing years, and this shift accelerated as pastel tiles came into vogue after World War II. In the postwar years, 4-inch by 4-inch and 6-inch by 6-inch tiles completely displaced the 3-inch by 6-inch subway tiles that had been the norm since the 1890s. The molded border tiles of the sanitary era were supplanted by sleeker-profiled edge tiles.

Tiles also became thinner during this period, as different installation methods took over. Earlier tiles were typically ½ to 5/8 inch thick and were set in a thick bed of wet mortar. This method is often called "mud set" tile. Later tiles are closer to ¼ inch thick and attached to cement backer board with a thin adhesive called thinset mortar. Today, this is generally a latex material rather than a cementitious mortar. With either method, white or colored grout was applied to fill the spaces between tiles after the mortar set.

Historic ceramic tile is a character-defining feature and should be retained whenever possible. Companies now reproduce Victorian-era tiles and historic subway tile using the historic Associated Tile Manufacturers standards to match damaged tiles that have to be replaced. Some companies offer custom coloring to give you a precise match between new and hundred-year-old tiles. Replacement tiles need to be set with the same method that was used for the surrounding historic tiles.

These art tiles of the 1880s, installed as a fireplace surround in Massachusetts, were produced by the Providential Tile Company of Trenton, New Jersey. They were removed in the 1980s and stored in a garage for decades before turning up in an online classified ad in Maine in 2015. They were dirty and had a fair amount of mortar attached to the backs; two were broken, and several others had small edge chips. But the price was right, and I bought them. After cleaning and repairing them (as in the next two photos), I resold them.

Another Aesthetic Movement fireplace surround tile set showing Japanese influence at the Eustis Estate.

It is not usually possible to replace a historic 5/8-inch-thick subway tile with a thin new tile from a big-box home improvement store. One alternative is to consolidate good historic tiles in the most visible areas and cover the areas the tiles were "harvested" from with plaster or cabinetry. This is the approach used for restoring the wonderful high wainscot tiles

One of the broken Providential art tiles is shown here after being glued back together with Araldite epoxy and having the gaps filled with setting plaster. The glue and plaster can both be removed by soaking in water if a future owner wants to undertake a better restoration treatment.

A close-up view of the repaired area after the plaster was colored with acrylic paints and the paint covered with an acrylic gloss coating. This photo was taken with a flash to make the touched-up area obvious. The goal was not to make the repaired area invisible, but simply to make it blend in when the seventeen tiles are installed together. Had one of the portrait tiles been broken, I would probably have undertaken a careful restoration rather than a quick repair.

in the kitchen of the Chicago bungalow (the Kowalski-Policht House) featured on page 453.

As discussed in Chapter 3, mimicking historic tile styles with new tiles can help to harmonize a new kitchen or bathroom with the character of an old house. See, for example, the subway tile tub surround in the "Example Project: A Compatible Bathroom" section in Chapter 3, where the goal was to suggest that the bathroom might have been added to the house in the 1890s.

Thirty years ago, the options for replacing damaged historic tiles were limited to a handful of companies, but today there is a plethora of historic tile styles being reproduced, and these are augmented by the continued availability of salvaged historic tiles. Use the internet to research what is available and to

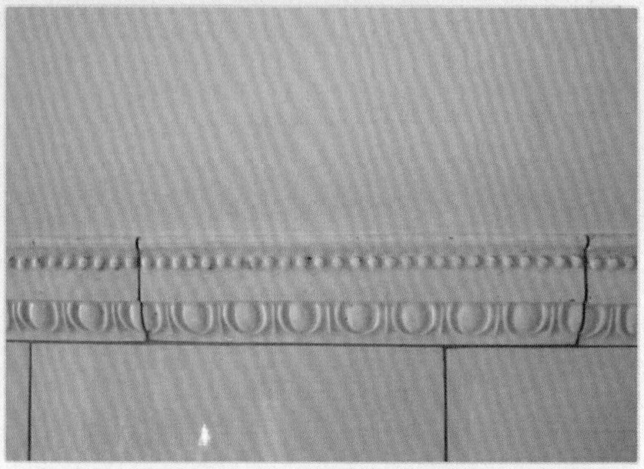

A wide range of trim tiles were produced for subway tile in the early twentieth century. Missing pieces can be a challenge to match. Possible sources include reproduction tile companies, local architectural salvage companies, and online auction sites.

A typical early-twentieth-century installation of sanitary, or subway, tile in a bathroom. These tiles were thicker than the subway tiles widely available today, but specialty reproduction tile companies produce historic thicknesses and can often do custom color matching.

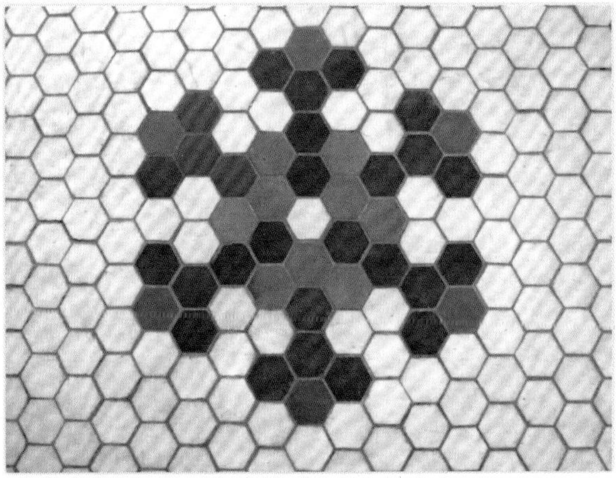

Multicolor 1-inch hex tile creates a pattern in a field of white tiles. Patterns like this were available preassembled on a paper backing that could be removed with water once the tile was set. Such patterns are again being produced by reproduction tile companies.

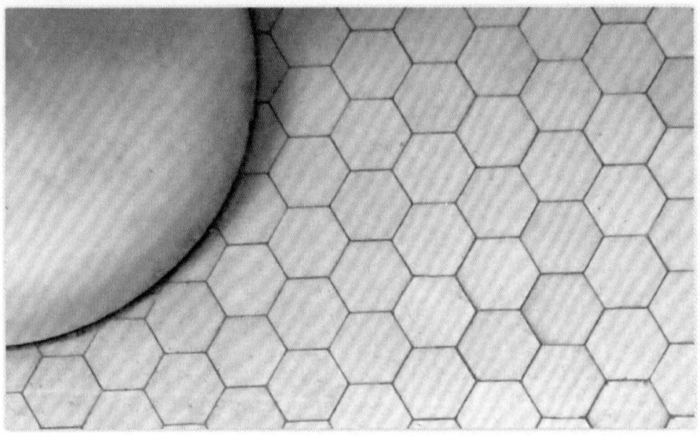

Hex tile was popular for bathroom floors and often used with subway tile wainscoting or walls. It was available in a variety of sizes from 1 to 3 inches. Often, white marble hex tile (like that shown here) was used with ceramic wall tile, but ceramic hex tile was also produced. Note the very tight spacing of the tiles, with minimal grout between. This was typical of the period and quite different from the 1/8-inch spacing typically used in modern hex tile sheets from the major manufacturers.

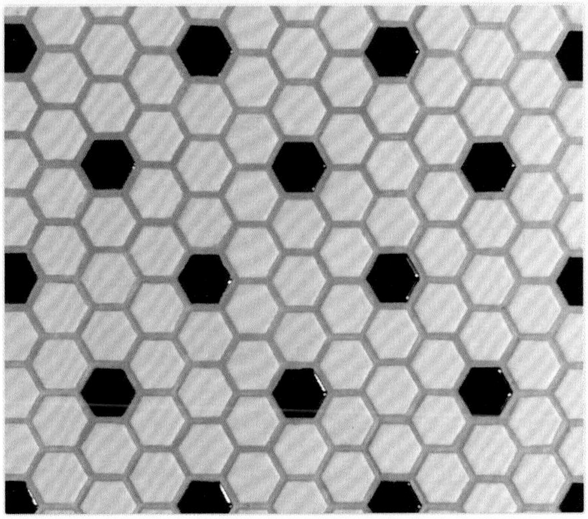

Modern 1-inch hex tile from a major manufacturer. These tiles come mounted onto 12-inch by 12-inch sheets for fiberglass mesh, ready to be laid in a thinset adhesive mortar, and are also available in solid white sheets. Similar sheets are available for rectangular tiles in basket-weave and other common historic floor tile patterns.

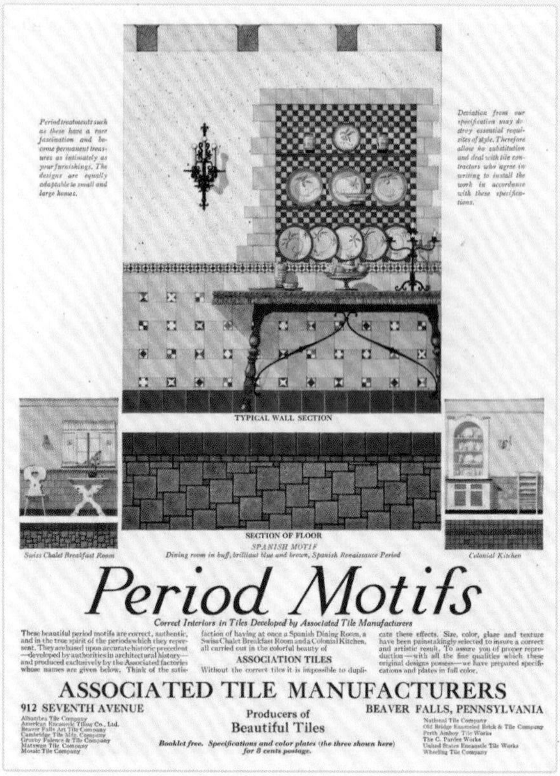

This 1920s magazine advertisement shows the effect of the Arts and Crafts movement on tile design and colorways. The ad shows "Spanish," "Swiss Chalet," and "Colonial Kitchen" motifs. The Associated Tile Manufacturers industry group included most of the major American tile producers of the early twentieth century and established standards for tile sizes that were adopted across the industry, making it possible to mix tiles from several manufacturers on a single job.

find examples of tile work in other historic homes for inspiration when you design a new kitchen or bath. Historic print advertising can also be an excellent source of period design ideas. Keep in mind that larger tiles can be cut down to match the dimensions of historic tile if necessary.

The National Park Service Preservation Brief #40, *Preserving Historic Ceramic Tile Floors,* can be found at

Glazed structural tile was produced in the early twentieth century. Laid up like brick or block, the tile didn't just finish the wall but was the wall. Here we see a "wainscot" of glazed tile, above which a hollow structural tile serves as a base for plaster. A new opening has been cut in the wall, and pieces of tile salvaged from elsewhere in the building are being used to replace broken tiles. The electrical box for the new sheathed cable has not yet been cut in.

This kitchen from 1967 uses 4-inch-square wall tiles with small white and green square and rectangular floor tiles in a seemingly random pattern. The sizes and colors of the tile are typical of the period.

This 1957 magazine advertisement for Hermosa Tile shows the then popular 4-inch-square tiles in a solid color with accent tiles to "insure PERMANENT BEAUTY for your home."

www.nps.gov/tps/how-to-preserve/briefs/40-ceramic-tile-floors.htm. Much of the information presented on ceramic floors is applicable to walls as well.

Quarry Tiles

Quarry, or terracotta, tile is made from clay, like porcelain, but without the added materials and high firing temperatures required to turn clay into porcelain. The red clay used for terracotta tiles is usually left unglazed, creating a matte red tile that is less dense and softer than porcelain and more susceptible to damage. These tiles are most associated with the

This image from a 1954 magazine advertisement shows unglazed quarry tile used for an indoor/outdoor living space. Often used for Spanish Colonial and Craftsman houses at an earlier period, quarry tile was frequently used in postwar modernist houses.

Southwest and the Spanish influences that came into the United States from Mexico long before those areas were part of the U.S. With the development of the Spanish Colonial Revival style and the adoption of terracotta as a favored material by the Arts and Crafts movement in the early twentieth century, terracotta tile made its way across the country.

Historic terracotta tile is a character-defining feature of a house and should be preserved following the recommendations for porcelain tile.

Marble Tiles

Marble tile is found in many historic houses, generally on floors. Larger slabs of marble might also be used for tub surrounds, wainscots, or countertops. Black and white 1-inch hexagonal or round tiles were used to create patterned floors as with ceramic tiles of the same shapes, often in rooms with ceramic tile

Marble used for a mantel and in a checkerboard pattern for the hearth of a re-created eighteenth-century fireplace at Colonial Williamsburg.

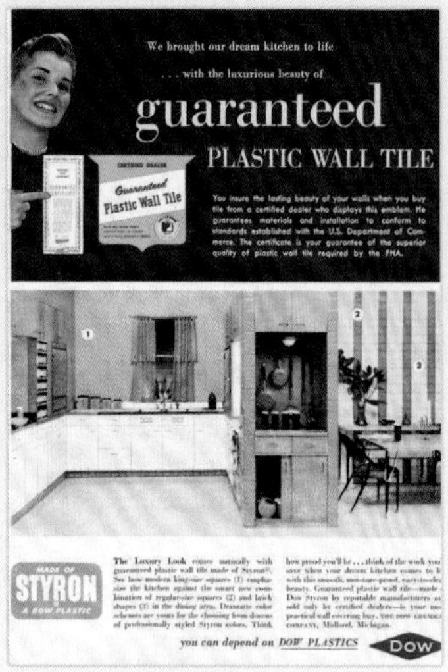

A 1956 magazine ad for plastic tile. A recent search of online auctions turned up a lot of this material in unused condition, some still in boxes, providing hope for those who need to replace damaged tiles.

on the walls. At a glance, marble tile of this type can appear to be unglazed ceramic, but a closer inspection will show that it is stone. As with ceramic tile, marble tile (and slabs) should be preserved if at all possible and repaired with matching replacement tiles. Because it is a natural material, an exact color match may require some searching.

Plastic Tiles

Plastic wall tile is a postwar product that had a short commercial life, primarily in the 1950s and 1960s. These tiles had a tendency to pop off the wall due to the minimal area of contact between the thin tile edge and the adhesive on the wall surface, so they were being replaced by the mid-1970s (as my parents did in 1977 when they bought a 1953 house with failing marbleized pink plastic tile in the bathroom). I am not aware of this tile being reproduced today; if you have a house with it and want to replace missing or broken tiles, your best bet may be eBay or other online sales sites. An exact color match may be impossible due to the effects of ageing on tile exposed to light in the bathroom versus tile that has remained in a box for fifty years. If tiles can't be matched, replacing it all with a ceramic tile in the same size and color will help to maintain the character of the space.

Glass Tiles

Although glass tile has roots in the third century BC, its use in American interiors was limited to interiors of the twentieth century, beginning in the 1920s. It was never as popular historically as it is today. As with other tile types—perhaps even more so due to its historic rarity—its preservation should be a priority. Matching historic glass tiles is more feasible today than twenty years ago. Follow the recommendations above for porcelain tile.

Mirror tiles came into popularity in the 1950s. Typically 12 inches square, they were used on walls in bathrooms, entry halls, and sometimes behind a home bar in a rec room. Reflecting (literally?) the spirit of the time, they were occasionally applied to bedroom ceilings in the 1960s and 1970s. For areas where they were more decorative than useful, they could have a smoky or marbled effect. Mirror tiles are still made and relatively easy to find.

Ebenezer Alden House, 1797

Transitional Georgian to Federal Style with Greek Revival Alterations

Built by a Massachusetts-trained finish carpenter and wood carver in the frontier region of Down East Maine (then a Massachusetts territory), the Ebenezer Alden House was exceptionally stylish and up to date for its remote location. Descended from John and Priscilla Alden of *Mayflower* fame, Ebenezer came to coastal Thomaston to carve interior trim for Revolutionary War General Henry Knox's mansion, Montpelier. Built by Alden's employer, Boston housewright Ebenezer Dunton, Montpelier was the grandest and most stylish residence in Maine when it was completed in 1795. Upon completion of the house, Alden purchased property nine miles inland on which to build his own home.

Thirty years after the end of the French and Indian War had removed the threat of violence that stymied settlement of interior Maine for 150 years, small towns were appearing throughout the territory, settled primarily by young people from Massachusetts who were descended from seventeenth-century Puritan immigrants. Twenty-four-year-old Ebenezer Alden first built a small house and store to serve the growing community, then undertook construction of his main house.

The large house featured a rather old-fashioned plan, with four large rooms on each floor flanking a wide center hall. The home's three entrances were centered on the north (street-facing), west, and east elevations, respectively. Of these, the west entrance was architecturally the principal one, opening into the center hall. A small vestibule inside the north door opened into a sitting room to the left and the north parlor to the right. The four rooms on the west side of the house, two down (north and south parlors) and two up (bedrooms), received the finest interior treatment, with elaborate Federal-style cornices, mantels, and door and window trim on the first floor and a somewhat simpler treatment on the second. The staircase and front hall also received fine, up-to-date detailing. The east-side rooms were simpler, with Georgian-style paneled fireplace walls and simple beaded-edge board partitions separating the rooms from the halls. A lean-to shed off the rear of the house was likely built at an early date as a woodshed. A one-and-a-half-story gable-roofed ell extended from the shed. The original house and store built by Alden stood nearby and likely continued in use as a store.

This impressively large house built by a man not yet twenty-five years old suggests that he came to Maine with substantial resources and wanted to show off his carpentry skills to potential clients in the rapidly growing community. In this he was apparently successful, as several other houses built by Alden survive in the region.

Like most rural Maine residents at the time, Alden was a farmer as well as a craftsman and entrepreneur. He is recorded as having the first wool carding machine in town in 1806, operated by waterpower from the river in a small wooden mill building. Alden was appointed postmaster for the community in 1813, a title he retained for thirty-two years, and he also served as the local coroner from 1809 to 1827. For several years in the early part of the nineteenth century, he manufactured five or six tons of potash annually in a building that he erected for that purpose. Used for bleaching wool and making soap and glass, potash was produced by pouring water through hardwood ashes to leach out the lye, then boiling down the lye to the potash residue. Frequently produced in areas being cleared and settled, it was an important cash export to England. While undertaking all these activities, Ebenezer and his wife, Patience, had nine sons and three daughters.

Like the Knox mansion, details in the house appear to have been derived from William Pain's *Practical Builder, or, Workman's general assistant, shewing the most approved and easy methods for drawing and working the whole or separate part of any building*, published in London in 1774 and in Boston in 1792. In addition to interior details, both houses have door surrounds based on Pain's drawing of a "Frontispiece in the Dorick Order." Remarkably, Ebenezer Alden's tools and tool chest remain in the house. As built, the house likely had a low-hipped roof, which was fashionable at the time. The current high-gabled roof in the Greek Revival style would have been added to update the house, possibly by Alden's son, Augustus, who inherited the house upon Ebenezer's death in 1862. His son also built the existing barn around 1868. The house remained in the Alden family through several more generations until 1965, with additional interior alterations during the Victorian period.

The next owners were Hazel and Joseph Marcus, who were in their early fifties when they moved from Great Neck, New York, and spent three years restoring the house. Joe Marcus worked in the textile industry in New York City, and the couple had been antique collectors since shortly after their marriage in 1938. Through their collecting in New York shops, they got to know some of the nation's most notable collectors, including Henry DuPont and Ima Hogg—founders of the Winterthur Museum in Delaware and the Hogg Museum in Texas. The Marcuses first saw the Alden House on a trip to Maine in the early 1960s while shopping for antiques. It was falling into neglect, but even from the outside they sensed it was more than a typical farmhouse. Several years later, Hazel noticed a "For Sale" sign when passing by the house. She immediately called Joe in New York, and he flew to Maine the next morning. As Joe remembered it thirty years later, when he walked into the dining room and saw the carved mantel, he said "I'll take it!"

Many photographs survive from the 1960s restoration project, making it possible to see that the work was done with care to preserve as much original material as possible and accurately replicate what was missing. Fortunately, original doors, paneling, and other elements had been stored in the barn when they were removed for Victorian-era updates to the interior. These included the upper half of the original exterior door surround for the west entrance, which was based on Pain's illustration. These elements were returned to their original locations, and missing elements were replicated, in some instances with Ebenezer Alden's own tools. The recreated door surround for the north elevation was slightly simplified from the original design to differentiate it. Lath and plaster finishes were removed from the original vertical-board walls in the secondary rooms of the house. The boards had been painted before the lath was installed, indicating that they had been the original exposed finish surface. The accordion lath used suggests that it was probably added within fifty years of construction, as sawn lath was the norm by 1850.

The original chimneys had been replaced and fireplace openings sealed up when smaller chimneys were built to serve the cast iron stoves that heated the house beginning in the mid-nineteenth century. The new chimneys had typical Italianate tops, further supporting a mid-nineteenth century date for this alteration. During the Marcuses' restoration, the fireplaces were reconstructed and the chimneys rebuilt at their original size and shape above the roof. Throughout the house, painted pine floors were carefully stripped to expose the wood, then varnished for protection. Paint analysis was undertaken to determine original colors, and the interior trim was repainted as it had been during Ebenezer's time. Wallpapers were selected from eighteenth- and nineteenth-century reproduction papers catalogued by historic wallpaper authority Nancy McClelland, who had been a friend of the Marcuses before her death in 1959. The shed and ell were rehabbed to accommodate a modern kitchen and bathrooms on each floor. The entire house was rewired, and electric radiant heat was installed in the new skim-coat plaster ceilings.

Once the restoration was complete, the Marcuses moved their notable collection of furniture and household objects from New York and settled in, happy to have found an appropriate setting for their antiques. Ebenezer Alden's store next to the house became an antique shop, allowing Joe and Hazel to pursue their passion in another way. The shop became a nationally known destination for discerning buyers for twenty-five years.

In 1995, the Marcuses prepared to retire and downsize by auctioning off their fifty-year collection of antiques with a two-day on-site sale at the house. They then moved to a small house on the coast. Current owners Dave and Suzy Shaub bought the Alden House in 1997 after satisfying Joe and Hazel that they would be good stewards, a promise kept to this day.

Dave and Suzy had each discovered a love of antiques growing up in Rhode Island and Vermont and began collecting together on their honeymoon. Settling in the Midwest, they shopped for antiques during annual vacations to New England. Over the years they filled their 1957 house with nineteenth-century interiors while raising a family and pursuing careers in education. As a teacher, Dave painted houses during the summer for a time and then moved into furniture repair. This work added to his knowledge of furniture construction and his appreciation for the qualities of handmade antique furniture. Upon early retirement, Dave and Suzy decided it was time to give their collection a more appropriate home in New England. Following their move to the Alden House, they became involved in the community, particularly the local historical society and Maine Preservation, the statewide preservation organization. A descendant of Paul Revere (and a *Mayflower* descendant like Alden), Suzy became active in the local chapter of the D.A.R. They also volunteered at the Knox Museum in Thomaston, the 1929 re-creation of Henry Knox's Montpelier.

Because the house and outbuildings had been so lovingly restored, the new owners did not have to undertake any significant work on them. Shortly after moving in, Dave decided to put his own "stamp" on the house with more contemporary interior paint colors, a decision he now calls the worst mistake he made with the house.

It just never looked or felt right to him after seeing the house in its original colors. Eventually, the rooms were repainted in colors of the period. Reflecting the scholarship and expectations of the 1960s, the rebuilt fireplace surrounds were left exposed during the Marcus restoration. With more recent scholarship making clear that the brick would not have been exposed in a house of this quality when it was built, Dave and Suzy had them appropriately parged with plaster.

The electric radiant heating system installed during the restoration was expected to be an affordable way to heat the house without any physical or visual impact to floors or baseboards. It was the mid-1960s, and experts were declaring that nuclear power would make electricity so cheap it wouldn't be metered. This, of course, never happened, and Dave and Suzy decided to install an oil-fired hot air system for the first floor to lower the heating bills. They also added eight inches of insulation in the attic to keep the heat in the inhabited spaces. Heat for the second floor relied on rising warmth with occasional backup from the radiant heat in the ceiling when needed. Out of sight of the road, in the angle between the main house and shed and just off the modern kitchen, they had a large screened porch built to extend their enjoyment of the views from the back of the house through "bug season."

Beyond the house itself, they rescued a rare surviving eighteenth-century cobbler's shop from a nearby town, which was slated for demolition, and had the diminutive structure disassembled, moved, and reassembled inside the barn, where it would be well protected from the elements. Near the barn, they built a small structure resembling a traditional privy, or outhouse, which is used to store gasoline and other flammable liquids and materials that could be a fire hazard in the historic barn.

After two decades of ownership and careful stewardship, Dave and Suzy have decided it is time to downsize and move closer to family. Thanks to their care and efforts, the 222-year-old Ebenezer Alden House will be passed on to new stewards in excellent condition, ready for another century.

The south parlor fireplace mantel with carved acanthus leaf detailing was the element that led Joe and Hazel Marcus to buy and restore the house.

The restored northeast bedroom. Before restoration, the fireplace and its paneled surround had been removed and the beaded board walls covered with plaster.

This kitchen—built in the shed addition during the 1960s restoration—opens into the original kitchen at left and the ell at right.

The restored south parlor.

The restored west façade door from the hallway. Because the original interior door trim was lost, a simple compatible trim was installed during restoration.

The south elevation in 1965.

Ebenezer Alden's branding iron remains in the house. Heated in a fire, this iron brand would have been used to mark the barrels of potash made in Alden's manufactory.

Ebenezer Alden's tools and tool chest remain in the house.

A 1965 pre-restoration view of the north and west façades with mid-nineteenth-century Greek Revival door surrounds..

The shop was the first structure Ebenezer Alden built on the property.

The north parlor features reproduction wallpapers hung during the 1960s restoration.

This view of the house in about 1890 shows the Greek Revival door surrounds and corner boards that were added in the mid-nineteenth century. *(Courtesy Maine Historic Preservation Commission)*

The front stairs.

Looking east toward the 1868 barn and earlier shop building. The ell of the house is at right.

A nineteenth-century clock keeps time in the upstairs hall.

The northwest bedroom.

A 1968 view of the restored door surround on the north façade, a simplified copy of the original door surround on the west façade.

One of the fireplaces restored in the 1960s shows the black-painted plaster parging added by the current owners to reflect current scholarship on how surrounds were finished historically.

The southeast bedroom.

The upstairs hall as seen from the northwest bedroom.

The east elevation in 1965, before restoration.

The Alden House was originally heated by eight fireplaces.

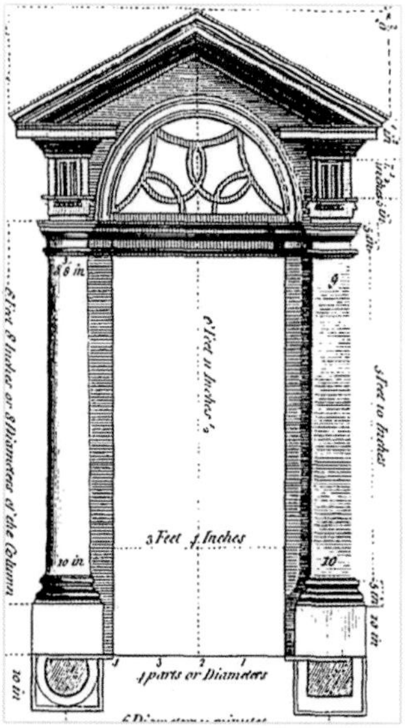

The west-elevation door surround was derived from this illustration in William Pain's Practical Builder, published in Boston in 1792. *(Wikimedia Commons)*

The original west elevation door surround was removed in the mid nineteenth century. The upper portion was stored in the barn and returned to its proper location during the 1960s restoration.

The carving on the south parlor mantel was done by Ebenezer Alden when he built his house in 1797.

The door to the left of the fireplace in the north parlor opens into the small hallway that serves the entrance on the north façade.

The Work

The reconstructed cooking hearth and brick oven in the original kitchen during the 1960s restoration.

The restored pine paneling around the original kitchen fireplace before the mantel was installed.

The restored original kitchen fireplace in 1968.

The south parlor at the start of the 1960s restoration. The later-nineteenth-century chimney has been removed for restoration of the fireplace.

The carved ornament on the south parlor mantel during restoration.

Stripping the paint from the south parlor mantel during restoration. The carved acanthus leaves have been temporarily removed for paint stripping. Note the restored fireplace.

The new ceiling in the north parlor before plastering, showing the electric radiant heat wires.

The north parlor cornice being stripped of paint.

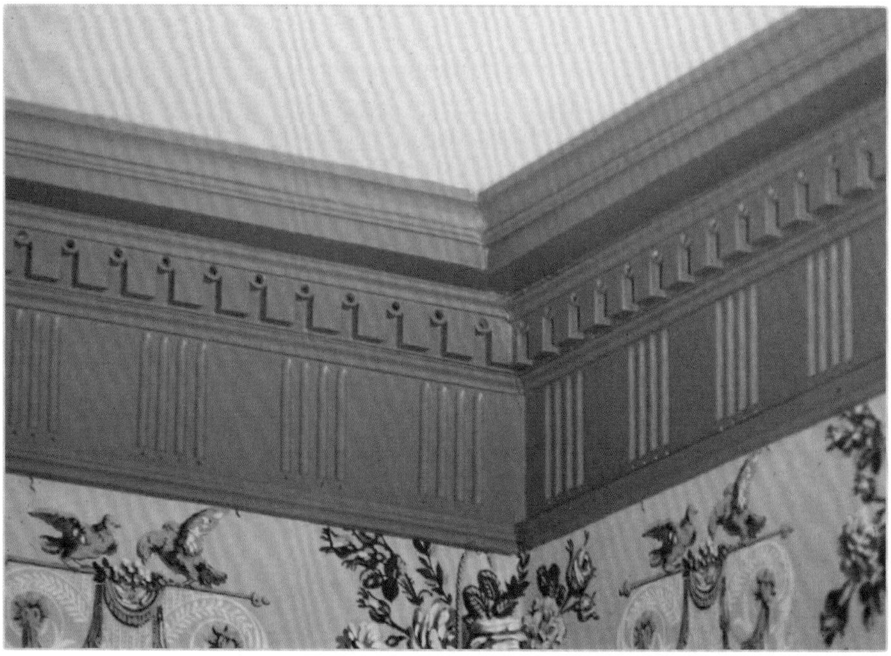

The restored south parlor cornice. Paint buildup had obscured much of the fine detail now on display.

Masons working on the reconstruction of the two chimneys and eight fireplaces during the 1960s restoration.

One of the second-story fireplaces during restoration. The clay flue liner is for a first-floor fireplace.

Section of an original paneled fireplace surround that had been stored in the barn since the late nineteenth century.

The fireplace wall in the northeast bedroom during restoration. The surviving original panels over the doors had been hidden behind wallpaper.

The same wall after being restored.

Removing plaster and accordion lath from a beaded-board wall during the 1960s restoration. The board wall was originally exposed and was covered with plaster within fifty years of construction.

The same wall with the plaster fully removed, showing the beaded boards and a reconstructed fireplace at left.

Exposed framing during restoration. Electric radiant heat was incorporated into the new plaster ceilings.

The restored fireplace wall in the southeast bedroom.

CHAPTER SIXTEEN

Ceilings

Historic ceilings are most often plaster but sometimes wood or metal, particularly pressed tin. Postwar ceilings generally use gypsum board of some sort, either as a base for skim-coat plaster or as the finish surface. Acoustical paper tile and sheet paneling also appeared as ceiling materials in the postwar period. It is not uncommon to find several ceiling layers in a historic house, each the result of a past effort to disguise problems with the previous layer. Sometimes a ceiling has been lowered in a misguided effort to reduce heating or cooling costs or to hide new electrical work or mechanical systems.

The ceiling is the most gravity-challenged surface in a room. Since the moment it was installed, gravity has worked night and day to pull it down. Sometimes this will have caused some or all of the plaster to loosen from the lath. Peeling paint is not uncommon, the result of multiple layers of incompatible paints.

Sometimes there is a layer of ceiling paper that may have been painted over.

The ceiling is usually the most visible element of a room, and any flaws are fully exposed. A historic ceiling should be repaired and preserved unless it is beyond help—an entire plaster ceiling lying on the floor; an extensively rusted tin ceiling—or there is some other compelling reason (such as gaining access to serious structural issues in the framing above) to replace it.

Plaster Repairs

As noted in Chapter 15, traditional plaster is superior to modern gypsum board for a number of reasons. For ceilings even more than for walls, the dead flatness of gypsum board sheets and their all-too-regular seams at 4-foot intervals are at odds with the character of a historic house. There is always a subtle modulation in a hand-plastered surface that speaks to the handmade qualities of an old house, and this character is particularly evident on the broad expanse of a ceiling. For this reason, it is nearly always preferable to preserve and repair a historic plaster ceiling rather than replace it.

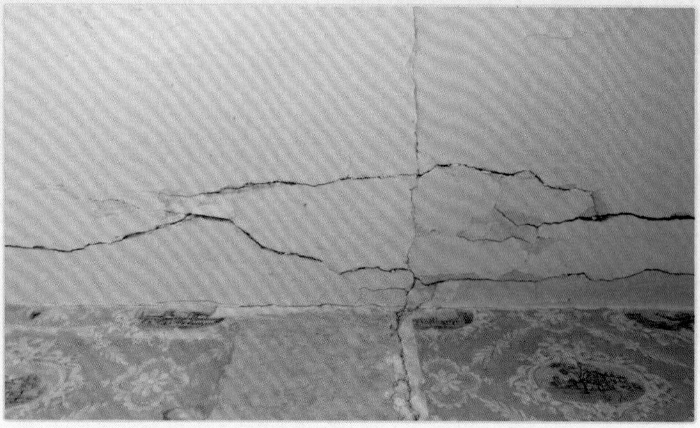

Movement in the framing often leads to cracked or loose plaster, as seen here. This is repairable.

The two common issues with historic plaster ceilings are cracking and detachment. A detached ceiling is likely to crack, but a cracked ceiling may not be detached. To determine whether a cracked ceiling is also detached from the lath, gently push up on the plaster on either side of each crack and at random intervals between cracks. If the plaster is solid and does not move upward when pressed, it is still keyed to the lath and solidly attached.

Please note that we are not discussing repair with historic lime plaster here. Lime plaster is best repaired with fresh lime plaster, but the material is neither readily available to the do-it-yourselfer nor easy to mix and apply. The following approaches work with readily available patching plaster, setting plaster, or joint compound. You should not attempt to repair historic ceiling (or wall) plaster with plaster of Paris, a very hard and fast-setting plaster that is suited to casting ornament and similar work but not to this type of repair.

If you would like to learn traditional plastering with lime or natural plaster, there are books available on the topic. Such finishes are receiving renewed attention as sustainable and natural products that are better for the environment (and home occupants) than many modern alternatives.

Cracked Plaster

Cracks in ceiling plaster generally result from movement in the framing above, which causes movement in the lath. Wood is somewhat flexible, whereas plaster—which is made up primarily of a binder (lime or gypsum) and sand—is more rigid and will crack if subjected to more than minimal movement. Fiber is added to wet plaster when it is mixed to help hold it together and prevent cracking. Historically, the fiber was horse or cow hair or some other organic fiber. Modern plaster often uses a synthetic fiber. Thin skim-coat plaster generally does not contain a fiber.

The cracks in this ceiling will be easy to repair once the peeling paint is dealt with.

To repair a crack and prevent it from reopening, you need to reinforce it with a new layer of binder. The modern fiberglass mesh tape used on seams in gypsum board does this job well. If a ceiling has a few widely spaced cracks or a group of cracks in a given area, the mesh can be applied just to those areas and covered with patching plaster or even drywall compound. If the ceiling is a generalized network of cracks, however, it may make sense to cover the entire surface with wide strips of mesh—which is available in wide sheets and rolls as well as the standard 2-inch joint tape—and apply a new finish coat of plaster. This approach is shown in the "Example Project: Plaster Ceiling Restoration" in this chapter.

For a localized repair, the mesh tape should be applied smoothly against the plaster for the full length of the crack after any loose material is removed. Traditionally, cracks were undercut on the sides to allow the patching plaster to key into the original plaster, but this is not absolutely necessary with fiberglass mesh holding everything together. The mesh is manufactured with a slight tack on one side to hold it in place until it is bedded and then covered over with plaster or compound. If the existing paint is well adhered and not calcimine, you can apply the repair over the paint. If the paint is peeling, see the section on "Dealing with Calcimine and Other Problem Paints" later in this chapter.

Once the tape is smoothed over the cracks, use a flat plaster trowel or 6-inch taping knife to apply a very thin coat of setting plaster (such as Durabond) or joint compound over the tape, filling the little holes in the mesh and crack behind it. Setting plaster cures quickly (via chemical reaction) whereas joint compound can take 16 to 24 hours to dry by evaporation and leaves a softer finished surface that is more susceptible to impact damage. Whichever you use, apply only enough to accomplish your task, building it up in thin layers. Every bit of excess "mud" you apply to the surface will have to be removed before painting—by you. Sanding joint compound or plaster overhead is a miserable task that you can minimize by careful application of the material.

Once your first application of plaster or compound, the "bedding layer," is cured or dried, drag a flat 6-inch joint compound knife along the tape to remove any bumps or dried gobs of extra material. Then switch to a 16-inch joint compound trowel—which has a very slight curve in the blade—and apply a second layer of mud over the tape. Make this application wider than the tape, covering it and tapering to nothing at the edges. The curving blade allows you to taper to the edges without scraping off all the mud over the tape, where you want it. Keep the application thin. You want to end up with a coating of mud that *just* covers the mesh tape before tapering away to nothing. Make this layer wide—a width of 8 or 10 inches over 2-inch tape is not excessive. Remember, ceilings are often seen with raking light from windows or a light fixture, which highlights changes in thickness. The wider your taper, the less noticeable it will be once the ceiling is finished.

Depending on the conditions of your ceiling and your prior experience with the process, it may take three or four full or partial coats to achieve a thin, smooth covering of mud that completely hides the tape and tapers to nothing. Scrape over the dried mud with the 6-inch taping knife after each application to remove any bumps or ridges. The smoother your original plaster, the more important it is to blend in your repair with a wide, smooth taper. Original plaster that exhibits more movement and texture can be more forgiving of repairs, as long as your repairs do not create a new texture in opposition to the existing one. Fortunately, subtle historic "ripples" or texture in a ceiling generally run parallel to the lath, which is the direction cracks most often run as well.

If a ceiling is being entirely covered with fiberglass mesh, as in the "Example Project: Plaster Ceiling Restoration" in this chapter, there will be no need to taper the new plaster into the old surface. Apply the new plaster with a flat 16-inch plaster trowel, working areas you can comfortably reach from one position at a time. Work adjoining areas in a line along one wall, then make another pass to coat the adjacent area, and so on until you have coated the entire ceiling. Use a plaster with a long enough setting time to allow you to smoothly cover each wet edge before it starts to set up. Joint compound is more forgiving in this respect, since it takes hours to dry, but it also results in a softer, less durable finished surface.

An alternative method is to apply 16-inch-wide strips of plaster or compound from wall to wall, parallel with the lath, leaving a gap of just under 16 inches between the leading edge of one strip and the trailing edge of the next. Once these strips cure or dry, apply plaster or compound to the gaps.

Regardless of the method, additional applications of mud will be needed to blend areas or strips together and create a surface that looks and feels like a single field of plaster. With careful application, little or no sanding should be necessary when it is completed.

Detached Plaster

Plaster is generally applied in several layers. The first layer, called the scratch coat, is applied to a thickness of ¼ to ½ inch (usually thinner on a ceiling than a wall) as a relatively "wet" mix that squeezes through the gaps between laths and spreads out a bit on their back sides, forming "keys." An additional thin layer (or two) of plaster is then applied to the cured scratch coat, creating the finish surface. In some historic

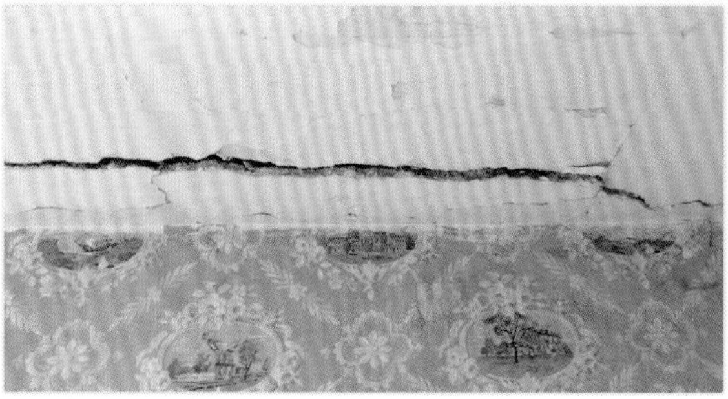

This small area of detached plaster along the edge of a ceiling will be simple to reattach.

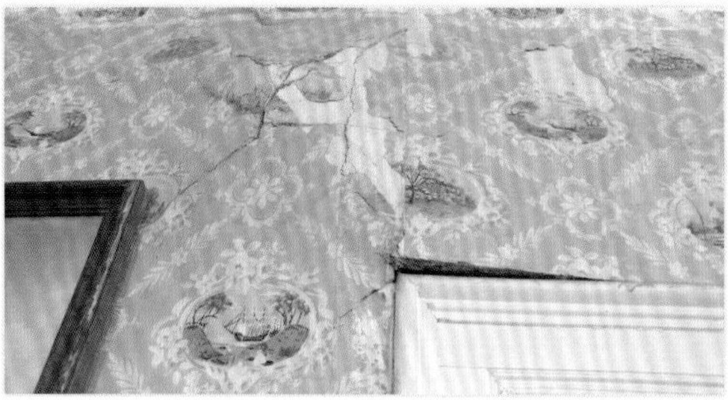

This bulging area of detached plaster on a wall may be more challenging to reattach. If hand pressure on the bulge will move it back against the lath, you're in luck. If not, loose plaster from the keys may have fallen into the gap between the plaster and the lath. The text describes what to do in this case.

construction, only one layer was used, but this is rare on ceilings except in the earliest or simplest houses.

Ceiling plaster detaches when the keys that hold it tight against the laths break off. The traditional method to reattach loose plaster is to anchor it mechanically with plaster washers and screws, which are then covered over with patching plaster. These can be difficult to install without cracking the plaster but are effective for areas of limited scope. They are less effective, however, if much or all of a ceiling is loose. Often in the past, there was no choice in this case but to take down and replace the ceiling, but now there are simple and effective methods of reattaching loose plaster with flexible adhesives.

This can be accomplished by drilling a series of $^{3}/_{16}$-inch holes through the loose plaster but not the lath, then vacuuming out any loose material

This setup is used by decorative paint restorer Tony Castro to apply pressure to loose plaster, either to temporarily stabilize it or to support it while adhesive is setting up. It is constructed of scrap lumber, old lath, drywall screws, and inflated plastic shipping pads. Here it is attached to the top of a door trim, but it could also be screwed to a stud or joist through the plaster. The arrangement of laths varies with the area that needs to be secured.

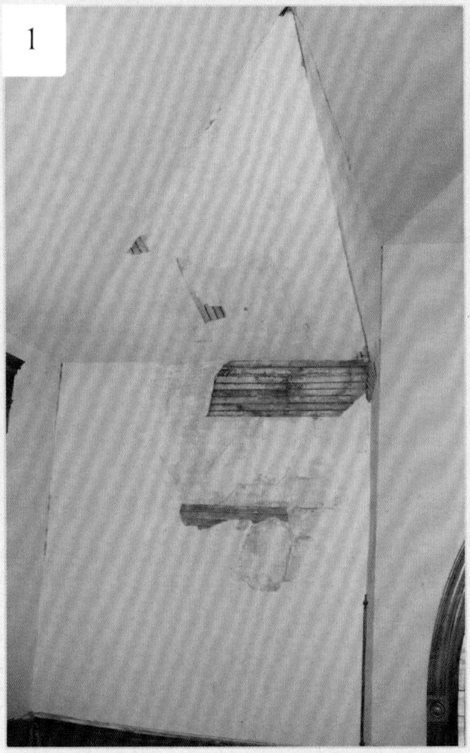

This is the first of eleven photos showing a repair with Plaster Magic. Here we see cracked, detached, and missing plaster. *(Courtesy Plaster Magic)*

Clean out any loose plaster you can see between laths or around the perimeter of a hole. To repair a crack, drill a line of holes with a 3/16-inch masonry bit 1 to 2 inches from the crack on either side, spacing the holes 3 inches apart and drilling through the plaster but not the lath. For loose or bulging plaster, space the holes evenly across the area, 3 to 4 inches apart. Vacuum plaster dust from the holes. *(Courtesy Plaster Magic)*

To wet the plaster and lath with conditioner, adjust the sprayer nozzle tip to the stream setting and apply three to five pumps of conditioner into each hole. *(Courtesy Plaster Magic)*

Wipe the excess with a wet sponge and wait 10 minutes for the conditioner to soak in. *(Courtesy Plaster Magic)*

5

Trim the tip of the adhesive tube close to its end so it fits securely into the drill holes. Apply one squeeze per hole by gently pulling the trigger of the adhesive gun until a little comes back out. There is no need to fill any hole that missed a lath. Wipe away the excess with a wet sponge. *(Courtesy Plaster Magic)*

6

Screw plaster clamps into some of the existing holes on each side of the crack, every 8 to 12 inches. Screwing in the clamps will draw the plaster and lath into "soft" contact with the adhesive. Wipe away excess adhesive with a wet sponge and allow to dry 24 to 48 hours. Unscrew the clamps and remove them with a putty knife. *(Courtesy Plaster Magic)*

7

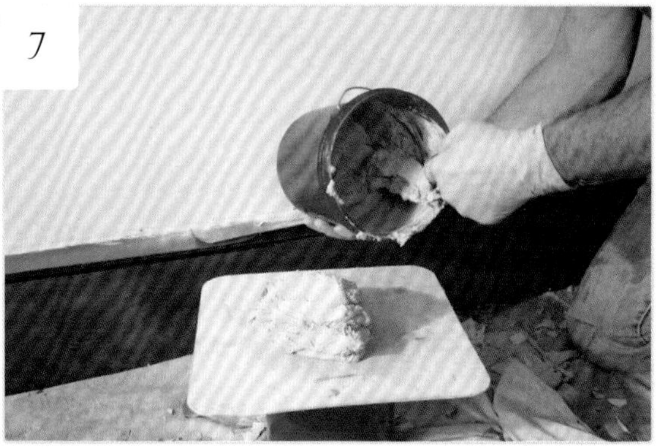

To deal with an area of missing plaster, once the perimeter of the hole has been stabilized, mix up the Plaster Magic patching plaster. The goal is to make enough mix to fill the hole halfway with the first layer and flush with the second or, if the hole is smaller than your hand, to fill it completely with one application. Pour some of the dry mix into the plastic mixing container. Add a small amount of water, stirring it into the patching plaster to create a peanut butter–like paste. (It's much easier to add more water than to take it out!) *(Courtesy Plaster Magic)*

8

Tuck the patching plaster under the edges of the existing plaster and into the keyways (the gaps between laths). Scratch the surface of this first layer with the corner of the putty knife to provide texture for the next layer to bond to. Let the first layer set up, approximately 1 to 2 hours. *(Courtesy Plaster Magic)*

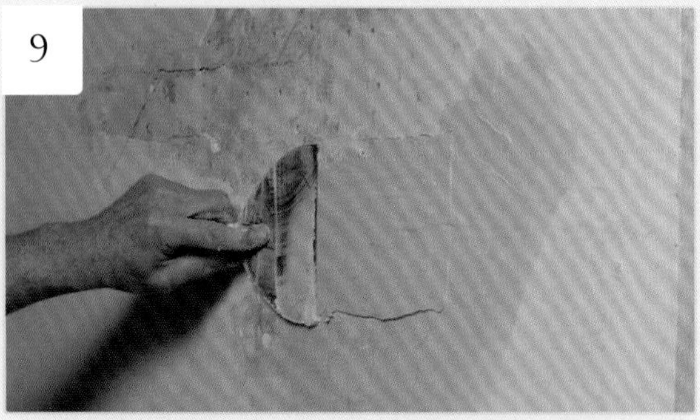

Mix and apply the second layer of plaster to fill the hole. *(Courtesy Plaster Magic)*

Remove any dried patching plaster from the surface of the existing plaster with the putty knife. Apply a topping of joint compound to fill any cosmetic defects. Let dry and lightly sand. *(Courtesy Plaster Magic)*

And here are the completed repairs. *(Courtesy Plaster Magic)*

that might prevent the plaster from moving back into contact with the laths. Inject a liquid bonding agent through each hole, then follow that with a viscous liquid adhesive applied with a caulking gun or squirt bottle. Once the adhesive is applied, you need to press the plaster tightly against the laths and hold it there with plastic support discs screwed through the drilled holes or with pieces of plywood propped or screwed in place. A sheet of plastic over the plywood prevents the adhesive from binding the plaster to the plywood.

If the back sides of the laths are accessible, as in an attic without a floor, it is preferable to drill the application holes through the lath instead of the plaster. A vacuum cleaner nozzle should be held next to the holes as they are drilled (like a dentist applying suction while drilling a tooth!) to prevent wood debris from lodging between a lath and the plaster. In the case of a detached ceiling with intact decorative painting, it may be worthwhile to temporarily remove flooring above to reattach the ceiling from the back.

In an ideal world, all the plaster will move easily back into its original position. In the real world, bits of broken key plaster or other detritus will likely have lodged in spaces between the laths and the detached plaster. In such an instance, cut a small hole with a hole saw directly beneath the obstruction, then suck out the debris on either side of the hole using a vacuum cleaner with a fine nozzle. Work carefully, and don't make any more holes than necessary, because they will all need to be patched. The discs of plaster removed to access trapped debris can be glued back in place with the adhesive, simplifying the patching.

Once the recommended curing time for the adhesive has elapsed, the temporary supports can be removed and the holes filled with patching plaster.

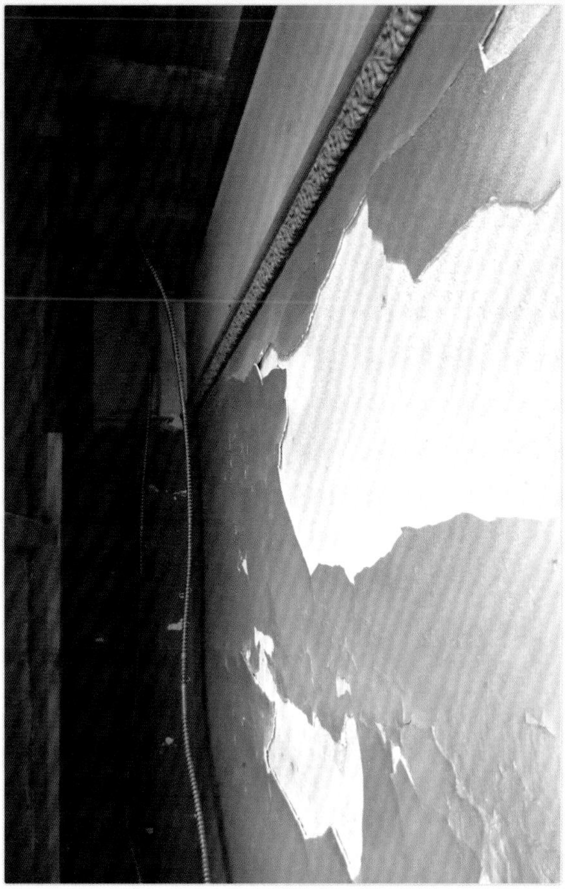

This is an example of a finish layer of plaster delaminating from the scratch or base coat. Skim-coating a new finish layer after removing the rest of the loose material is the best option in this instance.

Before plastering, I apply fiberglass mesh to any hole larger than ¾ inch to prevent future cracking around the filled hole. It may take several applications of plaster to fill the holes flush and bury the mesh in a thin layer that tapers gradually to nothing around the edges.

There are a number of bonding agents and adhesives that will work for this, and there are also prepackaged kits. One of the latter, Plaster Magic, was developed by a highly qualified conservator with many years' experience in historic preservation and plaster conservation. The Plaster Magic website includes all the technical bulletins and material safety data sheets you would expect from a serious conservator.

Note that occasionally sections of a finish layer of plaster will separate from the scratch coat while the scratch coat remains firmly keyed to the lath. If holes can be drilled in the detached layer without causing it to crumble, it may be possible to reattach it like fully detached plaster. If the finish layer can't be reattached, it will have to be skim-coated.

New Plaster

If a ceiling has lost its plaster entirely, or if the plaster must be removed to access framing for structural repairs, the recommended approach for replacing it is to skim-coat plaster on a new blueboard base. Blueboard is a paper-covered gypsum board that has been treated to bind chemically with plaster. The board is screwed up as in a typical drywall installation, joints are taped, and a thin layer of plaster is spread over the entire surface to cover the screws, joint tape, and board surface. With careful application, the surface can mimic the subtle textures of the historic plaster in a house. Unless your house has exceptionally smooth and level historic plaster, some degree of imperfection is acceptable and even desirable. The goal is to replicate the appearance and feeling of a hand-applied plaster over wood lath on a subsurface that is smoother and more even than any wood lath could be. Practice skim-coating in a secondary space before tackling your parlor ceiling. A portion of the ceiling in the "Example Project: Plaster Ceiling Restoration" section later in this chapter was new blueboard applied over the historic lath where original plaster had been removed in the 1960s. The application of the skim coat was the same as on the remainder of the ceiling.

If you're installing a new plaster ceiling where noise might be an issue—such as a corridor outside bedrooms, a large room that will not have soft furnishings and will host large gatherings, or an oval dining room (ask the folks who rehabbed Dark Harbor House, featured on page 213!)—acoustical plaster is a useful option. This consists of a very thin (1/16-inch-thick) application of skim-coat plaster over sound-absorbent backing boards. The plaster is spread on the backer board with a flat trowel or wide plaster knife, smoothed out, and then ridged with a 1/16-inch toothed trowel to set the depth. Inexperienced do-it-yourselfers are advised to let the ridged plaster set and then go back and fill the grooves flush with the ridge peaks. An experienced plasterer may have sufficient skill with a trowel to fill the grooves while the ridged plaster is still wet.

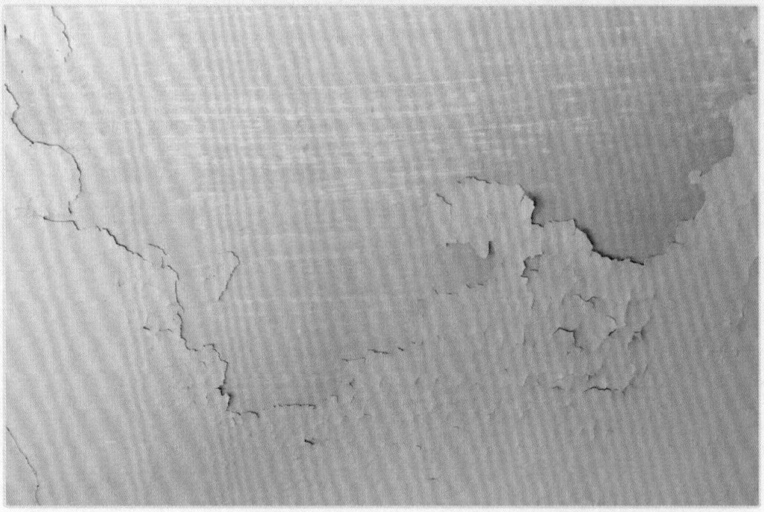

This ceiling was painted with a latex paint over an oil-based paint, which covered the original calcimine paint. Over time, the latex paint has exerted tension on the oil-based paint, which in turn has been breaking the bond between the calcimine paint and the ceiling. Note the coarse brush marks remaining from the original application of calcimine with a wide brush. The paint needs to be completely removed from this ceiling to get back to clean plaster for a new paint job that will stay attached.

Dealing with Calcimine and Other Problem Paints

Calcimine, or Kalsomine, paint was an inexpensive, highly popular distemper ceiling paint during much of the nineteenth century. It was water-based and could be mixed and applied by a homeowner—unlike the oil paints of the time, which had to be made with white lead, linseed oil, and hand-ground pigments by a professional painter on the job site. Calcimine ingredients include whiting (powdered chalk), glue-size binder, and water. Because it contained no oil, it would not yellow as oil-based white paint often did. Its popularity continued well into the twentieth century in many areas.

While inexpensive and easily mixed and applied, calcimine also had limitations. It was water soluble, did not bind firmly to surfaces, and consequently was easily damaged by water or abrasion. Because of its weak bond to the substrate, it could not be recoated with another coat of water-based paint, as the water in the second coat would loosen the first and cause it to peel. Recoating required washing off the existing paint before the new was applied.

If you have a calcimine ceiling that has not been covered with other types of paint, you will need to wash off the calcimine before undertaking repairs or repainting. Calcimine is still available, and if you choose to repaint with new calcimine, you do not need to remove every last trace of the old. If you plan to use any other type of paint, however, you must remove *all* of the calcimine. You will need a bucket of hot water mixed with TSP, a stiff scrub brush, and goggles. Protect the floor with plastic. After scrubbing off the calcimine, rinse the surface thoroughly with clean water. A well-soaked sponge mop can work effectively for rinsing. Use lots of water. Rub a dark piece of cloth here and there on the ceiling; white blotches on the cloth are an indication of remaining calcimine residue, and the process must be repeated until it is gone.

When premixed oil paints became readily available, people began to use them on ceilings. Oil paints covered calcimine with some success because, lacking water content, they did not loosen the bond of the calcimine to the plaster. Sometimes the calcimine was washed off completely and the ceiling sealed with varnish prior to painting. More often, however, the washed ceiling would retain a lot of calcimine, which the oil paint then covered. The common result was a reasonably well-bonded surface that would last as long as no water-based paints were applied over it.

With the increasing popularity of latex and acrylic paints since the mid-twentieth century, many of these ceilings have been coated with water-based paints, causing them to peel. Unfortunately, once a ceiling begins to peel in earnest, the only lasting solution is to remove all the layers of paint and start over. The ceiling shown in the "Example Project: Plaster Ceiling Restoration" section below had residual areas of calcimine under a layer of varnish, which had been covered by multiple layers of oil paint and finally a layer of latex paint. The paint had already peeled off large portions of the ceiling, and much of what remained showed a network of cracks promising to peel in the future. There was no choice but to remove all the paint, a decision rendered easier in this case because the plaster, too, had extensive cracking and areas of detachment, requiring extensive repairs and a new finish layer of plaster. Removing the remaining paint would ensure a good bond between new plaster and old.

Old oil-based ceiling paint over calcimine can often be removed with a sharp paint scraper, working carefully to avoid removing plaster with the paint. Some gouging of the plaster is inevitable, but the nicks are easily patched once the paint is off. After scraping, the ceiling must be washed to remove calcimine residue as described above.

The plaster in the "Example Project" ceiling below was so cracked and detached that much of it would have been dislodged by dry scraping. A less destructive method of removing the multiple layers of paint and varnish was needed, and a few experiments suggested that steaming was the best solution. By propping a wallpaper steamer into position with a stick for about ten minutes, I could sufficiently soften the paint in an 11-inch by 14-inch section to remove all the paint layers easily with a 3-inch flat-bladed wallpaper scraper. Some particularly stubborn areas of varnish required the use of a sharp paint scraper.

The best approach depends on the composition and combination of old paint layers on a ceiling. Experiment as necessary, starting with the least destructive method (such as hot water) before escalating to something more difficult.

Medallions

Plaster ceilings of the nineteenth century often feature medallions—usually at the center, but occasionally elsewhere on the ceiling in combination with

A cast plaster medallion of the mid-nineteenth century. Note the hole at center where the pipe for a gasolier once entered.

The polychromatic paint treatment of this plaster medallion of the 1880s was originally complemented by decorative painting on the ceiling around it.

plaster moldings. Medallions are typically cast plaster and were attached with screws and wet plaster after the ceiling plaster cured. Most often, a lighting fixture hung below the medallion. Oil and kerosene fixtures hung from an iron hook at the center of the medallion, whereas gas (and later electric) fixtures hung from a pipe that projected downward through the center of the medallion. The sculptural forms frequently found on medallions and the shadows they cast disguised the smoke and soot of oil or gas lighting, which would otherwise have been very apparent on a flat white ceiling. A polychromatic paint treatment on the medallion made its masking function even more effective.

Damaged medallions can often be repaired. If pieces have come loose, a countersink drill bit can be used to create holes for screws to reattach them. The screws should go through the plaster ceiling into the lath or joists. Remaining cracks can be filled with setting plaster or joint compound and smoothed with sandpaper. If replacement is necessary, reproduction plaster medallions are available; these are recommended over the hard foam versions, which generally lack the crisp detail of real cast plaster.

Decorative Paint on Ceilings

Decorative painting on ceilings was quite common in the second half of the nineteenth century for middle- and upper-class homes. In lucky cases, this painting remains exposed. More often, it has been painted over or covered by a later ceiling material but may be discovered during restoration. Approaches to dealing with decorative painting on ceilings is covered in Chapter 18, Paint and Paper.

A bit of historic decorative painting can be seen where later paint has come off this ceiling, which is the same ceiling shown above with the medallion. See Chapter 18 for more about decorative painting.

Example Project: Plaster Ceiling Restoration

This project involved the removal of multiple layers of paint and varnish before installing new lighting, repairing cracks, securing detached plaster, and applying a new coat of plaster over the entire ceiling that replicates the original texture and will resist cracking in the future. The ceiling—only 7 feet 2 inches above the floor—was installed in 1827 in the original kitchen of Whitten House. (See "Example Project: Restoration of an 1827 Kitchen" in Chapter 17 for more on this room.)

As with all open-hearth kitchens, the ceiling got dirty from smoke and grease and was recoated annually with calcimine paint. Ideally the old calcimine would have been washed off first, but that did not always happen. At some point, likely after cooking moved to a new kitchen in the adjacent woodshed, the ceiling was varnished to seal the calcimine and painted with an oil-based paint. It then received a coat of latex paint in the 1980s or 1990s.

When I bought the house in 2003, there were cracks in the plaster, particularly around the edges of the room, and paint was beginning to peel in spots. Quite unintentionally, the latex-on-oil-on-calcimine paint layers on the ceiling had been protected from moisture rising from the wet cellar by the floor's wall-to-wall carpet and foam padding. When I removed the carpet and pad, inadvertently allowing more humidity into the room, the deterioration of the ceiling paint accelerated rapidly as the latex overcoat created tension on underlying weak areas of attachment. By the time I got around to tackling the ceiling in 2018, it was a mess.

The accompanying photos and captions show the restoration of this ceiling.

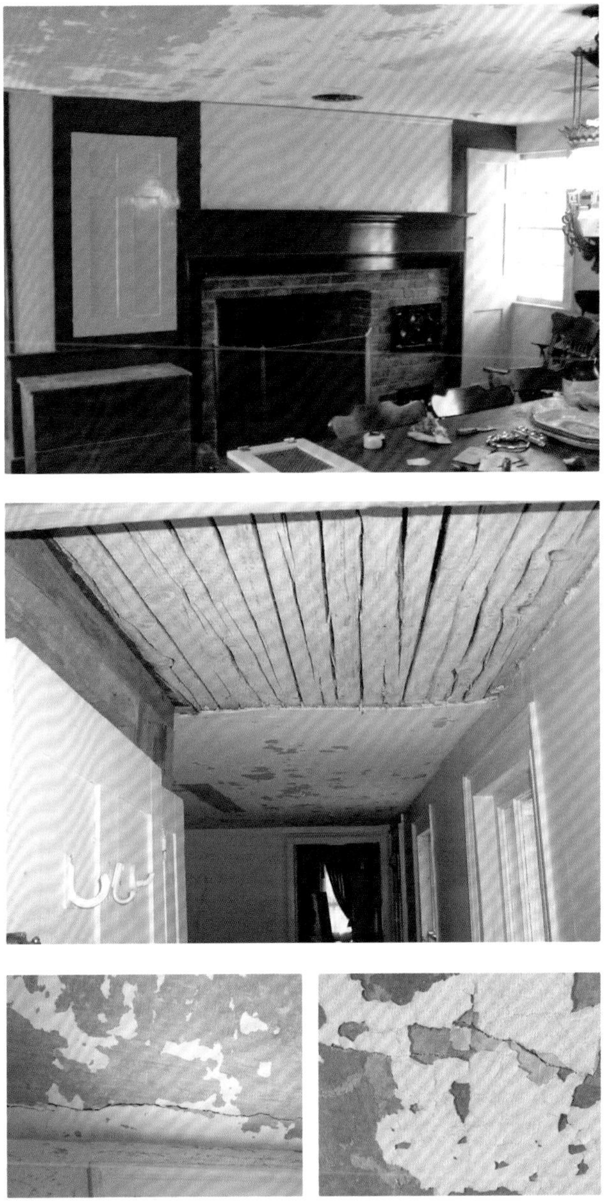

Prior to its restoration, the ceiling was afflicted by several conditions, none good. In addition to the paint and plaster issues described in the text, a portion of the ceiling plaster had been removed in the 1960s and replaced with gypsum board that was thicker than the surrounding plaster, creating an awkward transition. The gypsum board was removed when the chimney was restored, leaving the wood lath exposed. I installed gypsum blueboard for skim-coat plaster over the lath (as described in this chapter) as a starting point.

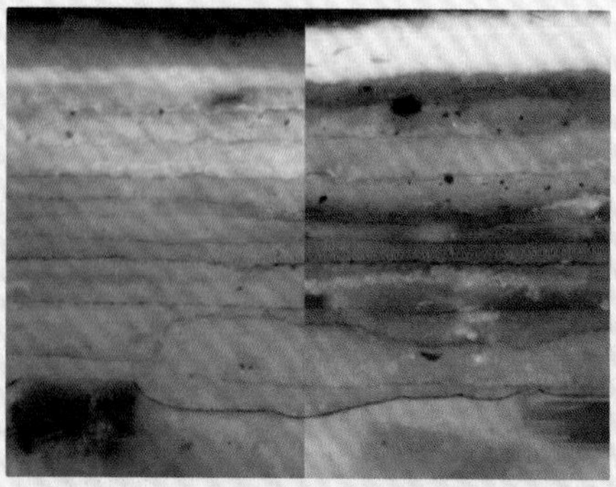

My co-worker, conservationist Amy Cole Ives, cast a paint chip from the ceiling in resin in order to examine it under a microscope. She found at least twelve paint layers in the chip. These images are cross-section photomicrographs of the paint layers at 200x magnification with reflected UV light (left) and reflected visible light (right). At this magnification you can see dirt between the layers as well as the inclusion of some pigments for color (in this case, blue pigments for a whitening effect). The bottom of the cross section shows traces of the lime plaster with its fine sand grains; that layer is overlain by a number of water-based paint layers and finally a layer of oil-based paint.

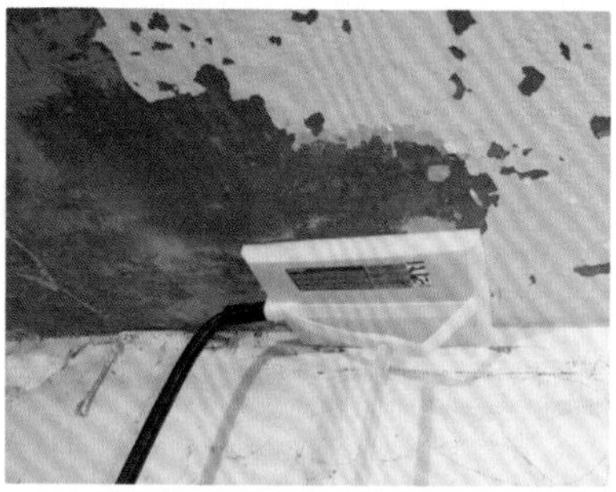

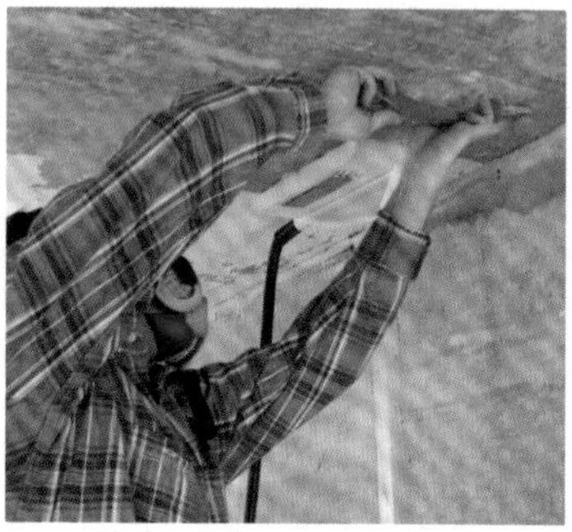

Steaming off the many paint layers turned out to be the best option for removal. Using an inexpensive wallpaper steamer, I propped the tray into position with a flexible length of wood and let it steam for ten minutes before moving it to an adjacent spot and immediately scraping the loosened paint, first pushing with a flat-bladed scraper, then pulling with a sharp paint scraper where needed, using enough pressure to remove the paint residue without digging excessively into the plaster. This close contact with the entire ceiling let me know where the plaster was well attached to the lath and where it was loose.

Following paint removal, I scrubbed the entire ceiling with a mixture of TSP in hot water, using a nylon scrubber sponge where necessary to remove stubborn residue. Finally, I rinsed the entire ceiling twice with a sponge mop and a lot of hot water. I was determined that there be no remaining calcimine residue.

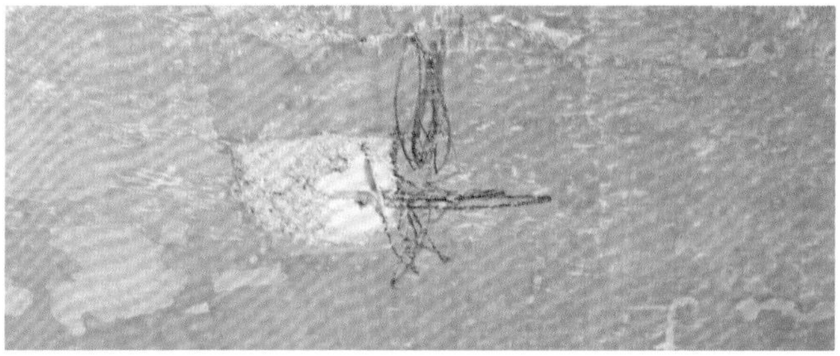

Electrical work came next. The ceiling would have one antique hanging fixture over the table and seven 3-inch recessed lights. When I marked the ceiling above the planned table center, I noticed a very old patch of plaster of Paris in almost the same spot—clearly the former location of an iron hook for hanging a candle or oil fixture. This confirmed that my hanging fixture would be where the Whittens had once hung theirs.

I do not usually recommend recessed lighting in historic houses, but the low ceiling in this room steered me to that option for task lighting over the fireplace (for hearth cooking), even lighting over the table, and pathway lighting between the adjacent modern kitchen and the rest of the house. Drilling the 3-inch holes with a hole bit in a right-angle drill and fishing through the wires was straightforward. The existing large hole in the ceiling once allowed a woodstove pipe to pass through before entering the chimney in the attic room above. A sheet metal collar between the ceiling and the floor above provided fire protection. When the fireplace was reconstructed, I had a second fireplace built in the room above, and the hearth for that fireplace now extends over the former pipe hole location. Strapping was screwed in place from above before the new hearth was laid; this allowed me to screw a disc of gypsum blueboard in place from below during this ceiling project. I wanted the recessed cans in place before replastering the ceiling, but they could not be allowed to "click" all the way into the holes. I used plastic zip ties to restrain the spring clips and prevent them from locking.

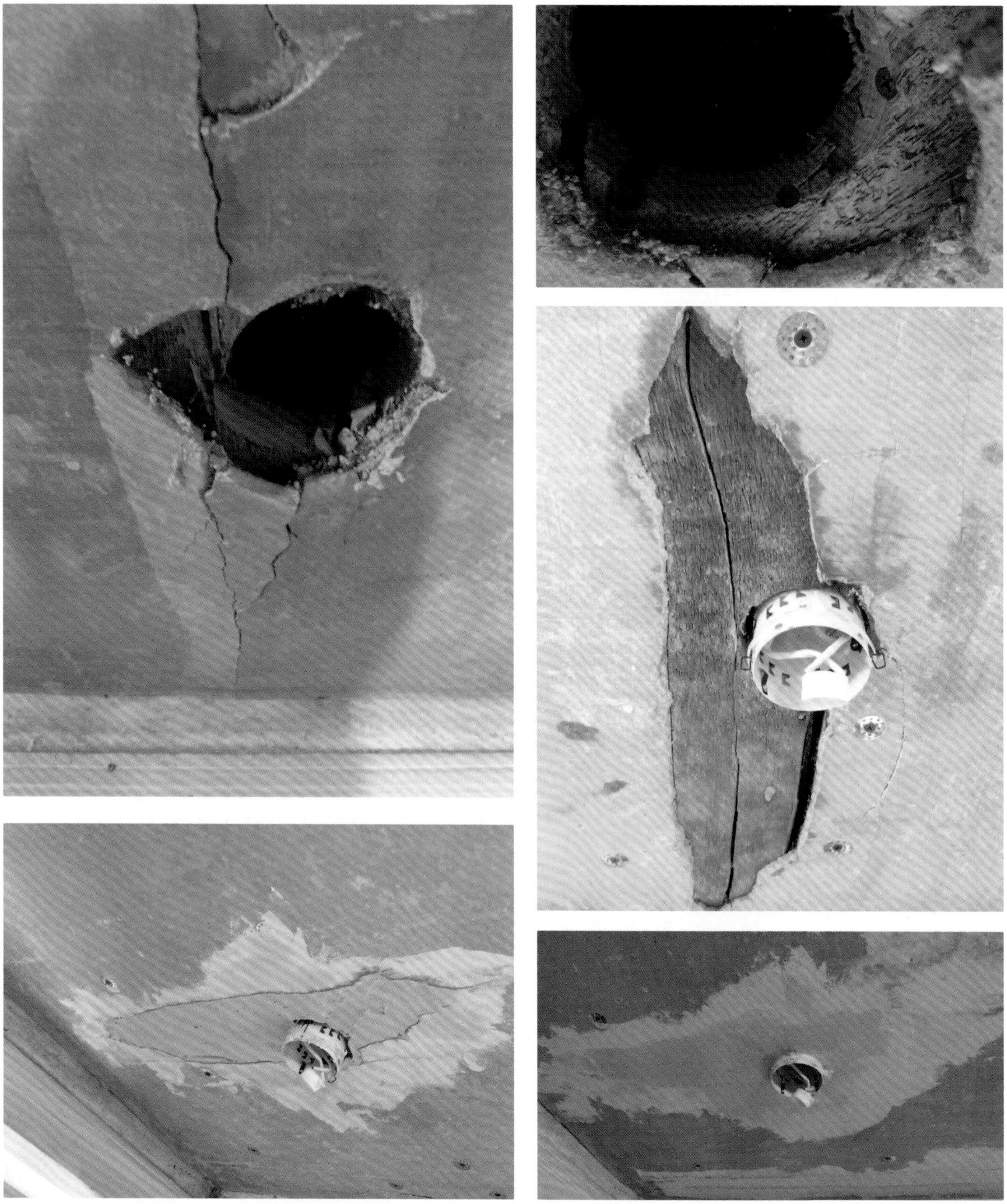

When you drill large holes in a lath-and-plaster ceiling, you need to worry about inadvertently loosening a piece of lath with consequent damage to the adjacent plaster. Fortunately, that only happened once in this project, as shown here. I secured the loose ends of the lath with drywall screws after predrilling the screw holes to prevent the lath from splitting. Then I removed the remaining loosened plaster and reinforced the surrounded solid plaster with plaster washers. Finally, I thoroughly wetted the wood lath and replaced the missing plaster with two layers of Durabond setting plaster.

The two worst areas of plaster damage were along the outside walls. Though I could have reattached the loose plaster with adhesive in both areas, I had no adhesive on hand and was working on a tight deadline. I chose instead to remove the loose plaster and apply Durabond (which is fast-setting) as a base for a new surface coating of plaster over the entire ceiling. I reinforced other areas along the edges of the room with plaster washers and applied strategic coats of Durabond to minimize roughness before replastering. I cleaned out and filled all cracks in the plaster as described in the text, and I filled the large hole over the fireplace with gypsum blueboard as described above, then filled around it with Durabond.

My product choices: Plaster Weld bonding agent to ensure a good attachment of the new plaster to the old and DAP Plaster Wall Patch for the new skim coat of plaster.

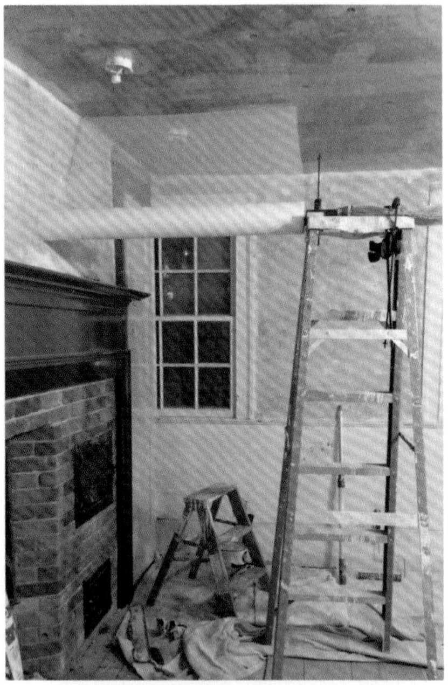

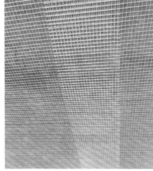

After scraping the ceiling with a 6-inch taping knife to remove any ridges and bumps from the preparatory patching, I coated the entire ceiling with the Plaster Weld bonding agent, the color of which makes it quite easy to tell what you've covered and what you haven't. Once the bonding agent was completely dry, I applied a 36-inch-wide fiberglass mesh reinforcement to the ceiling in strips. One side of the mesh is tacky to self-adhere to surfaces. Installing it on a ceiling is a little tricky when you're working alone; I clamped a broom to a 6-foot stepladder to serve as an axle for spooling the mesh. I lapped the mesh 2 inches down onto the walls and also overlapped the strips 2 inches to guard against future cracking along the joints.

Thanks to the low ceiling height, I was able to improvise staging with an old door and six paint cans. I mixed the plaster in a bucket using a rotary mixer in a cordless drill. The size of my staging limited how much ceiling area I could cover with each batch. I used any leftover plaster from a batch to patch holes in the walls.

Working one area at a time and moving the staging for each batch, I worked my way along one wall and then back along the line of the first pass, back and forth until the ceiling was covered. I applied the plaster at a thickness of 1/16 to 1/8 inch with a 6-inch taping knife, then smoothed it down and blended it into adjacent areas with a 16-inch flat knife, ending up with a thickness of about 3/32 inch, with some variation. A perfectly flat surface was not the goal; the historic plaster in the rest of the house isn't perfectly flat, and I wanted a similar modulation in this ceiling.

With the primary coat on the ceiling set up, I worked around the room, plastering over the fiberglass mesh that lapped onto the walls, tying the ceiling and walls together. This would be finished with a second coat when I skim-coated the walls.

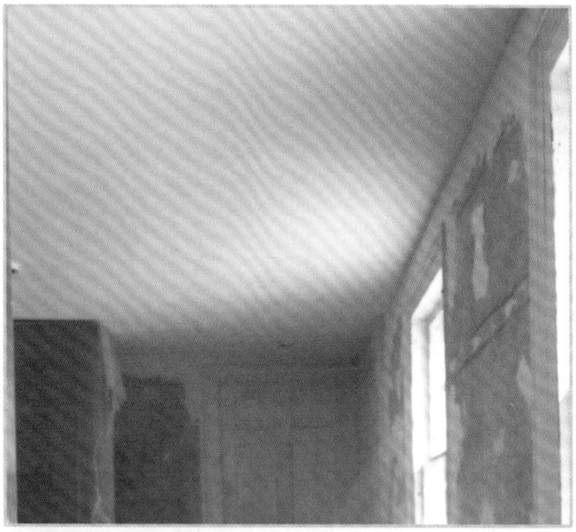

After a quick scrape with a 6-inch taping knife to remove any ridges or bits of dried plaster on the ceiling, I went back over it with joint compound and the 16-inch straight knife, filling minor defects and building up the surface a bit in one spot where the mesh was "reading" through the plaster. Joint compound is softer than plaster and not well suited as a primary coat, but it worked well for this touchup coat. Because it was applied very thinly to dry plaster, it dried quickly.

Here the finished ceiling has been painted with one coat of Fine Paints of Europe alkyd (oil) primer and one coat of their matte alkyd paint. To avoid the "orange peel" texture given to paint by modern paint rollers, I rolled the ceiling in strips (for speed of application) and followed with a semi-coarse wallpaper paste brush to give it a more traditional texture before the paint could set up. On a high ceiling this might have been unnecessary, but with a ceiling height just over 7 feet, the texture is highly visible.

Pressed Tin Ceilings

Decorative ceiling panels and cornices made by stamping sheets of thin metal between cast iron molds became widely available at the end of the nineteenth century and remained popular until the 1930s. Intended to mimic the appearance of expensive molded plaster, the ceilings were comparatively inexpensive and were marketed as being fireproof. Although most common in commercial and institutional buildings, they were used in many homes as well, often to cover plaster ceilings that had developed cracks. Typically, 2-foot by 2-foot or 2-foot by 4-foot panels are nailed into 1 x 3 strapping installed perpendicular to the ceiling joists on 2-foot centers. Overlaps are tapped tight with a ball-peen hammer and punches as needed, then caulked before painting. They were often installed with stamped sheet-metal cornices and may have medallion panels above light fixtures.

This late-nineteenth-century pressed tin ceiling has been soda blasted to remove paint in preparation for repainting.

In general, the greatest threat to tin ceilings is moisture. If water has gotten into the ceiling from plumbing above or a leaking roof, rust is likely to be present. Portions of the ceiling may have rusted away entirely. When the damage is extensive, the repair

This photo shows how tin panels were nailed to wood strapping installed below an original plaster ceiling.

can be complicated. The next most common threat is holes made for plumbing or to hang a suspended ceiling, but these are relatively easy to deal with.

If portions of the tin were removed to access plumbing and not reinstalled, the repair can be as complicated as a severely rusted ceiling. A lesser problem, but no fun to deal with, is peeling or chipping paint. Some surface rust may be present where paint is missing, but this is relatively easy to remove with a wire brush before priming with an oil-based metal primer. Water-based paints will cause more rust.

When portions of a ceiling are missing, the challenge is finding matching pieces. Other buildings in the area may have used the same ceiling design, in which case local architectural salvage shops might have some stock. Alternatively, several companies continue to make pressed tin ceiling panels from the same dies that were used a century ago. F.W. Norman Company, for example, still reprints a catalog from the early twentieth century. With luck, you can find your exact design. Failing that, you may find a design close enough in appearance to be undetectable to a casual observer.

Because panel sizes were largely standardized in the early twentieth century, pieces from one company usually fit with pieces from another. Sharp tin snips, heavy

This tin ceiling is in a house that remained vacant and unheated for more than a decade, allowing moisture to condense on the painted tin (which remains colder than the air when the space warms up). The result is failing paint and rusting tin, which, if allowed to continue, will eventually become unrepairable.

gloves, and safety glasses are essential for working with this material. A hand or bench grinder with a wire brush wheel will help remove burrs along new cuts.

Patching small holes is like doing autobody repair work. A Bondo-type putty (i.e., a two-part thickened polyester resin) and fiberglass mesh will take care of holes in flat areas. More putty may be needed to build up missing design elements. Mix the goop thickly and apply it with whatever tools allow you to build up the shapes you need. Once it cures, use sandpaper to smooth it and feather the edges into the metal. Once the repair is painted, you may be the only one who ever knows it is not a perfect match. Perfection is not required except in the most highly visible locations.

Wood Ceilings

Ceilings made entirely of wood are uncommon but not unknown. Early examples were usually painted. They became more popular during the Victorian era, when they were generally stained and varnished. Frequently these have a coffered effect, with panels of bead board framed by moldings and false beams. In secondary spaces, beadboard alone might be used on the ceiling as well as the walls.

This wood ceiling is in Historic New England's 1878 Eustis Estate in Massachusetts, a masterpiece of the early Queen Anne style designed by Boston architect William R. Emerson. *(Photo by Eric Roth)*

This Weldtex ceiling was installed when the attic of a kitchen ell on an 1808 Federal-style house was rehabbed as a rec room in the 1960s.

When this Victorian-era pantry was restored, the historic plaster ceiling had been damaged by leaking plumbing in the bathroom above. The ceiling was removed to replace the plumbing, and a new beadboard panel ceiling was screwed in place. This is compatible with the original pantry cabinetry and allows easy access to the plumbing above to address any future issues.

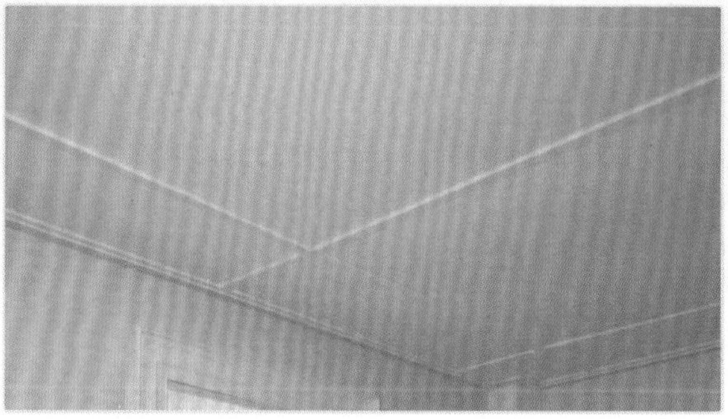

This parlor ceiling in an 1808 Federal-style house has been covered with Masonite panels. Wood lattice hides the joints. This was a common cover-up for cracked plaster ceilings in the second half of the twentieth century.

Wood ceilings are repaired like any other wood trim (see Chapter 17). Modern wood ceiling types include sheet paneling, usually with color-matched wood strips to cover the joints (see Weston House feature on page 243), and Weldtex striated plywood in natural or painted finishes. Weldtex was manufactured by U.S. Plywood from the 1940s until the 1970s, then disappeared from the market until 2012, when Eicher Siding in Southern California began to reproduce it.

Masonite is the best-known brand of manufactured fiberboard (hardboard, as it was traditionally known). Made of small wood fibers and resins, it has a smooth, dense finished surface and often an unfinished back. Thin (1/4-inch) 4-foot by 8-foot hardboard sheets were often used to cover cracked ceiling plaster in the later twentieth century, usually with ¼-inch by 1½-inch wood lattice stripes to hide the seams and create a paneled effect. Typically, the lattice created a symmetrical pattern on the ceiling. The finished ceiling would be painted, usually white. These hardboard ceilings are often removed during a rehab to restore the original plaster above them.

Fiberboard remains readily available in a variety of thicknesses and densities. Medium-density fiberboard (MDF) is widely used in modern construction.

Acoustical Tile Ceilings

Acoustical tile ceilings were another popular cover-up for cracked historic plaster ceilings and a widely used ceiling treatment in postwar new construction. Made of pressed paper fiber with a painted face and beveled edges, these were historically available in sizes from 12 inches by 12 inches to 24 inches by 24 inches. Only 12 inches by 12 inches is readily available today; fortunately, that was always the most commonly used size in residential applications. (As of this writing, at least one company is producing 24-inch by 24-inch tiles as well.) Acoustical tiles were usually installed by nailing up 1 x 3 strapping at intervals matching the width of the tiles and stapling the interlocking tiles to the strapping. Occasionally the tiles were glued up with a big dab of mastic or other adhesive on the back.

The cover of a 1960s Johns-Manville brochure for acoustical ceiling tiles and wall panels.

Where these tiles cover an earlier ceiling, it is usually desirable to remove them and restore the original ceiling. In more recent historic houses with original

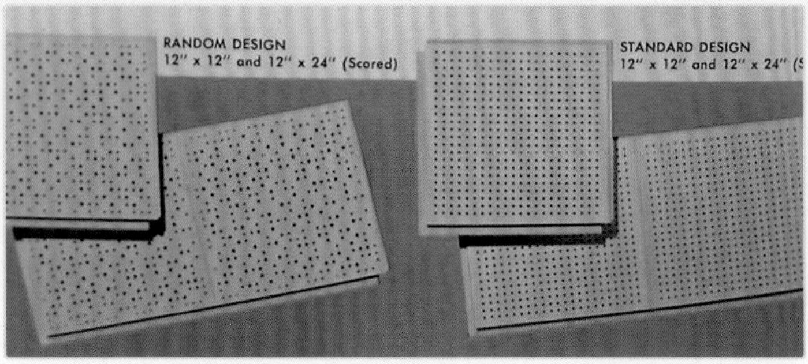

Two styles of ceiling tile were available in the Johns-Manville brochure. The plethora of manufacturers for these tiles over many decades makes it a challenge to find a match.

This image from the Johns-Manville brochure was intended to show that a do-it-yourselfer could install an acoustical tile ceiling. This is Dad. In another image (not shown here), we see Mom handing Dad a wall panel to staple in place.

acoustical tile ceilings, it may be necessary to undertake some repairs or replacements of tiles that are dangling or sagging, water-stained, or damaged.

I have successfully reattached sagging or dangling tiles by pushing them back into place and gently countersinking small finish nails at an angle through the tiles and into the strapping. Most acoustical tiles have many tiny holes in the surface to absorb sound (hence the "acoustical" part of the name), and the nail holes often blend right in. When nail holes are jarringly larger than the tiles' original holes, a small amount of lightweight spackle can be applied over the countersunk nail heads. Spackle can also be used to fill small divots and scrapes from impact. A small nail can be used to mimic holes in the tile when a solid patch stands out. To blend in the patches, it may be necessary to paint the ceiling as described in the next paragraph.

Water stains can often be sealed with commercially available stain sealers or even traditional shellac if soaking has not caused the tile to lose its structural integrity. Once sealed, the entire ceiling will have to be painted. This is an instance where a paint sprayer is the best solution, as you can easily cover the angled joints between the tiles as well as their faces with a thin, even coat. Painting with a roller is more challenging. A brush will probably have to be used in conjunction with the roller to make sure all the angled joints are covered without applying so much paint it ends up dripping from the joints.

Badly damaged acoustical tiles may need to be replaced. Try to match the texture or pinhole pattern of the original tile as closely as possible, but don't be surprised if you can't find an exact match. There have been many manufacturers of these tiles in the past half century and lots of variation in details.

Suspended Ceilings

Like acoustical tile ceilings, suspended ceilings have been a popular way to hide cracked plaster in old houses and were also used as original ceilings in

THE WORN CEILING
THE WARM CEILING
BE A CEILING WATCHER:
See how a dingy, old ceiling can be made new again with an Armstrong Ceiling.
If you have a ceiling that's unfinished, cracked, peeling, just plain dingy or dull, look into an Armstrong Ceiling. Today, you have a choice of dozens of patterns as well as many types. There are Armstrong Ceilings that soak up noise. Ceilings with built-in lighting that hang below your existing ceiling. Ceilings that have a rich inlaid effect. And lots more.
NEW CEILINGS!
In the picture above at right, you see one of our newest ceiling tiles. It's called Brunswick Temlok®. The deeply beveled edges and the wood pattern create a rich inlaid effect. And it costs only about $50 for a 12′ x 14′ room. You simply cement or staple the tiles right over your old ceiling.
Another new style is called Dover Temlok—sample at right.
ANOTHER KIND OF CEILING
There's a new kind of ceiling that hangs below an old ceiling. It's called an Armstrong Suspended Ceiling, and it's a great way to cover the pipes and beams in an unfinished area. To install it, you hang a metal grid by wires at any chosen height. Then you take large ceiling panels and just slip them into the grid. It's easy to do yourself, and the whole job takes just a few hours. Armstrong offers lots of patterns in suspended ceiling panels, and the materials for a suspended ceiling cost as little as $55 for a 12′ x 14′ room.
NEW KIND OF LIGHTING
A special kind of lighting fixture, called Gridmate™, has been developed by Armstrong to fit an Armstrong Suspended Ceiling. It attaches right to the grid and, combined with a luminous panel, gives you mod ern, recessed lighting. (Shown with luminou panel removed.)
JOIN THE CEILING WATCHERS!
Write for a complete Ceiling Watcher's kit. You'l get color pictures of all the Armstrong Ceilings detailed information about how to install them, and a Ceiling Watcher's button. Send ten cents to Armstrong, 6611 Sun Street, Lancaster, Pa. 17604.
LOOK UP!
Ceiling Watchers' Ceilings by
Armstrong

A 1966 magazine ad for Armstrong suspended ceilings specifically targeted the owner of an old house. The less common hidden-grid type of tile is shown at top right being admired by a happy housewife. The more common 2-foot by 4-foot exposed grid is shown at bottom.

postwar houses. The typical configuration is a grid of interlocking painted steel frame pieces (having an inverted T cross-sectional shape) hung from wires to support 24-inch by 24-inch or 24-inch by 48-inch pressed paper or fiberglass tiles. Most commonly the tiles are dropped into and rest atop the inverted Ts, leaving the grid exposed. Less commonly, the tiles abut one another and hide the grid, resembling an acoustical tile ceiling. (Similar grids have been used to support pressed tin panels, or plastic panels imitating pressed tin, in a usually unsuccessful attempt to mimic a traditional tin ceiling.)

Suspended ceilings are even less appropriate than acoustical tile for houses that did not have them when built. In a contemporary house, however, it is appropriate to maintain, repair, or replace in-kind as needed.

Textured Ceilings

Among the mostly doomed-to-fail attempts to hide cracked plaster instead of repairing it are various types of added texture. These can be sprayed ("popcorn ceiling"), rolled (striated ceiling), troweled ("stucco ceiling"), or brushed ("swirl ceiling"). None of these is appropriate in a pre–World War II house. Some forms of "popcorn" are fairly easy to scrape off with a putty knife. If the house you want to buy has textured ceilings, hope it's this stuff. Other textures are created with joint compound and are less easily removed. Often the most effective approach is to knock off the high points with a putty knife or straightedge scraper and then skim coat over the mess to get a new smooth surface. You will first need to deal with any cracks or detached plaster as described earlier in this chapter.

In a postwar house with original textured ceilings that are damaged or detaching, you have another challenge. I have no personal experience with the

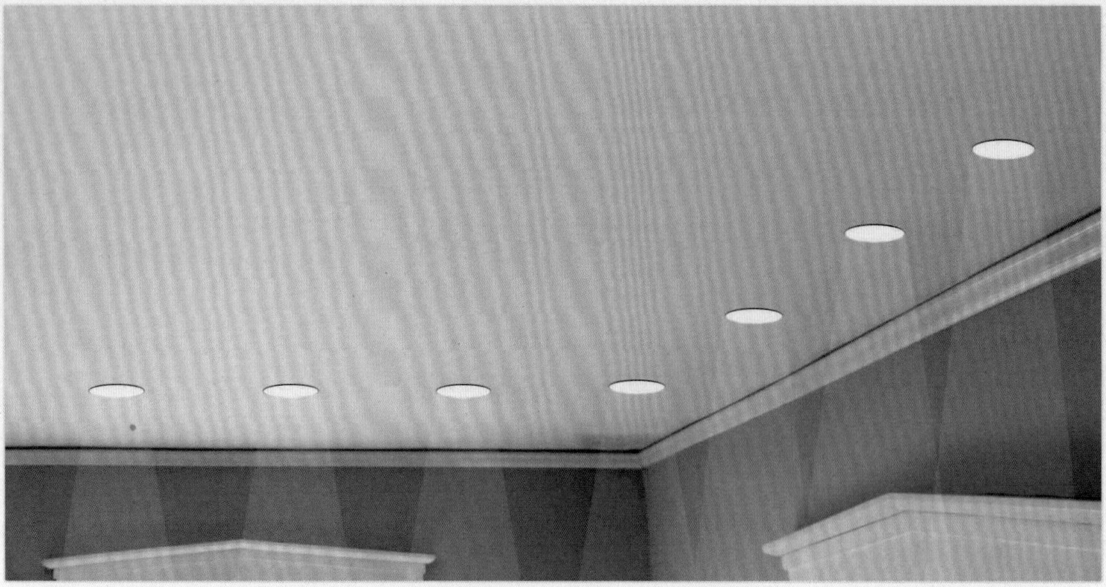

Recessed lights have become ubiquitous in new construction, and rooms with fifteen or twenty 6-inch holes in the ceiling are not uncommon. Recessed lighting appeared in the 1950s and can be entirely appropriate for postwar modern houses. Its use should be minimized in earlier houses, but sometimes it's the best option. The original kitchen in Whitten House has a ceiling that is barely more than seven feet above the floor, so ceiling-mounted light fixtures are impractical anywhere but over the table. The restored hearth and brick oven are occasionally used for cooking, and the room serves as the dining room. Several years of cooking in my own shadow made clear to me that some sort of task lighting was needed over the hearth, for safety as well as convenience. The solution was three 3-inch recessed LED lights operating on a dimmer. Two identical fixtures, also on a dimmer, were installed over each end of the eight-foot-long dining table, flanking the Victorian cast iron fixture at the center. These help with presentation of food and for task lighting at the table. Finally, two recessed lights (operating on a third dimmer) were installed inside the doors to the back hall and the modern kitchen, allowing the pathway to be lit without lighting the whole room. Because they are small and usually on a dim setting, these three groups of recessed lights do not greatly impact the character of the room. I replaced the "bright white" LED bulbs that came with the fixtures with "warm white" bulbs that better suit the space.

spray popcorn repair kits available at home improvement centers and hardware stores. Textures created with joint compound can be patched with a careful mimicking of the surrounding texture. Practicing on a gypsum board panel might be a good idea. Joint compound is inexpensive, so don't be hesitant to waste a bit while you figure out the best approach for a repair you may be looking at for decades to come.

Be aware that asbestos was sometimes used as an additive both in gypsum plaster and spray-on popcorn ceilings, primarily between 1900 and 1975. With this in mind, wear personal protective gear and clean up and dispose of these materials appropriately when removing or disturbing them. If you're unsure about the presence of asbestos, testing can be done for a positive determination.

Record Farm, 1844

Greek Revival Style

The Greek Revival–style Ezekiel and Mariam Record house was built shortly after the Records bought the land it sits on in 1844. Both Ezekiel and Mariam had deep roots in the rural western Maine community. Ezekiel Courtney Record's grandfather and three great uncles had been among the town's first settlers in the late eighteenth century; three of the four brothers (from Plymouth County, Massachusetts) built homes on 100-acre lots granted for their service in the Revolutionary War. Mariam Irish Record's family were also among the town's early settlers, and the house built for the young couple in 1844 was close to the homes of many Irish and Record relatives.

Ezekiel purchased the first portion of the land he would farm from Hiram Hall in 1838; this plot was part of the "Settling Lot" of Enoch Hall. He purchased additional acreage from Hiram Hall in January 1844, at which time, according to the deed, Ezekiel was a resident of Lewiston. This second plot included the site of the homestead, and the two plots combined to form a property of approximately 43 acres, divided into three sections by roads. Ezekiel is identified as a bootmaker in the 1850 US Census—work that very likely took place in the ell. The census also records that Ezekiel's younger brother, Clark Record, age twenty-three and also a bootmaker, was in residence in 1850. Given the challenging agricultural conditions of the region, secondary (and often seasonal) sources of income were critical to the local economy. Ezekiel owned the house for thirty years before selling it to his son-in-law in 1874, three years after Mariam's death.

The adoption of Greek architectural forms between the 1820s and 1860s symbolized the ideal of democratization over republicanism, the importance of Classical studies in the school curriculum, and a national pride in developing a new, non-British identity. The Greek Revival style became popular in Maine during the Greek war for independence (1821–1832), and as the Federalist postures of the early century gave way to a more populist concept of democracy, Greek architectural forms came to represent increasingly accessible democratic ideals. In building their house, the Records clearly sought to make a statement with boldly expressed detailing derived from Asher Benjamin's builders' companion books. In particular, the trim on the front façade and in the two parlors make this house noteworthy despite its modest size.

Architectural Historian Joyce Bibber, in her book *A Home for Everyman: The Greek Revival and Maine Domestic Architecture*, discussed the various builders' companion books available to Maine carpenters in the first half of the nineteenth century, stating:

> When the books offered similar drawings and builders freely adapted them, perhaps using a mantelpiece design to inspire a doorway or a doorway element to adorn an exterior pilaster, the results are charmingly varied, if frustrating for researchers! … It is not always easy to attribute a particular architectural

> detail to a specific design in a builder's book. While an occasional element may have been created, line for line, from a given plate, most appear to represent composites or compromises. Parts of one design were combined with parts of another, as details which looked too demanding for the carpenter's skills were either omitted or replaced by something he already knew how to do.

As Bibber suggests, it can be challenging to apply strict Classical terminology to all Greek Revival trim, since much "classical" trim of the period was freely adapted from published sources. The architectural detailing of the Record homestead embodies the creative expression of carpenters working with hand tools and builders' guides before steam-powered mills produced identical ready-made building components and rail lines shipped those products over long distances. The gradual shift in a carpenter's work from "making" to "installing" began in the house building trade in the mid-nineteenth century. The trim on the Record home predates this and is characteristic of the creativity of its style and period.

The Record Farm is composed of a one-and-a-half-story Greek Revival–style Cape dwelling with original kitchen ell, attached shed, and attached barn. The main house is typical in form for a rural Maine farmhouse of the 1840s but has somewhat more than the usual trim detail. This is particularly apparent in the elaborate corner pilasters on the front and side façades (repeated in a reduced form on the kitchen ell) and in the interior trim of the north parlor. The interior trim of the entry hall and south parlor are less exceptional but are still fine examples of Greek Revival detailing. The use of different molding profiles in the several rooms contributes to the character of this small house. The panel-within-a-panel Greek Revival door in the recessed entry is unusually "high style" for a rural home, where traditional six-panel doors were the norm during the Greek Revival period. As the home of a presumably prominent resident, a grandson of one of the early settlers of the town, the E. C. Record homestead shows clear evidence of a conscious effort to make a statement about style and status.

The connected farmstead plan may not appear exceptional to natives of northern New England, but it is quite exceptional when viewed in a national context. Commemorated in the nineteenth-century children's rhyme "big house, little house, back house, barn," the form was a mid-nineteenth-century development centered in a relatively small area of Maine and New Hampshire, with a little spillover into Massachusetts and eastern Vermont. The densest concentration of such farmsteads is in southwestern and central Maine, including Oxford County. These farmsteads are a unique characteristic of the northern New England landscape and heritage.

Most connected farmsteads were created between 1850 and 1890, often by relocating separated buildings to create the connected complex. This was part of a movement intended to make New England's older farms more competitive with newer Midwestern farms (where the soil was often better) and thus more attractive to sons who might be tempted to move west to farm or to the growing urban centers to try their luck in other trades. A connected farmstead was efficient and functional, placing the small-industry workrooms between the barn and kitchen while providing a winter windbreak for the dooryard and farmyard on the south or east sides of the buildings.

Connected farmsteads have become increasingly threatened as agricultural activity decreased and barns and sheds lost their utility and fell victim to deferred maintenance. The preservation of remaining intact examples is critical to ensuring that future generations can understand the historic system of mixed husbandry, home industry, and small-scale family farming that characterized life in much of rural northern New England for more than a century.

When the current owners of the Record house, Dave and Pat, purchased it in 1977, the house had been divided into two apartments and the property subdivided into several large lots. Despite suffering the usual alterations of an "old house" not valued as a "historic house," it remained relatively intact. During their decades of ownership, Pat and Dave have carefully restored elements of the house that had been altered over time. Among the many projects undertaken by the couple were installing new wood windows that replicate the six-over-six sashes of the original windows in place of later two-over-two sashes; re-shingling the barn; and replacing missing trim with pieces cut with molding knives to match original profiles. They also installed a compatible new kitchen and two bathrooms and rehabbed the second story space in the ell as an office for Pat's bookselling business specializing in books about conservation biology, environmental science, and natural history. Prior to his retirement, Dave was a professor of chemistry at Bates College when he wasn't working on the Record House and outbuildings. He continues to make fine reproduction furniture in his shop in the barn and cultivates hundreds of varieties of daylilies on the property.

From the time they bought the house, Dave and Pat hoped to reassemble the original acreage. They were able to acquire the adjacent wooded lot to the north of the house in 1980, and in 1997 they acquired another portion of the original farmstead across the road. Their sense of stewardship of the property led to the creation of a protective easement on the significant historic elements of the house and outbuildings as part of the donation of the house and land to a local land trust. It is anticipated that the trust will eventually sell the parcel containing the house and outbuildings, but the historic integrity of the buildings will be protected through the preservation easement. As described in Chapter 2, their thoughtful process included documentation of the significant historic features of the house and identification of areas where future changes can be accommodated without harm to the historic integrity of the buildings. To further emphasize the historic significance of the house to future owners, in 2011 they had the property listed on the National Register of Historic Places.

The trim is simple on the back of the house, which is not on public view.

Looking into the kitchen ell from the main house. The kitchen is modern but compatible in character.

Nineteenth-century landscape paintings in the south parlor depict the nearby White Mountains region.

Stone walls document the taming of a wilderness by European settlers to America.

Dave's collection of router bits, many of which were used in the 30-year restoration of the house.

Cast iron stoves like this one in the south parlor were likely installed shortly after the house was built.

Overhead doors in the east side, which faces away from public view, allow part of the barn to be used as a garage.

After completing the restoration, Dave began reproducing antiques like this tall case clock in the north parlor.

A corner of the north parlor mantel is reflected in a reproduction mirror made by Dave.

The house in about 1880 shows the original open porch and six-over-six windows.

The porch had been enclosed in this photo of about 1940, and the original six-over-six windows had been replaced by one-over-one sashes.

Dave's wood shop occupies one side of the barn.

The connected farmstead form suited the rural economy in this region of New England, which often combined farming with home occupations like Ezekiel's bootmaking.

The door on the front stairs was likely added when the house was divided for use by two families at one point in its history.

The connected form of the farm is clear in this view from the north of the house, ell, shed, and barn.

Pat operated a mail-order specialty bookshop from this office space in the attic of the ell for many years.

The north parlor received the most elaborate trim in the house and clearly was always intended to be the "best" room.

The rural farmstead in winter.

Dave did the period-appropriate grain painting on this swirled handrail, another exceptional detail for a rural house of the time.

Flowerbeds fill the property with color during the warmer months.

The Work

Dave gets some assistance while re-shingling the barn.

Stripping paint from the woodwork in the front parlor.

The shingling assistant points out a knothole in the barn sheathing.

Priming the new trim pieces around the restored fireplace.

CHAPTER SEVENTEEN

Interior Trim

Detail from an 1840s Greek Revival mantel.

The interior trim of a house identifies it with a style or period and—like exterior trim—is almost always a character-defining feature. If the goal is to respect and preserve a home's historic character, it is seldom acceptable to replace existing trim with off-the-shelf modern moldings. Here more than anywhere else in or on a house, details are important. A Federal-style molding is as problematic as a modern molding in a Victorian house, and the converse is equally true. Inappropriate moldings confuse a home's identity.

The trim in a very old house may have been handmade on site. As time went by, trim was more and more often produced in a mill and transported to the building site. In either case, it is important to match the dimensions, profiles, and materials as closely as possible when replacing a missing element or installing new trim. Hand-molded trim was produced with hand planes containing shaped blades. It is possible to find historic hand planes with blades that match common molding profiles, and it is also possible to have a millwork shop cut a shaper blade with a matching profile to run new molding on a shaping machine. Once installed and painted, such machine-molded trim will be nearly impossible to tell from the handcrafted trim it mimics. A local millwork company that produces custom moldings for old houses in the region may already have blades cut to the molding profiles you need. This will save you money, since

Stained and varnished trim in an 1890s Colonial Revival house. This finish was common during the late nineteenth and early twentieth centuries but was often painted over in the later twentieth century.

A grain-painted door in a mid-nineteenth-century house.

The capital on this pilaster inside the front door of a Colonial Revival house is likely cast composition material. Paint stripping is the primary danger to compo, as the chemicals in some strippers will soften the material to such a degree that it is removed with the paint. A soy stripper may be safe, but try it first in an inconspicuous spot.

cutting a custom blade generally runs around $100 or more.

If your house has historic alterations that are significant for documenting changes over time, it is important to match new work to the trim of the room you are working in. A Federal-style house with a Victorian-era addition should get matching Federal moldings in the original house and Victorian moldings in the addition.

Houses of the Victorian era and the early twentieth century often have trim with stained and varnished finishes that should be matched by any new or restoration work. If the original wood is mahogany or quarter-sawn oak, the new work should be as well. If that is financially unfeasible, or if matching wood is simply unavailable, it may be possible to mimic a finish by grain-painting on another wood. Such faux finishes were used historically to mimic expensive hardwoods, even in very expensive houses. Thomas Jefferson had softwood doors at Monticello grain-painted to mimic mahogany, and graining was frequently painted on inexpensive softwood trim in the rear portions of Victorian houses that had hardwood trim in their principal rooms.

The skills required to repair or replicate historic interior trim vary with the style, materials, and detail but are often easily mastered. Even elaborate trim usually consists of simple elements built up in layers—the exception being hand- or machine-carved ornament.

Much of what appears at first glance to be hand-carved ornament is in fact "compo" or plaster that has been painted or faux finished to look as though carved from the wood it is attached to. "Compo" is a period term for "composition" ornament; there were many variations, generally a mixture of a binder-like hide glue with a filler like sawdust, chalk, or gypsum and some sort of fiber to help hold it together. It could be cast in semiliquid form and retained a degree of flexibility, unlike cast plaster ornament. Some compo ornaments could be softened in hot water for application (usually with hide glue) to curved surfaces. Composition and plaster ornaments are still produced by a few surviving companies from the same molds they were made in historically.

It is sometimes possible to find a match for missing ornament that is identical or at least close enough to be undetectable to most people. It is also possible to cast a replacement piece from a mold that you make yourself from surviving ornament (as described

This hand-carved ornament from 1719 is on the stairs of the Brush-Everard House in Colonial Williamsburg. Few of us will ever find ourselves the steward of an early-eighteenth-century house, but hand-carved ornament continued to be created for the homes of those who could afford it into the early twentieth century. The options for replacing damaged or missing carved ornament include carving a replacement piece yourself; having someone carve it for you; substituting a commercially available carved ornament that is close in size and design; substituting a commercially available composition ornament; or casting a replica from a surviving original piece.

later in this chapter). Elaborate wood carving is a skill that requires significant time and experience to master. Local millwork shops may have a carver among their employees or may be able to direct you to an independent carver who can help.

Historic Trim Details

Window and Door Trim

Post-and-lintel trim features right-angle cuts, whereas picture-frame (or mitered) trim uses angled cuts. Historic window and door trim sometimes exhibits both forms—for example, a post-and-lintel base to which mitered moldings are applied on top. A variation on the post-and-lintel form is corner-block trim, which

Simple post-and-lintel trim. Some traditional carpenters slightly angled the ends of the lintel (the horizontal piece), making it imperceptibly wider at the top to offset an optical illusion that makes a square cut appear to taper the other way.

became popular during the Greek Revival period and carried right through the numerous Victorian styles. This type, too, is sometimes made more elaborate by an applied molding with mitered corners. Frequently the same trim elements were used in the fireplace mantel—something to bear in mind if you need to design a compatible replacement for a missing mantel.

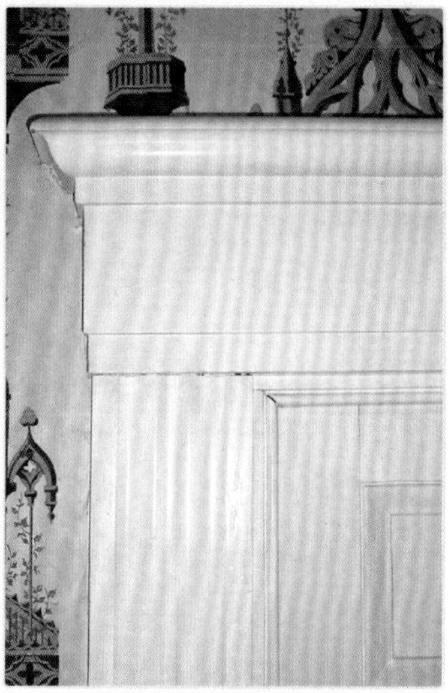

This more developed example of post-and-lintel trim features a molded post and a built-up lintel with a cap.

Another post-and-lintel trim piece with a molded post and a "Greek Peak" lintel with cap.

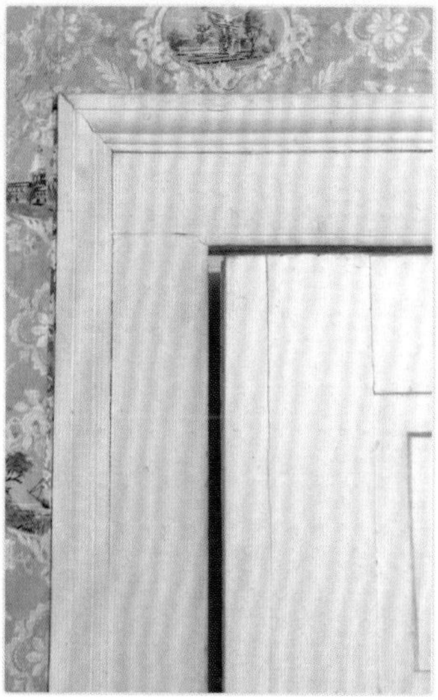

A post-and-lintel trim treatment with an added band molding at the outer edge, mitered at the corner.

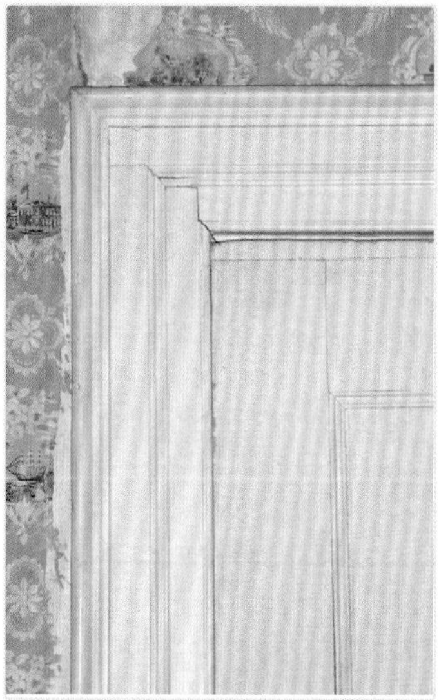

This post-and-lintel trim with a mitered band molding also features an intermediate post-and-lintel step with mitered molding.

An example of corner-block trim. Generally, the same molding(s) is used for the sides and top.

An example of mitered trim, which can be a single molded board or built up in layers like some of the previous examples.

This example combines corner-block and mitered trim. The floral element on the corner block is applied composition ornament.

Looking down the front stairs at Whitten House.

Staircase Trim

Primary staircases are almost always important architectural features and frequently received some of the most detailed millwork in a house. A back stair, by contrast, is often strictly utilitarian, with minimal detailing. If your staircase(s) are badly damaged, do your best to save them. A new staircase will have to meet code (few historic ones do) and may not fit where the old one was. The winding staircases that were once common are not allowed at all in many jurisdictions, which means that you could face a radical reworking of your floorplan if you can't save the original winding stairs. If your code official is making things difficult, you might offer to build a modern, code-compliant staircase in the back of the house while restoring the nonconforming stair in the front.

Stair-building was historically considered one of the most difficult tasks in finish carpentry. In larger communities there were craftsmen who only built staircases; later there were millwork shops that did the same, selling staircase "kits" to builders.

A homeowner repairing a historic staircase must often contend with loose balustrades (railings). When this happens, determine how the newel posts and railings are secured to the framing. A variety of methods were used historically, depending on where, when, and who built the staircase. In most cases, bolts or iron rods are placed inconspicuously to provide strength and solid connections to framing. Finding these and making sure they are still well connected is step one. In Victorian-era stairs, there is often a large nut on a threaded rod under the cap of the newel post, and the rod is attached to the floor framing below with another nut or screw plate. Other attachment methods are shown in the accompanying photos.

When a full second story was added to Whitten House around 1850, the center chimney left little room for a front staircase. The solution was this amazingly tight swirl of a stair that is not remotely close to meeting modern code, though it functions just fine. This stair could not legally be replicated, and a code-compliant stair would require drastic alterations to the floorplan on both stories. If at all possible, a staircase like this should be repaired, not removed.

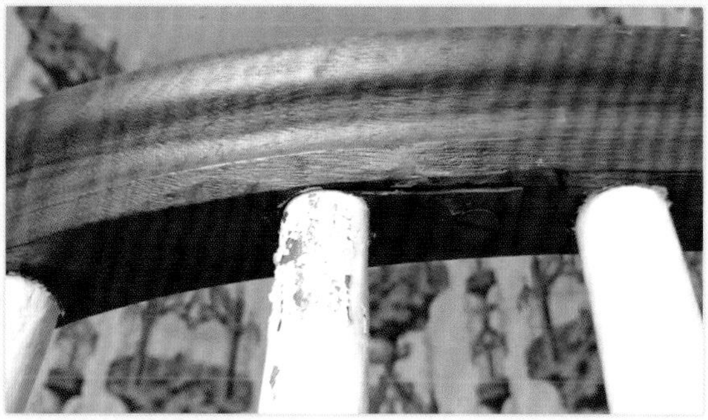

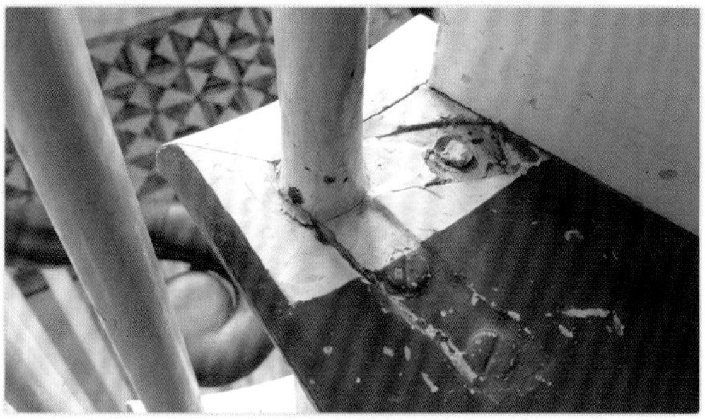

Hidden among the round wood balusters on the Whitten House front stair are three iron rod balusters that are attached by plates and/ or screws to the railing and floor or steps.

The solid turned newel post at the second floor of Whitten House's back stair is attached to the floor framing with three hand-forged carriage bolts. The nuts for the bolts are under the floor and can only be accessed by taking up a portion of the floor above or removing a portion of the ceiling below.

The turned newel post for the attic stair at Whitten House has no visible attachment. The post is solid without a cap, so a threaded rod isn't hidden there. Very likely the post is bolted to the floor framing with carriage bolts and nuts, accessing which would require removal of flooring or the ceiling below.

This mid-eighteenth-century stair shows no visible attachment points for the newel posts. Given its early date, the posts may be secured with wooden pegs or with hand-forged ironwork under or behind the finish surfaces.

Railings are attached to newel posts and walls with bolts or large screws, again placed as inconspicuously as possible. If a railing was lag bolted through a wall stud or blocking into the railing before the plaster was installed, it will be impossible to expose the attachment without some selective demolition. Other types of through-wall bolts may have a nut accessible in a recess under the railing. Once you've identified the major points of attachment for the railing system, tighten them. If some degree of reinforcement with new iron or steel is necessary, it should be done as inconspicuously as possible. Consult a friendly local blacksmith.

The balustrades on primary staircases often feature turned or even carved balusters and elaborate newel posts. Unfortunately, balusters are a common target of vandals and bad tenants, and many historic houses are missing some by the time restoration gets underway; finding or making replacements then becomes part of the task. As the staircase is a character-defining feature, it is important to match the original balusters as closely as possible, using one of the approaches outlined below. If a good match cannot be had, consider consolidating the surviving original balusters in the most visible location (usually the front hall) and keeping the new balusters inconspicuous (ideally upstairs).

This Victorian-era newel post probably has an iron rod attached to the floor framing and running up to the top of the post, where a large nut is accessible under the post cap.

This nicely detailed Greek Revival baseboard of the 1840s continues around the base of the fireplace mantel. Baseboard moldings and board thicknesses and heights vary with the period, style, and luxury of the house. The right molding on the right board helps make a restored room feel "right." Thin, off-the-shelf modern baseboard rarely does the trick.

Baseboards

Baseboards have been used in American houses from earliest colonial times. Plain or elaborate, they protect the wall surface where it meets the floor. Early baseboards often have a bead or molded profile at their top edge. Later baseboards usually have an applied baseboard cap molding. Baseboards in utilitarian spaces may lack any molding.

Until the middle of the nineteenth century, baseboards were typically installed before the finish flooring, so the ends of the floorboards butt against the baseboard. It became more common after the middle of the nineteenth century to install the finish flooring first and place the baseboard over it. This practice can allow a gap to open under the baseboard, because a board will shrink across the grain when it dries. To deal with this crack around the base of the room, shoe molding is sometimes found along the bottom of the baseboard.

Doors

Interior doors of the eighteenth through mid-twentieth centuries were constructed of rails, stiles, and panels except in the most utilitarian houses or rooms. The sizes and numbers of panels changed over time, with later doors typically having fewer and larger panels. Any added or replacement doors should be appropriate to the style and period of the space in which they are installed. A two-panel door from the

This wonderful paint-decorated door is in the collections of the Dewitt Wallace Decorative Arts Museum at Colonial Williamsburg. The paint decoration is exceptional, but the eight-panel stile-and-rail door construction is not. Stile-and-rail construction has been used for a wide range of door designs with anywhere from one to twenty panels. Among the most common for American houses are six-, four-, five-, and two-panel designs.

1930s will not look right in an eighteenth-century room that originally had doors with six raised panels. (On the other hand, a Colonial Revival house of the 1930s might have been built with two-panel doors despite its attempt to achieve an eighteenth-century feel; in other words, Colonial Revival houses often failed to be Colonial in all details.)

Door-panel configurations may vary within a house due to the hierarchy of spaces. For example, the most significant rooms might have six-panel doors while secondary spaces have four-panel doors. Alternatively or in addition, door panels in significant spaces might have applied moldings, whereas doors in secondary spaces do not. By this time it is probably unnecessary to remind you that doors are character-defining features and that modern, hollow-core, flush doors are not a desirable addition to an eighteenth- or nineteenth-century house.

Repairing or Replacing Wood Trim

Reproducing Damaged or Missing Moldings

Thanks to previous alterations to a house—especially if walls or architectural elements have been removed—it is not unusual to have to reproduce missing moldings. How best to do this will depend on the quantity needed, the availability of local millwork shops, the tools in your shop, and/or the availability of antique hand planes with appropriately profiled blades.

Always start with a search of available stock from local lumberyards or mail-order sources. It is usually considerably less expensive (in time and/or money) to buy an existing molding than to make one. Elaborate moldings usually comprise several simpler moldings that are pieced together; the newer your house, the better your chances of finding those component moldings in stock somewhere. Most of the commonly available wood moldings today have been around since the postwar period, and many Victorian-era moldings are now reproduced by mail-order millwork companies, a few of which also make Colonial- and Federal-style profiles.

When you cannot find matching moldings for sale, the next-best option is to make them. If you have or can locate an antique hand plane with a blade profile that matches what is missing in your house, and if the quantity you need is modest, you may wish to reproduce the molding as it was produced originally.

Reproduction moldings for this project were obtained from a custom millwork shop, the shelves of a home-improvement center, and on-site work with a router table.

With a sharp blade, clear stock, and a suitable work surface, it is straightforward and simple to plane a molding. A profile of any depth will require multiple passes with the plane, but the work can be enjoyable if you adopt a meditative frame of mind and decide that you are not in a hurry.

Custom mechanical millwork is another option, especially when you need a large amount of molding. (The cost of fabricating a custom cutter blade—upwards of a hundred dollars—can result in a very high cost per foot if you only need a little.) You can do this work yourself with a power shaper or have it done by a local millwork shop using a surviving sample of your original molding for the new cutter blade profile. If your molding profile was commonly used by a historic builder in your community, you might find a millwork shop that already has the blade on hand from a previous customer.

Another option is to reproduce your molding at home using a router in a stand or table. It is sometimes possible to combine several available profile bits to create the overall profile needed. For the simplest moldings, a single bit might provide the correct profile. Though it is less common to find modern profile bits that match early profiles, this possibility is certainly worth looking into.

The cost of a custom cutter blade or even a router bit is hard to justify when you only need a foot or so of molding. In a case like that you can cast the missing piece or pieces instead. This approach also enables you to bend the mold around a round object to create a curved molding, as described in Chapter 13, Exterior Trim and Windows.

If you're inexperienced with finish carpentry but want to learn to do it well, *Fine Homebuilding* magazine from Taunton Press is an excellent resource. Although not focused exclusively on restoration work, the magazine includes many articles on finish carpentry (and other work) that apply to the restoration of historic homes. Even better, the entire magazine archive since 1981 (280 issues as of this writing) is available in searchable digital form. The full archive flash drive is currently priced around $100. It goes on sale occasionally, and you'll know when if you join their informative email newsletter. For the record, I have no connection with Taunton Press or *Fine Homebuilding*; I just appreciate an excellent source of useful information.

Architectural Salvage

In many communities there are architectural salvage shops that sell materials removed from historic buildings. With luck, you may find exactly what you need in one of these shops, whether a door, molding, or mantel. Many salvage shops have knowledgeable staff who can help you find appropriate elements for your house, but others have staff who are focused on moving material out the door, whether or not it is appropriate to the house it is going to. Resist high-pressure sales tactics if you are unsure or need to double-check measurements or profiles before you buy. The item you're looking at may actually be "the last one," and there may actually be "someone coming to get it this afternoon" if you do not buy it now, but don't let that stop you from taking the time to be sure. Fortunately, in this age of smart phones with cameras and note-taking apps it is easy to carry the information you need about the things you are seeking. Preparing yourself ahead of time can greatly simplify the shopping.

This is the architectural salvage shop where the early-twentieth-century step-back cupboard used in the new kitchen at Whitten House was found (at center).

Resist the urge to replace missing elements with pieces that are fancier than the original trim in your house. An elaborate carved marble mantel will look out of place in a simple vernacular Victorian house, as will an exuberant mahogany newel post on a stair with simple round, painted pine balusters. Conversely, a slender Federal-style newel post will look odd in an Italianate house. Be true to the character of the house. If the character of a house is simpler than you would prefer, add interest with elaborate furniture, window, and wall treatments rather than altering architectural details.

Similarly, resist the temptation to install elements that are older than the house. "Earlying up" was one of the sins of the early Colonial Revival movement, when many eighteenth-century houses were "restored" to a fantasy seventeenth-century appearance. These works of imagination have a certain charm today, but their creation erased a lot of good eighteenth-century work. It is best to respect the history of your house and avoid fakery.

On the other hand, furnishings of an earlier period are always appropriate. After all, the original owners of a historic house likely had older pieces as well, brought from a previous home or inherited as heirlooms.

Repairing Damaged Woodwork

It is not uncommon for trim in an old house to have been damaged over time. Cosmetic damage may simply offer evidence of age and authenticity, but serious damage should be repaired so as not to distract from the historic character of the house. There are more repair options for painted trim than for stained and varnished pieces; matching wood patches are often the only option for the latter, though it may also be possible to blend in a repair material using faux grain painting. A number of books detail how to create effective faux finishes, including wood graining.

The simplest woodwork repairs are just good paint prep: filling nail holes, feathering the edges of chipped paint with sandpaper, filling small dings and dents with wood putty. The next level involves patching more serious damage such as large holes, broken edges, and rot with Dutchman patches or a high-quality epoxy wood filler (as described in Chapters 13 and 14). Elements that are missing or damaged beyond repair will have to be replaced with fabricated or salvaged pieces that match the missing items as closely as possible in style, detail, and dimensions.

Balusters on staircase railings are a common missing element, easy for a vandal to kick out. Usually the majority remain in place, so you know what to match. Start your search for matching balusters in the attic; I have spotted left-over original balusters in the attic of a house on a number of occasions. If you go in search of architectural salvage replacements, take one of the remaining ones with you. The same applies if you have a millwork shop replicate balusters. If you have a lathe, you can replicate turned balusters yourself.

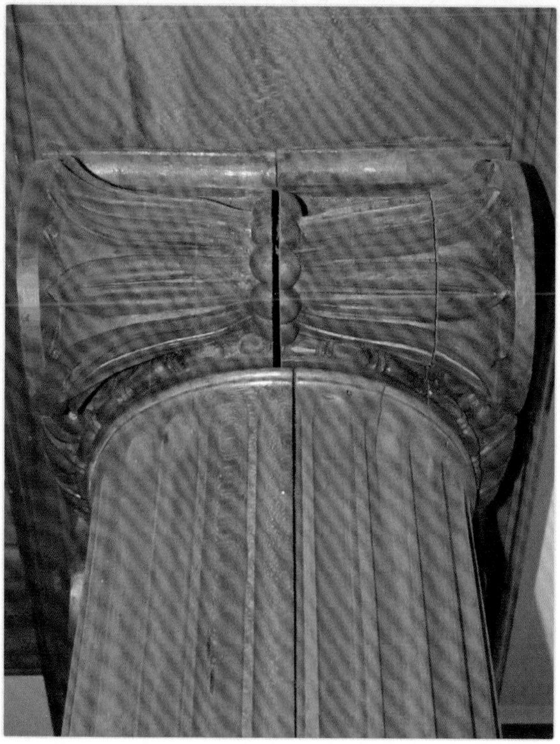

This Ionic capital and fluted wood column have split as a result of structural issues below the floor, which have caused uneven pressure on the largely decorative column. Once the underlying causes are resolved, the column will have to be repaired. Fortunately, it appears to have split along glue joints where its component pieces of wood were put together; a relatively straightforward repair with clamps and glue and perhaps a few countersunk screws will restore it.

Replacing Missing Doors

Doors are another commonly missing element. Fortunately, they are also commonly similar among many houses in a vicinity, because they were all supplied from the same local door and window mill. I have pulled matching doors for my house from the burn pile at the local solid waste facility (aka, the dump). As always, local architectural salvage shops are worth checking.

Sometimes a door has an uncommon detail, such as a panel molding, that you cannot find a match for. I encountered this problem in the c. 1850 second-story addition to Whitten House; when I rebuilt the missing wall between the master bedroom in the front and a small room in back (see Chapter 3, page 112), I needed to replace the door that had gone missing with the original partition. The doors in the back rooms lacked panel moldings but were otherwise identical to those in the front rooms. Having removed two closets from the back rooms to make room for two bathrooms and a walk-through closet, I had two extra doors without moldings. I simply milled a matching panel molding for one of these doors, making it indistinguishable from the front-room doors.

Other Trim Materials and Styles

Metal Trim

Along with pressed tin ceilings and ornamental exterior metalwork, metal interior trim was also produced in the first half of the twentieth century. Its great selling point was that it was fireproof; it was used primarily in institutional buildings but was advertised for use in homes as well. Typically, this material was installed before plaster went on, hiding the edges. If it is damaged, a repair will likely mean removing some plaster and patching afterward.

I recently looked for new chair rail stock to match that in the accompanying photo and was unable to find any in production. Instead we had to salvage material from some areas of the building to repair others. As an alternative, wood molding could be milled to a matching profile. Once painted, it would look much the same.

Pieces of formed-metal chair rail from another part of the building are being used here to fill in where a new opening has been cut in the wall. Salvaging from within the house is often the best option for obtaining obscure materials that are no longer produced.

Stonework

The most common use of stone within American houses was for fireplace surrounds and mantels. Early houses might have marble slabs for the surround between the firebox and a wooden mantel. Carved marble, slate, and soapstone mantels became popular in Italianate and Second Empire houses. These were often arched, and many had cast iron surrounds for a coal grate. Many "marble" mantels were actually cheaper slate with a faux finish.

Stone appeared in some grander houses for hallway floors and stair steps and risers, particularly toward the end of the nineteenth century and into the twentieth. Slate and soapstone sinks appeared in kitchens around the same time, sometimes with matching countertops. When bathrooms made their appearance, marble was commonly used for sink surrounds, usually with ceramic sinks set below a cutout in the marble. This was a logical progression from the marble-topped washstands that had long held basins and pitchers for washing in bedrooms. Although not common, showers were sometimes fabricated from large slabs of marble or even slate.

Any of these historic stone installations may be broken, chipped, or cracked. An experienced stone installer can repair or fabricate replacements for most damaged pieces using stone-patching compounds like Jahn Restoration Mortars (from Cathedral Stone Products) to custom-match colors and textures (including aggregate within the stone). Masons have

A simple marble fireplace surround.

to be certified by the manufacturer to buy and use these products; the results can be amazing, but this is not a do-it-yourself project.

Cracked stone in a slate or soapstone kitchen sink can often be made watertight with a marine caulking compound such as Life-Calk, made by Boat Life. This polysulfide adhesive-sealant comes in both liquid (for tight cracks) and traditional thick caulking in either white or black. Once cured, it can be sanded. I use it for all the seams in my slate kitchen sink, applying a thin bead along the joint and working it in with a plastic putty knife, then scraping off the caulking that remains on the surface. Before it sets up, I use 220-grit sandpaper to remove the residue from the surface and push slate dust into the caulking that remains in the seams. Once cured, there is no evidence of the caulking, and the sink is watertight.

When stone is merely stained and dirty, as can happen easily in the white marble beloved by Victorians, a homeowner can clean it. Homemade poultices are often effective, and there are many stain-removal products for stone on the market thanks to the recent popularity of stone countertops. Poultice recipes for cleaning stone can be found in many of the popular nineteenth-century housekeeping guides for women as well as in twenty-first-century online sources.

Postwar Trim

A number of materials came into use in home construction after World War II that had not previously been common. Some, such as aluminum and plywood, had experienced greatly expanded production during the war, and producers sought new peacetime markets. Plywood quickly became a standard home-building material, and this included exterior grades as well as varieties with birch or other veneers for interior

This built-in bar from 1957 is typical of much post-war work, incorporating plywood, laminate counter-top, and aluminum hardware for the shelves.

finish work. Production of waterproof laminate materials also took off during the war, followed by rapid adoption of these materials in postwar homes. Aluminum appeared in cabinet hardware and as extruded moldings to finish the edges of laminate countertops and sheets of Masonite "tile" in bathrooms and kitchens, as shown in the previous chapter.

Twenty-First Century Substitute Materials

Substitute materials are nearly as old as architecture. The Romans used scored stucco on rubble stone or brick walls to imitate ashlar stone blocks; they used trompe l'oeil painting to mimic architectural materials and features. Wallpaper was developed to mimic fabric and leather wall coverings. Cast compo ornament was developed to mimic hand-carved wood. In the nineteenth century, cast iron replaced stone for many structural applications and was often cast in forms imitating stone or wood.

It is never appropriate to rip out good historic material and replace it with a modern substitute. It is always appropriate to replace a missing element with a new element that matches it in design, dimensions, quality, and, if at all possible, materials. Still, there is nothing inherently bad or inappropriate about substitute materials, provided they are used with care.

Many companies now produce moldings in polyurethane and other modern synthetic materials. Such moldings vary widely in the quality of their design and construction and in the longevity of the material. When considering the use of substitute materials to replace missing historic elements, ask several questions:

1. Is the piece dimensionally correct for the period of the house? (Many of these materials are thinner or flatter than the historic materials they claim to replicate.)
2. Is the profile of a molding clean, crisp, and accurate to the period? (Many are not.)
3. Is the material as durable as the original material it's replacing?
4. Will the material accept appropriate period finishes? (Not all modern materials accept paint or stain well.)

If the answers are all yes and the original material is unavailable or unaffordable, the synthetic substitute may be a reasonable choice. The bottom line is that the work should look no different from the surrounding historic trim when completed and into the future.

The marketing claims made by product manufacturers should be investigated and independently confirmed before being believed. Traditional materials and methods are backed by a long track record, ample evidence of quality and durability, and the experiences of generations of tradesmen, salespersons, and customers. No such history buttresses the new-product claims of a manufacturer or related industrial marketing group. Proceed with caution, question everything, and make use of the remarkable research tool of the twenty-first century (the internet!) to read about peoples' real experiences with the material or product. It is impossible to know how a newly developed material will behave after a century in use, but if it was developed in a lab ten years ago and hit the market five years ago, you can at least find evidence of how it has behaved in the real world over that brief time. When considering the use of a new material in your home, intending to pass it along to future generations, take the time to learn all you can about it.

This wonderful doorknob and plate from the 1880s is still in working condition. Removing the works long enough to clean it up and give the moving parts a few drops of oil would be time well spent. The missing screw should be replaced to keep the plate from moving and marring the door finish, and a bit of metal polish would give it some luster. A locksmith should be able to replace the missing key if one is needed.

Center-mounted door knobs are a classic of the 1950s and 1960s but are virtually unknown before or since. This one from 1957 remains in use.

Hardware

Hardware can be an important character-defining feature in any historic building. Changing technologies, styles, and industrial capabilities have led to a vast range of available hardware over time. Earlier homes generally have simple forged or cast iron or brass hardware. The availability of decorative hardware in cast iron, brass, bronze, silver-plate, and other metals exploded in the Victorian age. In the twentieth century, nickel plating and, later, chrome became popular as styles evolved back toward simpler designs.

If your house is missing historic hardware, look carefully for evidence of what it might have been. Be aware that it was common in the nineteenth century to install high-quality hardware in primary rooms and less expensive or less stylish hardware in the rest of the house. Whitten House, for example, has porcelain door knobs on the parlor and sitting room doors, wood knobs in the front bedrooms upstairs, and iron

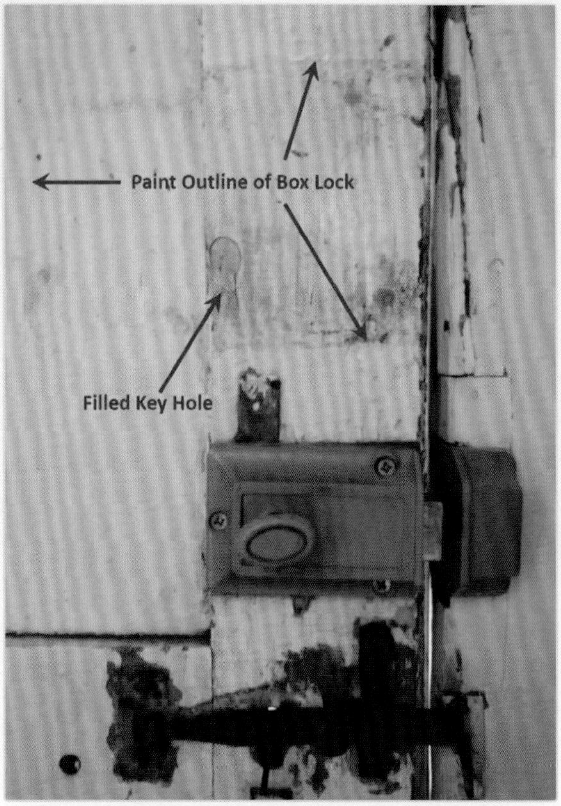

Here is evidence of a former box lock on the side door into the original kitchen at Whitten House. Such locks were common in the early nineteenth century but don't often survive in place. After noticing this evidence, I spent more than a decade looking for a box lock of the right size and orientation (they are left- or right-handed and are not reversible). Reproduction box locks are often cast in brass, which is too fancy for this location (and also expensive).

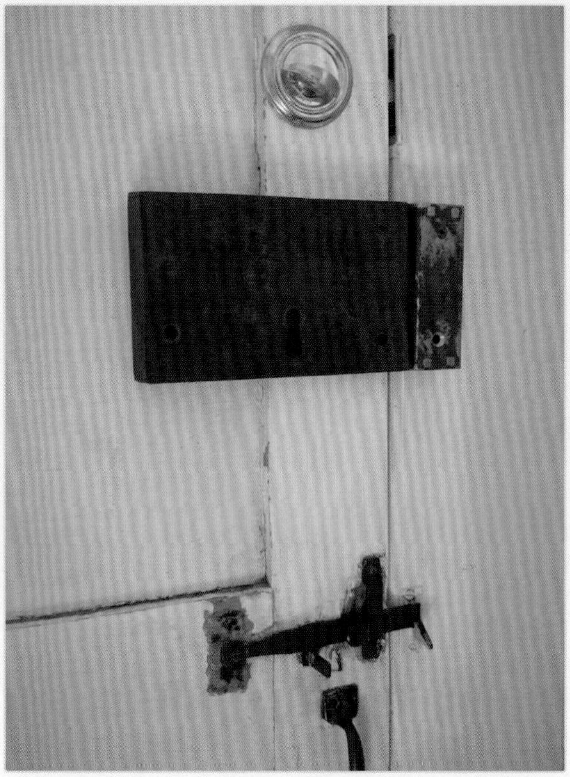

In a last-ditch effort before giving up the search and spending a lot of money on eBay for an antique lock that was bigger than the original, I checked a local architectural salvage shop one more time and found one of exactly the right size for under $50. It is simple and utilitarian, as I suspect the original would have been. The "catch" was missing (along with the key) but my local blacksmith was able to fabricate a matching catch. Here the lock and catch are placed temporarily with drywall screws to check the fit. The hardware will be painted black and mounted permanently once the door is repainted.

thumb latches on all other doors. If you live in a neighborhood of houses of similar age and style, one of your neighbors' homes may have original hardware that is missing from yours; this can show you what to look for as you seek replacements.

Antique and vintage hardware can be found in architectural salvage shops, flea markets, some antique shops, and online auction sites. Condition matters; make sure the item will still be functional once the 27 coats of paint are removed. A local blacksmith can repair iron and sheet-metal hardware. A remarkable amount of new old stock (NOS) hardware from as far back as the nineteenth century is out there. There is something wonderful about opening a box packed in 1880 and taking out a "brand-new," unused 140-year-old piece of hardware to install in your house of the period.

The availability of reproduction hardware was limited thirty years ago; the few companies making it sold primarily by mail order from catalogs. The range and variety of reproduction hardware available today would have been unimaginable to home restorers of the 1970s and 1980s. Similarly, the spread of architectural salvage shops has made antique hardware much easier to find than in the flea markets and antique shops of a few decades ago.

Stay true to the character of your house, matching original hardware as closely as possible in character-defining spaces. For hardware that will see regular use, such as cabinet and door hardware, buy the best quality you can. A kitchen cabinet door can get an amazing amount of use over time. Some of the major reproduction manufacturers and retailers are listed in Chapter 20.

Example Project: Restoration of an 1827 Kitchen

As described in Chapter 2, the original 1827 kitchen of Whitten House was altered in the late nineteenth century when the cooking hearth, fireplace, and brick oven were removed, then altered again in the 1960s when a half bath was installed in the former oven area and the room was opened into a new addition that replaced the original woodshed and late-nineteenth-century kitchen. I undertook a project of several years' duration to restore the 1827 appearance of the room, with a re-created functional cooking hearth and brick oven.

Maggie enjoys a fire in the restored fireplace in the 1827 Whitten House kitchen.

Before

These photos show the room as it looked when I first saw the house in December 2002. The mantel shelf visible below the corkboard was installed in the late nineteenth century when the original chimney and cooking facilities were removed and a much smaller

single-flue chimney to serve woodstoves was built in their place. The hearth bricks were also removed, and wide shelf boards recycled from the other side

of the original chimney were turned over and used as floorboards to cover the hearth base. Fortunately, the massive brick and granite base for the masonry remained in the cellar, providing critical information about the missing elements. By 2002, commercial carpet covered the floor except for an exposed portion of the chimney base, which was covered with painted concrete. The half bath used the original oven closet door for access. The only surviving element of the original chimney above the base in the cellar was the cleanout door at the base of the single-flue chimney, for which the cast iron firebox door from the original set kettle on the back of the chimney had been repurposed.

Demolition

I carefully stripped away the features added while the house was a public library and when the original chimney was removed in the late nineteenth century.

Carpet, library shelving, and the half bath were the first things to go, followed by the 1960s gypsum board on the ceiling. Removal of the latter exposed the 1827 lath, showed the original wall layout around the 1827 chimney, and revealed the size of that

chimney where it went through the ceiling. Next, I removed the late-nineteenth-century work, including the mantel, the plaster and lath where the fireplace opening had been, and the floorboards covering the hearth base.

What remained of the 1827 finish details were two vertical boards, the closet door and a horizontal board above it, and the plaster above the original mantel location. These few elements were enough to guide an accurate replication of the original room's other elements. At this point, the masons arrived to remove the single-flue chimney.

Reconstructing the Fireplace, Hearth, Brick Oven, and Chimney

Restoration mason Richard Irons and his team reconstructed the missing masonry elements, starting with the fireplace. For exposed surfaces they used the bricks from the single-flue chimney, which had been reused from the original chimney; for work that would be hidden, they used concrete block.

The cast iron firebox door from the original chimney (which had been repurposed as the cleanout door at the base of the single-flue chimney) provided

a model for the style of oven door needed. The oven door needed to be larger and have a flue and damper incorporated into its frame. In an antique shop I found an oven door of the right size with a similar fan design, though it is slightly more recent and has cast hinges and handle in lieu of the firebox door's hand-forged hinges and handle. Since the cast handle was broken, I had a local blacksmith make a hand-forged replacement copying the spiral detail of the firebox door handle. The oven door with frame is 17¾ inches wide, compatible with the width of a typical cooking fireplace of the period within the original vertical wood elements that contained both. The surviving firebox door was installed on the ash pit below the oven, as I was not having the set kettle reconstructed on the back of the fireplace.

Once the masonry reached the right height, the oven floor was laid over the ash pit and the "beehive" dome of the oven was constructed using wet sand to create a form for the top of the dome. The sand was removed through the oven door once the mortar set. A cast iron damper was installed above the fireplace (something the 1827 chimney would not have had), and the chimney was tapered down to the size of the opening through the ceiling, using clay tile to line the flues. All that remained on the

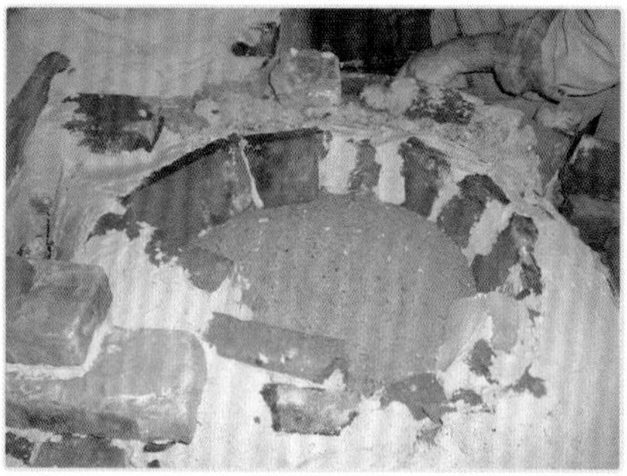

first floor was to lay the brick hearth and install an antique iron cooking crane in the fireplace.

Another fireplace was added on the second floor and from there, the chimney continued through the roof. Because there were not enough original bricks remaining for reconstruction of the larger chimney, the masons used matching historic bricks that had been salvaged from a demolished brick house. Building a chimney in Maine in January can be challenging, but an unusually warm spell of weather provided an opening, and the masons went for it. The work was wrapped in insulation at the end of each day to prevent the mortar from freezing before it cured, and things progressed well in spite of being interrupted by a snowstorm. To determine the height of the new chimney, I counted the brick courses on the late-nineteenth-century chimney in a photo taken in 1939, assuming the masons who built that chimney matched the height of the original they had just taken down. Once the work was complete and the mortar given time to cure thoroughly, the first fire was lit.

Finish Carpentry

The two surviving original vertical boards framing the fireplace and oven door preserved good information about the original trim work, now long gone. The boards' inside edges—visible once the later plaster was removed—had mortises showing the location and width of the horizontal board above the fireplace. On the faces of the vertical boards was a thick buildup of paint, the result of the many coats applied to cover the soot and grime of cooking on an open hearth. Recorded in this paint was the location of a band molding that had once surrounded the opening and the location and profile of the original mantel, which had been removed in the late nineteenth century. Because the original band molding remained on the door and window trim in the room, I knew the profile as well as the location of the molding and was able to have a local millwork shop replicate both it and the mantel. Twelve-inch-wide old-growth clear white pine is no longer readily available as it was in 1827, so I substituted two boards, joined edge to edge with a biscuit joiner and biscuits, for the wide horizontal board below the mantel, locating the joint under the band molding to hide it.

To the left of the fireplace, where a wide opening into the room had been added in the 1960s, plaster marks on the original ceiling lath documented the former location of a chimney cupboard, which I decided to reinstate. The bottom of the shallow cupboard had been above the wainscoting in order to leave headroom over the original cellar stairs, which had descended there from the woodshed. While this project proceeded over several years, I pondered what to do for a chimney cupboard door. One day I noticed a circular wear mark on the vertical board to the left of the fireplace, the type of mark made by a simple wooden turn catch for a door. It reminded me of a similar wear mark on a small door stored in an upstairs closet, and suddenly I realized that the original chimney

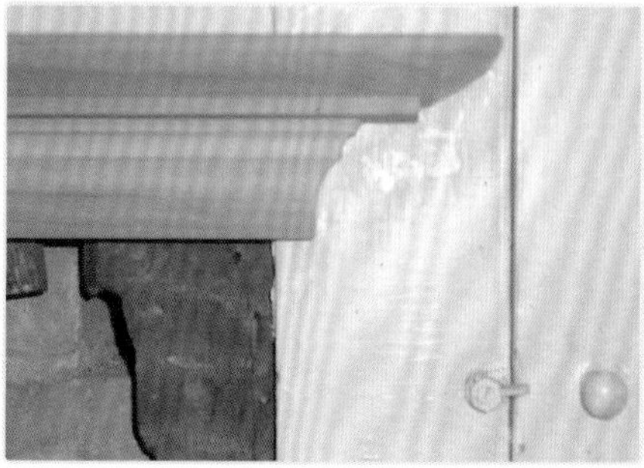

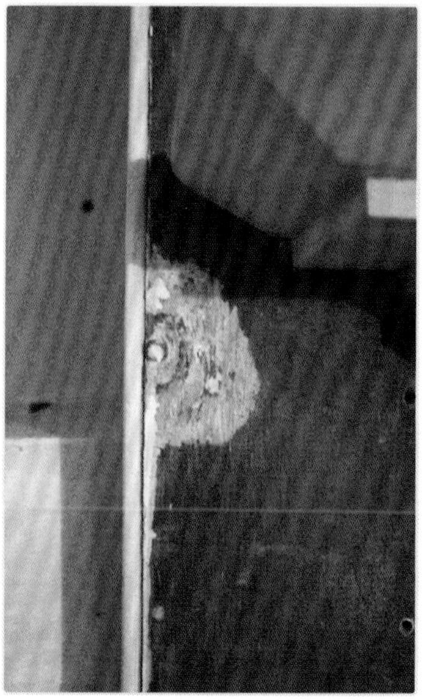

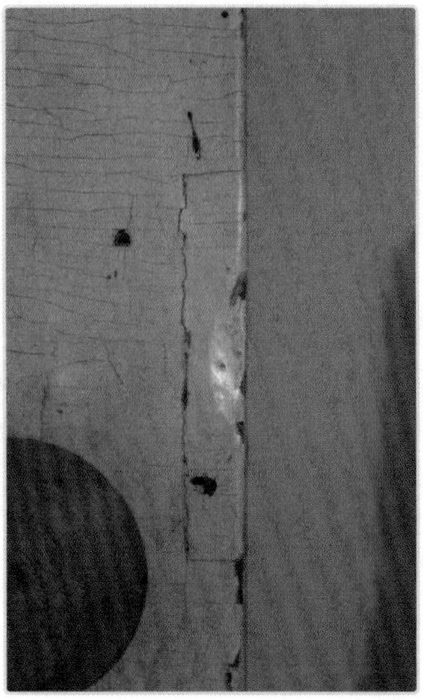

cupboard door was still in the house, waiting to be returned to use.

Since 22-inch-wide pine is even less available than clear 12-inch boards, I bought a 24-inch by 48-inch glued-up panel of clear pine and a #2 pine 1 x 10 to build the wainscot below the cupboard and around the corner toward the new kitchen, mitering the joint at the outside corner. The knotty #2 board filled the gap between the floor and the bottom of the panel, to be covered with 10-inch poplar baseboard. I made the wainscot cap by ripping down the width of off-the-shelf window stool cap (sill) molding, which happened to match the profile of the original wainscot cap in the room.

I had to install the cupboard door a bit above the wainscot cap in order to align the wear marks on the door with those on the vertical board to its right. I cut a base trim piece to fill the resultant gap under the door; then I cut side and top pieces to fit the door, beading their edges in the router to match the detail on the surviving door surround for the oven closet on the other side of the fireplace. This is a less "finished" look than was used in the more formal rooms in the house, where the beads

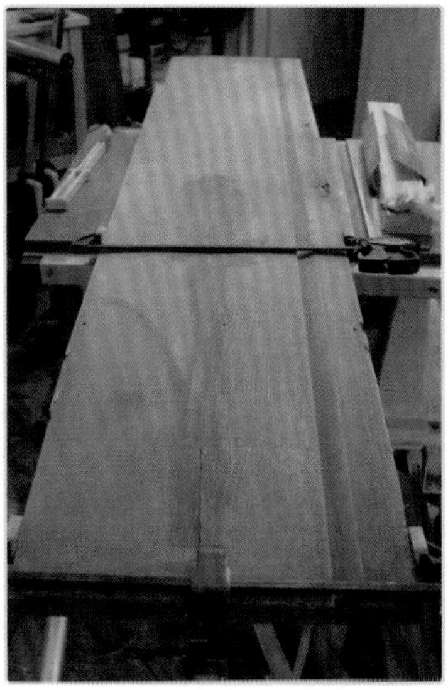

from upstairs (the closet was removed for a bathroom), and a shelf from the library-era bookcases. The set kettle

are mitered at the corners of door trim. The joints between these pieces were biscuit jointed and test fitted dry with the door, as seen in the photo.

Once the door and frame were ready, I made the shelves for the interior of the cupboard and installed them before putting on the face frame. The shelf materials were recycled from elsewhere in the house and included the 1827 shelf above the set kettle that was turned over and reused as flooring when the original fireplace was removed, a c. 1850 closet shelf

shelf was partially split and needed to be glued before being cut into two new shelves. It already had a plate groove planed in it, which I replicated on the former closet shelf. The closet shelf had an elliptical arch cutout in the front, to improve access to the shelf. This was duplicated on the shelf below. The shelves were mounted to molded cleats nailed into the plywood structure of the cupboard.

Once the shelves were in place, I nailed the face frame in place. I finished the narrow slice of wall between the door frame and the outside

corner with skim-coat plaster, reinforcing the corner with a narrow bead-edged board as was used elsewhere in the house in 1827. All that remained then was to match the profiles of the top of the baseboard and the molding under the wainscot cap. After stripping the paint buildup off a section of these moldings to reveal their precise original profiles, I was able to find a poplar molding made for dollhouses to match the wainscot piece and a router bit to match the baseboard top.

The final bit of carpentry was to piece in the missing floorboards around the work area, which I did using some of the boards removed at the start of the project.

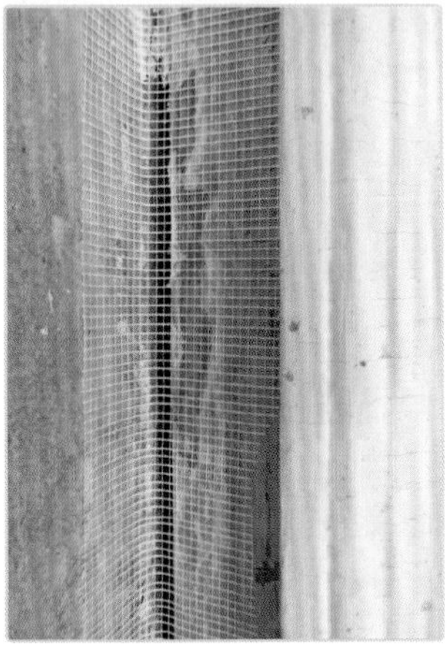

Plaster, Paint, and Paper

With the trim work completed, I stripped and replastered the ceiling as described in Chapter 16. The wallpaper was easier to strip than the ceiling paint, and it would have been even easier if it hadn't been painted over at some point. I used a scoring tool to pierce the paint and allow steam to reach the paper. With the steamer pan in one hand and a single-edge razor scraper in the other, I steamed a section, then moved the pan to a new section while scraping the section just steamed. The paint came off in the first pass; then I reversed direction and made another pass to remove the wallpaper. Interestingly, the back side of the peeling paint showed the wallpaper design in mirror image, providing a modicum of entertainment in an otherwise tedious task.

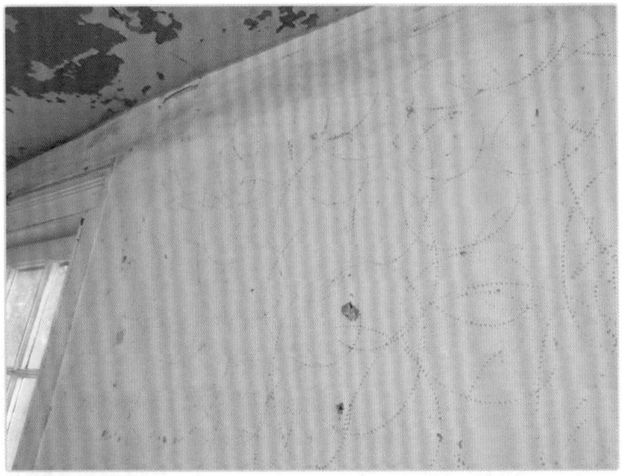

Once the wallpaper was gone, green paint marks from 1827 showed where hanging rails had originally been mounted on the walls. Hanging rails in Shaker buildings were equipped with turned wood pegs to hang things from, but the ones in this room probably had projecting square nails like those in a surviving piece of hanging rail in the closet under the back stairs. That one is beaded along both edges, and these probably were too. Having found evidence of their existence, I intend to replicate them before the wall covering goes up, so I photographed them with a tape measure held in place to document the locations.

The wall plaster was in reasonably good condition, but a few areas needed repair as shown with fiberglass mesh and Durabond setting plaster. In one corner a sizable crack had opened up behind the wallpaper. Door trim within 2 inches of this corner made access difficult; having lapped the mesh onto both walls, I applied the Durabond to the more accessible wall, allowed it to set, and only then applied it in the narrow space next to the door trim. This is easier than trying to form a clean corner in a tight spot with both sides wet.

After patching the major defects, I lightly skimcoated the walls with joint compound to fill any small surface cracks and divots, then finished the plaster work with a quick once-over with a 150-grit screen on a pole sander.

After the removal of lead in the 1970s, paint quality declined until the 1990s as manufacturers tried to keep up with changing regulations. Paint has improved since then, however, and the higher-quality American paints—or "coatings," as they are now called—are better today than they have been in a long time. That said, however, none of them come close to the quality of Fine Paints of Europe's (FPE) alkyd paints imported from Holland. This brand reminds me of the oil paints of the 1970s and is the only one I use in Whitten House. It is available by mail order.

The first step in painting the 1827 kitchen at Whitten House was to prime and paint the ceiling. As described in Chapter 16, I rolled the paint in strips and immediately brushed with a coarse wallpaper paste brush to create a more accurate historic texture on this highly visible surface.

Next I sanded all the trim and applied a coat of FPE's brushing putty to fill and smooth the crackled, alligatored old-paint surfaces. Once the putty was dry, I sanded it lightly, wiped it down with a tack cloth, and primed the trim with FPE's alkyd primer.

The oven closet door to the right of the fireplace appeared to have been repainted from green to an off-white/tan shade early on, probably the first time the woodwork in the room was repainted. I decided to

return the room to this scheme, with dark trim and light doors and window sashes. I returned the floor to the yellow ocher that was common for painted pine floors during the period. I applied a dark-red primer on the trim that was to be finished dark green, since priming with a complementary color creates a more visually "lively" surface.

The finish paint is FPE's Hollandlac Brilliant alkyd gloss. The company claims you cannot apply this paint too thinly, and they are right. It needs to be thin to avoid drips and sags, and even a very thin coat will not streak. If you are using it for the first time, use less than you think you need and keep checking your work to be sure it isn't running. Once you get the hang of it, it is a wonderful paint to work with. The wall color seen here is a primer sizing coat for a wall-covering that hasn't been hung yet. It is tinted to the base color in the covering to disguise any gaps in the butted joints of the covering once it's up. The floor has been painted its original ocher yellow.

The Result

Although still awaiting its wallcovering, the 1827 kitchen at Whitten House would be immediately recognizable to any of the Whittens who lived in the house. Re-creating the cooking hearth and oven set

me on the path to learn traditional cooking. Early on I learned that working in your own shadow when cooking on a hearth, while historically accurate, is inconvenient and even somewhat dangerous. Three 3-inch recessed LED lights on a dimmer over the hearth solved this problem. I am not generally a fan of recessed lighting in a historic house, but here they provided the solution with the least impact on the room's character. Although the room is used primarily as a dining room, not a kitchen, we use the cooking hearth and brick oven for Thanksgiving dinners and occasional dinner parties, serving food on the Whittens' china pattern.

CHAPTER EIGHTEEN

Interior Paint and Paper

Courtesy of Bradbury & Bradbury Art Wallpapers

Historically, the character of a room was often expressed as much in paint and wallpaper as in architectural features. In paintings of eighteenth- and nineteenth-century interiors—and even in black-and-white photos of the latter—it is often the combination of colors or wallpaper patterns that is most noticeable. A room will always best evoke its period of creation if the paint and wallpaper are right, and appropriate colors and patterns go a long way toward restoring character to rooms that have lost historic trim or other architectural features.

Historic Interior Finishes

Entire books have been written on historic architectural paints in America, and several are listed in Chapter 20. An understanding of how paint was used in historic houses can guide your color choices and placements.

Paint is less expensive and easier to change than architectural features. If you want to introduce a more contemporary feel to a historic house without making permanent changes to character-defining features, you can do so with paint. A common example is painting walls a solid color where patterned wallpaper would have been used historically. You could choose a period color for a slightly more contemporary treatment or a modern color for a distinct statement. Whatever you do will be readily reversible; you or another owner can always hang wallpaper over the paint at some future time.

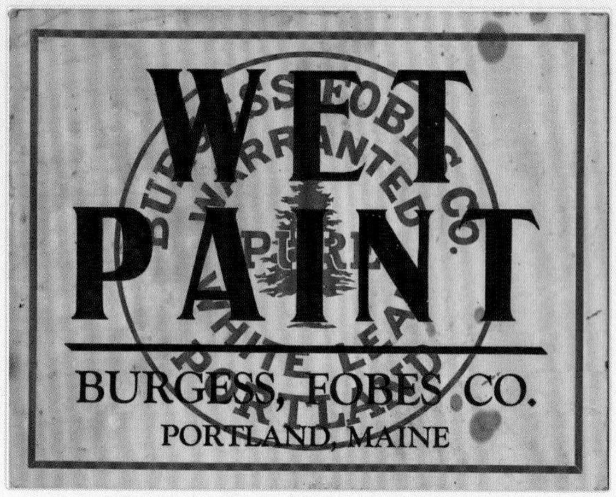

This antique "Wet Paint" sign is from a Portland, Maine, paint manufacturer founded in 1858, at the start of the era of pre-mixed paints.

Painting historic stained and varnished trim is a different matter, however. I have tried to avoid saying "never" in this book, because every house is unique and every owner has his or her own priorities and preferences, but I have to make an exception here. Think of the tens of thousands of tedious, laborious hours that have been spent by restorers stripping paint from beautiful woodwork that was never meant to be painted. Please don't do it. If the original finish of your woodwork has darkened and the grain of the wood is lost in the gloom, refinish it to restore its intended appearance; don't paint it.

Period Colors

Period-appropriate colors almost always complement historic spaces better than this year's "in" colors or the ever-popular beige on beige. "Safe" colors are rarely

Before and after photos of the hall at Historic New England's Eustis Estate illustrate the effect of returning historic spaces to their original colors. The warm red wall paint built up in four layers of hand-ground pigments and gold leaf cove molding harmonize with the wonderful woodwork to create a unified whole. *(Before photo courtesy Sutherland Conservation & Consulting; after photo by Eric Roth)*

what historic character calls for unless you're working in a postwar house. Eighteenth- and nineteenth-century homes used a lot of rich color in their interiors. Victorian-era homes frequently featured dark-toned woodwork and tile that loses much of its richness when walls are painted in white or pastels.

There is good guidance available in books and online for selecting period-appropriate colors. Museum houses are another good source of information when their interpretation is based on objective research rather than the subjective tastes of historical society members. In small volunteer-run museums, preserved historic interiors are often more reliable than restored spaces; even so, colors can fade or darken depending on their original composition and ongoing exposure to light, soot, dirt, and other environmental factors.

Because historic paint research has advanced significantly in the past several decades, recently restored interiors in well-run museum houses provide the best opportunities to see accurate historic finishes. Many older museums, like Colonial Williamsburg, have worked hard to reinterpret interior colors based on recent research. Colors of the eighteenth and early nineteenth centuries are now known to have been bolder and more vibrant than was previously supposed.

Before and after photos of the dining room at Historic New England's Eustis Estate. The striking original painted finishes on the walls and ceiling of this room were identified with cross-section microscopy. *(Before photo courtesy Sutherland Conservation & Consulting; after photo by Eric Roth)*

Paint Analysis

Early historic paint documentation was generally of the "scratch and match" variety. Whatever could be seen of the earliest layer determined what color to restore it to. This method guided the 1920s restoration colors at Colonial Williamsburg, which became hugely influential in the interpretation of thousands of private homes. But the method does not account for changes in paint pigments and binders over time, and the result was a misinterpretation of the colors used by the eighteenth-century residents of Williamsburg. Over the past several decades, Williamsburg and many other museums have been correcting these scratch-and-match errors using modern techniques in which small paint samples are removed and studied under a microscope.

When a paint sample is magnified between 100 and 400 times, it is possible to identify individual layers and the pigments and binders in each. Differences in the amount of dirt between layers suggest the relative lengths of time between paint jobs, and differences in the number of paint layers between one architectural element and another can document changes in a room. For instance, if a corner cabinet has many fewer layers of paint than the rest of the room, it was probably added later; if most of its layers do not match the paint layers elsewhere in the room, it may have come from another room or even another house. It may also have been painted a different color intentionally. A good conservator will take samples where the cabinet meets the wall, baseboard, painted floor, or other trim to look for evidence of colors overlapping from one element to another. A skilled conservator with knowledge of how paints were used historically can generally make a clear determination on such questions.

The most common use of microscopic paint analysis in historic homes is for documenting early colors. A full-house analysis can be expensive, but a selective analysis of key elements or rooms can better fit a budget.

A detail over the mantel from the full-room mural in the parlor at the Rufus Porter Museum in Bridgton, Maine. Porter was an itinerant painter in New England whose mural work is highly regarded today.

Historic Decorative Painting

Decorative painting on walls, ceilings, or trim was rare in America prior to the nineteenth century. The exception was grain painting; for example, Thomas Jefferson had softwood doors at Monticello and Poplar Forest grain painted to resemble mahogany. Beginning in the early nineteenth century, however, stenciled designs were often painted on walls, frequently mimicking wallpaper patterns. Itinerant painters traveled around rural America doing stencil decoration and naïve murals.

An example of the sort of decorative painting that became common in middle-class homes in the second half of the nineteenth century. By combining striping and stencils in a polychromatic scheme with a molded plaster cornice, the craftsmen who did this work created a more sophisticated treatment than the typical folk-art mural and stencil work of itinerant painters.

More sophisticated decorative painting on walls and ceilings came into vogue by the middle of the nineteenth century, particularly in urban areas. Many experienced European decorative painters immigrated to the United States, bringing their talents with them. Homes of the middle class and wealthy were decorated with painted trompe l'oeil moldings

The remarkable shimmering metallic effect on the Eustis Estate dining room walls was created by first painting the walls with dark green paint into which coarse sand had been mixed, then applying bronze powder to gild the grainy surface, allowing light to reflect off the thousands of sharp facets of the sand. The bay window arch received a smoother gilding treatment. This decorative painting was the height of Aesthetic Movement sophistication in the late 1870s.

This illustration from Gustave Stickley's 1909 book *Craftsman Homes* (originally published in *The Craftsman* magazine) shows a typical early-twentieth-century Arts and Crafts decorative paint treatment on the frieze. During this period, the painting was often applied over burlap for a more "earthy" feeling.

This decoratively painted coffered ceiling in a Spanish Colonial Revival mansion in Phoenix, Arizona, is typical of work from the 1920s.

surrounding painted "damask" panels on the walls; painted moldings and medallions also adorned ceilings. Wood graining continued along with marbleizing, tortoise shell, and other faux painted finishes. Ornamental plaster moldings came into wide use at the same time and were often combined with trompe l'oeil work.

Gold gilding, either with leaf or bronze power, was often used with decorative painting. Decorative painting sometimes involved creating textures on walls and other architectural elements with sand or impasto paint. Stenciled ornamental painting returned to vogue during the Arts and Crafts period in the early twentieth century, when it was sometimes applied to burlap-covered walls.

Restoring Decorative Painting

Decorative painting often survives under wallpaper or layers of paint. Sometimes a pattern can be seen through later paint or emerges behind peeling paint or wallpaper. When it is discovered, a homeowner must decide whether to attempt to uncover the pattern or at least document it before repainting or papering. The best documentation is high-resolution photos that include a tape measure or ruler to indicate the sizes and positions of decorative elements.

Many do-it-yourself home restorers have uncovered and successfully restored historic decorative painting. There are, in addition, professional conservators and restorers who do this work using recommended approaches, materials, and methods. A key principle of conservation work today is "reversibility," the ability for future conservators or restorers to undo restoration work that ages poorly or can be improved with better treatments. Reversibility is generally achieved by coating the exposed and stabilized historic paint with museum varnish, then doing the repair and restoration work over the varnish layer. It can be very helpful to pay a professional to assess your decorative painting and suggest appropriate treatments even if you plan to attempt the work yourself. A full examination with microscopic paint analysis and a written report can be pricy, but a several-hour visit with you taking notes on what the professional sees and says might be enough to meet your needs. Video is a great way to record a site visit and recommendations.

Before and after photos of the parlor at Historic New England's Eustis Estate. Mid-twentieth-century latex paint was covering an unusual cloud-like decorative paint treatment in warm, earthy colors. A treatment like this could never be truly reproduced, as it is a work of imagination and creativity by the craftsmen who did it. Fortunately, it was possible to remove the latex paint and restore the damaged areas of the original work with careful in-fill painting, using reversible conservation methods and materials. *(Before photo courtesy Sutherland Conservation & Consulting; after photo by Eric Roth)*

Decorative painting is generally easier to expose when hidden by wallpaper than when covered by later paint. Explore carefully and test the decorative paint for water solubility before using steam or water to remove wallpaper. Dry removal of wallpaper is generally possible; work a razor scraper behind the paper at a low angle to the wall and gently slice between the paper and paint. Removing some paint is unavoidable. On rough or damaged plaster, a 2-inch scraper will follow the contours more easily than a 4-inch scraper. This work can be slow and tedious, but it can also be exciting if you are exposing interesting decorative painting.

Many methods exist for removing later paint without destroying earlier decorative painting. It is not possible to explore all of them here, but I will discuss several. If you are lucky, the bond between the decorative painting and the first coat that covered it is weak. Cleaving, or separating, that bond is the goal. If you can't expose the entire area of historic decorative painting, you want to expose enough of it to allow duplication of the pattern atop the later paint.

Here I am working with acetone and cotton balls to dissolve layers of modern latex paint and expose an area of the original red wall color in the hall at Eustis Estate. Removing all of the modern paint by this method would have been impractical and was never considered, but exposing the color in a sample area confirmed what we were seeing in tiny paint chips under a microscope. The four-layer red paint treatment was replicated with hand-ground pigments to restore the original appearance of the hall.

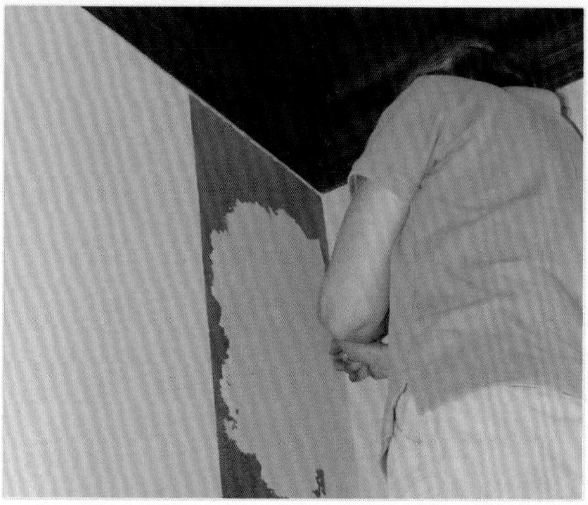

Conservator Amy Cole Ives of Sutherland Conservation & Consulting is using duct tape to create an exposure on the parlor wall at Eustis Estate. Microscopic paint analysis had indicated an unusual paint treatment that could only be understood by uncovering an area. Fortunately, a later layer of varnish did not bond well with the original varnished surface, which meant that the tape could pull off the later paint and varnish while leaving the original paint. The alternative would have been to clear an area with a scalpel, one flake at a time. Sometimes you get lucky.

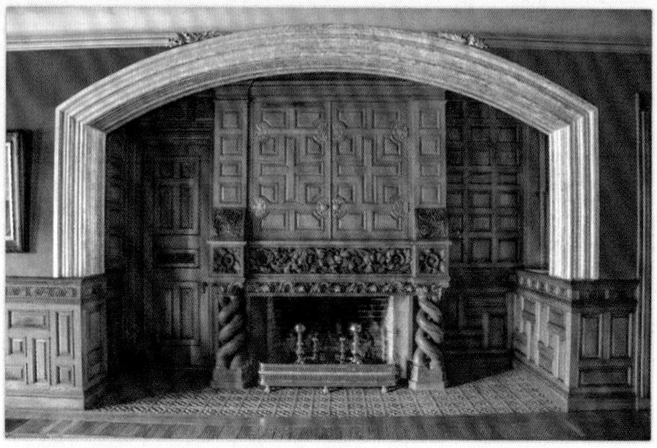

Metallic gilding was often used in the late nineteenth century to add shimmer and richness to interiors. Sometimes this was done with real gold leaf, but often it was done with bronze powders that resembled gold leaf when new but tarnished with age, turning green or brown. Before restoration, this arch in Historic New England's Eustis mansion was a dark olive-brown color without any of the shimmer and glow of the original treatment. New imitation gold leaf was applied over the original textured surface during the restoration, returning it to its intended effect, highlighting and showing off the spectacular terracotta fireplace and hearth and wood paneling of the alcove. *(Photo by Eric Roth)*

Large canvas paintings on the ceiling of an elaborate Victorian hotel dining room—in poor condition due to prolonged neglect and water leaks—required attention when the room was converted to a community space for the residents of an affordable senior housing project. The project budget did not permit full conservation and restoration of the canvases, but they were stabilized and cleaned in place. (And a sprinkler system support bracket, installed decades earlier, was removed from the eye of an angel!) The room's original colors were determined by microscopic paint analysis, and a cost-effective method of recreating the gilded surfaces was devised by decorative paint restorer Tony Castro and me. We would have preferred to fully restore the room and canvases using historic methods and materials, but an 85 percent restoration of historic character within the project budget was nevertheless a good outcome. A similar balancing of aspiration with realism can apply to the interiors of homes, where budgets are often limited.

A good method to try first is using an adhesive tape (our firm has had luck with generic duct tape) to pull off later paint layers. In an inconspicuous area, press a small piece of tape against the paint, leaving a raised end or ends to grasp, and pull quickly downward. With luck, you will remove the unwanted layers and leave the decorative paint. Look closely at the back of the paint you remove (the part facing you when you look at the peeled paint on the tape) to see if any decorative paint is being removed as well. Some removal is likely, but hopefully not too much. If you are pulling off much of the decorative paint with the overburden, this is not the right approach. Often, decorative painting was coated with varnish or shellac historically, either to protect it or to give it a gloss. The clear finish will sometimes remain on the decorative painting with the later paint cleaving from it. Old varnish tends to yellow with age, so if you have a varnish layer it may be discoloring the painting under it. Unless you have professional help, it is probably best to live with it.

Mechanical scraping is another method for removing later layers of paint. One of the owners of the Smith House (page 327) removed later paint from the entire parlor ceiling by scraping carefully over a period of months. Experiment on small areas (not in the middle of a room) with different scrapers to see what works best without removing too much of the decorative paint. If the later paint is calcimine and the decorative paint is oil, it may be possible to wash off the calcimine layers with hot water, TSP, and a rough-textured rag. (A scrub brush may be too aggressive.) Make sure the decorative painting is not also water-soluble, or you'll remove it as well.

Latex and other modern paints can sometimes be dissolved with mineral spirits or acetone and wiped away, though this method is likely to leave a "film" of dissolved paint that can be difficult or impossible to remove. If you can't remove it, tell yourself it is evidence of age and adds character to the room.

If exposing an entire ceiling or wall is too difficult or time-consuming, try to expose enough historic decorative painting to extrapolate the full design—probably with several localized exposures around the room—and then replicate the design on the areas not exposed. I recommend leaving areas of the original exposed as a record of what you copied.

Even if you remove all the later paint, you will likely want to touch up areas of decorative paint that were removed accidentally. You should seal the historic paint with reversible museum varnish first, then color match carefully in water-soluble paints for the inpainting. Try not to cover remaining areas of original paint; just fill in voids in the pattern. Where the plaster is scratched or was gouged by scraping, apply patches over the varnish before touching up with paint.

Sometimes painting was done on canvas in a studio and then pasted onto a plaster wall or ceiling. These are generally rondels or panels intended to be surrounded by plaster or trompe l'oeil moldings. Unlike most decorative painting, canvas panels were usually intended to be seen as "art," not merely as decoration. The conservation and restoration of painting on canvas is a specialized field. Depending on the condition of the pieces, work can range from in situ cleaning with appropriate methods to removal to a conservation studio for extensive (and expensive) restoration.

New Decorative Painting

Traditional decorative painting techniques can be used to blend new work into a historic space when the original materials or finishes are unobtainable or

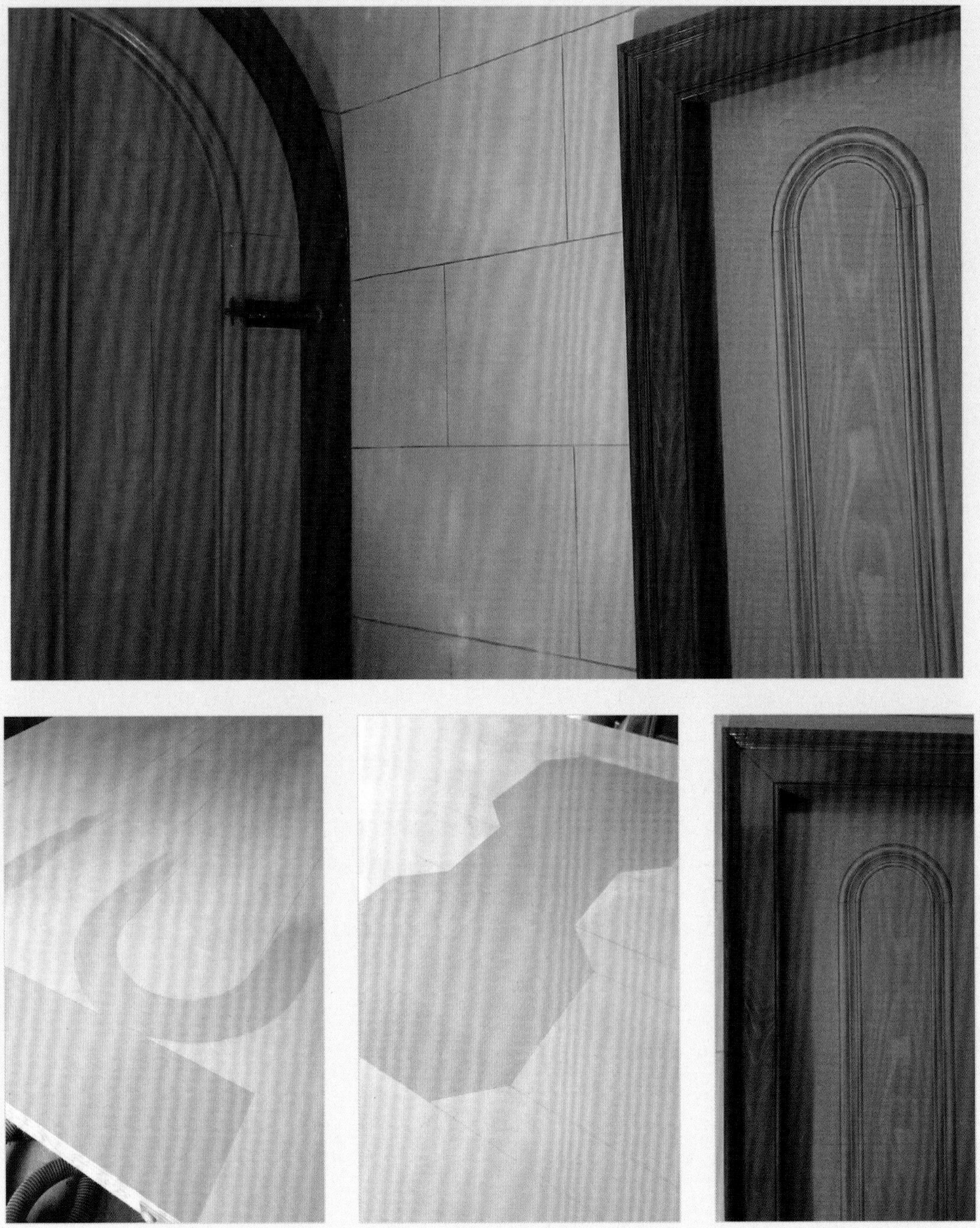

When a new lavatory was installed next to this small nineteenth-century entryway, there was only room for a flush pocket door. To make the door compatible with the space, decorative painter Tony Castro reproduced the appearance of the historic doors elsewhere in the house with a trompe l'oeil wood-grained treatment on the door and trim. Here we see the design for the paint laid out in pencil on the flat door and the first coat of paint being laid in, matching the brush stroke direction to the "grain" of the illusionary wood.

unaffordable. Say, for example, the mantel and the trim from one door are missing from a parlor with naturally finished historic chestnut trim. It would be nearly impossible to replicate the missing elements in chestnut, since that once-common species was virtually eliminated from the U.S. by fungal disease in the first half of the twentieth century. Instead, the elements could be replicated in another wood and matched to the existing chestnut trim with paint-graining techniques. Graining and marbleizing are relatively easy to learn, and there are many how-to books and videos available to help. Jocasta Innes's *Paint Magic* is a good introductory volume for non-specialist readers.

A talented and experienced decorative painter can produce results that are seemingly magical, particularly with trompe l'oeil effects to make three-dimensional features appear on flat walls and ceilings. Skilled artists can often be found in the theatrical set design field, the one area outside conservation and restoration where the art has been kept alive.

This pilaster and base are three-dimensional illusions created with distemper paint on flat plaster by a mid-nineteenth-century decorative painter. Such work by an experienced artist or determined amateur can be used to "replace" missing architectural elements.

Using paint to create the illusion of a feature not actually present is nothing new. Here, in a restored room in the colonial-era Governor's Palace in Santa Fe, New Mexico, an illusion of wainscoting has been created with flat paint applied to the lower wall.

Stripping Paint

For a long time, "restoring" an antique usually involved stripping off the paint, the assumption being that all wood must have been naturally finished when new. This was true of houses as well as antique furniture. Sadly, a lot of historic painted finishes were scraped and peeled away forever, with no record of the original finish. In recent years, the *Antiques Road*

This stripped and refinished woodwork is in the 1898 Pullman palace car *Gertrude Emma* at the Conway Scenic Railroad in New Hampshire. As a teenager, I spent a summer stripping paint in this car. Built as a sleeper-parlor-observation car for the Pennsylvania Limited express train from New York to Chicago, the car had an interior of carved mahogany and panels inlaid with exquisite floral marquetry in satinwood, rosewood, and other exotic woods. Retired from passenger service in 1938, it survived as a bunk car in work train service until the 1970s. When it was pulled from a scrap line for restoration by the Conway Scenic Railroad, its original interior finishes were hidden behind eighteen layers of paint. It took me and others several years to remove all that paint, and several years more for the woodwork to be refinished and other features restored. Since that formative experience, I have opposed the painting of any historically unpainted wood trim.

Show and other television programs have finally begun to get out the message that original finishes add value to antiques. The same is true of houses.

Most interior trim was painted before the mid-nineteenth century, and it should remain painted if the goal is a historically appropriate finish. Earlier restorers stripped and varnished a lot of knotty pine and other softwoods that were never intended to be seen. Returning such trim to a painted finish is appropriate.

The development of steam-powered mills for cutting, milling, and turning wood made hardwood trim more feasible, and the Victorians reveled in native and exotic hardwoods finished with glossy varnish. Consequently, stripping Victorian-era woodwork may well be the correct approach for returning it to its original appearance. Stripping (in preparation for repainting) can also be appropriate for originally painted surfaces *if* there are so many built-up layers that the original form and detail are lost or the paint is chipping. Leave one or more areas with all layers of paint intact as a record of the finishes in the room over time. If there were clearly different colors on different trim elements (doors, window sashes, baseboards, door and window trim, mantel, wainscots, crown molding, etc.) leave at least a 6-inch by 6-inch area of each in an inconspicuous location. Even after being painted over to blend in, this swatch can be identified by its thicker profile. It is a time capsule for future paint analysis.

Traditionally, paint has been stripped with either heat or a toxic chemical stripper. Open-flame torches have largely gone out of use—though not before inadvertently burning down more than one historic home—as have caustic lyes. Today's heat sources are guns, plates, or infrared heaters, and the chemical strippers have been joined by soy- and citrus-based products that are more environmentally friendly. Whatever method you use, you will need suitable scrapers to reach all areas of the surface you are stripping. A 6-inch putty knife will work fine on large, flat areas. For delicate carvings or cast ornament, you may need dental picks. Plastic putty knives of various widths can be cut to match molding profiles; fifteen minutes spent cutting and shaping a putty knife will more than pay for itself in scraping time if you have more than a small area of complex shape to strip.

Stripping Paint with Heat

A concern when stripping paint with heat is the release of toxic fumes if the temperature is high enough to vaporize lead. Wear a respirator when using a heat gun or heat plate, and don't allow children in the area. Infrared paint strippers claim to remove paint at a lower temperature so lead will not be vaporized, but it is still a good idea to be cautious.

The other danger of using heat to strip paint is the possibility of igniting wood shavings, sawdust, scraps of paper, or other detritus behind the material you are heating. There could be an unseen rodent's nest in the wall, or perhaps shavings from the hand-planing of boards or moldings. Keep a fire extinguisher at hand when stripping paint with heat and be prepared to cut a hole in the wall if you smell smoke and cannot see where it is coming from. Infrared heat strippers greatly reduce the likelihood of igniting a fire due to the way they generate and transmit heat to a surface.

Heated paint will soften on the surface, allowing it to be scraped off with a minimum of force. Heat each area long enough to soften all layers but not to scorch the wood behind the paint. Trial and error will teach you the right length of time for the job you are working on. A bit of scorching is inevitable and can be removed by sanding before repainting. The softened paint will be initially gooey, and you should protect the flooring if it is in good condition. The goo turns brittle upon cooling and is then easy to clean up with a broom or shop vac.

Liquid, Gel, and Paste Strippers

The chemical paint strippers in use twenty or more years ago were effective but vile, smelling strongly of volatile organic compounds (VOCs) and causing chemical burns on skin if left there for any length of time. And they really burned when spattered in your eye—trust me! Today's strippers are more user-friendly but still require a respirator, eye protection, rubber gloves, and fully covered skin. The chemical-infused paint scrapings are considered hazardous waste.

Citrus- and soy-based strippers produce much less odor and are far less toxic, some claiming to be safe even when applied without rubber gloves. In general, these strippers work more slowly than their more caustic forerunners but are equally effective. Be aware that some of these come in various formulations, with lower (and slower) concentrations being shelved in the big-box home centers. Track down the stronger stuff if you have much stripping to do.

The choice of a liquid, gel, or paste depends in part on the surface you are stripping. Gels and pastes stay on vertical surfaces better than liquids. Whichever stripper you use—chemical, soy, or citrus; liquid, gel, or paste—gently press clear plastic wrap over the wet surface as soon as you apply it. You want to slow the evaporation of its active compounds and give it time to do its work. Some strippers come with plastic-coated kraft paper for this purpose; the paint is supposed to (but may not) stick to the paper for easy removal.

Once stripped, the surface should be treated with a neutralizer to remove the chemical residue before refinishing. If the original finish was shellac, any residue should be removed with denatured alcohol and

0000 steel wood. Shellac is secreted by an insect from India, the lac bug. It is dissolved in alcohol for application, drying as the alcohol evaporates, and it can always be re-dissolved with alcohol.

Thin removable elements that are difficult to strip—such as louvered shutters—can be soaked in a bath of stripper. Build a shallow box of rigid foam insulation just slightly larger than the element to be stripped. The foam board can be glued together and the box lined with thick plastic sheet material. Close the box with a top the same size as the bottom while the stripper is working. You can reuse soy stripper if you periodically clean out loose paint by straining the emulsion through nylon screening or by scooping the residue from the box with a kitty litter scooper. Liquid strippers are more likely to raise the grain of the wood, especially softwoods. Any raised grain will need to be sanded smooth before refinishing or repainting.

Stripping Paint with Steam

Bagala Window Works introduced widespread use of steam stripping in 1998 with the development and commercial production of a paint-stripping machine for window sashes. No chemicals are used in this closed metal box with a built-in steam unit. A fully glazed window sash is placed in the box and steamed until the paint and glazing compound are softened and easily removed. The sash can then be repaired, sanded, primed, re-glazed, and painted. It is an effective method for removing lead paint from windows, as shown in Chapter 13.

Paint can be removed from plaster with a wallpaper steamer. This is particularly effective on calcimine paint, which is water soluble, and latex paint, which softens quickly in steam. I have used this method to strip a peeling ceiling with a stubborn combination of latex paint over oil paint over varnish over calcimine residue. (See "Example Project: Plaster Ceiling Restoration" in Chapter 16.)

Working on Lead-Painted Surfaces

The use of lead in paints dates to at least 300 BC. Although its dangers to the people who produced it were known by the Middle Ages, it remained in common use until the 1970s, when Canada, the U.S., the UK, and European countries began efforts to ban its use in household paints. It is still used in parts of the world with lax safety regulations. Of the several types of lead used as pigments in household paints, the most common was lead carbonate (white lead); this was followed by lead chromate (chrome yellow) and lead oxide (red lead), the latter being used principally as a primer for metal. White lead provided a white base into which other pigments could be mixed

Advertising pin for the National Lead Company.

This is an 1868 advertisement for Pure Snow White Zinc pigment. From the 1830s until the 1920s, white zinc was the most common pigment used for white paint and to lighten other colors. *(Courtesy the Library of Congress)*

to produce bright or light colors. Lead in paint sped up drying, increased durability, and resisted moisture while retaining a degree of flexibility—all desirable characteristics for a paint.

White zinc pigment was developed in Europe in the 1830s as a "clean" white well suited for mixing with other colors, which it will lighten and brighten without altering. It is nontoxic and is used today in sunscreen preparations. However, it dries to a brittle surface in oil paint unless mixed with another pigment, which was likely to be white lead in white paint manufactured before the twentieth century but might be titanium white after 1921.

Titanium white (titanium dioxide) began to compete with white lead in the 1920s and is sometimes called "the whitest white." It has twice the opacity of white lead and a very high reflective index. Like white zinc, titanium dioxide pigment is harmless to humans. In fact, it is used in food coloring, toothpaste, and cosmetics. However, titanium white does not dry to a hard yet flexible surface in an oil-based paint but remains somewhat gooey. It must be mixed with another pigment to offset this characteristic; the usual choice was white lead until it was banned from house paints, followed by white zinc.

Today's paints are much safer but do not match the quality of lead paint. Given the permanent damage lead ingestion causes to children, the trade-off is a small price to pay. Lead's threat to children is increased by the fact that it tastes sweet, encouraging kids to put paint chips in their mouths. If ingested by a child, it can cause nervous system damage, stunted growth, kidney damage, and cognitive developmental damage. The threat is somewhat smaller to those past their developmental years, but lead is still toxic to adults. It can cause reproductive issues and may be a carcinogen. Enough of it will kill you.

Since 2010, contractors must be certified in lead abatement and follow lead-safe practices when working in homes built before 1978 and disturbing more than six square feet of lead paint per room inside a house or twenty square feet of paint on exterior surfaces. Further information can be found at the EPA website at www.epa.gov/lead/renovation-repair-and-painting-program.

Homeowners are not required to be certified to work on their own homes, but the EPA provides safety guidance for do-it-yourselfers at www.epa.gov/lead/renovation-repair-and-painting-program-do-it-yourselfers. Among their recommendations:

Work Safely

- Remove all furniture, area rugs, curtains, food, clothing, and other household items until cleanup is complete.
- Items that cannot be removed from the work area should be tightly wrapped with plastic sheeting and sealed with tape.
- Cover floors with plastic sheeting.
- For a large job, construct an airlock at the entry to the work area. This consists of two sheets of thick plastic, one of which is taped along all four edges and cut down the middle. The second is taped along the top edge only and acts as a flap covering the slit in the first.
- Turn off forced-air heating and air-conditioning systems. Cover vents with plastic sheeting, taped in place.
- Close all windows in the work area.
- When using a hand tool on a lead-painted surface, spray water to keep dust from spreading.

Get the Right Equipment to Protect You and Your Family

- NIOSH-certified disposable respirator with a HEPA (high-efficiency particulate air) filter (N-100, R-100, or P-100).
- HEPA filter-equipped vacuum cleaner. (Regular household vacuums may release harmful lead particles into the air.)
- Wet-sanding equipment (e.g., spray mister), wet/dry abrasive paper, and wet sanding sponges for "wet methods."
- Two buckets and all-purpose cleaner. Use one bucket for the cleaning solution and the other bucket for rinsing. Change the rinse water frequently and replace rags, sponges, and mops often.
- Heavy-duty plastic sheeting and heavy-duty plastic bags.
- Protective clothing. To keep lead dust from being tracked throughout your home, wear clothes such as coveralls, shoe covers, hats, goggles, face shields, and gloves or clean work clothes and launder separately.

Additionally, EPA has published *Steps to Lead-Safe Renovation, Repair and Painting: October 2011,* a guide for contractors and homeowners on how to plan for and complete a home renovation, repair, or painting project in pre-1978 housing and child-occupied facilities. The 34-page PDF document can be found at www.epa.gov/sites/production/files/2013-11/documents/steps_0.pdf.

Prepping for Refinishing

Painting well requires only that you master several relatively simple manual tasks and do them in the correct order. First and foremost, you must properly prepare the surface for the chosen coating. The old saying is that "good prep is half of a good paint job," but in fact it is more like 75 percent, especially in an old building. No matter how carefully you apply new paint over a bad surface, you will never get a great result. The surface must be well prepped before the paint goes on.

A couple of professional painters from the days when walls were still painted with brushes, not rollers.

Surface Prep

Begin by determining whether or not the existing surface is well adhered to the substrate (base material). If paint is chipping copiously or coming off in sheets, you'll need to remove some or all of it to get back to a stable surface. Depending on the extent, that usually means scraping, sanding, or applying heat or steam as discussed above. (As a rule, grit blasting should be avoided except on a hard metal like cast iron or steel.) It is not generally necessary to remove paint that is still well bonded to the substrate. You will need to "feather" the transitional edges between bare wood and patches of remaining paint with sandpaper; a wide feathered edge is less apparent than a narrow one.

When woodwork has large areas of crackled paint or many small chips that would be time-consuming to feather, it is possible to smooth the surface with a filler. Wood putty and plaster-based fillers, the traditional choices for this work, can shrink while curing, requiring multiple applications to fill a recess or crack. Wood putties are also notorious for drying out, cracking, and ultimately losing their bond and falling out. Bondo-type polyester fillers are also popular, and I can recommend Brushing Putty, available from Fine Paints of Europe. Not well known in the U.S., Brushing Putty is an oil-based, high-build, sandable primer that has given me excellent results on old crackled and chipped painted trim. Fine Paints of Europe also carries Swedish Putty, a thicker, knife-applied putty that can be built up in thin layers to create an ultrasmooth surface for paint.

Although they are expensive, I am a big fan of the FPE line of paints and prep products. The quality more than pays for itself in better coverage and long-lasting finishes. It should be noted, however, that FPE paints comprise a system, and the compatible primer and finish paints must be used to get the best results (and meet the conditions of the warranty).

Correct preparation is equally important for new work, where it will likely involve filling nail holes and caulking joints. Again, corners cut at this stage will show in the finished surface. It is tempting to rush through tedious prep work to get to the satisfying sight of a freshly painted surface, but you will end up with flaws that you will notice forever. Looking at something you know you could have done better had you done it correctly, day after day, year after year, doesn't sit well. Trust me on this.

Choosing Paint

Before the late nineteenth century, house painting was a job for trained tradesmen who made their own paints, often grinding the pigments themselves from natural materials before mixing them with a water-based or lead and linseed oil–based binder. Water-based, or distemper, paints were less durable and formed a flat, or matte, finish. Linseed oil forms a

While the basic formulations for paint are the same, there are variations in the proportions of ingredients, and sometimes additional ingredients were used to adapt paints to particular purposes and degrees of gloss. Manufacturers sometimes advertised paints for specific uses, such as in this nineteenth-century paint store display card.

dense skin when it dries, which creates a more durable paint with a glossy finish. In addition to pigments, a distemper paint often included a binder such as milk casein or some form of glue. In oil paints, solvents were added to thin the mixture, improve its flow from a brush, and control the degree of gloss. Typical solvents were turpentine or other volatile organic compounds that evaporated from the paint mix, allowing the linseed oil to dry.

The development of premixed oil paints in the late nineteenth century allowed painting to be done by ambitious homeowners. Paint manufacturers increasingly catered to the homeowner market and formulated paints that were easier to apply and clean up after, particularly after World War II. Latex and acrylic waterborne paints earned increasing acceptance due to their ease of cleanup. When lead was banned from paint for household use in 1978, American manufacturers sought to develop equivalent quality in lead-free formulations, with mixed success. More recent regulations have required a significant reduction in the volatile organic compounds (VOCs) released from a paint's solvents as it dries (causing the sharp odors associated with traditional paints). This has required another round of reformulation by manufacturers.

The cumulative result of reformulations to satisfy regulations and optimize paint for do-it-yourself application is that modern paints are generally inferior to historic paints. Manufacturers will dispute this, but my experience is that paints are not as tough or long-lasting as they used to be, especially at the lower end of the market. However, some modern paints—or "coatings" as the industry now calls them—are far better than others.

You will work much harder to get a poor finish with cheap paint than to get a good surface with quality paint. And the poor finish will not last as long. Given that much of the labor in a paint job is in the prep work, and that poor paint is at least as hard to apply as good paint, and that a coat of inferior paint will always be the weak point of attachment beneath future coats, it is wise to buy the best paint you can afford.

Traditional oil paints provide the hardest, most durable finishes on doors, trim, cabinets, and staircases, which receive a lot of use and abuse. The recently developed waterborne alkyd (oil) paints are almost as hard and durable (and clean up with soap and water!). High-end latex/acrylic paints are less durable. Inexpensive latex paints are short lived.

In general, high-quality paints provide better coverage because they contain higher proportions of pigments to fillers. They also flatten better before drying, leaving fewer visible brush or roller marks. All good painting requires an appropriate primer coat and two finish coats. Surfaces need to be sanded with fine (220 grit) sandpaper between coats and wiped down with tack cloth to remove dust. Resist the temptation to apply one thick coat in place of two thin coats.

If you are painting dark colors, do not use a white primer. Paint stores will tint primer at no additional cost, and a tinted primer can save you from applying a third or fourth coat of dark topcoat. Gray will work, but for strong colors I usually tint primer to the complementary color of the finish paint. This is a technique from easel painting that was sometimes used by nineteenth-century house painters, especially for decorative painting. The undercoat of a complementary color will almost imperceptibly affect the finish color, making it "livelier" to the eye.

Clear-Coat Finishes

For wood trim as for hardwood floors (see Chapter 14), it is always appropriate to maintain or restore an original "natural" finish of stain and varnish or shellac. Shellacked or varnished surfaces may have darkened with age and may have lost finish from scraped or worn areas. Start by scrubbing the surface with a mild cleaning agent for wood like Murphy's Oil Soap. This will often brighten a surface quite a bit, particularly if it was subject to cigarette smoke over a long period of time. If you're lucky, you will only need to clean the surface, touch up a few worn or damaged spots, and give it a fresh coat of furniture polish. If you're unlucky, you'll encounter peeling varnish, white water stains, splattered paint, chunks of missing wood, or a shellac finish so impregnated with grime it appears black. In such instances you may have to strip the surface to bare wood, repair damage to the wood, and restore the original finish with stains or dyes and a clear coat. Most restorations fall somewhere between these extremes and require some combination of the approaches described previously for floors, ceilings, and walls.

Be cautious about applying polyurethane to surfaces originally finished with varnish or shellac. Refinishing with the original material is generally recommended. Traditional varnish is less available than it used to be, but marine varnishes are readily available and impart a more traditional—less "plasticized"—appearance than polyurethane. Shellac is still available and is an excellent choice for surfaces that will not be walked on or get wet. An insect secretion dissolved in alcohol, shellac is the most natural and sustainable of the clear-coat options. It is also the easiest to touch up; a scratch can often be filled with re-dissolved shellac by passing a cotton ball wetted with alcohol across the adjacent surface.

Matching stain colors for repairs is as much art as science and may require a mixture of several

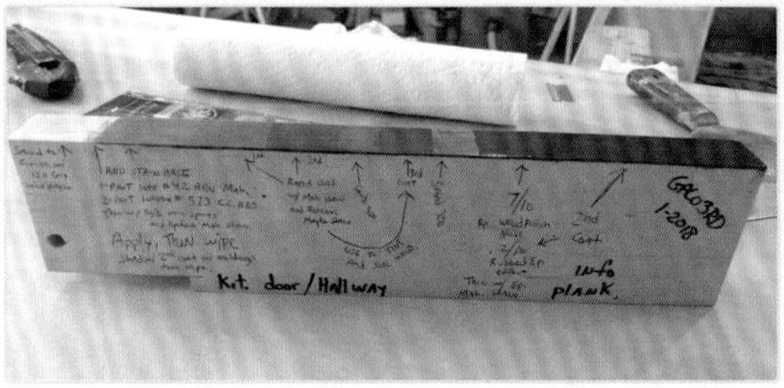

This formula board is for the "mahogany" finish in a restored kitchen. It includes all steps involved in creating the finish, which will allow any repairs or future work to match what has been done.

This remarkable finish is in the 1874 Boston & Maine Railroad station in North Conway, New Hampshire, the first building I ever helped restore. I was a curious eleven-year-old with a ton of questions; I'm not sure how much I helped, but I learned a lot. The alligatored finish is the result of oil paint being applied over varnish or shellac, probably after 1900. The oil paint may have had a high linseed oil content, or perhaps there were coats of both varnish and shellac under the paint. Whatever the underlying cause, the paint has pulled partially free of the varnish while hardening and shrinking yet remains well attached. The result, though unintentional, imparts a special character to this building. If you encounter unusual finishes like this in your house, think about preserving them. They tell a story (or at least hint at one) and will set your house apart.

commercial stain colors. Many restorers use wood dyes in place of stain with good results. Shellac's amber tint will subtly alter the color of the stain under it. It is always best to do a test on scrap material before applying your formula to the permanent work. If a finished repair is not matching as well as you'd like, try adjusting the color and tone with a tinted varnish, shellac, or polyurethane.

Brushes

As with any job, the right tools can greatly simplify painting and make it easier to get good results. Paint brushes are the most essential tool for painting once the prep work is done. Their price range is sweeping; a quick search for a 2½-inch sash brush on a major online retail site turned up options ranging from $1.99 to $135. In my experience, the least expensive brushes are not worth what they cost, and the most expensive brushes are unnecessary for good results. My favorite brushes cost around $20 each, twenty years ago, and I am still using them. I also have some expensive round sash brushes like those used historically, which I use occasionally. I like how they spread paint but prefer the handle shape of my modern brushes when I am doing a lot of painting. Authenticity of tools and techniques is nice, but hand cramps are not.

I know very good painters who prefer other brushes, some of which I have tried and was not at all satisfied with. Yet in the hands of these painters, those brushes produce work that is as good as or better than my own. Find the brushes that work best for you. A good brush will allow you to transfer a reasonable amount of paint from the bucket to the surface you are painting and spread that paint evenly across the surface with a minimum number of strokes while giving you good control of the paint and not causing excessive hand fatigue over time. Depending on the control needed for the work you are doing and your access to the surface, you may need to grip the handle close to the bristles or close to the tip. Do not struggle to paint every surface with a single grip. It is useful to have at least one brush with the handle cut down to a nub for reaching into spaces that do not have room to manipulate a full-length handle.

Cheap brushes have their uses, primarily for one-time application of stain sealers or other materials that are not easily cleaned out of a brush. Given a choice between throwing away a $1.99 brush after a single use or spending twenty minutes of my time cleaning a more expensive brush after a five-minute job, I know what I'm choosing. I do not use foam brushes for finish work, but sometimes they are the perfect tool for dabbing a small amount of paint onto a difficult-to-reach spot. They are also useful for applying primer to the cut ends of pre-primed boards used on the exterior, which are usually cut as they are installed. A sponge brush in a small can of primer next to the chop saw makes it easy to keep all sides of the board painted as it is installed.

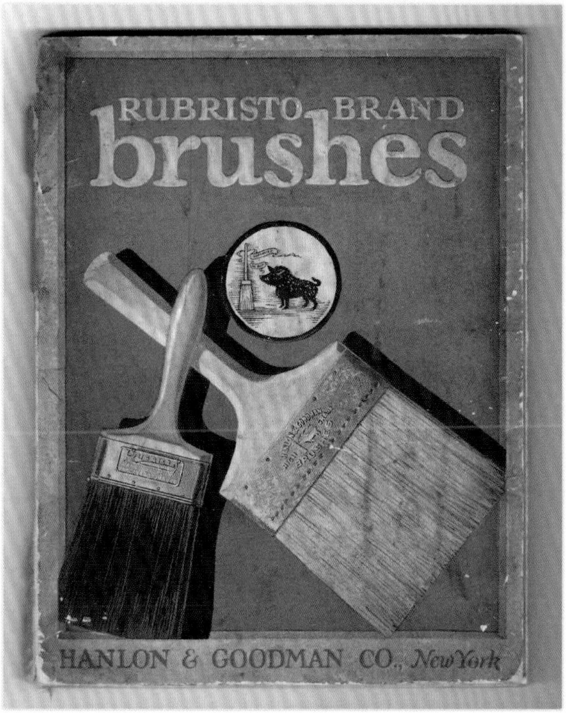

The cover of a 1930s paint brush catalog.

Among my favorite cheap brushes are the small "chip brushes" imported from China. These have unfinished wood handles and white natural bristles and come in widths from ½ inch to several inches. They can be found in bins at retailers for a couple dollars or less apiece. I find that ½-inch-wide, long-bristled chip brushes are handy for touch-up work and for applying small amounts of metallic paint. When I worked frequently with gold metallic paint, I kept one of these brushes wrapped in plastic in my freezer for years. I just pulled it out, let it warm up, and used it again without ever cleaning it. I cut the handle of one brush in half lengthwise and screwed the halves back together so I have an adjustable angled handle to reach difficult spots. This ninety-nine-cent custom brush has lived in my toolbox for twenty years and gotten a lot of use.

Painting Trim

Painted trim and cabinetry can take a lot of abuse in daily life, and they need paint that will stand up to it. High-quality paint is more important for trim than for walls or ceilings. Kitchen cabinets, in particular, need wear-resistant scrubbable finishes. I strongly recommend alkyd/oil-based paints for trim and cabinetry due to their durability and washability. The development of waterborne alkyd paints in recent years now makes it possible to use oil paint yet clean up your tools with soap and water. This is a seemingly magical product to those of us who have gone through gallons of paint thinner and mineral spirits cleaning brushes through the years. I am impressed with Fine Paints of Europe's version, called Eco Waterborne. It is not as hard or resistant to heat damage (on a mantel, for instance) as their Hollandlac Brilliant traditional oil paint, but it is superior to any latex/acrylic waterborne paint I have used. If you choose a latex/acrylic paint for trim, buy the best quality you can afford. There really is a difference between the cheap stuff and the good stuff. Save money on wall and ceiling paint if you must, but go with high-quality paint for trim and floors.

As with any surface to be painted, proper preparation of the surface is essential. Once prepped, a durable painted surface requires an appropriate primer coat and two coats of finish paint, sanded between coats. Many paint manufacturers claim to make products that are "self-priming" and cover with one coat, but there's a tradeoff in the quality of the paint. Such a paint needs to be applied in a very thick coat, requiring it to be full of fillers and drying agents to prevent runs and sags before the paint sets up, and this quick-drying characteristic causes it to set up before the brush marks can smooth out and disappear. It is also difficult to apply this paint well, as the leading edge of the area just painted frequently dries before the next strokes can be blended in. This is a particular problem on paneled doors, with their many overlapping joints. Quick-drying paints also cause brushes to become loaded with dried paint as you work, which eventually starts flaking off into the paint you're applying. In short, these paints are an exercise in frustration for anyone seeking a good result.

Painting trim is generally straightforward and requires only that one pay attention to the work. As with any task, practice improves both speed and results. Start slowly. In time, the brush will feel like a natural extension of your arm, and you will be able to apply paint as much by feel as by sight. Trim paint is nearly always applied parallel with the wood grain. When painting a long piece—a door casing, baseboard, crown molding, etc.—my method is to apply the paint with brush strokes away from the end I started at, then brush back over the just-applied paint, pulling it slightly into the previously applied section. This brushing of the overlap should be done with a light touch, creating a feathery edge—blending the most recently applied paint into the previous wet edge. Blending paint in sections like this removes the evidence of overlapping sections.

Painting Walls and Ceilings

Most people today use acrylic/latex paints for walls and ceilings, even when using alkyd/oil paints for trim and cabinetry. The standard approach is to cut in the edges with a brush, then apply paint to the rest of the surface with a roller. Done well, this will produce an even finish with a subtle stippled or "orange peel" texture imparted by the roller over the whole surface. Done poorly, it yields an uneven surface with brush marks around the edges and ridges where the edge of the roller left a thicker coat of paint than the middle.

After thousands of hours painting interiors, I've gained a reputation for doing it well. After prepping the surfaces, I use a 2½-inch angled sash brush for cutting in and a 9-inch roller with a ¼- or 3/8-inch nap. I paint trim first, lapping slightly onto the wall surface, and allow it to dry thoroughly. In historic houses, the junctions between walls and trim are seldom clean, smooth, right angles. More likely you'll encounter an uneven buildup of old paint or caulking or other irregularities that have appeared over time. If there is no clear, straight line to follow at the junction, I make one with the wall or ceiling paint. For long runs, such as baseboards or crown moldings, I use blue painters' tape to create the line. I always run the back of my thumbnail or the blade of a five-in-one tool along the tape's meeting edge to seal it to the trim paint. Elsewhere I usually "pull" the line freehand with a brush. Practice is the only way to get good at this. A do-it-yourselfer with a room or two to paint may find blue tape the better way to go, but painters' tape is expensive. If you are painting your entire house, you'll get good at cutting a sharp line on an irregular surface without tape.

I like to keep my brush well loaded with paint for cutting. It flows better for long strokes and reduces the number of trips the brush has to make back to the bucket. The exception is an inside corner with abutting trim. There, I use less wall or ceiling paint and make short strokes with the tip of the brush fanned out against the surface. With a light touch, it is possible to ease the paint up against but not onto the corner trim. Once the cut is made by feathering in this fashion, a wider band of paint can be applied with a more loaded brush, keeping a short distance away from the feathered edge.

I generally cut and roll one wall at a time so I can roll before the cutting paint is dry. This reduces the chances of a visible line of demarcation where brushed and rolled paint meet. I work with a roller on a telescoping aluminum pole, adjusting its length to the height of the work. If you are going to do a lot of painting, buy good-quality poles. Inexpensive wood poles that screw together in 24- or 30-inch sections will work but do not allow you to adjust pole length as you go. Wood poles are also smaller in diameter, providing a less comfortable grip. I have been using the same two roller frames for twenty years. They squeak as they rotate, giving an audible rhythm to the work that helps me maintain a steady pace. Paint applied at a steady pace and with even pressure will be applied evenly across the surface.

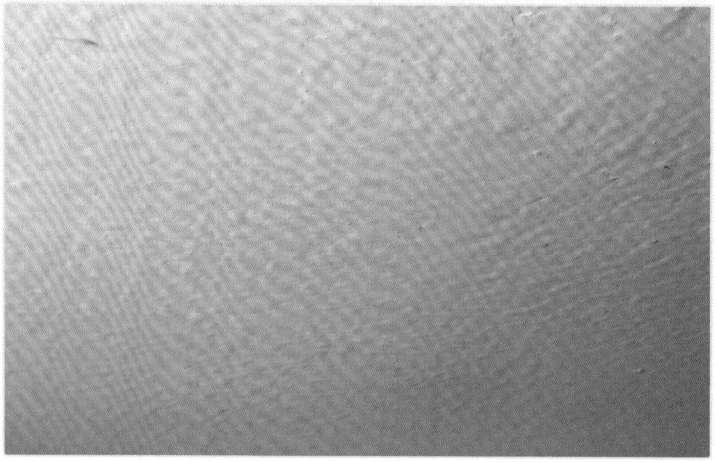

The "orange peel" texture of rolled paint on a ceiling.

I try to load as much paint from the tray onto the roller as possible without inviting drips. You want paint on the wall or ceiling, not the floor. There are online videos showing the steps for rolling paint on a wall. In brief, roll a tall, narrow W on the wall to transfer as much paint as possible from the roller, then spread the paint evenly with vertical strokes, filling the areas missed by the W. Then roll from top to bottom in parallel overlapping passes with light pressure to even out the coat. Once the area is filled evenly with paint, roll horizontally along the top and bottom boundaries, close to the trim but not touching it, to carry the orange peel texture over the cut paint. Any visible roller overlap marks need to be rolled again, gently, to spread the thicker paint causing the mark. If allowed to remain, those marks will dry into ridges that will remain visible forever in raking light.

Repeat this process, area by area, until you have covered the entire wall, overlapping sections slightly. You may need to roll horizontally rather than vertically over windows and doors. It is best not to take a break in the middle of a wall, so you are always blending the next section into the wet paint of the previous section.

Before applying the second coat, go over the wall quickly with a fine sanding screen on a pole sander, then wipe with a damp rag to remove any dust from sanding. It is not usually necessary to cut the second coat all the way to the trim edge unless the first coat is noticeably thin. I try to get within 1/8 to 1/16 inch of the edge, depending on the prominence of the junction. Where walls meet ceilings without an intervening molding, the junction is usually quite visible, and I cut it as close as I can without messing up the first cut.

The second coat is rolled like the first. If I am using blue tape for a cut line, I try to apply the second coat as soon as the manufacturer's instructions allow. This will be before the first coat is fully cured. As soon as the second coat is fully applied, I pull the tape off. Pulling it while the paint is still wet seems to produce a clean line most of the time. Folding the tape back on itself while pulling will reduce the pulling pressure on the surface as it is removed, decreasing the likelihood of pulling off trim paint with the tape. However, some touchup is virtually inevitable.

When rolling large areas with waterborne paints, especially ceilings, the most significant challenge can be maintaining a wet edge to blend the next section into. Rolling wet paint onto dry often creates a visible line at the overlap that is difficult to eliminate. Thinly applied latex/acrylic paint dries quickly, especially when the air is dry and hot. For much of the year in much of the U.S., the air near a ceiling is very dry from central heating, air conditioning, or a desert climate. If you can't paint on a humid day, additives like Floetrol will extend the drying time of a waterborne paint.

Some restorers of historic homes find the orange peel texture of rolled paint bothersome in a house built before World War II. Prior to the development of paint rollers with replaceable wool or nylon covers, paint was applied to walls with wide brushes, usually applied with the first coat brushed in one direction and the second coat brushed at a right angle to the first, producing a subtle cross-hatched texture. It is possible to brush over the rolled paint before it dries to impart this more authentic texture to the paint. This can be done with a 4- or 6-inch paint brush or a wider wallpaper paste brush.

Alkyd/oil paints are applied in nearly identical fashion, but you will have more time to work due to the much longer drying time of these paints. Oil paints run much more easily than latex/acrylic paints if applied too thickly, so keep the coats thin; the coverage will be fine provided the paint is of good quality.

As mentioned above, many modern paints promise one-coat coverage with no need of priming, but this is accomplished by formulating a gelatinous goop that is so full of fillers and driers that it can be applied very thickly without dripping or sagging. It is virtually impossible to apply such a paint thickly enough to cover well without leaving visible roller marks. When you roll over the just-applied paint enough to remove the marks, you make the coat too thin to cover the previous coat. Ignore the hype and buy good-quality paint to be applied in two thin coats over an appropriate primer.

If you need to touch up a wall or ceiling that has a rolled orange peel stipple, dabbing on the paint with a small brush will create a texture that blends in much better than if you apply the paint with brush strokes. For touching up tiny holes, such as from tacks or small nails, a dap of matching paint applied with a disposable bathroom cotton swab will work nicely.

Paint Sprayers

Many professional painters now use paint sprayers in place of brushes and rollers for walls, ceilings, and even trim. Sprayers require more prep—since every surface that is not to be painted must be carefully masked off—but apply paint much more quickly. In the hands of an experienced painter using high-quality paint, the result is a good finish, especially on walls and ceilings. The texture is a more subtle orange peel stipple than that imparted by a roller, and the very thin coat dries quickly, almost upon contact when flat latex/acrylic paints are sprayed. On a large area, such as a ceiling, it is not possible to spray the surface and then back brush it to change the texture before the paint dries, but this can often be done on trim or cabinetry, where smaller areas painted. Perhaps the greatest "flaw" of paint sprayers is that the finished surfaces are often too perfect, lacking the subtle evidence of the human hand. I have sent out cabinet doors to have their primer and first finish coats sprayed in a paint shop, applying a final coat by brush once they are hung.

The inexpensive paint sprayers available at a local home-improvement center are not the sprayers used by professional painters and will not produce equivalent results in the hands of a weekend warrior. Rather, they are likely to leave all too much evidence of the human hand, in ways that are not at all characteristic of historic finishes.

Wallpaper

Wallpaper has been used in America since the eighteenth century and became very popular in the nineteenth century. The earliest papers were imported from England and France, then the leading producer of high-quality papers. By the early nineteenth century, a number of American companies were producing wallpaper as well.

Reproduction Aesthetic Movement wallpaper in the B.J. Talbert room set by Bradbury & Bradbury showing patterns for the dado, border, and wall fill. A "room set" comprises ceiling papers with related wallpapers. *(Courtesy Bradbury & Bradbury Art Wallpapers)*

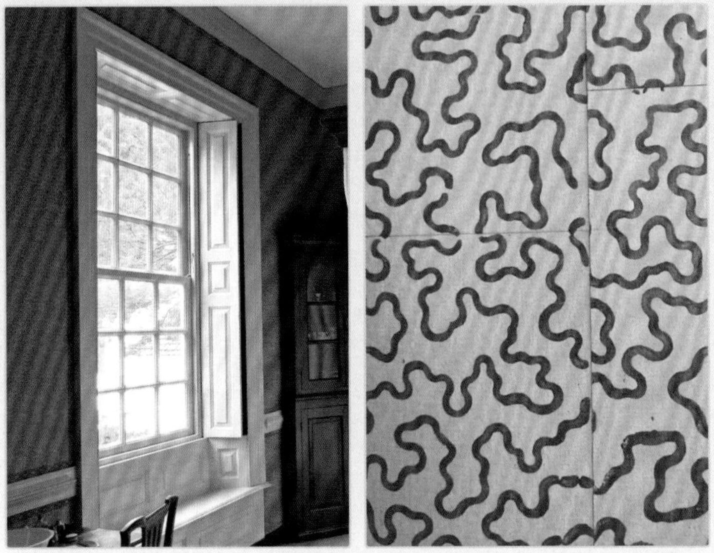

This reproduction vermicelli-pattern (vermicelli is Italian for worm) wallpaper used at Colonial Williamsburg has been printed on lengths of paper glued together from smaller sheets, as was done in the eighteenth century before paper was produced in long strips. In the detail shot, the horizontal joints were pasted before the paper was printed, and the vertical seam is where two strips met when hung on the wall. This installation used overlapping vertical seams, which only requires trimming one side of the paper, the common practice when all papers were purchased with untrimmed edges.

The earliest wallpapers were made by pasting together sheets of paper into long strips. In time, advancing technology made possible the manufacture of pieces long enough to reach from baseboard to ceiling; later still, paper was manufactured in much longer rolls. Early paper was made of rag fibers, primarily from recycled clothing. After painting the strips with a ground color, the design was printed on the paper with carved wood blocks. Each color of the pattern was applied with a different block, as shown on page 673. Many papers were quite simple, some simply painted a solid color to be hung with a patterned border. One- or two-color prints were very common. The most elaborate block-printed papers could have twelve or more colors. Companies like Zuber, in France, specialized in block-printed scenic papers, producing murals that could cover all four walls of a room.

Developed by the mid-nineteenth century, roller printing allowed much faster production of wallpapers. Wood pulp paper was developed around the same time, which was less expensive to produce than rag paper. These two developments made inexpensive wallpaper and ceiling paper available to the masses, though expensive papers were still imported and made by some American firms.

Preserving and Conserving Historic Wallpapers

If your house has historic wallpapers on the walls, it may be worthwhile to research what they are and determine whether they can and should be preserved. Often, earlier rag wallpaper will be in better condition than later wood pulp paper, which contains acids that cause the paper to turn brown and brittle over time. This is true of nearly all wood pulp paper produced before modern acid-free paper was developed.

Most of the research on historic wallpapers in America has been conducted in the Northeast, in large part because the climate there is more conducive to wallpaper longevity than in the humid South. However, research from New England can be considered representative of wallpaper use throughout much of the United States up to at least the Civil War. Most early American wallpaper manufacturers were between Boston and Philadelphia, selling nationwide,

This is a portion of a leftover partial roll of the French Gothic Revival wallpaper hung in the front hall of Whitten House in 1852. It remains in pristine condition because it was printed on cotton rag paper, which does not turn brown and brittle with age as wood-fiber paper does.

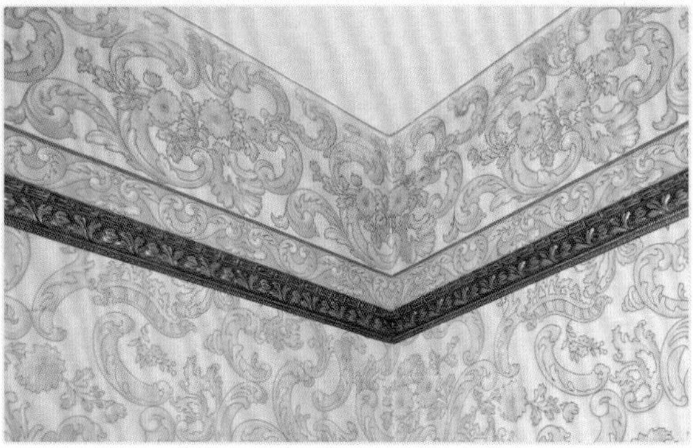

This Victorian-era wallpaper, border, and frieze survive in excellent condition with a period silver-gilded picture rail with composition ornamentation. The monochrome colorway of this pattern is less florid than many papers of the period, a plus for some tastes in the twenty-first century. Even if a paper like this offends your personal taste, please don't destroy it. Surviving examples are rare, and you are just passing through.

and papers were imported from England and France for distribution throughout the U.S.

In my opinion, historic wallpaper surviving on the walls it was hung on fifty or more years ago should be preserved if it is in reasonably good condition—especially if it is pre-1850. Such papers are rare in their original locations. Dirty or damaged wallpapers can be conserved and restored; it is not simple or inexpensive, but it is worthwhile if you have a surviving rare wallpaper.

Historic wallpapers are often buried under layers of later wallpapers and sometimes paint as well.

This before-and-after image of a conserved nineteenth-century William Morris wallpaper in Speke Hall, Liverpool, UK, shows what is possible in the hands of a skilled conservator. *(Courtesy Kendal Anderson)*

This bit of Victorian wallpaper and border was exposed when a 1940s shelf unit was disassembled in a small bedroom in Whitten House. The paper remains on the rest of the walls but was painted over sometime after the shelves were installed. Always document discoveries like this with photos, even if you have no intention of duplicating the treatment as part of your project. Once you remove the evidence, no one will ever again know what was there historically.

Wallpaper historians have spent many hours steaming apart layers of wallpaper to identify the different patterns. Depending on the period of significance you have identified for your house, you may be interested in the earliest layers or later layers. It is, of course, possible that the earliest layer you see was preceded by even earlier layers that were removed rather than covered. Historic wallpapers also turn up behind architectural elements that were added later, such as the trim around a door that was cut into a wall at a later date.

Papered Ceilings

Ceilings in the nineteenth century were often papered like walls, sometimes in elaborate patterns built up of "field" papers and various borders and medallions. While most wallpapers have a distinct orientation in their patterns—stripes being the most obvious example—field papers for ceilings have omnidirectional patterns that look the same from any point in the room.

Reproduction Victorian-era ceiling papers with related wallpapers have been available since the 1980s. Some of these have been developed beyond what was historically available to make it possible to reproduce the effect of a decoratively painted or stenciled ceiling with papers. Bradbury & Bradbury is the leading firm producing these high-quality hand-screened room-set papers.

A ceiling can be prepared for papering in much the same way it would be prepared for painting. Delaminated plaster needs to be reattached and cracks need to be repaired, because movement in the plaster will cause the paper to wrinkle or tear. Most high-end wall and ceiling paper manufacturers recommend installation of lining paper before the printed paper. As with wallpaper, it is a good idea to have your sizing primer

A papered ceiling in Bradbury & Bradbury's Fenway room set. *(Courtesy Bradbury & Bradbury Art Wallpapers)*

tinted to the background color of the printed paper to minimize the visibility of any gaps at the seams.

Papering a ceiling is not much more difficult than papering a wall, except that gravity fights your efforts all the way instead of assisting as it can on a wall. Good, solid staging is essential. (See Chapter 19 for a discussion of options.) Long runs of field paper are best installed on a ceiling by two people working together. I have done it alone, but I don't recommend it. Borders and ornamental pieces of ceiling paper, on the other hand, can readily be installed singlehanded.

Do not attempt to cut miters for angle joints before pasting up borders. Instead, run both lengths of the border, allowing the ends to cross, then lay your straightedge across the overlapped pieces, aligned with the crossing points on each edge, and cut through both pieces at once with a sharp blade. Remove the scrap pieces and pat the joint in place with a damp rag, wiping gently to remove residual paste on the surface of the paper. This approach will always give you a perfect miter, whether or not your pieces meet at exactly 90 or 45 degrees.

Lincrusta and Anaglypta Wallcoverings

During the Victorian period, several English manufacturers developed embossed, linoleum-like or heavy-paper products for imitating embossed leather wallcoverings. American manufacturers copied the products, and they became popular for hallways and other high-traffic areas as well as libraries and dining rooms, which were associated with "masculine" leather wallcoverings. These materials could be painted in flat colors or treated in any number of faux finishes or gilding.

Where these coverings survive, they can usually be repaired with paintable materials that can be shaped or sculpted to replicate missing portions. Consolidation of original material in the most visible locations is also an option. Accurate reproductions are available, as are less accurate embossed wallcovering materials. Avoid obviously bad copies; a coat of paint won't disguise them.

Reproduction Wallpapers

Few historic houses have surviving wallpapers that were not covered by later papers or paint. Excavating old layers of wallpaper can provide a sense of the patterns that were used historically in the house, guiding decisions about appropriate reproduction papers. Until quite recently, it was very expensive to reproduce historic papers from surviving fragments. The full design had to be drawn on paper by hand, then color separated in order to carve wood printing blocks for each color or make silk printing screens. Thousands of dollars would have to be spent before any printing could begin, and the printing itself, whether by wood block or silk screen, was labor intensive and expensive. But the recreation of patterns has moved

Part of the team at Adelphi Paper Hangings reproducing a wood block–printed flocked wallpaper. *(Courtesy Adelphi Paper Hangings)*

This portion of a leftover roll of nineteenth-century machine-made wallpaper was found in Whitten House. While there are several period wallpapers still in place in the house, this one was apparently removed in 1941, when the house became the town library and several rooms were repapered. This partial roll in excellent condition is a prime candidate for digital reproduction, as it will not need extensive (and expensive) digital restoration before being printed.

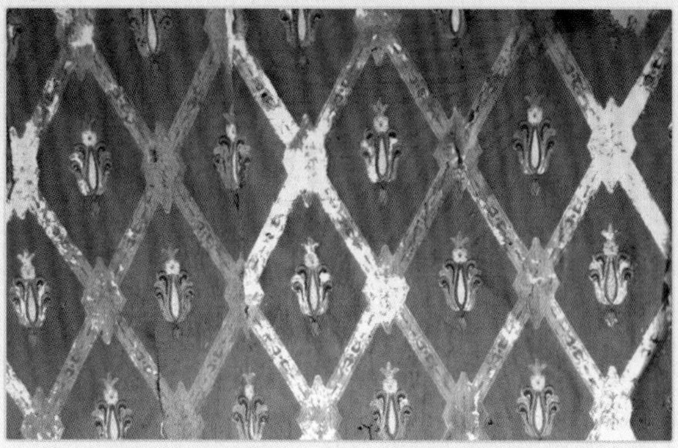

This historic paper survived into the twenty-first century beneath another layer of wallpaper in an attic room. Reproduction of a paper with this degree of damage will require the missing portions of the design be recreated by hand or digitally and the original colors to be determined based on how historic pigments react to UV light and other environmental factors. Some pigments are far more stable than others. Once the pattern and colors are established, the paper can be reproduced as a block print, screen print, or digital print.

from paper to computers in recent decades, providing some efficiency in the color separations and other preparatory steps. Wood printing blocks are now usually carved by computer-guided laser.

Wood block–printed paper is identical with its historic antecedents, recreating the characteristics of the original with great fidelity. Screen printing is not a replication of a historic process but does a fine job of recreating historic roller-printed wallpapers. Each of these printing processes provides a particular character. In my opinion, wood block printing is better for reproducing papers that were printed by that method historically. Both printing methods can create exquisite wallpapers. Both are illustrated here with photos from two of the leading reproduction wallpaper manufacturers.

Custom printing remains very expensive, but once the blocks or screens exist, it is much less expensive to produce additional rolls with the same patterns. The companies that make reproduction wallpapers by these methods offer existing patterns in their catalog at prices well below those of custom papers. There is information on reproduction wallpaper (and ceiling paper) companies in Chapter 20.

Hand Block Reproductions

These images, courtesy of Adelphi Paper Hangings, show the process for reproducing wallpapers with hand block printing. The paper shown is their "1776" pattern. This pillar-and-arch pattern, inspired by British architectural patterns of the 1760s and 1770s, is probably the first American design featuring allegorical symbols. Like most architectural papers, it was most often hung in a hallway, though it may also have been used in a parlor or bedchamber. The allegorical characters include France—handing a Declaration of

Individual sheets of paper are pasted together to form 33-foot-long rolls.

The background color, or ground, is painted onto the paper by hand.

The first color for printing is spread onto the fabric pad with a brush, and a carved pearwood block is pressed into the pad to transfer the paint onto the carved design. Each color of the pattern requires a separate block.

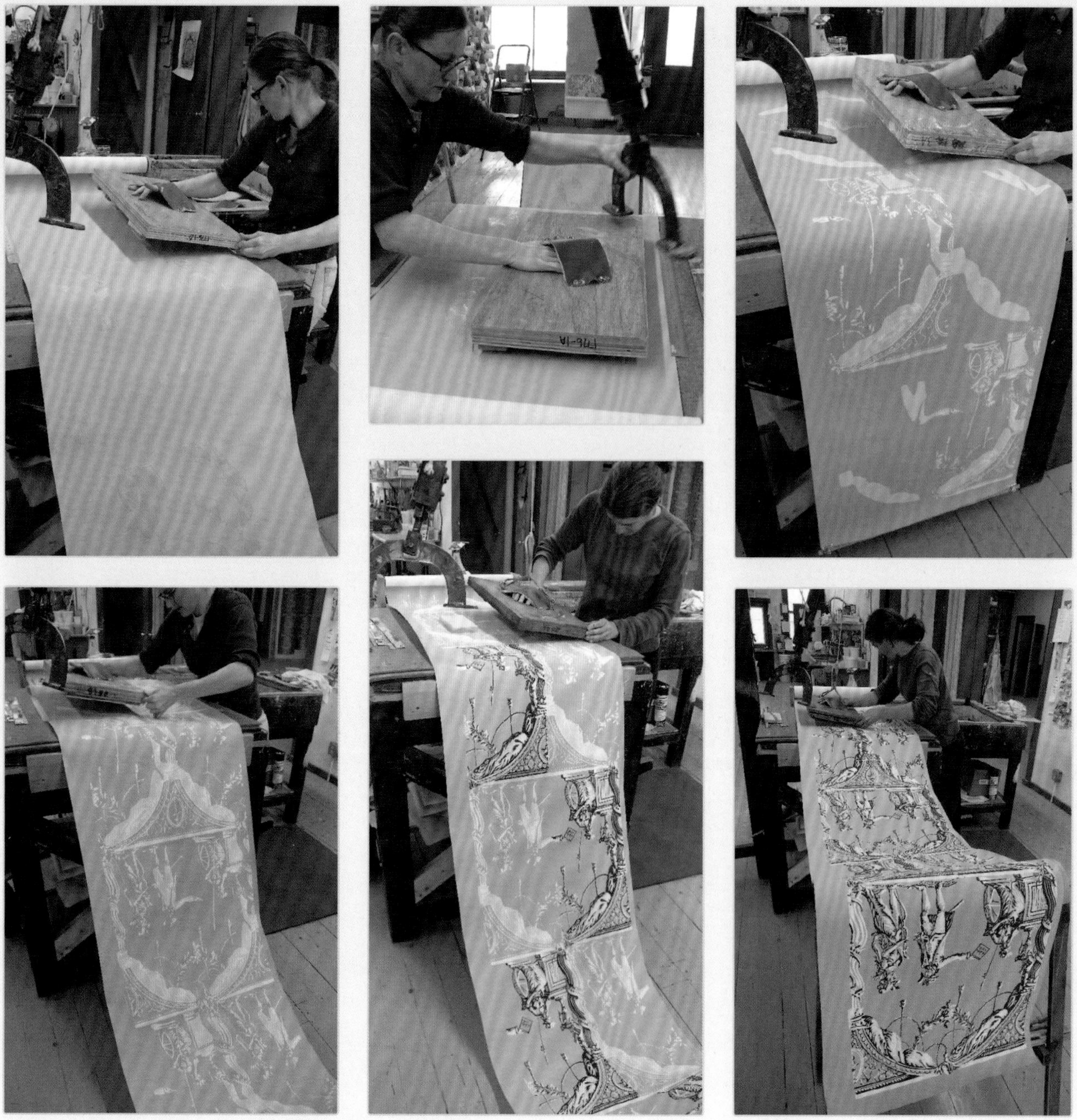

The block is positioned on the paper by hand, and a lever controlled by a foot pedal depresses the arched iron bar to push the block against the paper, transferring the paint. Then the block is lifted, the paper is advanced, and the process is repeated to the end of the roll. Once dry, the paper returns to the starting point and the process repeats for the next block and color, and so on until the full pattern is printed.

During the drying stages, the paper is hung from the ceiling on this rack.

Independence to a weeping Britannia—and America, represented by a Native American princess.

Screen-Printed Reproductions

Screen-printed reproduction wall and ceiling papers are made using a process similar to block printing, but with more modern materials as shown in these images courtesy of Bradbury & Bradbury Art Wallpapers.

A screen is prepared for each color of the design as follows: A polyester fabric is tightly stretched over a frame and coated with a photosensitive emulsion. Next, opaque artwork in red, on clear film, is laid over the coated screen, sandwiched against it in a vacuum frame, and exposed to strong ultraviolet light for several minutes.

The screen is rinsed with water. The areas of emulsion that were shielded from the light under the artwork wash out of the screen, but the exposed emulsion surrounding the artwork remains, creating a stencil in the mesh.

To ensure proper printing, each screen is then carefully aligned, or registered, to the 30-yard-long print table and to subsequent screens by attaching an L-shaped guide. The guide is used in conjunction with "stops" that control the pattern repetition along the length of the table.

The ink is poured across the bottom edge of the screen, and a rubber squeegee is used to pick up some of the ink and transfer it to the top of the screen. The ink is pressed through the screen as the squeegee is pulled by the printer. The screen is then carefully lifted, revealing the print. This is repeated for the full length of the table, and again for each subsequent color. Each color has a different screen and must be printed in the proper sequence.

Each ink color is carefully mixed by hand.

The paper is checked and touched up by hand if needed. After final drying, it is rolled.

Digital Reproductions

Digital printing is a new option for reproducing historic wallpapers that can provide custom work at a lower cost than the traditional methods allow. Computer design programs are used to recreate and color correct historic patterns from photographs of old wallpaper—or even from wallpaper fragments—which can then be digitally printed onto rolls of paper. The process is even easier for a pattern that survives in a partial roll of wallpaper, as the digital image will need few if any corrections. To overcome the limitations of digital printing—such as the inability to print metallic colors—some innovative companies use screen printing or other traditional techniques on top of the digitally printed paper to complete the process.

The following images are courtesy of Evergreen Architectural Arts.

Fragments of historic wallpapers were all that survived in the Clara Barton House in Washington, DC, like the strip that was preserved behind a baseboard in this hallway (now behind Plexiglas).

In the foreground, an artist at Evergreen Architectural Arts is working on a wallpaper design on her computer while a technician is printing on the digital printers in the background. With digital printing, all colors are printed at once and do not require drying time unless additional layers need to be added for metallic elements or flocking.

From the surviving fragment, the entire design was recreated digitally and then printed to restore the walls of the hallway to their post–Civil War appearance.

Artists work on a digitally printed base pattern to recreate a raised metallic design.

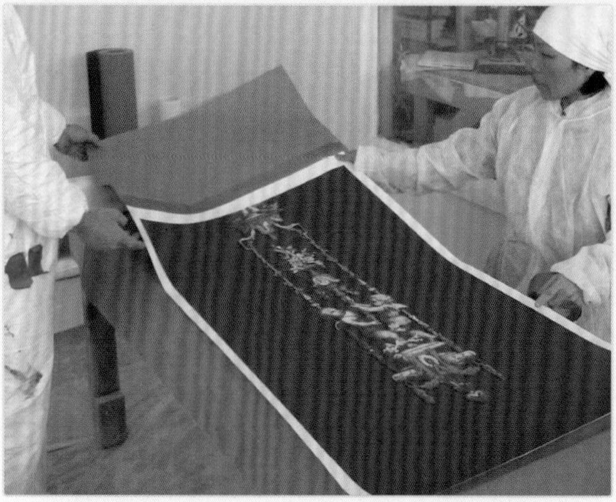

This is the base print from the digital printer for a panel that is to have the background color flocked. Flocking is a technique for attaching finely chopped felt to parts of the paper to create the effect of velvet that was very popular for a time in the nineteenth century.

This screen-printed stencil will mask the printed elements of the design while allowing adhesive to be squeegeed onto the background.

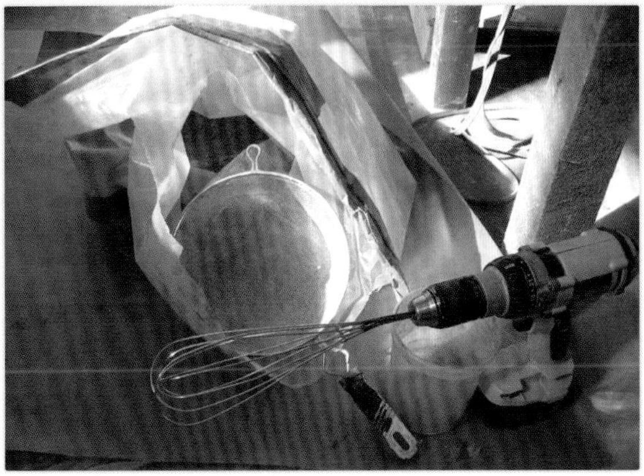

Tools for flocking include a bag of flock, a strainer to sift the flock, and a beater in a hand drill to vibrate the paper, which causes the flock to bounce around and settle lightly into the adhesive.

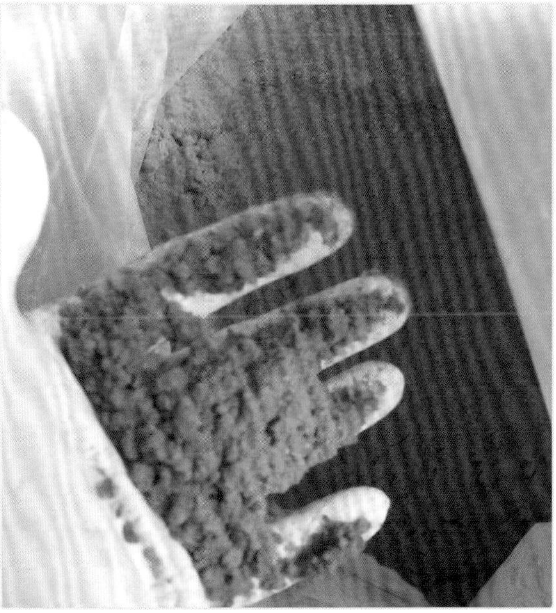

A close-up view of the flock.

The color-tinted adhesive.

Applying the adhesive through the screen with a squeegee before lifting off the screen to begin the flocking.

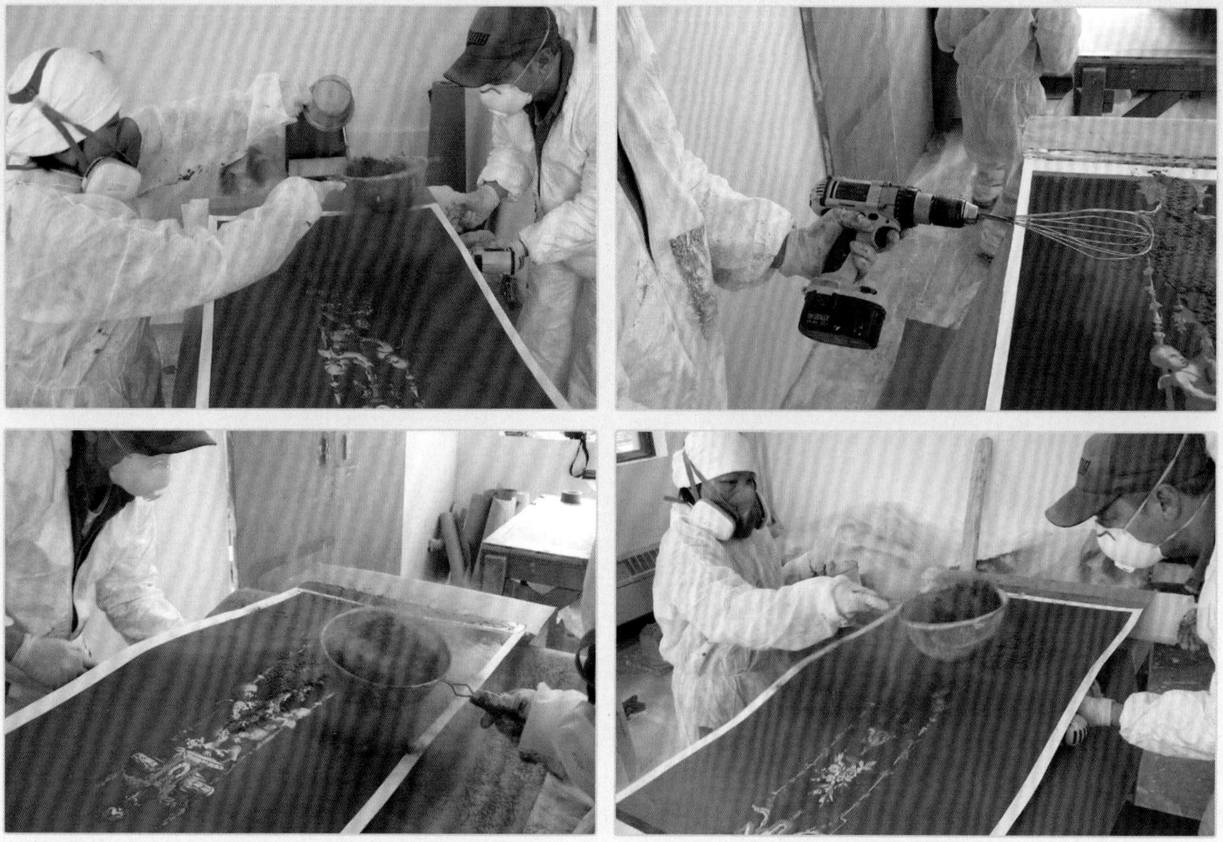

The flocking.

Machine-Printed Reproductions

Machine-printed wallpapers were developed in the mid nineteenth century using printers with multiple rollers to print different colors nearly at the same time. These metal rollers had thin metal outlines filled with an absorbent felt for the areas of color. Ink was absorbed by the felt and transferred to the paper as it passed over the roller. The expense of making these printing rollers, one for each color in the design, means that a lot of wallpaper must be sold to turn a profit. Some historic wallpaper patterns are popular enough today to be produced by machine printing.

Exercise caution when buying from "historic" collections of the major wallpaper manufacturers. Such patterns are based on historic designs but not necessarily on wallpaper designs. Instead, they might be taken from needlework, fabric prints, carpet designs, or even bed quilts. Sometimes, historic wallpaper patterns are reproduced in colors that were not used during the period. If your goal is to hang wallpapers that are accurate to your home's period of significance, spend some time researching what patterns were available then. There are several good books on historic wallpapers listed in Chapter 20, along with several excellent online collections of historic papers.

PART SIX

Tools and Resources

Measure twice, cut once.

To use a tool is to engage in an act of creation or destruction. Restoring a house will require both, and you'll need tools appropriate to the job at hand. The tools you use will affect the results you achieve. As Lord Byron wrote in *Don Juan*, "Good workmen never quarrel with their tools."

And good tools do their best work when you have good information to work with and good materials to work on. That's why the last chapter of this book is a compilation of sources that restorers of historic houses depend on.

So yes, measure twice before you cut once. And to take the adage a little further, plan three times before you measure twice.

HARTVILLE TOOL
Pure Air
Dust Collection System
16" AUTO

CHAPTER NINETEEN

Tools

Rehab projects require tools, and good rehab work requires the right tools. Anyone beginning a big project without a collection of the necessary tools will have to acquire them—a significant expense. Buy the best you can afford, especially those that will be used a lot. Used and reconditioned tools can often be found at much-less-than-new prices and represent a good value if they are in good condition. Used tools are sold by an individual, a pawnshop, or other dealer, usually without warranty. Reconditioned tools are used tools that have been returned to the manufacturer, inspected, repaired as necessary, and resold, generally with some sort of warranty attached. There are multiple online sources for used tools, including auction sites, retailers, and classified ad sites. Be aware that many stolen tools are sold via classified ad sites; a seller who seems sketchy may well be exactly that.

Power versus Hand Tools

Prior to the mid-nineteenth century, virtually every component of a house was made on the building site with hand tools. Through the nineteenth century, however, window sashes, doors, and other millwork components were increasingly fabricated off site using big, stationary tools powered first by water, then steam, then electricity, and shipped to the building site for installation.

The evolution away from hand tools and skilled hand crafting continued in the twentieth century as power tools became increasingly portable and common on building sites. The electric drill—usually credited with being the first portable power tool—was invented in 1895, and tools powered by small electric motors proliferated in the first decades of the twentieth century with the spread of home electrification. Table saws, radial arm saws, circular saws, and jigsaws were all available by the 1920s. Even so, hand tools remained in wide use until after World War II. The postwar housing boom, with its demand for speedy construction, accelerated the shift to power tools, and by the 1960s the whine of circular saws dominated building sites. In recent decades, pneumatic and battery-operated tools have chased corded tools from most contractors' toolboxes.

Today, many carpenters seldom touch a hand-powered tool. The reduction in time and physical effort required to accomplish a given task makes the use of power tools hard to resist for most applications.

Still, some historic-house restorers—both professional and do-it-yourself—choose to work primarily with hand tools. Repeating the physical movements of the carpenters who built a home can be deeply satisfying. The work is slower and quieter, and hand tools are sometimes the only way to accurately reproduce the surface character of missing or damaged elements in historic hand-built houses. The subtle ripples of a hand-planed floorboard cannot be reproduced with power tools.

Such considerations made me a slow adopter of power tools, attached as I was to the *idea* of replicating the actions of historic builders when working on old buildings. My attitude changed, however, when I realized that power tools can be far less impactful on the historic fabric of a house. Pounding nails into trim next to plaster causes significant vibration. Shooting a nail into the same trim with a pneumatic nailer causes almost no vibration. Multiplied by hundreds or thousands of impacts, the modern tool is far easier on a building. (When visible nail heads are important to the historic character of a surface, however, I use traditional—usually square-headed—nails. Pre-drilling the nail holes reduces the force needed to sink the nails and the consequent vibrations. It is not fast to pre-drill every hole, but it is faster than repairing damaged or detached plaster.)

Today I usually choose a hand tool when it will produce a noticeable and desired distinction in the appearance of a finished surface. Otherwise I'm likely to reach for a power tool. Reality being what it is, time constraints urge flexibility on this point. The bottom line is that most people restoring houses in the twenty-first century use power tools for most of the work.

Tool Needs

Essential tools for home renovation include pry bars, a reciprocating saw, a circular saw, a screw gun, a chop or miter saw, a table saw, hand saws, at least one hand plane, a router, a compressor, nail guns, at least one hammer, and jacks. Most professionals and experienced do-it-yourselfers have preferred tool brands, but their preferences vary.

In addition, specialized tools are needed for specialized renovation jobs such as cabinetry, electrical work, plumbing, tiling, drywall and plaster work, painting, and papering. These tools are referenced elsewhere in this book, where the jobs are discussed.

When you begin your rehab project, set up a tool room. You need a place to keep your tools organized and secure. If you have a space that will serve as a permanent shop, set it up from the start, installing decent lighting and whatever else is needed to work safely and efficiently. If you need to use a room that will ultimately be put to another use, choose one you can leave alone until your project is nearly completed. You do not want to move your tool room multiple times during a project if you can avoid it. The room does not have to be large enough to work in; its purpose is for tool storage and security.

Tool theft is common, especially for power tools. There can be thousands of dollars' worth of tools on site for a rehab project, and they are easily pawned or sold through classified ads, making them attractive to criminals who need cash quickly. You need a space that can be kept locked, and you need to lock it, especially if the house is unoccupied while the work is underway. It has become easy and reasonably inexpensive to install video cameras with motion detectors, and some systems can be set to notify your smart phone when the video is triggered, allowing you to see if the intruder is a thief or a curious opossum. Write your name on every tool in large black letters using permanent marker. It will deter some thieves and might make recovery easier if the tools end up in a pawn shop or for sale online. Consider putting small stick-on GPS trackers in discrete locations on expensive tools, and make sure your insurance will cover tool theft.

Try to keep your tool room neat, with a place for everything and everything in its place. Most of us working on our own houses push to accomplish as much as we can in the available time; we fail to leave adequate time for cleanup, and tools are simply dropped quickly in the tool room when we rush home for dinner. Eventually we find ourselves spending as much time finding a tool as using it. Putting large and heavy tools on wheels will save effort and increase the likelihood that you will put them away properly at the end of the day. Periodically reorganize your tool space as it gets cluttered. Tools that are used as intended and kept clean and sharp will serve your needs far better than tools that are misused or treated poorly.

At Whitten House, the shed and carriage house were torn down long ago, leaving only finished living spaces. My tool room is the original pantry, a 10-foot by 10-foot room opening onto the original kitchen and the sitting room. Its cabinetwork was stripped and replaced with bookshelves when the house served as a public library; I use the library shelving on three walls for tool storage and have fastened a hanging board for hand tools on the fourth wall. A repurposed rolling kitchen cart serves as a chop saw bench with tool storage underneath. For a long time I moved tools as needed into the adjoining original kitchen and worked there. Since restoring that space, I work in the backyard or in the room undergoing restoration. Eventually—probably about the time the house rehab is complete—I will have the missing carriage house rebuilt with a real shop space in it. It will be great for making and restoring furniture.

Ladders and Staging

Working on or in a house will require you to reach things you cannot reach from the floor. Sooner or later, probably sooner, you will need a ladder or staging of some sort. Six-foot stepladders are standard, but you'll need a taller ladder if you have 12-foot ceilings. I prefer fiberglass stepladders to wood or aluminum. Adaptable ladders that can be configured in different settings to stand solidly in a variety of circumstances (such as on a stairway) are now readily available and can be very useful.

If you have very high ceilings and will be doing a lot of work up there, it may be worthwhile to invest in a 4-foot by 6-foot rolling staging platform with safety railings. These units are not inexpensive but pay off in added safety and increased productivity. When you finish the rehab, you can sell the platform to another homeowner. For temporary interior staging—say, to hang wallpaper—a heavy (2-inch-thick) plank between two step ladders will often work. Exercise due care when working up high on any structure, whether factory- or home-made.

When plastering or papering a ceiling requires me to work the full width of a room, I have found it worthwhile to make a platform of ¾-inch plywood laid across sawhorses built to the appropriate height. You want to be able to touch the ceiling comfortably while standing on the platform. Plan on three horses for a 4-foot by 8-foot sheet of plywood, or horses placed every four feet across the room. A few coarse-thread drywall screws through the plywood into the tops of the horses will add stability, but don't overdo it; you'll have to take them all out when you move the platform for the next section of ceiling. If the room is just a bit too high for working from the floor, sheets of plywood or a solid-core flush door placed atop paint cans or five-gallon buckets might do the trick. Just make sure you support the platform in multiple places.

Exterior work often requires elaborate and expensive ladders and staging. I recommend modern aluminum or fiberglass extension ladders over old wooden extension ladders. Wooden ladders over 16 feet long are very heavy and difficult to maneuver; if stored outdoors, they may also harbor rot that is not obvious on the surface. Having a rung snap underfoot when you are 20 feet off the ground can make for an exciting day, to say the least.

Spend a few dollars for ladder top pads to protect your siding. Someday that siding will be a beautiful painted surface that you don't want to scratch.

Ladder jacks, which allow level planks to run between extension ladders, are the least expensive option for creating a narrow working platform well above the ground. Heavy (2-inch-thick) wood planks are increasingly yielding to much lighter aluminum planks these days. On such a narrow platform, fall protection is critical. Working alone, it is nearly impossible to set up ladder jacks more than about 10 feet off the ground.

Modular metal staging provides a wider and more solid working platform but is expensive to rent for any length of time. If your exterior walls need extensive work requiring weeks or months of labor, look into buying used staging. Make sure all staging is solidly footed and level. At heights above 8 feet, the staging should be firmly enough attached to the building to remain stable and be unable to tip over.

Mechanical lifts—scissor lifts or boom lifts—are another option for reaching high exterior sections. A scissors lift has to be maneuvered close to and parallel with the surface you are working on and must sit on level ground. A boom lift can be based farther from the building and will adapt to a sloped site with careful base placement. Both types can be rented short or long term. The owner of the Hench House, featured elsewhere in this book, rented a boom lift for several years to complete the exterior restoration of his house. The elaborate Queen Anne–style forms of the building—with multiple dormers, bay windows, and balconies—made the boom lift the only practical option. Follow all safety recommendations when working from a lift or staging. This generally means being tied in if you are more than a few feet above the ground. It may be a pain to do this, but falling will hurt a lot more.

Safety

I have touched on safety repeatedly—above and in earlier chapters—because it cannot be over-emphasized when working on a house. Whether it means wearing hearing protection, providing fall protection, or simply noting what direction a right-angle drill will swing if it binds up and keeping your head on the opposite side, stay safe. Safety was not a priority in the building trades when I grew up, but concussions, cuts, bruises, and hearing loss have taught me that safety matters.

Buy a hardhat for everyone who will be on the jobsite regularly, and keep an extra one or two for visitors. They should be worn whenever work is being done overhead or in areas with low clearances. The same is true for safety glasses or goggles. Buy them and use them, all the time.

Respirators are essential. Demolition work in an old house fills the air with fine particles of plaster and who knows what else. You do not want it in your lungs. Respirators are equally important when working with materials that produce toxic fumes. Most respirators will accommodate different filter heads for different hazards. Make sure you are using the right one for the task you are doing.

A Tyvek suit is useful and inexpensive. If will keep your clothes clean if you are working in a crawlspace, sanding joint compound, or painting; if you are removing lead paint or working with any other hazardous material, a Tyvek suit will keep your clothing toxin-free.

Read the instructions that come with power tools. Really.

Keep chisel blades sharp with regular sharpening and honing. A sharp blade will go where you intend it to go, a dull one may not. Most chisels need to be sharpened when you buy them. Sharpening instruction videos are readily available online. A few drops of machine oil or mineral oil rubbed onto a blade will prevent rust from humidity while it is idle.

Worksites are dangerous places. To stay safe, be careful.

Don't Do This!

There are many ways to disrespect a historic house. These half-dozen photos show just a few.

This ungainly dormer addition is out of scale with the house and detracts from its historic character. Previously the door surround (a transitional Federal-to-Greek Revival style) was the most prominent feature, as intended by the early-nineteenth-century designer; now the dormer dominates the façade. A shed dormer on the opposite plane of the roof would have added as much interior space with little impact to exterior character.

This Craftsman-style bungalow looks in danger of being swallowed by its outsized addition in an incompatible contemporary style. The tall, narrow windows in the new gable, which show off the interior chimney, do not help.

These two houses started life looking nearly the same. The one at right has lost its corner boards behind vinyl siding and has had the slate on the front plane of the mansard roof replaced with a less-than-ideal roofing material—but at least it is roofing material! Its companion on the left now has vinyl siding on its mansard roof. Siding belongs on walls, not roofs! I won't even get started on the windows . . . .

Whoever built this portico seems to have tried to relate it to the Federal-style house but lacked sufficient knowledge of the rules of Classical design to pull it off.

This contemporary door hood design was apparently intended to respect the elliptical arch of the Federal door surround, but it doesn't. The ramp and railing could have been handled in a much more compatible way as well.

CHAPTER TWENTY

Resources

For your convenience, here is a compilation of information sources, appropriate materials, and many other things you may need for your project.

Preservation Briefs

The Technical Preservation Services department of the National Park Service has published forty-eight *Preservation Briefs* on multiple aspects of historic preservation work. These documents are well researched and provide a great deal of useful information. They can be found at www.nps.gov/tps/how-to-preserve/briefs.htm.

The Briefs are as follows:

1. **Cleaning and Water-Repellent Treatments for Historic Masonry Buildings**
2. **Repointing Mortar Joints in Historic Masonry Buildings**
3. **Improving Energy Efficiency in Historic Buildings**
4. **Roofing for Historic Buildings**
5. The Preservation of Historic **Adobe Buildings**
6. **Dangers of Abrasive Cleaning to Historic Buildings**
7. The Preservation of Historic Glazed Architectural **Terra-Cotta**
8. **Aluminum and Vinyl Siding on Historic Buildings: The Appropriateness of Substitute Materials for Resurfacing Historic Wood Frame Buildings**
9. The Repair of Historic **Wooden Windows**
10. Exterior **Paint Problems** on Historic Woodwork
11. Rehabilitating Historic **Storefronts**
12. The Preservation of Historic Pigmented **Structural Glass** (Vitrolite and Carrara Glass)
13. The Repair and Thermal Upgrading of Historic **Steel Windows**
14. New **Exterior Additions** to Historic Buildings: Preservation Concerns
15. Preservation of Historic **Concrete**
16. The Use of **Substitute Materials** on Historic Building Exteriors
17. **Architectural Character—Identifying the Visual Aspects of Historic Buildings as an Aid to Preserving their Character**
18. Rehabilitating **Interiors** in Historic Buildings—Identifying Character-Defining Elements

19. The Repair and Replacement of Historic **Wooden Shingle Roofs**

20. The Preservation of Historic **Barns**

21. Repairing Historic **Flat Plaster**—Walls and Ceilings

22. The Preservation and Repair of Historic **Stucco**

23. Preserving Historic **Ornamental Plaster**

24. **Heating, Ventilating, and Cooling Historic Buildings: Problems and Recommended Approaches**

25. The Preservation of Historic **Signs**

26. The Preservation and Repair of Historic **Log Buildings**

27. The Maintenance and Repair of Architectural **Cast Iron**

28. **Painting Historic Interiors**

29. The Repair, Replacement, and Maintenance of Historic **Slate Roofs**

30. The Preservation and Repair of Historic **Clay Tile Roofs**

31. **Mothballing Historic Buildings**

32. Making Historic Properties **Accessible**

33. The Preservation and Repair of Historic **Stained and Leaded Glass**

34. Applied Decoration for Historic Interiors: Preserving Historic **Composition Ornament**

35. Understanding Old Buildings: The Process of **Architectural Investigation**

36. Protecting **Cultural Landscapes**: Planning, Treatment, and Management of Historic Landscapes

37. Appropriate Methods of Reducing **Lead-Paint Hazards** in Historic Housing

38. **Removing Graffiti from Historic Masonry**

39. Holding the Line: **Controlling Unwanted Moisture** in Historic Buildings

40. Preserving Historic **Ceramic Tile Floors**

41. The **Seismic Retrofit** of Historic Buildings

42. The Maintenance, Repair, and Replacement of Historic **Cast Stone**

43. The Preparation and Use of **Historic Structure Reports**

44. The Use of **Awnings** on Historic Buildings: Repair, Replacement, and New Design

45. Preserving Historic **Wooden Porches**

46. The Preservation and Reuse of Historic **Gas Stations**

47. **Maintaining the Exterior of Small and Medium Size Historic Buildings**

48. **Preserving Grave Markers in Historic Cemeteries**

Appropriate and Hard-to-Find Materials and Products

This list primarily consists of companies that serve the entire nation. You may find local or regional companies that offer similar products. Where possible, I have included companies in each category in multiple regions, as shipping costs may be lower when buying from companies closer to your location—particularly for heavy building materials. I have personal experience with the products and companies marked with an asterisk (*) and can recommend them based on my experience. Other products and companies are listed based on their long-established reputations in the preservation community or at the recommendation of other preservation professionals.

Exterior

Radial-sawn Clapboards

Bear Creek Lumber: Winthrop, Washington. Producer of vertical-grain red cedar clapboards since 1977; www.bearcreeklumber.com.

Katahdin Clapboard Company: Maine-based producer of durable vertical-grain radially sawn clapboards; www.katahdinclapboardcompany.com.

Ward Clapboard Mill: Vermont producer of durable, vertical-grain radially sawn clapboards made on vintage machinery for over 100 years; www.wardclapboard.com.

Glass

Curved

B&L Antiqurie, Inc.: Michigan-based supplier of curved glass including antique glass, serpentine bends, J-bends, irregular bends, one-of-a kind bends, varied tints and thicknesses, etc. Ships nationwide; www.bentglasscentral.com.

Standard Bent Glass Corp.: Since 1936, Pennsylvania-based company supplying custom flat- and bent-glass products including flat or curved laminated glass, curved tempered glass, compound-curved glass, decorative laminated glass, and bullet-resistant glazing; www.standardbent.com.

Wavy

Bendheim Restoration Glass: Established in 1927 in Wayne, New Jersey. Reproduces the cylinder antique window glass of the seventeenth through early-twentieth centuries. The architect and designer's first choice for replacement of broken or missing antique window glass, as well as the detailing of reproduction homes and furniture; www.restorationglass.com.

Hollander Historic Restoration Window Glass: Established in 1956, with five locations across the United States. Reproduces period glass using the same manufacturing methods used in America from the seventeenth to mid-twentieth centuries; www.restorationwindowglass.com.

Pioneer Glass: Massachusetts based and family owned for more than 50 years. Offers a wide range of restoration glass types for different periods; www.pioneer.glass.

Stained

Hollander Historic Restoration Window Glass: Established in 1956, with five locations across the United States. Antique glass, drawn glass, rolled glass, opal and cathedral glass, restoration window glass of all types, decorative patterned glass, warm glass, accessories, supplies, etc.; www.restorationwindowglass.com.

Glazing Putty

***Sarco Putty Dual Glaze:** Family owned since 1943. Elastic glazing compound for sealing a watertight bond between glass and wood, glass and metal sash, or glass and doors. Made of non-drying oils that set into a firm, tough cushion that adheres tightly to glass and sash; www.sarcoputty.com.

Mortar

*__Edison Coatings:__ Connecticut-based manufacturer of lime and natural cement products including lime putty and natural cement mortars; www.edisoncoatings.com.

Freedom Cement: Massachusetts-based producer of natural cement mortars for a variety of uses, including casting; www.freedomcement.com.

Lancaster Lime Works: Pennsylvania-based supplier of lime putty, plaster, mortar, and related tools and materials; www.lancasterlimeworks.com.

Lime Works US: Pennsylvania-based importer of French lime plaster, mortar, and related tools and materials; www.limeworks.us.

U.S. Heritage Group: Illinois-based producer of quality bedding and pointing mortar in several classes for different applications. Can be color- or custom-matched based on mortar analysis;
www.usheritage.com/repointing-mortars.

Masonry Repair

Abatron: Founded in Wisconsin in 1959, a leading supplier of flexible epoxy concrete repair products for historic preservation; www.abatron.com.

*__Jahn International:__ Dutch company, producer of a leading line of masonry repair and restoration products used worldwide, available from Jahn trained and certified masons nationwide in America;
www.jahn-international.com.

Tile Roofing

Ludowici Tile: Ohio based since 1888, produces a wide range of roofing tile styles, shapes, and colors. Custom work available to match historic tiles;
www.ludowici.com.

Patterned Steel Roofing

F.W. Norman Co.: Extensive line of pressed metal ceilings, cornices, siding, shingles, and more, still produced with early-twentieth-century molds and equipment. This company is a national treasure; www.wfnorman.com

Shutters

Beech River Mill: New Hampshire-based manufacturer since 1851, specializes in the design, manufacture, and installation of custom wooden blinds, shutters, and doors. All products made to order and finished using original Victorian-era machinery along with twenty-first-century technologies; www.beechriver.com.

Colonial Shutterworks: Massachusetts-based manufacturer of custom wood-louvered and paneled shutters; www.colonialshutterworks.com.

New England Shutter Mills: Massachusetts-based manufacturer of wood exterior and interior shutters since 1999; www.newenglandshutter.com.

Kirtz Shutters: Oklahoma based since 1987, manufacturer of hardwood and softwood plantation shutters; www.kirtz.com.

Storm Windows

Adams Architectural Millwork Co.: Founded in 1981, Iowa-based family-owned company supplying custom millwork including wood storm windows nationwide; www.adamsarch.com.

*__Allied Window:__ Ohio based since 1950, offering a wide range of storm windows that "disappear" in the opening, whether mounted inside or outside, and have been used on many of the nation's most important landmarks; www.alliedwindow.com.

Indow: Oregon-based company founded in 2010, supplying interior storm windows across the United States and Canada; www.indowwindows.com.

Steel Windows

Crittall Windows: United Kingdom based, available in the United States and Canada. New steel windows from "the original steel window manufacturer," founded 1849; www.crittall-windows.com.

Optimum Window: New York based since 1985, produces new custom steel windows including fire-rated units; www.optimimwindow.com.

Seekircher Steel Window Corp.: New York based since 1977, restores and refurbishes historic steel windows on site or at their shop. Large selection of restored steel windows available for sale; www.seekirchersteelwindow.com.

*__Universal Window:__ Massachusetts based for nearly 50 years, produces modern custom insulated aluminum-framed windows, including Steel Replica windows that mimic the appearance of historic steel windows; www.universalwindow.com.

Wood Windows

Adams Architectural Millwork Co.: Founded in 1981, Iowa-based family-owned company supplying custom millwork including wood windows nationwide; www.adamsarch.com.

Window Restoration

*__The Steam Stripper:__ Developed and sold by Bagala Window Works in Maine, holds sashes up to 44 inches by 72 inches for fast and complete paint and glazing removal; www.bagalawindowworks.com/steam-stripper.

Window Hardware

*__Pullman Manufacturing Company:__ Since 1886, New York-based manufacturer of spring window balances in a range of sizes for different window weights; www.pullmanmfg.com.

*__SRS Hardware:__ Produced initially for their own use at Smith Restoration Sash, founded in 1985, manufacturer of high-quality forged brass window locks, ball bearing sash pulleys, stackable cast iron and lead sash weights, stop bead adjusters, spring bolts, storm sash hardware, sash lifts, and more available to the trade and public; www.srshardware.com.

Interior

Online General Sources of Hardware, Fixtures, Etc.

American Historic Hardware: Source for unused, original stock builders' hardware for doors, windows, and furniture with stock from circa 1860 to 1940; www.ahhardware.com.

Historic House Parts: Extensive online catalog of architectural salvage and restored hardware, lighting, plumbing, etc.; www.historichouseparts.com.

Kennedy Hardware: Zionsville, Indiana. Large line of reproduction hardware including Hoosier Cabinet hardware; www.kennedyhardware.com.

*__Lehman's Hardware:__ Longtime mail-order source of traditional hardware and housewares. Product lines have expanded into other areas, but still a good source for oil lighting lamps, parts, and accessories and traditional kitchen items; www.lehmans.com.

*Signature Hardware: Wide range of plumbing, hardware, etc., much of it appropriate for historic homes; www.signaturehardware.com.

*Van Dyke's Restorers: Large selection including hardware, kitchen and bath fixtures, carved wood trim elements, wire grilles and panels, and lighting; www.vandykes.com.

Reproduction Hardware

Ball & Ball: Exton, Pennsylvania. High-quality true reproductions of antique hardware using eighteenth-century craftsmanship; www.ballandball.com.

E.R. Butler & Co.: Manufacturer and supplier of custom architectural, builders', and cabinetmakers' Early American, Federal, and Georgian period hardware; www.erbutler.com.

House of Antique Hardware: Producer of high-quality reproduction hardware with one of the largest selections of original and authentic reproduction hardware on the web; www.houseofantiquehardware.com.

*Horton Brasses: Founded in 1936 to reproduce antique furniture hardware, the company still uses actual antique hardware for dies and patterns and employs many of the same painstaking methods used by historic craftsman. Produces more than 1,000 items of architectural and furniture hardware; www.horton-brasses.com.

Nostalgic Warehouse: Founded in 1980, specializes in reproduction door hardware; www.nostalgicwarehouse.com.

Phelps Company: New Hampshire-based designer and manufacturer of traditional window hardware including sash pulleys, chain, weights, locks, and lifts; casement window hardware; and transom hardware; www.phelpscompany.com.

*SRS Hardware: Produced initially for their own use at Smith Restoration Sash, founded in 1985, manufacturer of high-quality forged brass window locks, ball bearing sash pulleys, stackable cast iron and lead sash weights, stop bead adjusters, spring bolts, storm sash hardware, sash lifts, and more available to the trade and public; www.srshardware.com.

Vintage Hardware and Lighting: Port Townsend, Washington. Reproduction Victorian and early-twentieth-century hardware and lighting since 1978; www.vintagehardware.com.

Lime Plaster

American Clay: Natural earth clay plaster produced in New Mexico, nontoxic and available in 224 colors for paint-free plaster surfaces; www.americanclay.com.

*Edison Coatings: Connecticut-based manufacturer of lime and natural cement products including lime putty and plaster; www.edisoncoatings.com.

Lancaster Lime Works: Pennsylvania-based supplier of lime putty, plaster, mortar, and related tools and materials; www.lancasterlimeworks.com.

Lime Works US: Pennsylvania-based importer of French lime plaster, mortar, and related tools and materials; www.limeworks.us.

US Heritage Group: Producer of bagged dry lime plaster with aggregate, and supplier of fiber for plaster; www.usheritage.com/plaster-stucco.

Renaissance Lime Putty: Massachusetts-based importer of Italian lime putty, fiber, and plaster-working tools; www.limeputty.us.

Cast Plaster

Boston Ornament Company, Inc.: Massachusetts-based manufacturer of ornamental plaster medallions, moldings, brackets, and custom plaster work, shipping nationwide; www.bostonornament.com.

J.P. Weaver Company: Since 1914, California-based manufacturer of extensive lines of plaster, composition, and Petitsin ornament for architectural applications; www.jpweaver.com.

Ornamental Plasterworks: Virginia-based manufacturer of an extensive line of moldings, door pediments, niches, pilasters, mantels, and a wide variety of custom items; www.ornamentalplasterworks.com.

Plaster Repair

*__Plaster Magic:__ Plaster reattachment and repair system developed by a building conservator and offered in do-it-yourself kits with step-by-step instructions; www.plastermagic.com.

Composition Ornament

Bomar Designs: Since 1984, Kansas-based producer of an extensive line of traditional composition ornament for buildings and furniture; www.bomardesigns.com.

Decorators Supply Corp.: Established 1883, this Chicago firm produces composition and wood fiber ornament from molds carved in the nineteenth century. Also plaster and carved wood ornament. Extensive selection; www.decoratorssupply.com.

J.P. Weaver Company: Since 1914, California-based manufacturer of extensive lines of composition, plaster, and Petitsin ornament for architectural applications; www.jpweaver.com.

Wood Repair

*__Abatron:__ Founded in Wisconsin in 1959, a leading supplier of flexible epoxy wood repair products for historic preservation; www.abatron.com.

Historic Plumbing Fixtures and Repair

Old School Plumber: Available online and from New England showroom; www.oldplumbfixer.com/home.html.

Plumbing Geek: Online and mail-order source of antique fixtures, information, and repairs; www.plumbing-geek.com.

Vintage Bathroom: Online source for antique bath and kitchen fixtures; www.vintagebathroom.com.

Reproduction Plumbing Fixtures

*__American Standard:__ Since 1929, a leading supplier of bath and kitchen fixtures and accessories available through stores and dealers nationwide; www.americanstandard-us.com.

*__Kohler:__ Since 1873, a leading supplier of bath and kitchen fixtures and accessories available through dealers and stores nationwide; www.us.kohler.com/us.

Sunrise Specialty: Accurately reproduced cast iron tubs, fixtures, and other bath products available through dealers; www.sunrisespecialty.com.

Textiles for Floors

Design Village Floorcloths: New Jersey-based company producing reproduction painted canvas floor cloths. Designs are based on historic floor cloths *and* quilts and wall stencil designs. Documented floor cloth designs; www.floorcloth.net.

Dunberry Hill Designs: Traditionally crafted floor cloths made in Vermont; www.dunberryhilldesigns.com.

Family Heirloom Weavers: Since 1983, Pennsylvania-based family-owned company specializing in reproduction ingrain carpet with designs dating from the late eighteenth to early twentieth centuries; www.familyheirloomweavers.com.

*__J. R. Burrowes & Company:__ Founded in 1985, this Boston-area company reproduces nineteenth-century wallpapers and imports reproduction lace curtains and carpeting from the United Kingdom; www.burrows.com.

Thistle Hill Weavers: Cherry Valley, New York. Specializes in accurate historic reproduction textiles and carpets, working from surviving examples, documented patterns, and period weavers' drafts covering a wide range of periods; www.thistlehillweavers.com.

The English Wilton Company: Located in Kidderminster, England, with a US office in California, reproduction of historic carpeting on historic looms; www.historic-carpet.com

The Weavery: Virginia-based company specializing in hand-woven textiles, including popular nineteenth-century carpet types such as striped Venetian carpets; www.theweavery.com.

Ceramic Tile

American Restoration Tile: Arkansas-based, facilitates the restoration of historically significant buildings and residences. Uses modern manufacturing technology and decades of ceramic engineering experience to exactly reproduce the sizes and colors of old ceramic tile installations including patterns, borders, corners, and medallions; www.restorationtile.com.

*__Heritage Tile:__ Illinois-based family of companies producing traditional ceramic tiles in several collections including subway tile, Subway Mosaics, and Batchelder Tile; www.heritagetile.com.

*__Subway Ceramics:__ Illinois-based, manufacturing subway ceramics to the original tile specifications and standards of the 1920s. Part of the Heritage Tile family of companies; www.subwaytile.com.

Wallpaper

Vintage

Hannah's Treasures: Founded in 1990, an online retailer for unused vintage wallpapers dating from 1900 to the 1970s, including reproductions of eighteenth- and nineteenth-century papers; hannahstreasures.com

Reproduction

*__Adelphi Paper Hangings:__ Since 1999, the leading producer of fine hand-block-printed reproduction wallpapers from documented sources, 1750–1930; www.adelphipaperhangings.com.

*__Bradbury & Bradbury Art Wallpapers:__ Founded in San Francisco in 1979, produces an extensive line of reproduction nineteenth- and twentieth-century wall and ceiling papers; www.bradbury.com.

*__Evergreene Architectural Arts:__ Founded in 1979, New York City-based decorative arts firm that has pioneered reproduction of historic wallpapers using digital printing combined with more traditional methods; www.evergreene.com.

*J. R. Burrowes & Company: Founded in 1985, this Boston-area company reproduces nineteenth-century wallpapers and imports reproduction lace curtains and carpeting from the United Kingdom; www.burrows.com.

Mason and Wolfe: New Jersey-based, reproducing period wallpapers from the Arts & Crafts movement and the late Victorian era. Collections are developed using historic design sources and rare survivors of vintage wallpaper. Hand-screened in authentic colors; www.mason-wolf.com.

Inside and Out

Cut Nails

*Tremont Nail Company:** Founded 1819, Massachusetts-based manufacturer of a wide range of historic nail types and sizes in cut steel; www.tremontnail.com/tremont-index.htm.

Paint

*Benjamin Moore:** Founded in Brooklyn, New York, in 1883, manufactures an extensive line of paints and coatings with stores and dealers nationwide. Offers lines of historic colors; www.benjaminmoore.com.

*Fine Paints of Europe:** Founded in 1987 in Vermont to import high-quality Dutch paints to America. Carries an extensive line of oil- and waterborne paints of exceptional quality. Sold through dealers and mail order; www.finepaintsofeurope.com.

*Sherwin Williams:** Founded in Cleveland, Ohio, in 1866, manufactures paints and other coatings, with stores nationwide. Offers lines of historic colors; www.sherwin-williams.com.

Reproduction Lighting

Arroyo Craftsman: California-based manufacturer of high-quality Arts & Crafts–inspired interior and exterior lighting; www.arroyo-craftsman.com.

Authentic Designs: Vermont based through three generations. Custom, handmade lighting based on eighteenth- and nineteenth-century designs; www.authenticdesigns.com.

Ball & Ball Lighting: Pennsylvania based, founded in 1932. Manufactures a broad product line that includes furniture and door hardware, and began reproducing eighteenth-century and turn-of-the-century American Revival period lighting in the 1950s; www.ballandball-us.com.

Faubourg Lighting: Alabama based, founded in 1978, manufactures handmade gas and electric lighting in copper including wall sconces, porch lighting, and lanterns influenced by traditional New Orleans gas fixtures. Styles include Colonial, Federal, and Italianate; www.charlestonlighting.com/faubourg-lighting-designs.

King Chandelier Company: Since 1935, North Carolina-based manufacturer of high-quality Victorian lighting, including gas light fixtures; www.chandelier.com.

*Lehman's Hardware:** Longtime mail-order source of traditional hardware and housewares. Product lines have expanded into other areas, but still a good source for oil lighting lamps, parts, and accessories; www.lehmans.com.

Old California Lantern Company: California based since 1989, manufactures both interior and exterior Arts & Crafts-period lighting fixtures; www.oldcalifornia.com.

*__Rejuvenation:__ Founded in Portland, Oregon in 1977, reproducing light fixtures, now part of the Pottery Barn/ Williams Sonoma companies; www.rejuvenation.com.

Revival Lighting: Spokane, Washington. A supplier of antique and reproduction lighting from the nineteenth and early twentieth centuries; www.revivallighting.com.

Rinaudo's Reproductions, Inc.: California-based. Hand-crafts Victorian reproductions ranging in design from Gothic European to American Arts & Crafts, and restores or duplicates antique fixtures; www.victorianreproductionlighting.com.

Vintage Hardware and Lighting: Port Townsend, Washington. Reproduces Victorian and early-twentieth-century hardware and lighting since 1978; www.vintagehardware.com.

The Federalist: Connecticut-based. High-end reproductions of late-eighteenth- and early-nineteenth-century fixtures adapted for electric bulbs; thefederalistonline.com.

The Gas Light Company: Baton Rouge, Louisiana. Hand-crafts reproduction exterior gas light (and electric and candle) lanterns and fixtures; www.thegaslightcompany.com.

Ornamental Metal

Stamped Sheet Metal

American Tin Ceilings: Produces a line of traditional designs for tin ceilings; www.americantinceilings.com.

Brian Greer's Tin Ceilings: Produces a line of traditional designs for tin ceilings; htwww.tinceiling.com.

Chelsea Decorative Metal Co.: Extensive line of tin ceiling panels and cornices made in Texas; www.thetinman.com.

*__F.W. Norman Co.:__ Extensive line of pressed metal ceilings, cornices, siding, shingles, and more, still produced with early-twentieth-century molds and equipment. This company is a national treasure; www.wfnorman.com.

M-Boss, Inc.: Extensive line of tin ceiling panels and cornices made in Ohio; www.mbossinc.com.

Cast Iron and Aluminum

King Architectural Metals: Extensive line of cast and wrought ornamental iron, cast aluminum, and other hand-forged architectural metals. Locations in the east, south, and west; kingmetals.com.

Wrought Iron and Steel

King Architectural Metals: Extensive line of cast and wrought ornamental iron, cast aluminum, and other hand-forged architectural metals. Locations in the east, south, and west; kingmetals.com.

Wiemann Metalcraft: Oklahoma based since 1940, a full-service custom ornamental and architectural metal fabricator using traditional metalworking methods and finishes combined with state-of-the-art CAD-based modeling, prototyping, and machining; www.wmcraft.com.

Millwork

Adams Architectural Millwork Co.: Founded in 1981, Iowa-based family-owned company supplying custom millwork nationwide; www.adamsarch.com.

Kuiken Brothers Company, Inc.: New York/New Jersey-based since 1912, reproducing Classical moldings for Federal and Georgian buildings; www.kuikenbrothers.com.

Mad River Woodworks: California-based since 1981, reproducing Victorian gingerbread trim and moldings; www.madriverwoodworks.com.

The Wood Factory: Maine-based, reproducing Victorian millwork; www.thewoodfactorymillwork.com.

Vintage Woodwork: Texas based since 1978, reproducing a wide range of old-fashioned, solid wood, handcrafted millwork; www.vintage woodwork.com.

Columns

*Chatsworth: North Carolina based, founded in 1987, produces columns representing all five orders of classical architecture made of wood, fiberglass, and polyester resin along with other architectural elements; www.columns.com.

Periodicals

Old House Journal
Period Homes
Traditional Building
Fine Homebuilding

Historic Preservation Websites

www.traditionalbuilding.com: from *Traditional Building* magazine.

www.oldhouseonline.com: from *Old House Journal.*

www.nps.gov/tps: Technical Preservation Services at the National Park Service.

retrorenovation.com: Deemed by *The New York Times* as "the go-to destination for enthusiasts who want to restore houses built during the post-World War II boom."

www.YourHistoricHouse.com: The companion site to this book, with additional and updated information, video content, and more.

Books

Architectural Styles and History

Beecher, Catherine E., and Harriet Beecher Stowe. *The American Woman's Home* (1869), reprint. New Brunswick, NJ: Rutgers University Press, 2004.

Bibber, Joyce K. *A Home for Everymen: The Greek Revival and Maine Domestic Architecture.* Portland, ME: Greater Portland Landmarks, 2000.

Booker, Margaret Moore, and Steve Larese. *The Santa Fe House: Historic Residences, Enchanting Adobes, and Romantic Revivals.* New York: Rizzoli, 2009.

Cheek, Richard. *Selling the Dwelling: The Books That Built America's Houses, 1775–2000.* New York: The Grolier Club, 2013.

Downing, A. J. *Architecture of Country Houses: Including Designs for Cottages, Farm-Houses, and Villas, with Remarks on Interiors, Furniture, and the Best Modes of Warming and Ventilating.* Forgotten Books, 1851 (facsimile editions available online).

Friedman, Donald. *Historical Building Construction, Design, Materials, and Technology*, Second Edition. New York: W.W. Norton and Company, 2010.

Garrett, Wendell. *Victorian America: Classical Romanticism to Gilded Opulence.* New York: Rizzoli, 1993.

Garvin, James L. *A Building History of Northern New England.* Hanover, NH: University Press of New England, 2001.

Guter, Robert P., and Janet W. Foster. *Building by the Book: Pattern Book Architecture in New Jersey*. New Brunswick, NJ: Rutgers University Press, 1992.

Ierley, Merritt. *Open House: A Guided Tour of the American Home – 1637–Present*. New York: Henry Holt and Company, 1999.

Mayhew, Edgar De N., and Minor Myers. *A Documentary History of American Interiors: From the Colonial Era to 1915*. New York: Scribner, 1986.

McAlester, Virginia (illustrated by Suzanne Patton Matty and Steve Clicque). *A Field Guide to American Houses: The Definitive Guide to Identifying and Understanding America's Domestic Architecture*. New York: Alfred A. Knopf, 2017.

Pugin, Augustus. *Pugin's Gothic Ornament: The Classic Sourcebook of Decorative Motifs*. New York: Dover Publications, 1987.

Reiff, Daniel D. *Houses from Books: Treatises, Pattern Books, and Catalogs in American Architecture, 1738–1950: A History and Guide*. University Park, PA: Pennsylvania State University Press, 2001.

Sobon, Jack. *Historic American Timber Joinery: A Graphic Guide*. Becket, MA: Timber Framers Guild, 2014.

Ware, William R. *The American Vignola: A Guide to the Making of Classical Architecture* (1903). New York: Dover Publications, 1994.

Woodward, George E., and Edward G. Thompson. *A Victorian Housebuilders Guide: "Woodwards National Architect" of 1869*. New York: Dover, 1988.

Carpentry

Bealer, Alex W. *Old Ways of Working Wood: The Techniques & Tools of a Time-Honored Craft, Revised Edition*. Edison, NJ: Castle Books, 1980.

Ellie, George. *Modern Practical Joinery* (1902). Fresno, CA: Reprint Linden Publishing, 1987.

Hodgson, Fred T. *Modern Carpentry: A Practical Manual* (1917). Amsterdam, The Netherlands: Reprint Fredonia Books, Amsterdam, 2002.

Taunton Press. *Finish Carpentry*, For Pros by Pros series. Newton, CT: 2003.

Walker, C. Howard. *Theory of Mouldings* (1926), with Forward by Richard Sammons. New York: W.W. Norton, 2007.

Exteriors

Portland Porch Booklet: www.ci.portland.me.us/DocumentCenter/View/2705.

Historic Builders' Guides in Reprint

Benjamin, Asher. *American Builders Companion: Or a System of Architecture, Particularly Adapted to the... Present Style of Building*. S.l.: Forgotten Books, 2015.

Benjamin, Asher. *Architect, or Practical House Carpenter: Illustrated by Sixty-Four Engravings, Which... Exhibit the Orders of Architecture, and Other Elem*. S.l.: Forgotten Books, 2016.

Bicknell, Amos Jackson. *Victorian Architectural Details Designs for over 700 Stairs, Mantels, Doors, Windows, Cornices, Porches, and Other Decorative Elements*. Mineola, NY: Dover Publications, 2005.

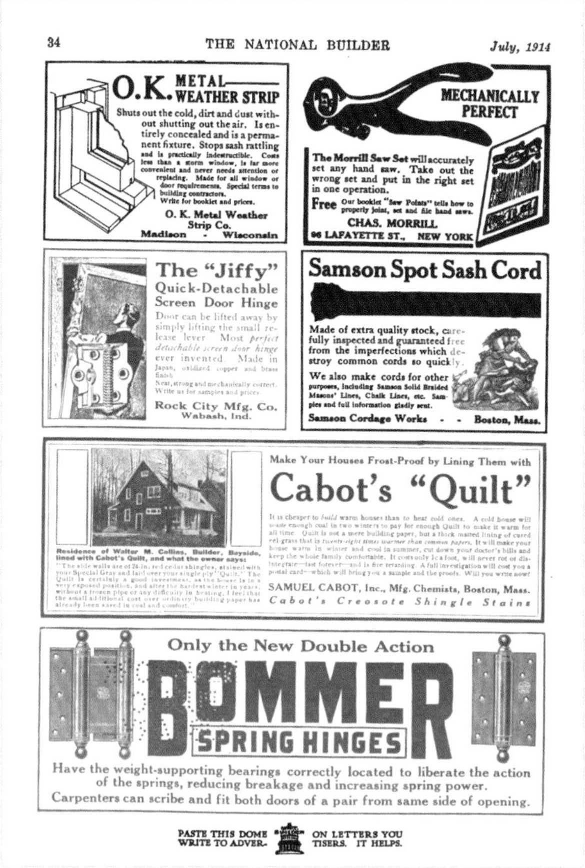

34 THE NATIONAL BUILDER *July, 1914*

O.K. METAL WEATHER STRIP

Shuts out the cold, dirt and dust without shutting out the air. Is entirely concealed and is a permanent fixture. Stops sash rattling and is practically indestructible. Costs less than a storm window, is far more convenient and never needs attention or replacing. Made for all window or door requirements. Special terms to building contractors.
Write for booklet and prices.

O. K. Metal Weather Strip Co.
Madison - Wisconsin

MECHANICALLY PERFECT

The Morrill Saw Set will accurately set any hand saw. Take out the wrong set and put in the right set in one operation.

Free Our booklet "Saw Points" tells how to properly joint, set and file hand saws.

CHAS. MORRILL
96 LAFAYETTE ST., NEW YORK

The "Jiffy"
Quick-Detachable Screen Door Hinge

Door can be lifted away by simply lifting the small release lever. Most *perfect detachable screen door hinge* ever invented. Made in Japan, oxidized copper and brass finish.
Neat, strong and mechanically correct. Write us for samples and prices.

Rock City Mfg. Co.
Wabash, Ind.

Samson Spot Sash Cord

Made of extra quality stock, carefully inspected and guaranteed free from the imperfections which destroy common cords so quickly.

We also make cords for other purposes, including Samson Solid Braided Masons' Lines, Chalk Lines, etc. Samples and full information gladly sent.

Samson Cordage Works - - Boston, Mass.

Make Your Houses Frost-Proof by Lining Them with

Cabot's "Quilt"

It is cheaper to *build* warm houses than to heat cold ones. A cold house will waste enough coal in two winters to pay for enough Quilt to make it warm for all time. Quilt is not a mere building paper, but a thick matted lining of cured eel-grass that is *twenty-eight times warmer than common papers*. It will make your house warm in winter and cool in summer, cut down your doctor's bills and keep the whole family comfortable. It costs only 1c a foot, will never rot or disintegrate—last forever—and is fire retarding. A full investigation will cost you a postal card—which will bring you a sample and the proofs. Will you write now?

Residence of Walter M. Collins, Builder, Bayside, lined with Cabot's Quilt, and what the owner says:

"The side walls are of 24-in. red cedar shingles, stained with your Special Gray and laid over your single ply 'Quilt.' The Quilt is certainly a good investment, as the house is in a very exposed position, and after the hardest winter in years, without a frozen pipe or any difficulty in heating, I feel that the small additional cost over ordinary building paper has already been saved in coal and comfort."

SAMUEL CABOT, Inc., Mfg. Chemists, Boston, Mass.
Cabot's Creosote Shingle Stains

Only the New Double Action

BOMMER SPRING HINGES

Have the weight-supporting bearings correctly located to liberate the action of the springs, reducing breakage and increasing spring power.
Carpenters can scribe and fit both doors of a pair from same side of opening.

PASTE THIS DOME ON LETTERS YOU WRITE TO ADVERTISERS. IT HELPS.

A page from The National Builder's July 1914 issue.

Biddle, Owen. *Biddle's Young Carpenters Assistant*. Mineola, NY: Dover Publications, 2006.

Ellis, George. *Modern Practical Joinery: A Treatise on the Practice of Joiner's Work by Hand and Machine, for the Use of Workmen, Architects, Builders, and Machinists,* 1902. (1908 edition from Linden Publishing, Fresno, CA, available on Google Books) reprint, 1987.

Shaw, Edward. *The Modern Architect: A Classic Victorian Stylebook and Carpenters Manual*. New York: Dover Publications, 1996.

Historic Catalogs in Reprint

Linoff, Victor M. *Universal Millwork Catalog, 1927: Over 500 Designs for Doors, Windows, Stairways, Cabinets, and Other Woodwork*. Mineola, NY: Dover Publications, 2003.

Ornamental Designs from Architectural Sheet Metal: The Complete Broschart & Braun Catalog, ca. 1900. Philadelphia: Athenaeum, 1992.

Roberts Illustrated Millwork Catalog: A Sourcebook of Turn-of-the-Century Architectural Woodwork. New York: Dover Publications, 1988.

Turn-of-the-Century Doors, Windows and Decorative Millwork: The Mulliner Catalog of 1893. New York: Dover Publications, 1995.

Historic Preservation

Page, Max. *Why Preservation Matters*. New Haven, CT: Yale University Press, 2016.

Tyler, Norman, Ted J. Ligibel, and Ilene R. Tyler. *Historic Preservation: An Introduction to Its History, Principles, and Practice*, Second Edition. New York: W.W. Norton and Company, 2009.

Interior Finishes

Eastlake, Charles L. *Hints on Household Taste: The Classic Handbook of Victorian Interior Decoration*. New York: Dover Publications, 1986.

Garrett, Elisabeth Donaghy. *The Antiques Book of American Interiors, Colonial and Federal Styles*. New York: Crown Publishers, 1980.

Garrett, Elisabeth Donaghy. *At Home: The American Family 1750–1870*. New York: Henry N. Abrams, 1990.

Hoskins, Lesley. *The Papered Wall: The History, Pattern and Techniques of Wallpaper*, Second Edition. London: Thames and Hudson, Ltd., 2005.

Jester, Thomas C. *Twentieth-Century Building Materials: History and Conservation*. Los Angeles, CA: Getty Publications, 2014.

Lynn, Catherine. *Wallpaper in America: From the Seventeenth Century to World War I*. Wayne, PA: Barra Foundation, 1980.

Mayhew, Edgar De N., and Minor Myers. *A Documentary History of American Interiors: From the Colonial Era to 1915*. New York: Scribner, 1986.

Moss, Roger W. *Lighting for Historic Buildings: A Guide to Selecting Reproductions*. Washington, DC: Preservation Press.

Moss, Roger W., and Gail Caskey Winkler. *Victorian Interior Decoration: American Interiors, 1830–1900*. New York: Henry Holt, 1992.

Nylander, Jane C., and Richard C. Nylander. *Fabrics and Wallpapers for Historic Buildings*. Chichester, CT: Wiley, 2005.

Nylander, Jane. *Fabrics for Historic Buildings, Revised Edition*. Washington, DC: The Preservation Press, National Trust for Historic Preservation, 1990.

Nylander, Jane C. *Our Own Snug Fireside: Images of the New England Home 1760–1860*. New York: Alfred A. Knopf, 1993.

Nylander, Richard C. *Wall Papers for Historic Buildings: A Guide to Selected Reproduction Wallpapers*. Washington, DC: The Preservation Press, National Trust for Historic Preservation, 1983.

Nylander, Richard C., Elizabeth Redmond, and Penny J. Sander. *Wallpaper in New England*. Boston: Society for the Preservation of New England Antiquities, 1986.

Rosenstiel, Helene Von. *Floor coverings for Historic Buildings: A Guide to Selecting Reproductions*. Place of publication not identified: Diane Pub., 1999.

Saunders, Gill. *Wallpaper in Interior Decoration*. New York: Watson Guptill Publications, 2002.

Shivers, Natalie. *Respectful Rehabilitation: Walls and Molding, How to Care for Historic Wood and Plaster*. New York: National Trust for Historic Preservation, Preservation Press, John Wiley & Sons, Inc., 1990.

Thornton, Peter. *Authentic Decor: The Domestic Interior 1620–1920*. London: Weidenfeld and Nicolson, 1985.

Wharton, Edith, and Ogden Codman, Jr. *The Decoration of Houses* (1897). Reprint, New York: Dover Publications, 2015.

Mechanical Systems

Ierley, Merritt. *The Comforts of Home: The American House and the Evolution of Modern Convenience*. New York: Clarkson Potter Publishers, 1999.

Ogle, Maureen. *All the Modern Conveniences: American Household Plumbing, 1840–1890*. Baltimore, MD: Johns Hopkins University Press, 1996.

Wooson, R. Dodge. *Radiant Floor Heating*. New York: McGraw Hill, 1999.

Paint

Baty, Patrick. *The Anatomy of Color: The Story of Heritage Paints and Pigments*. London: Thames and Hudson, 2017.

Cavelle, Simon. *The Encyclopedia of Decorative Paint Technique*. Philadelphia: Running Press, 1994.

Innes, Jocasta. *Paint Magic*. New York: Pantheon Books, 1987.

Innes, Jocasta. *The New Decorator's Handbook: Decorative Paint Techniques for Every Room*. New York: Harper Collins, 1995.

Lefko, Linda, and Jane E. Radcliffe. *Folk Art Murals of the Rufus Porter School: New England Landscapes 1825–1845*. Atglen, PA: Schiffer Publications, 2011.

Miller, Judith and Martin. *Period Finishes and Effects*. New York: Rizzoli, 1992.

Moss, Roger W., and Gail Caskey Winkler. *Victorian Exterior Decoration: How to Paint Your Nineteenth-Century American House Historically*. New York: Henry Holt, 1992.

Moss, Roger W. *Paint in America: The Colors of Historic Buildings*. New York: J. Wiley, 1996.

Phillips, Laura A. W. *Grand Illusions: Historic Decorative Interior Painting in North Carolina*. Chapel Hill, NC: North Carolina Office of Archives and History, University of North Carolina Press, Inc., 2018.

Van Herman, T. H. *Every Man His Own Painter and Colourman: A Complete System for the Amelioration of the Noxious Quality of Common Paint, Etc.* London: I. F. Setchel, 1829.

General Renovation

Barker, Bruce A. *Codes for Homeowners*. Minneapolis, MN: Quarto Publishing, 2015.

Johnson, Amy. *What Your Contractor Can't Tell You: The Essential Guide to Building and Renovating*. Burlington, VT: Shube Publishing, 2008.

Litchfield, Michael W. *Renovation: A Complete Guide*. Newtown, CT: Taunton Press, 2019.

Restoration/Renovation

Hewitt, Mark Alan, and Gordon Bock. *The Vintage Home: A Guide to Successful Renovations and Additions*. New York: W.W. Norton and Company, 2011.

Hutchins, Nigel. *Restoring Old Houses*. Toronto: Key Porter Books, 1997.

Nash, George. *Renovating Old Houses*. Newtown, CT: Taunton Press, 1998.

Shirley, Frank. *New Rooms for Old Houses: Beautiful Additions for the Traditional Home*. Newtown, CT: Taunton Press, 2007.

Stephen, George. *New Life for Old Houses*. Washington, DC: Preservation Press, National Trust for Historic Preservation, 1989.

Roofing

Jenkins, Joseph C. *The Slate Roof Bible: Everything You Wanted to Know About Slate Roofs Including How to Keep Them Alive for Centuries*. Grove City, PA: Jenkins Pub., 2000.

Windows

Jordan, Steve. *The Window Sash Bible*. Create Space Independent Publishing, 2015.

Leeke, John. *Save America's Windows: Caring for Older and Historic Wood Windows*. Create Space Independent Publishing, 2009.

Meany, Terry. *Working Windows: A Guide to the Repair and Restoration of Wood Windows*. Guilford, CT: Lyons Press, 2008.

INDEX

Page numbers in *italics* refer to illustrations and/or their captions.

ACKNOWLEDGMENTS

I owe thanks to many people who have contributed to this book in many ways. The most substantial contribution has been that of photographer David Clough, who has literally driven across much of America to shoot the wonderful homes featured in this book. Dave's photos capture the exceptional restoration and rehabilitation work done by more than two dozen historic homeowners who have generously shared their homes here. Others who have shared their work include architects Don Mills, Craig Whittaker, Nancy Barba, and Cynthia Wheelock. Images to illustrate the text were graciously supplied by Abatron Building and Restoration Products, Adelphi Paper Hangings, Bagala Window Works, Bradbury & Bradbury Art Wallpapers, Building Sciences Corporation, Tony Castro & Company, Evergreene Architectural Arts, Plaster Magic, Pullman Manufacturing Company, The Unico System, and my employer Sutherland Conservation & Consulting.

My publisher and editor, Jonathan Eaton of Tilbury House Publishers, has been endlessly patient and supportive of this project. When he approached me about the possibility of doing another book after the publication of *Homes Down East* (2014), I don't think he had any idea what he was getting into! He has been a pleasure to work with, and this book owes a great deal to his careful editorship and publishing expertise.

Any book based on a lifetime of experience owes thanks to a long list of people who contributed to that experience. I am grateful to have grown up with a father who built things, and who had friends who worked in the trades and helped with remodeling projects at our house. These men showed me that building things involves a series of steps taken in a certain order. Once you know the steps and the order, you can build things. Once you know how things are built, you can repair them. These were valuable early lessons for the path I have followed.

My introduction to historic preservation and restoration came at age eleven, when Dwight A. Smith recognized my interest and willingness and welcomed me as a volunteer on the restoration projects involved in the startup of the Conway Scenic Railroad in my hometown in New Hampshire. I got to spend some of my formative years helping to restore buildings, railroad cars, and a steam locomotive under the tutelage of experienced carpenters, painters, mechanics, and railroad men. They were all great teachers and very tolerant of the kid with a million questions.

As a young adult, I had the very good fortune to find myself working on a rehabilitation job with Donald Essman, an extraordinary restoration carpenter. Among other things, he showed me that it was possible to build a life by pursuing one's passion for preservation. After more than 30 years of friendship, I am still learning from Don.

When I made the transition from working on houses to working as a preservation professional, I had the good fortune to work under Deb Andrews. For more than 30 years, Deb has worked to preserve the historic buildings and character of Portland, Maine, first as executive director of Greater Portland Landmarks and, since 1990, as head of the city's historic preservation program. Working with Deb for nearly five years was a master class in preservation practices at the community level. It was also a lot of fun. I owe many thanks to Deb for her role in preparing me to write this book.

Moving from the Portland Planning Department, I became an architectural historian for Sutherland Conservation & Consulting and learned the intricacies of the state and federal historic tax credit program from Amy Cole Ives. Amy had long experience as historic tax credit reviewer for the Maine Historic Preservation Commission prior to founding SCC, and I could not have learned from anyone better. Needless to say, the lessons learned in working with Amy for eight years are reflected in this book in many ways.

Writing this book has been a four-year effort, during which I have often been distracted from domestic duties and pleasures by work on the book. This includes rehabilitation work on my own house. My husband, Andrew Jones, has tolerated my distraction with patience and good grace. I owe him thanks for that, and many other things.

SCOTT T. HANSON has thought deeply about the renovation of historic homes in his 40 years as a designer, carpenter, municipal historic district regulator, historic preservation consultant, and architectural historian. His own fifteen-year renovation of a historic house, doing the vast majority of the work himself, has allowed him to put his ideas to the test and to learn from his mistakes while accumulating the library of how-to photos that illustrates this book. Scott is Director of Preservation Consulting Services for Sutherland Conservation & Consulting. He has researched and written numerous National Register nominations and historic building documentation projects. He is the co-author of *Homes Down East: Classic Maine Coastal Cottages and Town Houses* (Tilbury House, 2014).

DAVID J. CLOUGH has had a lifelong fascination with images that tell a story. His architectural photography places exteriors and interiors within a context of spaces that the built environment does not encompass. He lived for ten years in Japan, where his work has been published in the Japan National Trust's *Official Guidebook for the Yasuda House*. A touring exhibit of his work in 2016 was exhibited in Venice, Yokohama, Milan, Suzhou (China), and Stockholm. His photography has appeared in *Built Heritage Journal, Coastal Home, Down East, Maine Home + Design*, and *Maine Boats, Homes and Harbors* magazines and is featured in *Homes Down East* (Tilbury House, 2014).